The
Rudolph J. Nunnemacher Collection
of
Projectile Arms

By

John Metschl

PART II
SHORT ARMS

Preface to the Greenwood reprint by
ELDON G. WOLFF
Curator, Department of History
Milwaukee Public Museum

GREENWOOD PRESS, PUBLISHERS
WESTPORT, CONNECTICUT

BULLETIN

OF THE

PUBLIC MUSEUM OF THE CITY OF MILWAUKEE

Vol. 9, Pp. 1-1017,   Plates 1-113,   Text figures 1-20,   May 25, 1928

The
Rudolph J. Nunnemacher Collection
of
Projectile Arms

By

John Metschl

PART II
SHORT ARMS

S. A. BARRETT
EDITOR
IRA EDWARDS
ASSISTANT EDITOR

Special Acknowledgment

Special acknowledgment is made to Dr. Paul B. Jenkins, the Museum's new expert on arms, for helpful editorial comments and for the addition of many items of great interest in the final preparation of this volume, especially for the section entitled "Origin, Purposes and Rise of Projectile Arms" and for the appendices and index.

Editor.

PART II, SHORT ARMS

FOREWORD

In treating of long arms in Part 1 of this work, occasion has arisen to discuss quite fully "the Origin, Rise and Purpose of Projectile Arms," the various systems of ignition and many other general topics which are equally as applicable to small arms as they are to long arms. Such matters will not, therefore, be again treated here, but the reader may refer to them for information on any of these subjects. Only such special discussions as particularly concern topics relating to small arms alone will receive consideration in this part of the work.

In the appendices attached to this, Part II will be found in ready reference form, a list of patents, a list of makers, and a glossary of terms which will be very useful to the student of arms. The index covers both parts of this volume and is made as full as possible without being too cumbersome.

PLATE 64.

HAND CANNON MATCH-LOCKS AND WHEEL-LOCKS

No. 1. (N3876). MATCH-LOCK ESCOPETTE. Caliber .50. Four-inch barrel. This specimen formerly had a match-lock hammer and a ball at the end of the tang which fitted into a socket on a ring carried about the neck of a man or horse. The specimen dates from about the year 1450.

England.

No. 2. (N3877). REPLICA OF A HAND CANNON. Caliber .60. Seven-inch, smooth-bore barrel. The whole arm is cast in one piece of brass. This is one of the most primitive of firearms. As its name implies, it was fired by simply applying by hand a torch or hot iron to the flash-pan which ignited the priming and so fired the charge.

Spain.

No. 3. (N672). WHEEL-LOCK PISTOL. Caliber .55. Seventeen-inch barrel. Full stock richly inlaid with ivory and mother-of-pearl. Upon the fore-end and butt-plate are carved ivory masks, while opposite the lock-plate is an inset carved in high relief showing a recumbent soldier within a fortress. Sixteenth century.

Germany.

No. 4. (N668). WHEEL-LOCK PISTOL AND BATTLE AXE. Caliber .66. Seven-inch, octagonal barrel. The ivory inlaid butt of the pistol forms the handle of the battle axe which is provided with a ring to hang it to the saddle. Below the specimen is a piece of pyrite, which mineral was frequently used in the early wheel-lock in the place of flint. Pyrite being relatively soft, did not injure the serrations of the wheel of the wheel-lock as a piece of flint was apt to do. Sixteenth century.

South Germany.

No. 5. (N670). DOUBLE WHEEL-LOCK PISTOL. Caliber .55. Fourteen-and-one-half-inch, octagonal, smooth-bore barrel, engraved with the letters "B.S.B.S." The barrel has a cannon-shaped muzzle and the butt is ball-shaped. The full stock is richly inlaid with bone. Trigger-guard and parts of the barrel are overlaid with gold leaf. There is a safety device on the lock for holding the trigger at rest. Two charges are placed in the barrel, one on top of the other, and are fired separately by means of the two cocks. Sixteenth century.

Saxony.

No. 6. (N609). DOUBLE WHEEL-LOCK PISTOL. Caliber .57. Fourteen-and-one-quarter-inch, half-octagonal, smooth-bore barrel, the breech of which is stamped with hall and maker's marks. The muzzle is cannon-shaped. The ornamentation upon the stock is similar to that of No. 5. There is a safety device opposite the lock for holding the trigger at rest. Two superposed charges are placed in the barrel. There is but a single trigger which serves both cocks, each charge is fired separately, if the first bullet properly performed its office as a gas check. Sixteenth century.

Saxony.

No. 7. (N671). WHEEL-LOCK PISTOL. Caliber .54. Sixteen-inch barrel, bearing maker's marks and the date "1564." The stock and butt are inlaid with engraved ivory.

Germany.

No. 8. (N3884). WHEEL-LOCK PISTOL. Caliber .50. Seventeen-inch, half-octagonal, engraved barrel. The stock and butt are made of engraved iron. There is a belt-hook upon the left side. The wheel of the lock is enclosed in a housing which indicates that this is one of the later types. A safety catch prevents the trigger from accidentally releasing the wheel.

Spain.

No. 9. (N666). HAND CANNON. Caliber .40. Two-and-one-half-inch brass barrel which is ornamented with a raised floral design.

Japan.

No. 10. (N667). HAND CANNON. Caliber .15. One-and-three-quarters-inch barrel. The butt is of wood and the top of it bears in gilt what is probably the crest of the Shimadzu family of the province of Satsuma.

Japan.

No. 11. (N3878). MATCH-LOCK PISTOL. Caliber .55. Ten-and-one-quarter-inch barrel which is held to the stock by means of two bands. The top of the barrel is ornamented with a dragon. This is said to have been a former Japanese government arm and came from Tokio in 1907.

Japan.

No. 12. (N475). MATCH-LOCK PISTOL. Caliber .31. Ten-inch, octagonal barrel, inlaid with a brass dragon and clouds. The muzzle is cannon-shaped.

Japan.

No. 13. (N3880). MATCH-LOCK PISTOL. Caliber .36. Four-and-one-half-inch, octagonal, brass barrel, engraved with a dragon and the crest of a Japanese noble house in relief. The muzzle is cannon-shaped.

Japan.

No. 14. (N3882). MATCH-LOCK PISTOL. Caliber .45. Two-inch, octagonal barrel, engraved with a dragon. Has a cannon-shaped muzzle. The barrel and stock are ornamented in gilt with a crest which is probably that of a Japanese noble house. The mountings are of silver.

Japan.

No. 15. (N3883). MATCH-LOCK PISTOL. Caliber .24. One-and-one-quarter-inch barrel, having a cannon-shaped muzzle. All mountings of this miniature pistol are of silver.

Japan.

No. 16. (N665). MATCH-LOCK PISTOL. Caliber .27. One-and-three-quarters-inch barrel, very lightly engraved with a design representing a dragon. The barrel is held to the stock by means of a band. Has a cannon-shaped muzzle.

Japan.

No. 17. (N3881). MATCH-LOCK PISTOL. Caliber .34. Three-and-one-half-inch, octagonal, bronze barrel, all parts of which are unornamented. The mountings are of brass.

Japan.

No. 18. (N3879). MATCH-LOCK PISTOL. Caliber .20. Four-inch, octagonal barrel. The barrel is unornamented and of the straight form. All mountings are of brass.

Japan.

PLATE 65.

MIQUELET LOCKS AND BLUNDERBUSSES

No. 1. (N687a, b). PAIR OF MIQUELET LOCKS. Caliber .63. Eleven-and-three-quarters-inch barrel, with inlaid design in gold and silver. The entire stock and fore part of the barrel are covered with cast metal ornamented with raised designs.

Albania.

No. 2. (N3893). MIQUELET LOCK. Caliber .70. Twelve-and-three-quarters-inch barrel. The entire stock is covered with a thin sheet of carved brass. No marks of identification are given.

Persia.

No. 3. (N2878). MIQUELET LOCK. Caliber .65. Thirteen-and-three-quarters-inch barrel. The stock is entirely covered with thin, engraved sheet brass. This specimen dates from the 18th century. No maker's name is given.

Arabia.

No. 4. (N690). MIQUELET LOCK. Caliber .65. Eleven-and-one-quarter-inch barrel. The stock is entirely covered with thin, engraved sheet brass. The comb is inlaid with lead. No marks of identification are given. The specimen dates from the 18th century.

Arabia.

No. 5. (N3894). MIQUELET LOCK. Caliber .60. Nine-inch barrel. The stock and comb are ornamented in a similar manner to No. 4. No maker's name is given.

Arabia.

No. 6. (N678). MIQUELET LOCK. Caliber .55. Double, super-posed, ten-and-one-half-inch, octagonal barrels which revolve by hand. The butt is painted in black and gilt.

Germany?

No. 7. (N4025). FLINT-LOCK. Caliber .55. Double, nine-and-one-quarter inch, half-octagonal, superposed barrels. There is but a single lock with a sliding pan which conveys the flash to either barrel separately or to both simultaneously. This pistol has engraved iron mountings and a brass lock-plate. The stock is inlaid with a silver wire.

Turkey.

No. 8. (N3898). MIQUELET LOCK. Caliber .72. Eleven-and-three-quarters-inch, half-octagonal barrel, marked with the military designations: "E X J H, 6016, F.M., 1814, R." All mountings are of brass.

England.

No. 9. (N2760). MIQUELET LOCK. Caliber .72. Eight-inch, half-octagonal barrel, which is marked "E X, A G, 1820, O M, 10246." The mountings are of red brass.

Spain?

No. 10. (N3899). MIQUELET LOCK. Caliber .65. Seven-and-three-quarters-inch, half-octagonal barrel, inlaid with silver and carved. This pistol bears an Italian hall-mark in gold-leaf. The lock parts and the brass and silver mountings are elaborately engraved.

Italy.

No. 11. (N5633). MIQUELET LOCK. Caliber .66. Six-and-three-quarters-inch, half-octagonal, fluted barrel. All parts of the lock are elaborately engraved. There is a belt hook upon the left side. The grip is cross-hatched and the ramrod has a horn tip. All mountings of this dagg type of miquelet lock are of brass.

Spain.

No. 12. (N7024). MIQUELET LOCK. Caliber .68. Six-and-five-eighths-inch, smooth-bore, tapering barrel having a cannon-shaped muzzle. This pistol has been converted to the percussion system. The mountings are of brass. There is a belt-hook upon the left side. The rear portion of the barrel is octagonal and is inlaid with floral designs and a hall-mark bearing the letters "P R" over "A? T."

Spain.

No. 13. (N3896). MIQUELET LOCK. Caliber .68. Five-and-one-half-inch, half-octagonal barrel, inlaid with silver. The muzzle is cannon-shaped. This dagg type of pistol is ornamented with engraved brass mountings. It is said to date from the 16th century.

Spain.

No. 14. (N681). MIQUELET LOCK. Caliber .64. Five-and-three-quarters-inch, engraved, half-octagonal barrel. The stock is ornamented with iron floral designs and carries a belt-hook upon the left side. This is another form of the dagg type of miquelet pistol.

Spain.

No. 15. (N3895). MIQUELET LOCK. Caliber .68. Seven-and-one-half-inch, half-octagonal barrel. The grip is short and has a ball butt. All mountings, including the elaborate front band, are of brass. This is another dagg type of miquelet lock pistol.

Spain.

No. 16. (N6335). MIQUELET LOCK. Caliber .63. Five-inch, smooth-bore barrel. The breech of the barrel and the lock-mechanism are engraved with scroll designs. A belt-hook is upon the left side. Trigger-guard, butt-plate and bands are of silver. The wooden ramrod is tipped with ivory. There are no sights. Upon the lock-plate the name "Lopez" is engraved.

Spain.

No. 17. (N6342). MIQUELET LOCK. Caliber .52. Three-and-one-quarter-inch, smooth-bore barrel. The rear portion of the barrel is octagonal and is inlaid with diamond-shaped pieces of brass. All mountings are of brass. A belt-hook is upon the left side. No marks of identification are given.

Spain.

No. 18. (N3885). MIQUELET BLUNDERBUSS. Eight-inch, iron barrel, having a diameter of one-and-three-eighths-inch at the muzzle. The grip and stock are ornamented with designs cut out of sheet brass. The butt is covered with a brass plate. There are no sights upon this blunderbuss. It is provided with a belt-hook and a brass trigger-guard. There are no marks of identification. This piece was probably used during the early part of the 16th century.

Spain.

No. 19. (N2766). MIQUELET BLUNDERBUSS. Nine-and-three-eighths-inch, iron barrel, having a diameter of one-and-one-quarter inches at the muzzle. This blunderbuss is practically devoid of ornamentation. The stock is apparently of recent origin and has a belt-hook on one side. The trigger-guard is of brass. There are no sights on the barrel, neither are there any marks of identification.

Spain?

No. 20. (N3886). MIQUELET BLUNDERBUSS. Six-and-five-eighths-in, iron barrel, having a diameter of one-inch at the muzzle. The wooden grip is plain and smooth. There are no sights. The trigger-guard is made of thin brass, and the stock is slightly engraved. Upon the top of the barrel is an undecipherable inscription. No other marks of identification are to be found upon the blunderbuss.

Spain.

No. 21. (N3887). FLINT-LOCK BLUNDERBUSS. Seven-and-one-half-inch, iron barrel, having a diameter of one-and-three-

quarters-inches at the muzzle. The rear portion of the barrel is octagonal and bears numerous proof-marks. Farther toward the muzzle is an
engraved design in relief. The grip is made of smooth wood and has an
iron butt-plate. The lock is ornamented, as is also the trigger-guard.
This was once an expensive arm but has been badly misused. It was
made in London.

England.

No. 22. (N3890). FLINT-LOCK BLUNDERBUSS. Ten-and-
one-quarter-inch, iron barrel, having an horizontally elliptical muzzle,
whose major diameter is one-and-five-eighths-inches and whose minor
diameter is seven-eighths of an inch. Blunderbusses having round bell-
shaped muzzles, scattered portions of the charge both too high and too
low for effectiveness. As an improvement the muzzle was made in the
form of an horizontal oval so that the charge, when leaving the gun,
would act like a scythe. This blunderbuss has a smooth grip and is
fitted with a belt-hook. The trigger-guard is of brass and all metal
parts are unornamented. Upon the lock-plate are the words "Constant,
Rotterdam."

Netherlands.

No. 23. (N2765). FLINT-LOCK BLUNDERBUSS. Seven-
inch, iron barrel, the rear portion of which is octagonal. The muzzle
is elliptical in shape, the large diameter being one-and-one-half-inches
and the small diameter one-inch. This blunderbuss has a smooth grip
and a heavy brass butt and trigger-guard. There are no sights. The
lock-plate and the rest of the pistol are unornamented. No marks of
identification are given.

England?

No. 24. (N3888). FLINT-LOCK BLUNDERBUSS. Six-and-
five-eighths-inch, steel barrel, having a diameter of one-and-one-quarter-
inches at the muzzle. The barrel is unornamented except for two engraved rings at the muzzle. The grip is smooth and has a carved face
at the butt. There is a concave rear sight but no fore-sight. The
trigger-guard is of wood, apparently of recent date. There is a safety-
device on the lock-plate, the latter also bears the name "Laurent
Gilles."

France?

No. 25. (N3889). FLINT-LOCK BLUNDERBUSS. Four-inch, damaskeened, steel barrel, having an elliptical muzzle whose large diameter is one-and-one-quarter-inches and whose small diameter is seven-eighths of an inch. This form of blunderbuss was also known as blunderbeusse. The grip is smooth and has a heavy iron butt-plate. All mountings are unornamented. There are no sights, or marks of identification. This blunderbuss is of French manufacture and dates from about 1773.

France.

No. 26. (N2057). FLINT-LOCK BLUNDERBUSS. Three-and-one-quarter-inch, brass barrel, having an elliptical muzzle whose large diameter is one-and-one-quarter-inches and whose small diameter is five-eighths of an inch. The grip is plain and smooth, and there are no sights. The hammer is placed on top of the barrel. The trigger-guard is of iron and the sides of the barrel are slightly engraved. This pocket blunderbuss was brought to Canada, in 1812, by Jean Baptist Dozois.

France.

No. 27. (N3892). FLINT-LOCK BLUNDERBUSS. Two-and-seven-eighths-inch, brass barrel, having a muzzle whose diameter is three-eighths of an inch. The rear half of the barrel is octagonal. Has a flat, smooth grip. The cock and flash-pan are set on top of the barrel, hence there are no sights. The trigger-guard and trigger are of iron. Upon the left side of the pistol are several proof-marks and the name "Reeve." This blunderbuss is said to have been made in London; however, there was also a gunsmith named "Reeves" in Birmingham, who was active about the year 1812.

England.

No. 28. (N3891). FLINT-LOCK BLUNDERBUSS. Two-and-five-eighths-inch, brass barrel, having a muzzle whose diameter is seven-eighths of an inch. There are no sights because cock and flash-pan are placed on top of the barrel. The grip is smooth and has two "cheek-pieces" inlaid with silver wire. The sides of the barrel behind the breech are engraved with flags, quivers and arrows. No marks of identification are given.

England?

No. 29. (N736). PERCUSSION BLUNDERBUSS. Four-inch, half-octagonal, brass barrel, having a muzzle diameter of three-quarters of an inch. There are no sights. There is a safety-catch back of the hammer. Upon the lock-plate is the word "Glasgow" and the barrel is engraved "George Washington, Anno Domini 1755." Washington never owned this pistol since the percussion principle was not known in his time.

Scotland.

No. 30. (N5790). PERCUSSION BLUNDERBUSS. Five-and-one-half-inch, steel barrel having a muzzle whose diameter is seven-eighths of an inch. The hammer is placed on top of the barrel. There is no ornamentation on any part of the blunderbuss. The number "31" is stamped on the under-side of the barrel near the trigger-guard. The left side of the barrel bears the Liége proof-mark.

Belgium.

PLATE 66.

EARLY FLINT-LOCK PISTOLS
CHIEFLY ORIENTAL

No. 1. (N696a, b). PAIR OF FLINT-LOCK PISTOLS. Caliber .63. Twelve-and-one-quarter-inch, steel barrel, the rear portion of which is inlaid with silver scrolls and a hall-mark. This pistol has a concave rear sight, a copper bead for a fore-sight and a ribbed flash-pan cover. The grip of each pistol is carved and inlaid with wire, silver and semi-precious stones. The butt-plate consists of a short, silver cylinder. Inlaid upon the top of the barrel is the word "DIDO."

Oriental.

No. 2. (N3913). FLINT-LOCK PISTOL. Caliber .68. Eight-and-one-quarter-inch, round, iron barrel, inlaid with silver designs just above the vent. There is a concave rear sight but no fore-sight. The cover of the flash-pan is ribbed. There is no ramrod. All wooden parts of the pistol are inlaid with silver wire in the form of scroll designs. The hammer, butt and lock-plates are also inlaid with silver. The barrel bears two unknown proof-marks but no other marks of identification.

Oriental?

No. 3. (N3904). FLINT-LOCK PISTOL. Caliber .64. Twelve-and-one-half-inch, smooth-bore, iron barrel. There are no sights on this pistol. It was made in England for trade in the near East.

England.

No. 4.   (N689a, b).   PAIR OF FLINT-LOCK PISTOLS. Caliber .66.   Ten-and-one-half-inch, smooth-bore, iron barrel.   The lockparts and barrel are inlaid with gold and engraved.   The bands and butt ornaments are of elaborate repoussé silver alloy.

Albania.

No. 5.   (N688).   FLINT-LOCK PISTOL.   Caliber .60.   Twelveand-one-quarter-inch barrel.   This pistol has a full stock and the broad fore-end band is completely covered with elaborately ornamented gilded, cast brass.

Albania.

No. 6.   (N3900).   FLINT-LOCK PISTOL.   Caliber .65.   Fourteen-and-three-quarters-inch, half-octagonal, engraved and ribbed barrel. The full stock is overlaid with a mosaic of diamond-shaped pieces of wood and ivory ornamented with small insets of metal mosaic in cloisons of brass.

India.

No. 7.   (N3905).   FLINT-LOCK PISTOL.   Caliber .60.   Thirteen-inch, smooth-bore, iron barrel, which is engraved and bears an inscription in Persian.   The front sight is of copper set in an ornamented brass inlay.   The flattened, bulbous butt is also inlaid with brass.

Persia.

No. 8.   (N3275).   FLINT-LOCK PISTOL.   Caliber .68.   Nineinch barrel.   The stock is carved and the butt is inlaid with a design in silver wire.   A silver alloy plate partly covers the butt.

Persia.

No. 9.   (N3902).   FLINT-LOCK PISTOL.   Caliber .65.   Nineand-one-quarter-inch, engraved, half-octagonal barrel.   The carved stock is inlaid with silver wire.   This pistol has a repoussé silver foreband and an embossed silver butt-plate.

Arabia.

No. 10.   (N686).   FLINT-LOCK PISTOL.   Caliber .62.   Nineinch, half-octagonal barrel, the rear portion of which is covered with a piece of sheet brass conforming to the steel of the barrel.   Lock-plate, trigger-guard and butt are made of brass.   The grip is somewhat broken and was formerly inlaid with a plate at the top.   The left side of the stock is inlaid with a brass scroll design. The ramrod is missing. No marks of identification are given.

Arabia.

No. 11. (N679). FLINT-LOCK PISTOL. Caliber .67. Eleven-and-one-half-inch, half-octagonal, smooth-bore, iron barrel. The full stock is elaborately engraved and inlaid with silver wire. The butt is of engraved brass.

Morocco.

No. 12. (N3903). FLINT-LOCK PISTOL. Caliber .64. Eleven-and-three-quarters-inch, engraved, smooth-bore barrel. Has a full carved stock which is inlaid with silver wire. The brass butt is engraved with line designs.

Morocco.

No. 13. (N3901). FLINT-LOCK PISTOL. Caliber .60. Thir-teen-and-one-half-inch, smooth-bore barrel. The lock-parts and barrel are inlaid with elaborate designs in gold, including the monogram of the sultan of Turkey. The silver bands are elaborately ornamented with repoussé work.

Turkey.

No. 14. (N682). FLINT-LOCK PISTOL. Caliber .74. Seven-and-three-quarters-inch, half-octagonal barrel, having a cannon muzzle. All mountings are of engraved brass. Has a ball butt. This pistol is of the "dagg" type and is said to date from the 17th century.

Spain?

No. 15. (N3907). FLINT-LOCK PISTOL. Caliber .60. Eleven-and-one-half-inch, engraved, smooth-bore ribbed barrel. This pistol has no sights and it is practically without ornamentation. No maker's marks are given. It is said to date from the 17th century.

Spain.

No. 16. (N685). FLINT-LOCK PISTOL. Caliber .50. Seven-inch, smooth-bore, fluted barrel. Engraved upon the top of the barrel is the name "Lazarino Cominazzo" who was one of Italy's most cele-brated gun barrel makers. The Cominazzo family specialized in the manufacture of gun barrels for several generations and the name given upon the barrel here illustrated probably refers to Lazarino Cominazzo, the younger, who died in 1696. All mountings are of iron and elaborately embossed. Upon the lock-plate is the name "Vicensso Mari, Brescia."

Italy.

No. 17. (N3911). FLINT-LOCK PISTOL. Caliber .46. Ten-and-three-quarters-inch, smooth-bore, half-octagonal barrel. The breech of the barrel, lock-plate and cock are engraved in relief. No marks of identification are given.

Italy.

No. 18. (N680). FLINT-LOCK PISTOL. Caliber .65. Nine-and-one-half-inch, smooth-bore barrel, richly engraved and ornamented with gold leaf. All mountings are finely embossed and plated with gold. This piece is said to date from the 17th century. No maker's name is given.

Italy.

No. 19. (N676). FLINT-LOCK PISTOL. Caliber .58. Four-teen-and-one-quarter-inch, smooth-bore barrel. The breech-end of the barrel is ornamented with horsemen and a seated woman in relief. Upon the flat butt is an iron plate which is decorated with a scene representing mounted and foot soldiers who evidently are attacking a camp. A pea-cock engraved in relief is shown upon the trigger guard. This piece was made in Venice, sometime during the 16th century.

Italy.

No. 20. FLINT-LOCK PISTOL. Caliber .69. Twelve-and-one-half-inch, smooth-bore, half-octagonal barrel which is finely engraved. Butt-plate, lock-plate and trigger-guard are finely engraved, while the butt itself is elaborately inlaid with silver wire. This is another Venetian pistol.

Italy.

PLATE 67.

EARLY FLINT-LOCK PISTOLS
CHIEFLY FRENCH AND GERMAN

No. 1. (N3906). FLINT-LOCK PISTOL. Caliber .65. Twelve-inch, smooth-bore, half-octagonal barrel. Upon the lock-plate is the name "I. Lecomte."

France.

No. 2. (N4038). FLINT-LOCK PISTOL. Caliber .56. Seven-and-one-half-inch, half-octagonal barrel, with rounded top inlaid with floral designs in gold. The fore-sight is of silver, as are also the engraved mountings. The mountings are stamped with hall-marks. Engraved upon the lock-plate are the words "Cassaignard à Nantes."

France.

No. 3. (N3928). FLINT-LOCK PISTOL. Caliber .60. Six-and-one-half-inch, half-octagonal, smooth-bore, brass barrel, having a bell-shaped muzzle. There is a belt-hook upon the left side. All mountings are of brass. The butt-plate is made in the form of a bird's head. This pistol dates from the 18th century.

France.

No. 4. (N4052 and N4053). PAIR OF FLINT-LOCK PISTOLS. Caliber .60. Eight-inch, half-octagonal, brass barrel. There are no sights. One of the pistols bears a small brass plate upon the left side which is stamped with the impression of an eagle. These pistols are probably of English manufacture, although no proof-marks are visible.

England?

No. 5. (N3912). FLINT-LOCK PISTOL. Caliber .58. Eight-and-one-half-inch barrel. This is a composite pistol. The barrel and other parts are from a Simeon North pistol, while the lock is from an English gun. This pistol has a full wooden stock ornamented with elaborate designs in brass wire. The gun was brought from Persia in 1901.

Persia.

No. 6. (N810). FLINT-LOCK PISTOL. Caliber .55. Seven-inch, half-octagonal, smooth-bore barrel. All parts of the gun are devoid of ornamentation and no marks of identification are given.

Germany.

No. 7. (N6781). FLINT-LOCK PISTOL. Caliber .74. Thirteen-and-one-half-inch, iron barrel, which is slightly flared at the muzzle. Has a fore-sight but no rear sight. The ramrod is missing. The trigger-guard and butt-plate are of brass. Upon the lock-plate are two marks, one in the form of a large crown over the letters "G.R.", and the other in the form of a small crown over the letter "V." In addition, the

lock-plate is marked "C. Armer, 1712." This pistol came from Olden-
burg, Germany, but the lock is probably of English manufacture.
Donor: Adolph Meinecke.

Germany?

No. 8. (N675). FLINT-LOCK PISTOL. Caliber .60. Four-
teen-inch, smooth-bore barrel, which is ribbed upon its exterior. The
stock formerly extended the full length of the barrel. Upon the lock-
plate is inscribed "Goh. Carl Morth, in Fürth." This piece came from
Bavaria.

Germany.

No. 9. (N2920). FLINT-LOCK PISTOL. Caliber .77. Twelve-
and-one-half-inch, smooth-bore, iron barrel. The stock is ornamented
with brass ferrules. It is considerably worm-eaten. The trigger-guard
and butt-plate are also of brass. The ramrod is missing. This pistol
has a concave rear-sight and a brass fore-sight. Barrel and lock-plate
bear the number "198." The top of the grip is inlaid with an oval brass
plate engraved with a cross and crown over the intials "C.F." No
maker's name or proof-marks are given.

Strassburg, Germany.

No. 10. (N2758). FLINT-LOCK PISTOL. Caliber .70. Seven-
inch barrel. The butt is flat and covered with a brass butt-plate. All
other mountings are also of brass. There is an undecipherable mark
upon the lock-plate.

Germany?

No. 11. (N2757). FLINT-LOCK PISTOL. Caliber .70. Seven-
and-three-quarters-inch, smooth-bore barrel. All mountings are of
brass. Upon the comb is a brass plate engraved with a crown and the
letters "F. W. A." The lock-plate is engraved "Potzdam Mag." This
is probably a Prussian military or police pistol.

Germany.

No. 12. (N702). FLINT-LOCK PISTOL. Caliber .55. Twelve-
inch, half-octagonal, smooth-bore barrel. All parts of the pistol are
made of iron, including the flat butt. Engraved upon the lock-plate are
the words "Jan Cloeter à Grevenbroich." Grevenbroich is a village near
Düsseldorf. This specimen came from the Judenburg at Nuremberg.

Germany.

No. 13. (N712). FLINT-LOCK PISTOL. Caliber .42. Three-inch, half-octagonal, rifled, iron barrel. The grip is smooth, made of wood, and has a brass butt plate. There are no sights. The lock-plate and cock are crudely engraved. Part of the brass trigger-guard is gone. No marks of identification are given.

Germany.

No. 14. (N691). FLINT-LOCK PISTOL. Caliber .24. Two-and-one-quarter-inch, steel, tapering barrel. The trigger-guard, ferrules and butt-plate of this vest pocket pistol are of engraved gold. This pistol is provided with front and rear sights and a wooden ramrod. Upon the top of the barrel are the words "Illig A Baureuth."

Bavaria, Germany.

No. 15. (N809). FLINT-LOCK PISTOL. Caliber .70. Nine-and-one-quarter-inch, half-octagonal, smooth-bore barrel. All mountings are of brass. No marks of identification are given. This probably is another form of the German military pistol.

Germany.

No. 16. (N705). FLINT-LOCK PISTOL. Caliber .65. Nine-inch, smooth-bore, engraved barrel, having a brass sight and a bell-shaped muzzle. All mountings are of embossed brass. The butt-plate is in the form of a monster's head. The pistol is marked with the name "A. Prion."

France.

No. 17. (N3914). FLINT-LOCK PISTOL. Caliber .64. Nine-and-one-half-inch, smooth-bore barrel. The butt is flat and bears a brass plate marked "19. L. W. R. 3. E." This pistol is said to have been captured by the American Colonists during the Revolutionary War from a Hessian soldier at or near Newburgh, N. Y. The pistol was made in Potsdam.

Germany.

No. 18. (N3966). FLINT-LOCK PISTOL. Caliber .70. Eight-and-three-quarters-inch barrel. The flash-pan and mountings are of brass. Upon the lock-plate are the words "G. Berleur, Liége."

Belgium.

No. 19. (N2695). FLINT-LOCK PISTOL. Caliber .55. Five-and-three-quarters-inch, half-octagonal barrel. The full wooden stock is ornamented with elaborate designs in ivory.   The mountings are of engraved brass.   No maker's name is given.

Germany.

No. 20. (N677).   FLINT-LOCK PISTOL.   Caliber .77.   Eight-and-one-half-inch, round, iron barrel.   The lock-plate, cock, hammer and spring are on the left side of the pistol.   Has a concave rear sight but no fore-sight.   All iron parts are badly rusted, and the grip has been repaired.   The trigger-guard is of brass.   No marks of identification are visible.   This pistol is probably of German manufacture.

Germany.

PLATE 68.

ENGLISH FLINT-LOCK PISTOLS

No. 1. (N1192).   FLINT-LOCK PISTOL.   Caliber .65.   Thirteen-and-three-quarters-inch, smooth-bore, half-octagonal barrel.   The stock formerly extended to the end of the barrel.   Barrel, lock-plate and butt-plate are engraved.   No maker's name is given.

England.

No. 2. (N714).   FLINT-LOCK PISTOL.   Caliber .65.   Twelve-and-one-half-inch, half-octagonal, smooth-bore barrel.   The stock is carved, while the barrel, lock-plate and butt are engraved.   No marks of identification are given.

England.

No. 3. (N3909).   FLINT-LOCK PISTOL.   Caliber .60.   Twelve-inch barrel, elaborately inlaid with silver.   All mountings are of heavy, embossed brass.   There are no marks of identification.

England.

No. 4. (N3918).   FLINT-LOCK PISTOL.   Caliber .60.   Nine-and-one-half-inch, smooth-bore, half-octagonal barrel.   The stock is tipped with horn while the butt is rather flat and covered with a brass butt-plate.   This pistol was made in 1769 but bears no maker's marks.

England.

No. 5. (N3916).   FLINT-LOCK PISTOL.   Caliber .66.   Eight-and-one-half-inch, smooth-bore barrel.   All mountings are of brass.   The butt is flat and bears an engraved, brass butt-plate.   The barrel is stamped with the English proof-mark of 1716.   The lock-plate bears the name "F. Horner."

England.

No. 6. (N3919). FLINT-LOCK PISTOL. Caliber .60. Nine-and-one-quarter-inch, smooth-bore barrel. All mountings are of engraved brass. No marks of identification are given.

England.

No. 7. (N3276). FLINT-LOCK PISTOL. Caliber .70. Five-and-one-half-inch, steel barrel, which is fitted with a fore-sight but no rear sight. This pistol has a smooth, oval, polished black grip and a wooden ramrod. There is no butt-plate. Ferrules and trigger-guard are of brass. The lug of the safety-catch upon the lock-plate is broken. Upon the barrel are two English proof-marks and the lock-plate is engraved with the words "Brander & Potts, 70 Minories, London."

England.

No. 8. (N3921). FLINT-LOCK PISTOL. Caliber .60. Twelve-inch, smooth-bore barrel. All mountings are of brass. There formerly was a belt-hook upon the left side. The lock-plate bears the word "Tower" and a crown over the letters "G.R.", together with the broad arrow which denotes British government possession. This pistol was made about the year 1750.

England.

No. 9. (N2755). FLINT-LOCK PISTOL. Caliber .65. Nine-inch, smooth-bore barrel. This is another "Tower" piece. Upon the lock-plate is a crown over the letters "G.R.", and the word "Tower." No maker's name is given.

England.

No. 10. (N3922). FLINT-LOCK PISTOL. Caliber .58. Ten-and-one-quarter-inch, smooth-bore barrel. All mountings are of brass. There is a large belt-hook upon the left side. The lock-plate is bordered with a line near its edge. The inscription upon the plate is the same as that of No. 9, as both are "Tower" pistols.

England.

No. 11. (N2761). FLINT-LOCK PISTOL. Caliber .75. Nine-inch, smooth-bore, half-octagonal barrel. This is a later model "Tower" pistol and is provided with a swivel ramrod. The cock is also of a later type, as it is not of the "goose-neck" variety. The lock-plate is marked in a similar fashion to that of No. 10.

England.

No. 12. (N3925). FLINT-LOCK PISTOL. Caliber .66. Nine-inch, smooth-bore barrel. This is another form of "Tower" pistol. It is provided with a swivel ramrod and with a safety-catch upon the lock-plate. Upon the lock-plate is a crown over the letters "G.R.", and the word "Tower," together with the broad arrow.

England.

No. 13. (N3926). FLINT-LOCK PISTOL. Caliber .64. Nine-and-one-quarter-inch, smooth-bore barrel. This is apparently a composite pistol, as the barrel bears a Belgian proof-mark, and the lock is of the "Tower" variety. The butt is, however, not similar to that of the usual "Tower" pistol.

England?

No. 14. (N3920). FLINT-LOCK PISTOL. Caliber .60. Twelve-inch, smooth-bore barrel. Upon the lock-plate is a crown over the letters "G.R." and a broad arrow. That portion of the lock-plate in the rear of the cock is marked "Vaughan 1744." It is not known in what part of England a gunsmith by this name worked.

England.

No. 15. (N3923). FLINT-LOCK PISTOL. Caliber .60. Twelve-inch, smooth-bore barrel. The left side of the stock is provided with a belt-hook. Upon the lock-plate is a crown over the letters "G.R." and the broad arrow. That portion of the lock-plate in the rear of the cock is marked "Edge 1759." This gunsmith was located in London.

England.

No. 16. (N3924). FLINT-LOCK PISTOL. Caliber .70. Nine-inch, smooth-bore barrel. There is a line surrounding the edge of the lock-plate, in other respects it is marked in the same manner as that of No. 15. Both pistols are made by Edge in London.

England.

No. 17. (N3935). FLINT-LOCK PISTOL. Caliber .60. Nine-and-one-half-inch, smooth-bore, half-octagonal barrel. All mountings are of engraved brass. This is an English officer's model pistol of 1772. There is a patch and flint-box in the end of the butt. No maker's name is given.

England.

No. 18. (N3897). FLINT-LOCK PISTOL. Caliber .70. Eight-inch, smooth-bore barrel. All mountings are of brass. The hammer is not of the "goose-neck" variety. This is another example of a British military pistol. No marks of identification are given.

England.

No. 19. (N3934). FLINT-LOCK PISTOL. Caliber .68. Nine-inch, half-octagonal, smooth-bore barrel. Engraved upon the lock-plate are the words "Ketland & Co.", while the top of the barrel is engraved with the word "London." The grip has a decided curve and has no butt-plate. An escutcheon is inlaid upon the top of the grip while a plate upon the left side is engraved with the former owner's name "A. Hammond."

England.

No. 20. (N3927). FLINT-LOCK PISTOL. Caliber .41. Seven-inch, smooth-bore barrel. The stock is slightly carved. Upon the lock-plate is the name "Mastrich" which is engraved in script.

England?

PLATE 69.

FLINT-LOCK PISTOLS, MOSTLY ENGLISH

No. 1. (N3930). FLINT-LOCK PISTOL. Caliber .60. Eight-inch, smooth-bore, half-octagonal, brass barrel, having a muzzle which is slightly bell-shaped. The butt-plate, which is also of brass, is in the form of a buffoon's head. This pistol was made by Ketland in London.

England.

No. 2. (N3932). FLINT-LOCK PISTOL. Caliber .56. Eight-inch, smooth-bore, brass barrel. All mountings are also of brass. This pistol was made by Ketland and Company of London.

England.

No. 3. (N3933). FLINT-LOCK PISTOL. Caliber .56. Eight-and-one-half-inch, smooth-bore, octagonal, brass barrel. The fore-sight is of silver. The pistol is marked "Ketland & Co., London."

England.

No. 4. (N3929). FLINT-LOCK PISTOL. Caliber .52. Eight-inch, smooth-bore, brass barrel. All mountings are of brass. There is no rear sight but there is a small fore-sight in the form of a bead. The name "Hampton" appears upon the lock-plate, while the barrel is engraved with the word "London" and bears two proof-marks.

England.

No. 5. (N3917). FLINT-LOCK PISTOL. Caliber .56. Nine-inch, smooth-bore, brass barrel. This pistol has no sights. All mountings are of brass. The name "I.P. Moor" appears on the lock-plate and the barrel bears the word "London."

England.

No. 6. (N3931). FLINT-LOCK PISTOL. Caliber .60. Eight-inch, brass barrel, having an iron back strap. The muzzle is slightly bell-shaped. Upon the lock-plate appears the words "Ketland & Co." The barrel bears the word "London" and two proof-marks in the form of the letters "V" and "P." The pistol has no sights and all mountings are of brass.

England.

No. 7. (N3915). FLINT-LOCK PISTOL. Caliber .54. Seven-and-one-half-inch, smooth-bore, brass barrel. There are no sights, and all mountings are of brass. Upon the lock-plate appear the words "W. Ketland & Co." There are no proof-marks upon the barrel. The Ketlands were in business in Birmingham, from 1750 until 1829; nothing appears to be known regarding the London Ketlands.

England.

No. 8. (N2867). FLINT-LOCK PISTOL. Caliber .67. Nine-inch, half-octagonal, engraved barrel. The muzzle is slightly bell-shaped. The stock and grip are cross-hatched. The lock-plate is slightly engraved but bears no maker's name.

England.

No. 9. (N3938). FLINT-LOCK PISTOL. Caliber .62. Nine-inch, smooth-bore barrel. This pistol has neither sights nor butt-plate. All mountings are of brass. Upon the lock-plate is the name "Red-fern's" who did business in Birmingham about 1790.

England.

No. 10. (N3936). FLINT-LOCK PISTOL. Caliber .66. Nine-inch, smooth-bore barrel. This is one of the later type of flint-locks and it has neither sights nor butt-plate. Upon the lock-plate appear the words "Blake & Co., London."

England.

No. 11. (N3937). FLINT-LOCK PISTOL. Caliber .60. Eight-inch, round, smooth-bore barrel. The brass trigger-guard is ornamented and the brass butt-plate represents a griffin's head. A portion of the stock is inlaid with silver wire. Upon the lock-plate is the name "Thomas," while the barrel is stamped with two Birmingham proof-marks.

England.

No. 12. (N2752, N2753). PAIR OF FLINT-LOCK PISTOLS. Caliber .75. Seven-inch, octagonal barrels. All mountings are of engraved iron. The grips are crudely cross-hatched. No marks of identification are given.

England.

No. 13. (N4026). FLINT-LOCK PISTOL. Caliber .58. Eight-inch, double, superposed, iron barrels, which are separated by a wooden rib. The grip is smooth and slightly curved near the butt. The butt-plate and fore-sight are of brass. The lock-plate is engraved with serpents and bears the name "Erttel Adresde."

France?

No. 14. (N4030, N4031). PAIR OF FLINT-LOCK PISTOLS. Caliber .54. Seven-inch, double, damaskeened, steel barrels. The stocks extend to the end of the muzzles where they are tipped with horn. The grips are inlaid with silver wire and slightly carved. The butts are covered with iron plates. Each pistol has a fore-sight and a concave rear sight. The trigger-guards are of iron and slightly engraved. The lock-plates are unornamented and bear the name of "L. Gilles." These pistols are sometimes known as English coach pistols.

England.

No. 15. (N4027, N4028). PAIR OF FLINT-LOCK PISTOLS. Caliber .48. Five-and-one-half-inch, double, brass barrels. The grips are smooth, flat and ornamented with carved roses and leaves and inlaid with silver wire. The rear sights are concave but there are no fore-sights. The trigger-guards are of engraved brass. Lock-plates and cocks are unornamented and bear no maker's name.

England.

No. 16. (N6754). FLINT-LOCK PISTOL. Caliber .55. Six-and-five-eighths-inch, double, damaskeened, iron barrels. The grip is finely cross-hatched and inlaid with silver wire. The butt is covered with an engraved iron plate, while the trigger-guard and ferrules are of polished iron. This pistol has a fore-sight and a concave rear sight.

The lock-plate is plain and provided with a safety-catch. Engraved beneath the spring on the lock-plate is the name "Berleur."

France?

No. 17. (N4029). DOUBLE-BARRELED FLINT-LOCK PISTOL. Caliber .54. Five-and-one-half-inch, damaskeened barrels. The grip is smooth and the top is inlaid with silver wire. Carved roses ornament the grip near the breech, while the butt is covered with an oval, engraved, iron plate. Provided with a front sight and a concave rear sight. The lock-plates are slightly ornamented and have safety-catches. Trigger-guard and ferrules are of engraved iron. No maker's name or proof-marks are given.

England.

No. 18. (N701). FLINT-LOCK PISTOL. Caliber .48. Three-and-one-quarter-inch, round, iron barrel, provided with a large, smooth grip. The stock extends to the end of the muzzle. The lock-plate of this pocket pistol is unornamented. Mountings are of brass. There are English proof-marks upon the barrel.

England.

No. 19. (N3939). FLINT-LOCK PISTOL. Caliber .31. Four-and-one-half-inch, smooth-bore, half-octagonal barrel, having a cannon-shaped muzzle. The butt-plate is of brass. This is another form of flint-lock pocket pistol. No maker's name or proof-marks are given.

England.

No. 20. (N6780). FLINT-LOCK PISTOL. Caliber .44. Three-and-five-eighths-inch, half-octagonal, brass barrel. The stock extends to the end of the muzzle. Has no sights. The top of the barrel is ornamented with scroll designs. The upper part of the long grip is carved. Has a smooth, brass trigger-guard and butt-plate. The lock-plate is slightly decorated with a crude line design. No proof-marks or maker's name are given.

England?

No. 21. (N3940). FLINT-LOCK PISTOL. Caliber .40. Four-inch, smooth-bore, half-octagonal barrel, having a bell-shaped muzzle. All mountings of this pocket pistol are of brass. No maker's name is given.

England.

No. 22. (N3945). FLINT-LOCK PISTOL. Caliber .50. Three-and-one-half-inch, smooth-bore barrel. A folding spring bayonet, which may be released by a catch, is located upon the top of the barrel. The trigger-guard is of engraved brass. A safety-catch is located upon the lock-plate. The words "Bunney & Son" appear upon the lock-plate, while the barrel bears the word "Birmingham."

England.

No. 23. (N3942). FLINT-LOCK PISTOL. Caliber .47. Three-inch, half-octagonal, smooth-bore barrel. Has neither fore nor back sights. This pistol is provided with a brass trigger-guard and butt-plate. The stock extends forward to the end of the muzzle. Has a smooth, round grip and an unornamented lock-plate. The ramrod is missing. No maker's name or proof-marks are given.

England?

PLATE 70.

FLINT-LOCK PISTOLS
CHIEFLY FRENCH MILITARY

No. 1. (N3964). FLINT-LOCK PISTOL. Caliber .70. Nine-inch, smooth-bore barrel, having no sights of any kind. All mountings are of iron. Upon the lock-plate is engraved "Mr. de Libreville," an abbreviation for "Manufactured in Libreville." The date "1763" appears upon the tang. During the French Revolution (1789-1799) when everything named after royalty, or that which in any way suggested it, was abolished, the name Charleville was changed to Libreville. Later the name of this city was again changed to Charleville. There was no regular French model military pistol before 1763, so this pistol represents one of the first models. Other models made their appearances in 1773 and 1777, although all three kinds were made long after the year 1777.

France.

No. 2. (N3962). FLINT-LOCK PISTOL. Caliber .70. Nine-inch, smooth-bore barrel. There are no sights and all mountings are of brass. Upon the lock-plate is engraved "Manufre. Nat. A. de Charleville," and upon the tang "M 1763." This is another example of the model 1763 Charleville model French military pistol.

France.

No. 3. (N3963). FLINT-LOCK PISTOL. Caliber .69. Nine-and-one-quarter-inch, smooth-bore barrel, having neither front nor rear sights. Stamped upon the lock-plate is a crown over the letters "HB" and engraved upon the same plate are the words "Mr. Roy. de St. Etienne." Several proof-marks appear upon the barrel. The tang is engraved with "M.1763." This is a French military pistol made at St. Etienne.

France.

No. 4. (N3961). FLINT-LOCK PISTOL. Caliber .70. Nine-inch, smooth-bore barrel, which has no sights, as is usual in this model of French military pistol. All mountings are of brass. Upon the lock-plate is a crown over the letter "R" and the words "Manufacture de St. Etienne." The figures "M 1773" appear upon the tang. This is the 1773, or the second model of the French military pistol.

France.

No. 5. (N3965). FLINT-LOCK PISTOL. Caliber .69. Eight-inch, smooth-bore barrel. The lock-plate is marked with a crown over the letter "S" and engraved with the words "Mre. imp. de St. etienne," the "i" of "Imp." and the "e" of "Etienne" not being capitalized. Upon the barrel are the figures "AN. 13" standing for the thirteenth year of the French Republic.

France.

No. 6. (N3952). FLINT-LOCK PISTOL. Caliber .68. Nine-inch, half-octagonal, smooth-bore barrel, having no sights. All mountings are of brass. No provision is made for carrying a ramrod on the pistol. The lock-plate is marked with the letter "P" and engraved with the words "Manufre. D. Essen." The letter "P" appears three times upon the barrel. This is another military pistol.

France?

No. 7. (N3908). FLINT-LOCK PISTOL. Caliber .75. Eight-and-one-half-inch, round, smooth-bore, iron barrel. Trigger-guard and butt-plate are of brass. There are no sights. No provision is made to carry a ramrod upon this pistol. Upon the lock-plate is a crown over the letters "ON." The butt is equipped with a ring.

France?

No. 8. (N3960). FLINT-LOCK PISTOL. Caliber .69. Eleven-inch, smooth-bore barrel, having no sights. All mountings are of brass, including the flash-pan.   The ramrod is of the swivel variety.   There are no marks upon the lock-plate but the barrel bears a Liége proof-mark.

France.

No. 9. (N3954). FLINT-LOCK PISTOL. Caliber .70. Eight-inch, smooth-bore barrel. Mountings and flash-pan are of brass. Upon the lock-plate is the letter "S" enclosed by a diamond and the words "Mre. imple. de St. etienne," the "i" of Imple. and the "e" of "St. Etienne" are not capitalized. The barrel is marked with the figures "M. AN. 13" and "B. 1812."

France.

No. 10. (N3956). FLINT-LOCK PISTOL. Caliber .72. Eight-inch, smooth-bore barrel. Flash-pan and mountings are of brass. Upon the lock-plate are the words "Maubeuge Manufre. Imple." The barrel is stamped with the figures "B. 1808," and engraved with "M. An. 13," the latter figures being in script. This is said to have been a 1777 model pistol which was altered in 1806.

France.

No. 11. (N3955). FLINT-LOCK PISTOL. Caliber .70. Eight-inch, smooth-bore barrel. All mountings and the flash-pan are of brass. The fore-band has a strap which connects with the lock-plate screw, as is usual with this model. Upon the lock-plate is stamped the letter "S" within a diamond and engraved upon the plate are the words "Manuf. Imp. de St. Etienne." On the barrel are the figures "G.75. M. AN. 13," and "B. 1813." The tang is stamped with a crown over the letter "G." This is another model 1777 which was altered in 1806.

France.

No. 12. (N3274). FLINT-LOCK PISTOL. Caliber .70. Eight-inch, smooth-bore barrel. The mountings are of brass. The lock-plate is marked with the letter "J" and "Mre. imp. de St. etienne." Upon the barrel are the figures "1808 B," and the tang is marked with a crown over the letter "P."

France.

No. 13. (N3957). FLINT-LOCK PISTOL. Caliber .68. Eight-inch, smooth-bore barrel. Mountings and flash-pan are made of brass. The lock-plate is marked "Manufre. Imple. de Charleville" and the barrel "M. an. 13 1808" together with the letter "B" within an oval. The grip is much thinner than on most pistols of this model.

France.

No. 14. (N3953). FLINT-LOCK PISTOL. Caliber .70. Nine-inch, smooth-bore barrel. The mountings and flash-pan are made of brass. There is a belt-hook upon the left side of the pistol. The lock-plate is marked with a crown over the letter "M" and with the words "Mre. Nle. de Tulle." The pistols made at Tulle were intended for sea service and for the American Colonies during the Revolutionary War.

France.

No. 15. (N3959). FLINT-LOCK PISTOL. Caliber .69. Eight-inch, smooth-bore barrel. Mountings and flash-pan are made of brass. The lock-plate is marked with an oval within which is the letter "S," and with the words "Manufre. Roy. de Charleville." Upon the tang are the figures "M 1816."

France.

No. 16. (N3958). FLINT-LOCK PISTOL. Caliber .70. Eight-inch, smooth-bore barrel. Mountings are the same as those of No. 15. The lock-plate is marked with the letter "S" within an oval and with the words "Mre. Rle. de Charleville." The tang is marked with the date "1822."

France.

No. 17. (N3951). FLINT-LOCK PISTOL. Caliber .68. Eight-inch, smooth-bore barrel, equipped with front and rear sights. Mountings and flash-pan are made of red brass. This is an officer's model pistol, having a finely checked grip and a brass-tipped ebony ramrod. The lock-plate is stamped with the letter "T" within a diamond and engraved with the words "Mre. Rle de Charleville." The tang is marked "M 1816."

France.

No. 18. (N3949). FLINT-LOCK PISTOL. Seven-and-one-half-inch, smooth-bore barrel. The mountings and body are made of brass and the flash-pan forms part of the body. There is a belt-hook upon the left side. The body is marked with the letter "F" over the number "80" and with the word "Charleville." The trigger is straight.

France.

No. 19. (N3948). FLINT-LOCK PISTOL. Caliber .69. Seven-and-one-half-inch, smooth-bore barrel. Mountings and body are of brass, and, as in the case of No. 18, the flash-pan is cast to the body. The body is marked with a crown over the letter "J" and with the word "St. Etienne." This pistol has no belt-hook. The barrel is marked with a crown and with "B.M. 85." The trigger is curved.

France.

No. 20. (N3950). FLINT-LOCK PISTOL. Caliber .69. Seven-and-one-half-inch, smooth-bore barrel. Mountings and body are the same as those of No. 19. The trigger is straight. The top of the barrel is marked with the year "1777" and with the numeral "18."

France.

No. 21. (N3947). FLINT-LOCK PISTOL. Caliber .64. Eight-and-one-half-inch, smooth-bore barrel. All mountings are of silver. There are front and rear sights. Engraved upon the barrel are the words "Peniet A Paris." This is said to be an 1809 government model pistol.

France.

No. 22. (N3968). FLINT-LOCK PISTOL. Caliber .60. Eight-inch, octagonal barrel, rifled with thirty-six grooves. Provided with front and rear sights. All mountings are of engraved brass. The butt is checkered and inlaid with silver wire. No maker's name is given.

France.

No. 23. (N3969). FLINT-LOCK PISTOL. Caliber .66. Eight-inch, octagonal barrel, rifled with sixty grooves. The muzzle is bell-shaped. Hammer, pan and frizzen are very elaborate. The barrel is provided with front and rear sights. This is probably an officer's model pistol. Engraved upon the top of the barrel are the words "F———ni Par Lgiage à Paris." This is said to be a 1763 model pistol made in 1802.

France.

No. 24. (N3973). FLINT-LOCK PISTOL. Caliber .62. Five-inch, smooth-bore barrel. All mountings are made of iron. This is a 1763 model pistol, the first model issued to the French gendarmerie. Although not intended for use in the army or navy, nevertheless it was a pocket pistol which was popular with officers of low and medium rank in these branches of the service. Upon the lock-plate is engraved "Manufacture de Charleville."

France.

No. 25. (N3972). FLINT-LOCK PISTOL. Caliber .60. Five-inch, smooth-bore barrel. The mountings are all of iron. This is another example of the pistol used by the French gendarmerie. It is probably a 1776 or 1777 model and in most respects is similar to No. 24. Engraved upon the lock-plate are the words "Maubeuge Manuf. Nle."

France.

No. 26. (N3974). FLINT-LOCK PISTOL. Caliber .60. Barrel and mountings are the same as those of No. 25. Engraved upon the lock-plate are the words "Maubeuge Manuf. N." This is another example of the French gendarme pistol and it is said to be of the 1802 model.

France.

No. 27. (N3971). FLINT-LOCK PISTOL. Caliber .64. Five-inch, smooth-bore barrel. The mountings are of iron and the flash-pan of brass. The lock-plate is engraved with the words "Maubeuge Manuf. Im.", and the barrel bears the date "1811."

France.

No. 28. (N3970). FLINT-LOCK PISTOL. Caliber .60. Five-inch, smooth-bore barrel, provided with a brass flash-pan. Engraved upon the lock-plate are the words "Manuf. Roy. de Maubeuge." The tang is marked with the figures "M an 9" and with "C. 1817."

France.

No. 29. (N3977). FLINT-LOCK PISTOL. Caliber .60. Five-inch, smooth-bore, half-octagonal barrel, having a bell-shaped muzzle. All mountings are of iron. The name "Duc" appears upon the lock-plate and the barrel is engraved with the words "F. par Duc à Paris."

France.

No. 30. (N3277). FLINT-LOCK PISTOL. Caliber .55. Five-inch, octagonal barrel, rifled with forty grooves. The muzzle is bell-shaped. The ramrod is of horn and has a brass head. This pistol is equipped with a rear sight, but no front sight. All mountings are of iron. Upon the barrel are three gold hall-marks.

France.

No. 31. (N3978). FLINT-LOCK PISTOL. Caliber .56. Five-and-one-half-inch, half-octagonal barrel. Has a front sight but none in the rear. All mountings are of iron. No marks of identification are given.

France.

No. 32. (N3975). FLINT-LOCK PISTOL. Caliber .54. Five-and-one-quarter-inch, half-octagonal barrel, having a bell-shaped muzzle. All mountings are of iron but the flash-pan is of brass. No maker's name is given upon the lock-plate. The barrel bears the Liége proof-mark.

Belgium.

No. 33. (N707). FLINT-LOCK PISTOL. Caliber .60. Four-and-one-quarter-inch, octagonal barrel. This pistol is provided with a rear sight only. The mountings are of iron. The piece is marked with the letters "L.G", or "L. C."

France.

No. 34. (N3944). FLINT-LOCK PISTOL. Caliber .56. Three-inch, half-octagonal barrel, having a cannon-shaped muzzle. All mountings are of iron. The grip is inlaid with silver wire. No marks of identification are given.

France.

No. 35. (N3941, N3943). PAIR OF FLINT-LOCK PISTOLS. Caliber .52. Three-and-one-quarter-inch, half-octagonal barrels, which are partly gilded. The muzzles are cannon-shaped. The ramrods are of horn and the butts are inlaid with silver wire. No maker's name is given.

France.

PLATE 71.

FLINT-LOCK CANNON BARRELS AND POCKET PISTOLS

No. 1. (N3278). FLINT-LOCK PISTOL. Caliber .55. Seven-inch barrel, having a cannon-shaped muzzle. All parts of the pistol are made of engraved steel, this being a Scotch Highlander Pistol. It has a side-hammer and the sear extends through the lock-plate and holds the hammer at full cock. There is no guard around the knob-trigger. A belt-hook is upon the left side of the pistol. The grip is flat and the butt is scroll-shaped. Upon the lock-plate is the name "John Campbell." One gunsmith by the name of John Campbell worked about 1860 and another of the same name about 1775; both working at Doune, Scotland.

Scotland.

THE SCOTCH HIGHLANDER PISTOL

The Highlander pistol is said to have developed from the all-steel German wheel-lock pistols. The earliest dated Highlanders known at the present time are in the collection of the Museum of Dresden, and bear the date "1598" and the armorer's initials "F. K." They are a pair of snaphaunces, one pistol being left-handed, the other right-handed.

Highlander pistols were commonly made of steel. Some, however, were made of copper, brass, bronze or combinations of these metals in the same arm. Those made of iron or steel were generally brightly polished, but sometimes they were also blued. The barrels were rarely rifled and they were usually single shot. Some true Highlanders are known which are multi-barreled.

Highlander pistols probably came into existence in answer to the demands of the Scotch chieftains for elegant and flashy weapons with which to decorate their sporrans and belts. The money for the purchase of these weapons was usually obtained by the sale of cattle on market days at such towns as Doune, Dundee, St. Andrew's and other places, where were located the gunsmiths who made these pistols. The price for the pistols ranged from fifty dollars per pair for the ordinary kind to six hundred dollars per pair for the more elaborate varieties, these prices being in terms of present day money.

Highlander pistols were in considerable vogue during the period 1700-1750 when English and Continental nobility and army officers used them to a great extent. This pistol also played a part in history and romance. The first shot fired at the battle of Lexington in the Revolutionary War, on April 19, 1775, was fired by Major Pitcairn at Captain Parker's men from a Highlander pistol. A pair of these pistols, captured by the Americans during this battle, were carried during the war by General Putnam. They may now be seen in the Hancock-Clark House at Lexington. Rob Roy, the outlaw made famous in Sir Walter Scott's novel, also owned a beautiful pair of Highlander pistols.

As weapons they were not very useful, however, though they made a showy appearance. Usually they were of large caliber and were light in weight, which combination produces a strong recoil and consequent inaccuracy. About the year 1800 these pistols were no longer made in Scotland, but English and Continental copies were produced during the first part of the nineteenth century.

No. 2. (N4020). FLINT-LOCK PISTOL. Caliber .55. Six-inch, smooth-bore barrel, having an octagonal muzzle. All parts are made of elaborately engraved steel. The hammer is of the side variety and the knob-trigger has no guard. There is a belt-lock upon the left side. The grip is flat and has a scroll-shaped butt. The name "Mc-Lauchlan" is engraved upon the lock-plate and the top of the barrel is engraved with the word "Edinburgh." McLauchlan worked in Edinburgh, about the year 1810 and usually made his Highlander pistols out of cast iron, they being intended for decoration only.

Scotland.

No. 3. (N4021). FLINT-LOCK PISTOL. Caliber .48. Five-and-one-half-inch, rifled barrel, having a cannon-shaped muzzle. The barrel may be unscrewed from the profusely engraved octagonal section of the body. All parts are made of steel. Has a top-hammer and a folding-trigger with a safety device which may be set at half-cock. The grip is flat and has a hole through the butt which serves as a barrel wrench, this pistol being made in pairs. This piece is engraved with the words "Ponsin, Liége."

Belgium.

No. 4. (N4024). FLINT-LOCK PISTOL. Caliber .35. Four, two-and-one-half-inch, rifled, cannon-shaped barrels, each of which may be unscrewed from the body of this multi-shot pistol. All parts of the piece are made of plain, unornamented steel. The gun has two top-hammers with a safety-device in the form of a slide which holds the cocks at full-cock. There are two triggers and four flash-pans and frizzens. The trigger-guard forms a lock for the barrels which are revolved as a block by hand. No maker's name is given.

Spain.

No. 5. (N4022). FLINT-LOCK PISTOL. Caliber .31. Two-and-one-quarter-inch, rifled, cannon-shaped barrels. The barrels of this double-barreled pistol unscrew from the octagonal section of the body. All parts are made of engraved steel. There are two top-hammers and two triggers. The siiding trigger-guard can be used as a safety lock. Engraved upon the pistol are the words "Richards, London." This gunsmith was active about the year 1700.

England.

No. 6. (N4023). FLINT-LOCK PISTOL. Caliber .35. Two-

and-three-quarters-inch, round, rifled barrels. This is another double-barreled pistol, the barrels of which can be unscrewed from the body. All parts are made of steel, most of which is engraved. The cocks are not attached to the side, but work through the top of the lock. The pistol bears the words "Richard, London"; probably the same gunsmith who signed his name "Richards" on No. 5 described above.

England.

No. 7. (N3279). FLINT-LOCK PISTOL. Caliber .35. Two-and-three-quarters-inch, smooth-bore barrel, the front section of which unscrews from the octagonal portion of the body. All parts are made of steel and the body is engraved. The grip is flat and the butt is rounded. Engraved upon the pistol are the words "Griffin, London."

England.

No. 8. (N6340, N6341). PAIR OF FLINT-LOCK PISTOLS. Caliber .44. Five-and-seven-eighths-inch, cannon-shaped, tapering barrels, each rifled with eight grooves. The rear portion of each barrel is octagonal and unscrews from the front part. Formerly there was a device which connected the front end of the stock to the ring around the barrel; the object of the device being to prevent the barrel from falling to the ground when it was unscrewed from the body of the gun. All mountings are of engraved, chiselled iron. Each trigger-guard is decorated with an engraved head. The stocks are of beautiful wood, which appears to be mottled maple. The barrels are stamped with the view and the London Gunmaker's Company proof-marks and with a private mark. Engraved upon the lock-plates and barrels is the name "R. Savage."

England.

No. 9. (N694 a, b). PAIR OF FLINT-LOCK PISTOLS. Caliber .78. Six-and-one-quarter-inch, round, smooth-bore barrels, having cannon-shaped muzzles which may be unscrewed from the bodies of the pistols. One reason for unscrewing the barrel from the pistol was to load the breech with a ball whose diameter was such as to prevent its rolling out of the barrel should the pistol be pointed muzzle downwards. The barrel was usually slightly chambered at the rear to receive the bullet. The pistols here illustrated are sometimes known as "Queen Anne Pistols." All types of cannon-shaped pistols were usually highly ornamented and inlaid; most of the butts terminating in

silver or brass masks. They were usually used by English gentlemen
and were expensive to make. The weapons shown here are ornamented
with embossed silver mountings; the butts being covered with silver
plates and the grips inlaid with silver wire. Each barrel is stamped
with three proof-marks which are in the form of a crown over the
letter "P". Below this mark is that of the maker in the form of an
oval enclosing the letters "W.H." and below this the viewer's mark in
the form of a crown over the letter "V". All these are early London
marks. Engraved upon the under-side of each barrel is the gunsmith's
name, "W. Heath." Bodies and barrels of each pistol are numbered
to avoid screwing the wrong barrel upon any certain body.

England.

No. 10. (N3981). FLINT-LOCK PISTOL. Caliber .45. Six-
inch, half-octagonal, rifled barrel. This is another pistol having a
barrel which is threaded to the body of the pistol, and which can be
unscrewed by the aid of a wrench. All mountings are of iron and
unornamented. This is said to be a model 1797 French officer's pistol.

France.

No. 11. (N2868). FLINT-LOCK PISTOL. Caliber .60. Six-
and-one-half-inch, smooth-bore, brass barrel, having a cannon-shaped
muzzle. The pistol has a top action and an enclosed lock. Hammer,
frizzen, trigger, and guard are of iron. The round wooden grip is
inlaid with elaborate floral designs in the form of silver wire. The butt
is covered by an embossed brass griffin's head. Engraved upon the
pistol are the words "F. F. Taylor, London."

England.

No. 12. (N6377, N6378). PAIR OF FLINT-LOCK PISTOLS.
Caliber .60. Seven-inch, smooth-bore, cannon-shaped barrels, which
may be removed by unscrewing them. The bodies are engraved and
bear several proof-marks. The round grips are inlaid with silver wire
forming floral designs, while the butts are covered by plates in the form
of embossed griffins' heads. These pistols are said to have been used
by a French general in the battle of Leipsic, October 16-18, 1813. The
bodies are engraved with the name "Walsingham." William Walsingham was a gunsmith who worked in Birmingham, before and after the
year 1770.

Donor: C. H. Holtz.

England.

No. 13. (N707). FLINT-LOCK PISTOL. Caliber .60. Two, interchangeable, smooth-bore barrels, one being five-and-one-quarter-inches in length, the other two-and-one-quarter-inches. This is an early form of cannon-shaped pistol. The peculiarities of this type were an action fitted into a box behind the breech; a hammer at the side; and a frizzen spring placed beneath the pan curving round to a point midway between the pan and hammer. All mountings upon the pistol are of silver. Upon the barrel are London proof-marks and the body is engraved with the words "David Wynn, London."

England.

No. 14. (N3982). FLINT-LOCK PISTOL. Caliber .60. Seven-and-one-quarter-inch, cannon-shaped barrel. The smooth, flat grip is elaborately inlaid with silver wire while the butt is covered with an embossed silver plate. The action is located on top of the barrel. There are no sights. The barrel is easily unscrewed. Stamped upon the breech are several proof-marks, probably London. One side of the lock-plate is engraved "Jones, London."

England.

No. 15. (N674). FLINT-LOCK PISTOL. Caliber .45. Four-and-one-quarter-inch, brass barrel, having a cannon-shaped muzzle which can be unscrewed from the square brass body. This is another pistol of the top action variety. It has a round wooden grip inlaid with conventionalized floral designs in fine silver wire. The butt is covered by an embossed, German silver plate. Several proof-marks appear upon the barrel. This piece was probably made by Joseph Bunny, who worked in Birmingham, about 1770 to 1812.

England.

No. 16a. (N3983). FLINT-LOCK PISTOL. Caliber .45. Three-and-one-half-inch, smooth-bore, steel barrel, with a cannon-shaped muzzle which can be unscrewed from the body. This pistol also has a top action. The body is engraved, the round grip elaborately inlaid with designs in fine, silver wire and the butt covered with an embossed German silver plate showing a grotesque human head. Three proof-marks are stamped upon the under-side of the barrel. Engraved upon the body are the words "W. Aston, Manchester."

England.

No. 16b. (N3984). FLINT-LOCK PISTOL. Caliber .41.

Three-and-one-half-inch, smooth-bore, cannon-shaped barrel, which may be unscrewed from the body. The general characteristics including the form of the butt-plate are the same as those of No. 16a. However, the present pistol has no silver wire inlaid into the grip. The three proof-marks stamped upon the under-side of the barrel appear to be of the early London type. One side of the lock-plate is engraved with the gunsmith's name, "Annely."

England.

No. 17a. (N3987). FLINT-LOCK PISTOL. Caliber .45. Three-and-one-half-inch, cannon-shaped, iron barrel, which can be unscrewed. There are no sights and the hammer is of the superposed kind. The flat grip is inlaid with silver wire in scroll designs and the lock-plate is engraved. The trigger-guard is ornamented with an engraved flower. London proof-marks are stamped upon the barrel. Engraved upon the lock-plate are the words "Barbar, London," Barbar being a gunsmith who worked about the year 1705.

England.

No. 17b. (N3988). Three-and-one-quarter-inch, smooth-bore, cannon-shaped barrel. The grip is smooth and flat and the top is inlaid with a silver escutcheon bearing the letter "V". An embossed silver plate covers the butt and a small amount of engraving adorns the lock-plate. There are two proof-marks upon the barrel. The words "Joiner, London," are engraved upon the lock-plate.

England.

No. 18. (N3986). FLINT-LOCK PISTOL. Caliber .45. Three-and-one-half-inch, smooth-bore, cannon-shaped barrel, which may be unscrewed from the body of the pistol by means of the wrench shown below the gun. It has a top action. The flat grip is inlaid with silver wire in scroll designs, the lock-plate is engraved and the barrel stamped with London proof-marks. The trigger-guard is ornamented with an engraved flower. Upon the lock-plate are the words "Barbar, London," a gunsmith who worked about the year 1705.

England.

No. 19. (N3985). FLINT-LOCK PISTOL. Caliber .45. Three-inch, smooth-bore, cannon-shaped barrel, which may be unscrewed from the body. The round grip is elaborately ornamented with inlaid silver wire and the butt is covered with a grotesque, embossed silver head. Upon the barrel are two proof-marks, one in the form of a crown over the letter "P" and the other with a crown over the letter "V". The left side of the lock-plate is engraved with the name "T. Davies," and the right side "Surgeon."

England.

No. 20. (N3980). FLINT-LOCK PISTOL. Caliber .50. Three-and-one-quarter-inch, smooth-bore barrel, having a cannon-shaped muzzle. The front section of the barrel unscrews from the body. This pistol has a side-hammer and battery. The smooth, plain grip is rather short. The body is engraved with the name "T. Hawley."

England.

No. 21. (N683). FLINT-LOCK PISTOL. Caliber .52. Two-and-one-half-inch, smooth-bore barrel, having a cannon-shaped muzzle which may be unscrewed from the body. This pistol has a straight trigger without a guard. The grip is short, smooth and rather clumsy. No marks of identification are given.

Germany.

No. 22. (N3989). FLINT-LOCK PISTOL. Caliber .50. Two-and-one-half-inch, cannon-shaped, steel barrel, which can be unscrewed with the accompanying wrench. The flat, checkered grip has the top inlaid with silver wire. The superposed hammer is provided with a safety catch. There are no sights. The trigger is of the concealed or folding variety. A slight amount of engraving adorns the lock-plate which is also inscribed with the name "Nicholson, London."

England.

No. 23. (N3979). FLINT-LOCK PISTOL. Caliber .55. Three-and-three-quarters-inch, octagonal, smooth-bore barrel. The grip is very finely checkered. A ramrod is carried beneath the barrel. This is a "Queen Anne" pocket pistol, the action being enclosed in a box behind the breech, the hammer being on the side, and the frizzen spring underneath the pan, curving round to a point midway between the pan and hammer. Barrel and body are slightly engraved and the butt is covered with an embossed lion's head. Engraved upon the weapon are the words "Mortimer, London," who may have been P. W. Mortimer who worked about the year 1789.

England.

No. 24. (N3991, N3992). PAIR OF FLINT-LOCK PISTOLS. Caliber .50. One-and-three-quarters-inch, smooth-bore, brass barrels, which can be unscrewed from the brass bodies. These pistols have concealed triggers and the top-hammers are provided with safety-devices in the form of slides. All grips are finely checkered. Proof-marks are stamped upon the under-side of each pistol. Each is engraved with the words "Goodwin & Co., London."

England.

No. 25. (N3993, N3994). PAIR OF FLINT-LOCK PISTOLS. Caliber .48. One-and-three-quarters-inch, smooth-bore, brass barrels, which may be unscrewed from the bodies. These pistols have flat, plain wooden grips, and top-hammers. Three proof-marks are stamped upon the under-side of each breech. Upon the left side of each lock-plate is the name "Sharpe" and upon the right side is the word "London."

England.

No. 26. (N4014). FLINT-LOCK-PISTOL. Caliber .46. Two-and-one-half-inch, cannon-shaped, brass barrel. Barrel and body are cast in one piece. The top-hammer is equipped with a safety-device in the form of a slide. The grip is flat and smooth. Stamped upon the barrel is the Liége proof-mark.

Belgium.

No. 27. (N3990). FLINT-LOCK PISTOL. Caliber .36. Two-and-one-half-inch, removable, steel barrel. The smooth grip was formerly inlaid with wire. Has a superposed hammer and consequently no sights. The lock-plate and all steel parts of this pistol are unornamented. No maker's name or proof-marks are given.

Belgium?

No. 28. (N4000). FLINT-LOCK PISTOL. Caliber .44. Three-and-one-half-inch, detachable, steel barrel. Has a flat, smooth grip and a superposed, engraved hammer. There are no sights. The steel trigger-guard is engraved with a flower. Two proof-marks are stamped upon the barrel. Engraved upon the lock-plate are the words "Booth, Sunderland."

England.

No. *29*. (N4003). FLINT-LOCK PISTOL. Caliber .41. Three-inch, removable, cannon-shaped, steel barrel. Has a flat, smooth grip, and a superposed hammer but no sights. The trigger-guard is of iron and all parts of the weapon are unornamented. Stamped upon the barrel are two proof-marks while the lock-plates bear the words "Thomas, London."

England.

No. *30*. (N4009). FLINT-LOCK PISTOL. Caliber .36. Three-and-one-half-inch, rifled, octagonal, steel barrel. Has a smooth, semi-oval grip and a superposed hammer, which formerly had a safety-catch. There are no sights. The breech and body of this pistol are of brass, while the trigger-guard is of iron. The gun is marked "Ketland, London," and in addition bears the Liége proof-mark.

England.

No. 31. (N3998). FLINT-LOCK PISTOL. Caliber .43. Three-and-five-eighths-inch, round, steel barrel. The superposed hammer has no safety-device. The body is of brass and the trigger-guard of engraved iron. Engraved upon the lock-plate are the words "Patrick, Liverpool."

England.

PLATE 72.

FLINT-LOCK POCKET PISTOLS
MOSTLY ENGLISH

No. 1. (N692). FLINT-LOCK PISTOL. Caliber .55. Five-and-seven-eighths-inch, round, steel barrel, with superposed hammer. The flat, smooth grip was formerly inlaid with wire. The butt is covered with the embossed brass head of an animal. Upon the left side of the lock-plate is the inscription "Speder á Liége." The right side of the pistol is provided with a folding bayonet which springs into position ready for use upon releasing a catch. This device was the invention of John Waters of Birmingham, who was granted patent No. 1284 covering this invention on March 9, 1781. The abstract of the patent reads:

"Pistols with a bayonet. The pistols have one or more barrels, and to each a knife or bayonet may be connected by a spring, slide, hinge or otherwise."[165]

Belgium.

No. 2. (N4006). FLINT-LOCK PISTOL. Caliber .45. Four-and-three-quarters-inch, round, steel barrel, which may be unscrewed. The pistol has a superposed hammer and consequently no sights. It is provided with a smooth, wooden grip, and a three-and-one-half-inch, folding bayonet which is released by pulling back the trigger-guard. The muzzle of the barrel is slightly ornamented and the lock-plate is engraved and bears the name "Jovir, London."

England.

No. 3. (N4010). FLINT-LOCK PISTOL. Caliber .50. Four-and-three-quarters-inch, octagonal, steel barrel. This pistol has a smooth, flat grip, an ornamented trigger-guard and a superposed hammer provided with a safety-catch. An iron ramrod is carried beneath the barrel. The lock-plate is engraved with wreaths and bears the name "H.I. Burg."

Germany.

No. 4. (N4011). FLINT-LOCK PISTOL. Caliber .50. Four-inch, round, steel barrel. This weapon has a flat, finely checkered grip, a superposed hammer provided with a safety-device and a trigger-guard engraved with a flower. A ramrod is fastened beneath the barrel. The lock-plate is engraved but no maker's name or other marks of identification are given.

England?

No. 5. (N4008). FLINT-LOCK PISTOL. Caliber .45. Three-and-three-quarters-inch, round, steel barrel. Has a flat, smooth grip, the top of which is inlaid with an oval silver escutcheon. The superposed hammer is fitted with a safety catch. There are no sights. The trigger-guard and lock-plate are slightly engraved and the latter is inscribed "Dally, Paris."

France.

No. 6. (N4013). FLINT-LOCK PISTOL. Caliber .50. Three-and-three-quarters-inch, round, steel barrel. This is another pistol having a top action and a smooth, flat grip. It also has the usual sliding

---

[165]"Patents for Inventions, Abridgments of the Specifications Relating to Firearms and Other Weapons, Ammunition and Accoutrements." London, 1859. p. 30.

safety device which locks the cover of the flash-pan to prevent the accidental escape of the priming while in the pocket. The trigger-guard is slightly engraved and the ramrod is carried in ferrules beneath the barrel. Lock-plate and barrel are devoid of ornamentation. No marks of identification are given.

England?

No. 7. (N4012). FLINT-LOCK PISTOL. Caliber .48. Three-and-one-half-inch, octagonal, steel barrel. The general characteristics of this gun are the same as those of No. 6. A ramrod is carried beneath the barrel. Stamped upon the left side of the barrel is the Liége proof-mark. No maker's name is given.

Belgium.

No. 8. (N3946). FLINT-LOCK PISTOL. Caliber .40. This is merely the body or frame of a pocket pistol; the barrel and stock are missing. The superposed hammer is provided with a sliding safety device. Engraved upon the lock-plate are the words "Rigby, Dublin."

Ireland.

No. 9. (N4007). FLINT-LOCK PISTOL. Caliber .52. Three-and-three-quarters-inch, round, steel barrel. Has a small, flat grip, a superposed hammer provided with a safety-device and a folding-trigger, which is brought down automatically into place by bringing the pistol to full-cock. The lock-plate is crudely engraved and bears the name "Grifin, London."

England.

No. 10. (N2677). FLINT-LOCK PISTOL. Caliber .50. Four-inch, octagonal, brass barrel. Has a smooth, semi-oval grip, a super-posed hammer, provided with a safety-catch, and a steel trigger-guard which is slightly engraved. An iron ramrod is carried in brass ferrules underneath the barrel. The lock-plate is engraved with spears, drums and flags. No maker's name or proof-marks are given.

Donor: Robert Nunnemacher.

England?

No. 11. (N698, b). PAIR OF FLINT-LOCK PISTOLS. Cali-ber .43. Two-and-one-half-inch, detachable, steel barrels. These pistols have flat, checkered grips, superposed hammers equipped with safety-devices but the weapons have no sights of any kind. Each lock-plate

is engraved with spears, flowers and banners. Upon the left side of
each barrel is the Liége proof-mark. No maker's name is given.

Belgium?

No. 12. (N3995, N3996). PAIR OF FLINT-LOCK PISTOLS.
Caliber .45. Two-and-three-eighths-inch, steel barrels, which may be
unscrewed from the bodies of the pistols by the aid of a wrench. Trigger-guards and hammers are slightly engraved. Upon the under-side
of each barrel are three proof-marks. The lock-plates are engraved
with the words "Goldring & Son, Kingston."

England.

No. 13. (N4005). FLINT-LOCK PISTOL. Caliber .47. Three-
and-one-half-inch, steel barrel, which is detachable from the body of
the pistol. Has a smooth, flat grip, a superposed hammer fitted with
a safety-catch to hold the pistol at half-cock and a folding-trigger. The
lock-plate and the rest of the pistol lack ornamentation and marks of
identification.

England.

No. 14. (N4004). FLINT-LOCK PISTOL. Caliber .50. Three-
inch steel barrel, which may be unscrewed. Has a smooth, flat grip and
a superposed hammer with the usual safety-device. There are no sights
and the pistol is unornamented. Stamped upon the barrel is the Liége
proof-mark, while the lock-plate bears the word "London."

England.

No. 15. (N3997, N3999). PAIR OF FLINT-LOCK PISTOLS.
Caliber .50. Two-inch, detachable steel barrels. The superposed hammers are provided with sliding safety-devices. There are no sights.
The flat, checkered grip has a smooth border. Each pistol has a folding-trigger and each lock-plate is slightly engraved but there are no
marks of identification.

England?

No. 16. (N699a, b). PAIR OF FLINT-LOCK PISTOLS. Caliber .55. Two-and-one-eighth-inch, steel barrels, which may be unscrewed from the frames. The flat, checkered grips are studded with
minute nails spaced equidistantly. The lock-plates are ornamented with
engravings of flowers and leaves. Upon the left side of each barrel is
the Liége proof-mark. No maker's name is given.

England?

No. 17. (N3280). FLINT-LOCK PISTOL. Caliber .48. Two-and-one-quarter-inch, detachable, steel barrel. Has the usual flat, smooth grip and the superposed hammer. The trigger-guard is engraved with a radiating star, and the lock-plate is engraved with banners and the name "Smith, London."

England.

No. 18. (N3396). FLINT-LOCK PISTOL. Caliber .48. Two-and-one-half-inch, detachable, steel barrel. Has a smooth, flat grip and a superposed hammer provided with a safety-device. All parts of the pistol are unornamented. The barrel is stamped with the Liége proof-mark and the lock-plate bears the words "Ketland, London."
Donor: Mrs. W. T. Cushing.

England.

No. 19. (N4001). FLINT-LOCK PISTOL. Caliber .48. Two-and-one-half-inch, detachable, steel barrel. Has a smooth, flat grip fitted with a ring in the butt. The trigger-guard is ornamented with an engraved flower and the barrel is stamped with two proof-marks. The lock-plate is stamped "Clarkson, London."

England.

No. 20. (N4002). FLINT-LOCK PISTOL. Caliber .45. One-and-three-quarters-inch, detachable, steel barrel. Has the usual top-hammer with a safety-device. Stamped upon the pistol is the Liége proof-mark and the number "42." The lock-plate bears the words "Thoms, London."

England.

No. 21. (N697). FLINT-LOCK PISTOL. Caliber .44. Two-and-one-quarter-inch, steel barrel, which unscrews from the body. Has a smooth, flat grip which is inlaid with silver wire. The trigger is of the folding kind, consequently there is no trigger-guard. The lock-plate is engraved but the barrel has neither proof-marks nor maker's name.

England?

No. 22. (N4019). DOUBLE-BARRELED FLINT-LOCK PISTOL. Caliber .35. Two-and-five-eighths-inch, round, steel barrels, having superposed hammers but no safety-devices or sights. The grip is smooth and flat and the trigger-guard is ornamented with a star. The inner back-strap is inscribed with the letters "N.I." and the lock-plate is engraved with quivers and arrows. No marks of identification are given.

England?

No. 23. (N700). DOUBLE-BARRELED FLINT-LOCK PISTOL. Caliber .35. Two-and-three-eighths-inch, half-octagonal cannon-shaped barrels. The bulbous grip is slightly carved on top. The trigger-guard is engraved with a rose and leaves and the lock-plate with scroll designs. No maker's name or proof-marks are given.

England?

No. 24. (N703a, b). PAIR OF FLINT-LOCK PISTOLS. Caliber .42. Three-and-one-half-inch, double, superposed, brass barrels. The grips are partly checkered and fitted with two cheek-pieces which are inlaid with silver wire. There is one top-hammer. The upper barrel is discharged first, after which a cylinder with a hole, sometimes known as a "faucet" or "tap" is rotated by means of a small lever outside of the barrel. The rotation of the cylinder brings the hole over the vent of the second barrel so that it may be discharged. The trigger-guards are of brass and engraved with a flower. The lock-plates are engraved with floral designs and stamped with the letters "EX."

England.

No. 25. (N4015, N4016). PAIR OF FLINT-LOCK PISTOLS. Caliber .45. Four-inch, double, superposed, brass barrels. The super-posed hammers are provided with safety-catches. Both barrels can be fired at once or each separately by the "faucet" arrangement described under No. 24. Each pistol is provided with a spring bayonet two-and-one-quarter-inches in length, which is released by drawing back the steel trigger-guard. The lock-plates are engraved with flags and bear the words "Lacy, London."

England.

No. 26. (N4018). DOUBLE-BARRELED FLINT-LOCK PISTOL. Caliber .45. Three-and-one-half-inch, half-octagonal, removable, steel barrels. The pistol has two top-hammers but no safety-devices or sights. All parts of the weapon are unornamented and no maker's name or proof-marks are given.

England?

No. 27. (N706). DOUBLE - BARRELED FLINT - LOCK PISTOL. Caliber .45. Four-and-one-quarter-inch, half-octagonal, brass barrels. One top-hammer serves both barrels. The left barrel is covered by a steel slide when the right one is fired. The slide is moved over the vent by means of a pulling device attached to the side of the barrel. The hammer is provided with a safety-device. There are no sights. The trigger-guard is of iron. Proof-marks are stamped upon the under-side of the barrels and the lock-plate bears the name "Wheeler, London."

England.

No. 28. (N3281). FLINT-LOCK PISTOL. Caliber .55. Five-inch, double, superposed, brass barrels, having one top-hammer provided with a safety-catch. Both barrels can be fired at once or each can be fired separately by means of the "faucet" arrangement described under No. 24. A three-inch, folding, triangular bayonet, which is released by pulling back the trigger-guard, is carried beneath the lower barrel. Engraved upon the lock-plate are the words "P. Bond, No. 45 Cornhill, London." The name may refer to Philip Bond, who worked in London about the year 1776.

England.

No. 29. (N4017). FLINT-LOCK PISTOL. Caliber .45. Four-and-one-quarter-inch, double, superposed steel barrels which can be unscrewed by inserting a special wrench into the muzzles. Both barrels can be fired at once or each separately as in No. 24, already described. The trigger-guard is engraved with a flower and the barrels are stamped with two proof-marks. Upon the lock-plate are the words "W. Ketland & Co., London."

England.

No. 30. (N6338). TRIPLE - BARRELED FLINT - LOCK PISTOL. Caliber .39. Three-and-three-quarters-inch barrels, each of which is rifled with eight grooves. One frizzen and one cock serves the three barrels, two of which can be fired at once or any one may be discharged separately by turning the "faucet" device by means of the lever on the left side of the gun. This "faucet" or "tap" serves as a false bottom to the pan and on being turned exposes a charge of priming powder lying in a longitudinal recess in the tap itself and communicating with the other barrels. If the pistol is fired in the usual way, the upper left-hand barrel is discharged first, next the upper right barrel, and lastly the lower barrel. A two-and-three-quarters-inch, triangular, spring bayonet, which can be released by pulling back the trigger-guard, is carried beneath the lower barrel. The grip is smooth and flat and the top of it is inlaid with an oval silver escutcheon. The sliding safety and the trigger-guard are slightly engraved. Stamped upon the lower barrel are the view and proof-marks of the Gunmaker's Company. The lock-plate is engraved with banners, drums and wreaths and is inscribed "Nicholson, Corn-Hill, London."

England.

No. 31. (N6337). FOUR-BARRELED FLINT-LOCK PISTOL. Caliber .40. Four, four-and-one-half-inch barrels, each of which is rifled with eight grooves. Each barrel can be unscrewed by inserting a special wrench into the muzzle. There are two top-hammers and two triggers. Each barrel has its own frizzen. By pulling back on a catch outside and below the trigger-guard the four breech-blocks and four barrels can be turned as a unit so that two barrels are presented to the hammers at a time. After the two barrels are discharged the catch is released and the other two brought into position. A sliding safety-device is fitted to the rear of the hammers. The smooth grip is made of mottled wood, is thick at the butt and inlaid with silver wire in the form of scrolls. Engraved upon the body are quivers, leaves and arrows. The trigger-guard, safety and other parts of the pistol are also slightly engraved. Upon the inner backstrap are the letters "L.A.L.", engraved in script. No other marks of identification are given upon the weapon which is said to be of French origin.

France?

PLATE 73.

PISTOLS IN CASES

No. 1. (N6333). PAIR OF FLINT-LOCK PISTOLS. Caliber .67. Seven-and-one-quarter-inch, smooth-bore barrels, fitted with five-inch, triangular, folding bayonets carried on top of the barrels. The caliber given above represents the average caliber, as one pistol has a caliber of .674 and the other one of .666. The barrels of both pistols are browned and both are provided with swivel ramrods. Trigger-guards and ferrules are of engraved brass. The stocks are finely checkered, tipped with horn, and inlaid on top with oval silver escutcheons. Rollers are attached to the frizzen and bayonet springs in order to reduce friction. The roller upon the frizzen spring came into being about 1790. Its use caused the flash-pan to open much more quickly and consequently the priming received many more sparks, thus providing a more rapid and certain ignition. Stamped upon the under-side of each barrel are the interlaced letters "G.P.", representing the Gunmakers' Company proof-mark. Also the letter "V" surmounted by a crown, denoting the same company's view mark. Each lock-plate is engraved with the words "T. Mortimer & Son," and upon the top of each barrel is the word "London." Safety-devices are attached to each lock-plate just behind the hammer. The case in which the pistols are contained is unusual in that the weapons and accessories are fitted into it in a horizontal, instead of the more usual flat position. Accessories consist of a powder-flask, wooden ramrod tipped with a heavy concave brass end, an iron bullet mold and a few bullets. The powder-flask also contains a compartment for a few balls. The bullet mold is provided with a pair of cutting blades at the rear of the mold. These cutters are for removing the projection of lead from the bullet after casting. Stamped near the cutting edges of the mold is the number "17" which denotes that the mold casts bullets of seventeen "bore"; or that seventeen bullets cast by this mold would weigh one pound. Bullets of seventeen bore had a diameter or caliber of .649 inches. This bore also corresponds to that of the English government ammunition commonly used during the period in which these pistols were used.

Upon the inside of the cover of the box containing the pistols there is a label bearing engravings of the British, United States, and several other coats-of-arms. About these engravings are the words:

"Patent Breeching, &c. Thomas Mortimer & Son, Gun & Pistol Manufacturers, To His Britannic Majesty, The United States of America, The Hon.ble East India Comp.y &c., No. 44 Ludgate Hill, London. Removed from Fleet Street. No connection with any other House. Merchants & Captains supplied with every article in the above Trade, on the most approved principles. Richly mounted in Gold, Silver & Steel, or set with Diamonds & suited for Europe, or any part of the Globe."

England.

No. 2. (N6336). PAIR OF PERCUSSION PISTOLS. Caliber .70. Eight-and-one-quarter-inch, octagonal barrels, each having a flared muzzle and each rifled with twenty-two grooves. The barrels are browned in mottled colors made to simulate damaskeening. All other metal work is covered with a plating of white metal which may be tin or zinc. Each trigger-guard is provided with a spur or auxiliary grip for obtaining greater steadiness in aiming. All parts of the stock are smooth and polished, the butt being rather wide and terminating in a swivel-ring. Both pistols are provided with front and rear sights. Stamped upon the left side of each barrel is the Liége proof-mark. No maker's name is given. The accessories are rather complete and consist of: a wooden ramrod, a cleaning rod, a wooden mallet for driving the bullets into the rifled barrel, a combination nipple wrench and screw-driver, two extra nipples, a powder-flask, a wooden patch-box for linen patches and grease, an oiler, a worm for removing wadding and bullets from the barrel, a hog's-hair bristle brush for attachment to the cleaning rod, and also a long sponge for the same rod, about a dozen bullets and a bullet mold marked with the number "16," indicating that it casts bullets of "sixteen bore," that sixteen bullets cast by this mold weighed one pound; or in other words, each ball weighed one ounce. The pistol case is made of mahogany lined with red plush. In all probability the pistols have never been fired.

Belgium.

No. 3. (N4049, N4051). PAIR OF PERCUSSION PISTOLS. Caliber .55. Five-and-one-half-inch, double, superposed, smooth-bore, steel barrels, with a rib between. The pistols are equipped with front and rear sights. Each weapon has two side-hammers and each hammer has a safety-catch. Each pistol also has a belt-hook upon the left side and a swivel-ramrod which is carried beneath the lower barrel. All metal parts, except the barrels, are finely engraved and all steel parts are blued, the barrels being in mottled colors. The grips are flat and checkered in a shallow fashion. Oval, silver escutcheons are inlaid in the top of each grip and a finely chiseled cap-box is located in the butt. Each barrel is stamped with the Gunmakers' Company proof-mark, and also with the view-mark. Engraved upon each lock-plate are the words "Parker Field & Sons." Upon the upper barrel are the words "Parker Field & Sons' High Holborn, London." The number "9383" appears engraved upon the inner backstrap of each pistol. The accessories consist of a wooden ramrod, a long-handled ivory brush, an oiler, a wooden box and a combination tool which is a screw-driver, nipple prick, nipple wrench and tongs.

England.

No. 4. (N663a, b). PAIR OF FLINT-LOCK DUELING PISTOLS. Caliber .53. Nine-and-one-eighth-inch, smooth-bore, octagonal, steel barrels fitted with front and rear sights. Each trigger-guard is provided with a large spur or auxiliary grip. Hammers are of the French type, having a curved under-spur. The vents are lined with gold. Lock-plates and hammers are slightly engraved and each of the latter is provided with a safety-catch. Each frizzen spring has a roller for reducing friction. All mountings are of engraved steel. Engraved upon each lock-plate is the name "Ja.ˢ Wilkinson," and upon each barrel the words "Jas. Wilkinson, London, Gun Maker to his Majesty." The former owner's name, "John D. Chauncey, U.S.A." is engraved upon a silver plate affixed to the cover of the case. The accessories consist of a wooden ramrod, having extra tips in the form of a worm and cleaner, a powder flask, several extra flints and bullets, and an iron bullet mold marked with the number "30," indicating that it cast balls of 30 bore or of .537 caliber. The case is made of polished bird's-eye maple.

England.

No. 5. (N6840). PAIR OF PERCUSSION DUELING PISTOLS. Caliber .52. Eight-and-five-eighths-inch, smooth-bore, octagonal, damaskeened barrels, fitted with front and rear sights. The grips are cross-hatched and have silver bands around the butts. Lock-plates, hammers and trigger-guards are engraved. Each lock-plate is inscribed with the names "Williams & Powell," and each barrel with the word "Liverpool." The accessories consist of a wooden cleaning rod, a nipple wrench and a powder flask. Other accessories are missing.

England.

No. 6. (N5615, N5616, N5617). PAIR OF PERCUSSION DUELING PISTOLS. Caliber .56. Eight-and-three-quarters-inch, smooth-bore, octagonal, damaskeened, steel barrels, fitted with front and rear sights. The saw-handle grips are finely cross-hatched and each butt is bound by an engraved silver band. The engraved, silver trigger-guard is equipped with a spur and the trigger is pro-vided with a small screw for regulating the trigger-pull. An ebony ramrod is carried beneath each barrel in engraved silver ferrules. Each lock-plate bears the name "Ino. Wadsworth," and each barrel the word "London." The balance of these pistols is delicate. The accessories consist of a spring ball-remover, several cleaning heads which can be attached to the wooden ramrod, a metallic cap-box, and a powder flask.

England.

No. *7.* (N661a b). PAIR OF FLINT-LOCK DUELING PISTOLS. Caliber .48. Three-inch, octagonal, smooth-bore, steel barrels. Barrels, lock-plates, triggers and back-straps are engraved. All metal work is left bright and is highly polished, there being not a speck of rust on any of it. Each flash-pan is equipped with rain drains. The pistols have concealed triggers which are revealed upon bringing the hammers to full cock. The rear sight is also a safety-device; pushing it forward prevents the pistol from being cocked at any position, the trigger also staying concealed in this case. The grips are of checkered ebony. Each barrel is engraved with the words "Boutet à Versailles." Boutet was director of the arsenal at Versailles, about the year 1800. Each pistol is inscribed with the number "316." The accessories consist of an oiler, a powder-flask, and a bullet mold, one leg of which serves as a ramrod, and the other as a screw driver.

France.

No. 8. (N660a, b). PERCUSSION PISTOLS. Caliber .48. Two-and-one-half-inch, rifled, blued, steel barrels, which are slightly engraved near the muzzle. The chambers are of engraved brass. The hammer being of the top-action variety, accounts for the fact that there are no sights upon the pistol. Each pistol is provided with a nipple-shield and a concealed trigger. These pistols are probably remodeled from the flint-lock system. Each weapon is provided with a small spring-bayonet which is carried along the right side of the barrel and which may be brought into play by releasing a catch. Each pistol bears the words "Jno. Jones & Son, London." Most of the accessories are missing, those in the box consist of a powder-flask and a bullet mold, the latter being marked with the number "38," which indicates that it casts 38 bullets to the pound, the diameter of the balls of this mold being 38 "bore" or .497 inches.

England.

No. 9. (N4413). PERCUSSION PISTOL. Caliber .28. Six-and-one-half-inch, octagonal, steel barrel, having front and rear sights. The grip is partly cross-hatched and the lock-plate and part of the back-strap are finely engraved. This pistol has a "faucet," in the rear of the barrel, and a stem entering the faucet, upon the end of which the cap is placed. The side-hammer striking this stem explodes the cap in the breech of the pistol. Engraved upon one side of the lock-plate is the name "Tanner," "In Hannover," appears upon the other side.

Germany.

No. 10. (N6334). PAIR OF PERCUSSION PISTOLS. Caliber .66. Eight-inch, smooth-bore, brass barrels. All metal parts of the pistols are made of brass except the sights, hammers, trunnion-screws and butt-plates which are of blued steel. This form of breech-loading percussion arm is sometimes known as the "Pauly" pistol. To load, the hammer is brought to full cock, the steel band in front of the trunnions is then pulled toward the muzzle, after which the breech of the barrel tips upward as shown in the pistol outside the case in the illustration. After this the cartridge, in the form of a very heavy brass shell, is inserted into the breech and the action closed. The cartridge consists of a slightly tapering cylinder bored out of one piece of brass. The rimmed head is rather heavy and contains a threaded recess in which an iron nipple, such as is commonly found upon percussion-cap pistols, is screwed. This nipple communicates with the bullet end of the cartridge by means of a small hole. The powder chamber of the cartridge is about .40 inches in diameter and half-an-inch in length. The upper rim is concavely beveled on the inside, forming a cup, so as to snugly receive and hold a one-ounce, spherical, lead ball. In order to remove the cartridge from the breech, a two-pronged iron fork, illustrated below the case, was applied to a groove around the head of the cartridge and the shell drawn out. This is in reality the first center-fire metallic cartridge, although the inventor was somewhat ahead of his time. Firing-pin, mainspring and other parts of the lock are concealed in the frame of the gun. The grip is of light wood, probably walnut, and is finely polished and checkered. The accessories consist of five extra cartridges, originally there were six; two nipple wrenches and nipple pricks, only one set of which formerly belonged to the gun; a wooden ramrod having a brass tip, which when detached converts the rod into a worm or charge extractor; a two-pronged fork for removing the cartridges from the breech; and a copper powder flask made by Sykes and stamped with his name. Two compartments are situated in the bottom of the flask, one of which contains five lead balls for the pistols. There are no proof-marks or other marks of identification upon the pistols.

England.

PLATE 74.

PISTOLS IN PAIRS
FLINT-LOCK AND PERCUSSION

No. 1. (N2894-5). PAIR OF FLINT-LOCK PISTOLS. Caliber .65. Eleven-and-seven-eighths-inch, steel barrels, the rear one-fourth of which are octagonal. The grips are smooth, and each butt is covered with a gold plate which is elaborately embossed. Each pistol has a concave rear-sight and a silver fore-sight. Lock-plates and hammers are finely engraved and the barrels are blued. The trigger-guards and ferrules are heavily gilded and embossed, as are also the mountings opposite the lock-plate. The name "Florkin à Liége" appears upon the lock-plate under the flash-pan.

Belgium.

No. 2. (N4043-4). PAIR OF FLINT-LOCK PISTOLS. Caliber .60. Eleven-and-one-half-inch, imitation damaskeened, steel barrels which are rifled. These pistols are very heavy and each is fitted with a belt-hook and a swivel ramrod. The grips are smooth and fitted with very heavy brass butts. There are neither front or rear sights. All parts of the pistols are devoid of ornamentation. Each lock-plate bears the words "Brander & Potts, Minories, London." This company did business at 70 Minories, London, before and after the year 1812.

England.

No. 3. (N2754, N2756). PAIR OF FLINT-LOCK PISTOLS. Caliber .62. Nine-inch, round, steel barrels, each of which is orna-mented with an engraved sunburst. The grips are checkered and the top of each is inlaid with a silver escutcheon marked with the letter "A". There are front and rear sights. Lock-plates, hammers and trigger-guards are engraved with floral designs. These pistols have no metal butt-plates. Each lock-plate bears the name "Manton." There were two Mantons, John Manton and Son, and Joseph Manton and Sons, all of whom worked in London. It is not known which of these made the above pistols.

England.

No. 4. (N4034-5). PAIR OF FLINT-LOCK PISTOLS. Cali-ber .58. Eight-and-five-eighths-inch, octagonal, steel barrels. The grips are cross-hatched but have no butt-plates. Each pistol has a brass fore-sight and a steel rear sight. The lock-plates are slightly engraved and each has a safety-catch. The trigger-guards are ornamented with engravings of flags and shields. Each barrel is stamped with private proof-marks and with the words "Strand, London." Engraved upon the lock-plates is the name "Iohn Richards." John Richards was in business at 55 Strand, London, before and after the year 1812.

England.

No. 5. (N4039-40). PAIR OF FLINT-LOCK PISTOLS. Caliber .57. Eight-and-one-half-inch, octagonal, steel barrels. The grips are smooth and each is protected by a solid silver butt which is slightly engraved. There are front and rear sights. Lock-plates and hammers are very slightly engraved. There is a safety-catch to the rear of each hammer. The trigger-guards are of embossed silver and the ferrules for the ramrod are of plain silver. The barrels are stamped with Birmingham proof-marks and each lock-plate bears the name "Blair." A gunsmith by this name was in business in Birmingham, about the year 1790.

England.

No. 6. (N4032, N4037). PAIR OF FLINT-LOCK PISTOLS. Caliber .72. Eight-inch, half-octagonal, steel barrels. The grips are finely checkered and inlaid on the top with a diamond-shaped escutcheon. There is a catch on each lock-plate which may be used to hold the hammer at half-cock. There is no rear sight. Each pistol bears the words "Carter & Heele."

England?

No. 7. (N4045, N4048). PAIR OF PERCUSSION PISTOLS. Caliber .48. Nine-and-three-quarters-inch, octagonal, damaskeened barrels. These pistols were remodeled from the flint-lock system. Lock-plates, hammers, barrels and trigger-guards are delicately engraved. Each of the checkered grips is inlaid with a silver escutcheon. There are back and fore-sights. All iron mountings were formerly blued. Stamped upon each barrel are the words "I. Blanch, London," and upon each lock-plate simply "I. Blanch." In 1812 John Blanch was in business at 39 Fish Street Hill, London, and this concern is still in business today under the same name.

England.

No. 8. (N2718-9). PAIR OF PERCUSSION PISTOLS. Caliber .57. Nine-inch, octagonal, damaskeened barrels. The grips are covered with checkering resembling fish-scales. A cap-box is fitted into each butt. Each pistol is provided with front and rear sights. The lock-plates and hammers are slightly engraved and the barrels are stamped with the Liége proof-mark. The pistols are numbered "1" and "2". No maker's name is given.

Belgium.

No. 9. (N4041-2). PAIR OF FLINT-LOCK PISTOLS. Caliber .62. Seven-and-three-quarters-inch, half-octagonal barrels. Lock-plates and hammers are engraved. Each grip is inlaid and ornamented with silver, and the butt-piece, also of silver, is in the form of a dog's head. The barrels have brass fore-sights but no rear sights. Each weapon bears the name "Appary" and the word "Paris."

France.

No. 10. (N704 a, b). PAIR OF FLINT-LOCK PISTOLS. Caliber .55. Eight-and-one-half-inch, ribbed barrels of damascus steel. The grips are smooth, and each butt is covered with a brass plate. The stocks are slightly carved. Each pistol is provided with a brass fore-sight but no rear-sight. Lock-plates and hammers are slightly engraved. The trigger-guards are of brass as are also the ferrules for the ramrods. No marks of identification are to be found upon either pistol.

Germany.

No. 11. (N2720-1). PAIR OF PERCUSSION PISTOLS. Caliber .43. Nine-and-one-quarter-inch, octagonal, rifled barrels. The barrels are damaskeened and each is provided with a two-leaved rear sight and a fore-sight. The back-strap of each weapon is handsomely engraved and ornamented with the raised figure of a woman holding a shield and spear. Lock-plates and hammers are also delicately engraved. The nipples are equipped with rain-drains. Each pistol has two triggers; pulling the rear one back converts the forward one into a hair-trigger. Each barrel is ornamented with gold floral designs near the breech and bears the name of "Jung. Teplitz."

Germany.

No. 12. (N695 a, b). PAIR OF FLINT-LOCK PISTOLS. Caliber .65. Seven-and-three-quarters-inch, half-octagonal barrels each of which is blued and inlaid with gold fleur-de-lis and other floral designs. There is a belt-hook upon the left side and all mountings are engraved. There are no sights upon either gun. Each barrel bears the name "BVSTINDVI" beneath a crown stamped in gold. There were two gunsmiths by the name of Bustindui, Jusepe and Santos, both of whom worked in Valencia.

Spain.

No. 13. (N4046-7). PAIR OF PERCUSSION PISTOLS. Caliber .50. Nine-and-five-eighths-inch, octagonal barrels. These pistols were remodeled from the flint-lock system. Lock-plates and trigger-guards are engraved. There is an auxiliary grip upon each trigger-guard to give greater steadiness in aiming. Each butt is inlaid on the top with an oval silver escutcheon. There are back and front sights upon each weapon. Stamped upon each barrel are the words "R. Constable, Philadelphia." The number of this set of pistols is "1062."

United States.

PLATE 75.

FLINT-LOCKS, CONVERTED FLINT-LOCKS, AND PERCUSSION ARMS OF VARIOUS COUNTRIES

No. 1. (N4036). FLINT-LOCK PISTOL. Caliber .68. Ten-and-one-half-inch, octagonal barrel. There are no sights upon this pistol. The grip is checkered but is not made with a metallic butt-plate. There is a safety-catch to hold the trigger at half-cock. Upon the lock-plate is the name "W. Parker." This is probably William Parker, who worked in London, before and after 1825.

England.

No. 2. (N5621). FLINT-LOCK PISTOL. Caliber .50. Nine-and-one-half-inch, round barrel. All metal parts except the trigger-guard and butt-plate are devoid of engraving. This pistol has front and rear sights and a plain, smooth grip. The flash-pan is lined with gold. The lock-plate is equipped with a safety-catch and bears the name "Wogdon." Upon the barrel are English proof-marks and the words "Wogdon, London." Wogdon was an English gunsmith who worked in London, about the year 1770.

England.

No. 3. (N4033). FLINT-LOCK PISTOL. Caliber .58. Eight-and-one-half-inch, octagonal barrel. Hammer and lock-plate are very slightly engraved and the grip is checkered. There are front and rear sights upon this pistol. Upon the lock-plate is the name "H. Nock," and the barrel is marked "London." Henry Nock, who made a considerable number of improvements in firearms, worked in London, about 1775.

England.

No. 4. (N5889). FLINT-LOCK PISTOL. Caliber .51. Ten-and-one-half-inch, steel barrel, which is slightly flared at the muzzle. The barrel is etched with scroll designs. The grip is checkered and has an engraved steel butt-plate. This pistol is provided with a front sight and a two-leaved rear-sight. The lock-plate is engraved, has a safety-catch, and is inscribed with the name of "Joseph Egg," the English gunsmith who apparently made the lock for this gun. The hammer is also engraved and all steel parts are blued. Upon the barrel is the name "Joh. And. Kuchenreuter," the celebrated Bavarian gunsmith of Regensburg.

Germany.

No. 5. (N2729). PERCUSSION PISTOL. Caliber .60. Nine-and-seven-eighths-inch, octagonal barrel. This pistol has been altered from the flint-lock system. It has an engraved hammer, lock-plate and trigger-guard. It also has front and rear sights and a safety-catch. A silver escutcheon is inlaid in the top of the grip. The name "William (?) Smith" appears upon the lock-plate and the barrel is marked "London."

England.

No. 6. (N4268). PERCUSSION PISTOL. Caliber .48. Nine-and-seven-eighths-inch, octagonal, rifled barrel. Has a grip which is checkered and a cap-box in the butt. There are front and rear sights upon the barrel. Lock-plate and hammer are slightly engraved and the trigger-guard is fitted with an auxiliary grip or spur. Upon the lock-plate is engraved the name "Jean Siber."

France.

No. 7. (N4260). PERCUSSION PISTOL. Caliber .38. Eleven-and-one-quarter-inch, rifled, octagonal, steel barrel. Has a smooth grip which is partly checkered, the butt of which is covered by a silver plate and the sides of which are inlaid with triangular bits of ivory.

The trigger-guards and ferrules are of engraved silver. The barrel is provided with front and rear sights and a wooden ramrod. This pistol was probably remodeled from the flint-lock system. Upon the lock-plate are the words "Ashmore, Warranted."

England.

No. 8. (N2706). PERCUSSION PISTOL. Caliber .50. Eleven-and-five-eighths-inch, octagonal barrel. This pistol was altered from the flint-lock to the percussion-cap system. The lock-plate and hammer are engraved with scroll designs and there are two gold bands back of the breech. Has front and rear sights. The grip is checkered and the stock is inlaid with silver escutcheons of various shapes. The pistol is marked "Walker, London."

Donor: E. E. Teller.

England.

No. 9. (N4265). PERCUSSION PISTOL. Caliber .45. Eight-inch, rifled, octagonal, iron barrel. Has a smooth, polished grip of curly maple wood. The butt is inlaid with an engraved brass plate and the top is inlaid with a diamond-shaped bit of ivory. The barrel is of Spanish make but bears no proof-marks, and the lock was made by Hutchinson. This pistol was also remodeled from the flint-lock system.

England.

No. 10. (N4261). PERCUSSION PISTOL. Caliber .40. Nine-inch, half-octagonal, iron barrel. Has a smooth, polished grip of a form sometimes known as the "fish-tail." This pistol is provided with front and rear sights the latter consisting of a groove. The ramrod is of wood and the trigger-guard of brass and slightly engraved. No maker's name or proof-marks are given. This pistol was remodeled from the flint-lock system.

Belgium?

No. 11. (N3286 a, b). PAIR OF PERCUSSION PISTOLS. Caliber .50. Nine-and-one-half-inch, octagonal, rifled barrels. Each pistol has a three stub damascus iron barrel which is fitted with front and rear sights. The grips are checkered and each has a cap-box in the butt. Lockplates and hammers are devoid of engraving. The trigger-guards are of brass. Each has an auxiliary grip. Inlaid with gold upon the top of each barrel are the words "Blaetterlein te Deventer."

Netherlands?

No. 12.   (N6767-8).   PAIR OF PERCUSSION PISTOLS. Caliber .50. Ten-and-one-quarter-inch, octagonal, rifled barrels, which are of scalp damaskeened steel. Each butt is inlaid with an engraved metal plate. There are rear and front sights upon each barrel. Lock-plates and hammers are engraved with scenes of the chase. The trigger-guards are of brass and are similarly engraved. Inscribed upon each lock-plate is the name "Muhlmeister."

Germany.

No. 13.   (N720 a, b).   PAIR OF PERCUSSION PISTOLS. Caliber .63. Seven-and-one-quarter-inch, octagonal barrels, having slightly flared muzzles. Each grip is decorated with carved leaves. These pistols have fore-sights but none in the rear. Lock-plates and hammers are engraved. Each barrel bears the name of "I. Adam Kuchenreuter," a noted gunsmith of Regensburg, Bavaria.

Germany.

No. 14.   (N2731, N2736). PAIR OF PERCUSSION PISTOLS. Caliber .64. Six-and-one-quarter-inch, octagonal, rifled barrels. Each grip is checkered and is fitted with a cap-box in the butt. There are front and rear sights upon each weapon. Lock-plates and hammers are crudely engraved and the trigger-guard is of German silver. Each barrel is stamped with the Liége proof-mark.

Belgium.

No. 15.   (N4253).   PERCUSSION PISTOL. Caliber .57. Eight-and-one-quarter-inch, octagonal, rifled barrel, having a slightly bell-shaped muzzle. The butt contains a cap-box hinged with an embossed brass cover. The hammer and lock-plate are but slightly engraved, while the trigger-guard is of brass and decorated with scroll designs. Upon the inside cover of the cap-box are the words "G. D. 25. Deposé."

France?

No. 16.   (N4249).   PERCUSSION PISTOL. Caliber .63. Seven-and-one-quarter-inch, octagonal barrel which is finely rifled. The checkered grip has an engraved, iron butt-plate. Equipped with front and rear sights, and formerly with a ramrod. Trigger-guard, hammer and lock-plate are engraved. Stamped upon the left side of the barrel is the Liége proof-mark. No maker's name is given.

Belgium.

No. 17. (N3288 a, b). PAIR OF PERCUSSION PISTOLS. Caliber .52. Three-and-one-quarter-inch, steel barrels. Lock-plates, hammers, trigger-guards and barrels are engraved. Each barrel is inlaid with a gold band near the breech. These pistols were altered from the flint-lock system. There is a safety-catch upon each lock-plate. Weapons of this kind were sometimes known as "carriage pistols." They are marked with the words "Bales in Ipswich."

England.

No. 18. (N3287 a, b). PAIR OF PERCUSSION PISTOLS. Caliber .52. Three-and-three-quarters-inch barrels. These pistols were remodeled from the flint-lock to the percussion-cap system. The stock extends to the end of the barrel. Trigger-guards, hammers and lock-plates are engraved. This is another set of "carriage pistols." They are marked "Jno. Jones & Co., London."

England.

No. 19. (N5770). PERCUSSION PISTOL. Caliber .65. Ten-and-one-half-inch, half-octagonal barrel. Has a brass fore-sight but no rear sight. The ramrod is carried in ferrules beneath the barrel. Has a smooth grip the butt of which is covered with a brass strip. All parts of the pistol are unornamented and no marks of identification are to be found upon it. This pistol also was remodeled from the flint-lock to the percussion-cap system.

England.

No. 20. (N717). PERCUSSION PISTOL. Caliber .69. Ten-and-one-half-inch, half-octagonal barrel. Has a smooth grip having the butt in the form of a dog's head made out of embossed brass. There is a front sight but none in the rear. A wooden ramrod formerly was carried beneath the barrel. The trigger-guard is made of brass, bearing figures in relief. The lock-plate and hammer are unornamented. No maker's name or proof-marks are given. This pistol was probably remodeled from the flint-lock to the percussion-cap system.

Germany.

No. 21. (N715). PERCUSSION PISTOL. Caliber .65. Twelve-and-one-half-inch, round, tapering barrel. Has a smooth-grip fitted with an embossed, brass butt. The fore-end of the stock is carved with scrolls in relief. The pistol has a brass fore-sight but no rear sight. It formerly carried a ramrod. The ferrules and trigger-guard are of embossed brass. The lock-plate is made of the same metal and the under-side bears the name "Michel Hetamaue Amaetric" which is crudely scratched upon the brass. No other marks of identification are given.

Italy.

No. 22. (N4252). PERCUSSION PISTOL. Caliber .63. Twelve-inch, round, tapering barrel, bound with three iron bands. Has a smooth grip having a large wooden knob at the butt fitted with an iron ring. Has no sights or ramrod. The trigger is in the form of an iron ball and is not protected by a trigger-guard. All parts are unornamented. The barrel is stamped with two unknown proof-marks near the nipple. No maker's name is given.

South Africa.

No. 23. (N718). PERCUSSION PISTOL. Caliber .60. Eight-and-one-half-inch, half-fluted, round barrel. Has a smooth, sloping grip having an embossed brass butt-plate. There is a crude, concave rear sight but no fore-sight. The trigger-guard and ferrules are made of engraved brass. The lock-plate also shows evidence of having been formerly engraved. No proof-marks or maker's name are given.

Germany.

No. 24. (N4262). PERCUSSION PISTOL. Caliber .62. Eight-and-one-half-inch, octagonal barrel. Has a checkered grip which has no butt-plate. Equipped with front and rear sights and a wooden ram-rod. All mountings are of steel and the trigger-guard and lock-plate are engraved. Upon the lock-plate is stamped "J.Bishop, Warranted." The barrel, which is not stamped with proof-marks, is engraved on top with the word "London." This pistol was remodeled from the flint-lock system.

England.

No. 25. (N2728). PERCUSSION PISTOL. Caliber .45. Five-and-one-half-inch, octagonal barrel. This pistol has a checkered and greatly curved grip which is fitted with a cap-box in the butt. Lock-plate, hammer and trigger-guard are engraved. The lock-plate is stamped with the name "J. Hishop," which probably is a misprint for "J. Bishop." Upon the barrel is the word "London."

England.

No. 26. (N4248). PERCUSSION PISTOL. Caliber .74. Five-and-one-quarter-inch, octagonal, brass barrel. The grip is checkered and has an engraved brass butt-plate. There is a front but no rear sight. An iron ramrod is carried beneath the barrel in brass ferrules. The trigger-guard is made of engraved brass and the hammer is engraved with floral designs. Upon the left side of the barrel is the Liége proof-mark. No other marks of identification are given.

Belgium.

No. 27. (N4267). PERCUSSION PISTOL (Pill-Lock). Caliber .38. Eight-and-one-half-inch, rifled, octagonal barrel. The grip is smooth and has a brass butt-plate. The stock extends to the end of the muzzle. The rear-sight is set about one-third the length of the barrel from the breech. Instead of using a percussion-cap, this pistol used a percussion-pellet or pill consisting of a detonating mixture which was placed in the cavity near the vent. The fall of the hammer exploded the pellet and discharged the gun. Locks of this type were common just before the invention of the metallic percussion-cap. No maker's name or proof-marks are given upon this pistol.

England?

No. 28. (N2167). PERCUSSION PISTOL. Caliber .70. Eight-and-one-quarter-inch, round, brass barrel. This pistol has a smooth, round grip having a brass butt-plate. There are no sights of any kind. A wooden ramrod is carried in brass ferrules beneath the barrel. The trigger-guard is of brass and is slightly engraved. Stamped upon the lock-plate are the words "T. Ketland & Co.", and upon the barrel the name "London." No proof-marks are given. This pistol was found in an old house in Holland, Sheboygan County, Wisconsin.

England.

PLATE 76.

*UNITED STATES SINGLE-SHOT MILITARY PISTOLS
FLINT-LOCK*

No. 1. (N3273). FLINT-LOCK PISTOL. Caliber .65. Eight-inch, round, steel barrel, which is stamped with the letters "U.S.P.", which may signify either United States Property or United States Pistol. This pistol is fitted with brass mountings, the flash-pan forming part of the body, that is, it is cast onto the body. This pistol is very likely a Charleville model of 1777. A number of these pistols were purchased in 1808 from the French and used in the United States army. The inside of the trigger-guard of the specimen here illustrated bears the letters "I.M.", probably the initials of a former owner.

United States.

No. 2. (N4055). FLINT-LOCK PISTOL. Caliber .70. Eight-and-one-half-inch, smooth-bore barrel. This pistol is fitted with a swivel ramrod and a brass trigger-guard, the latter having six file scratches upon it. The lock-plate is of steel and bears a Turkish or an Arabic stamp. There are no sights upon the barrel but there is an undecipherable stamp above the vent. This pistol is said to have been made for the American revolutionary army by an English gunsmith at Woodville, Jefferson County, New York. This man received a royalty upon each piece made. Later he fell into the hands of the British and was never heard of again by the Americans.

United States.

No. 3. (N4056). FLINT-LOCK PISTOL. Caliber .75. Eleven-and-three-quarters-inch, smooth-bore barrel. All metal parts are made of iron. This pistol has no sights. Upon the lock-plate are the words "Virginia Manufactory, Richmond, 1806." This manufactory was established to provide arms for the Virginia militia. The act of the Virginia legislature which authorized it was passed in December, 1797. The manufactory continued to make, alter and repair arms until the close of the Civil War. Probably the first output of pistols was in 1798 and doubtless many kinds were issued in subsequent years. The variety here shown is the only kind known at present. Its odd design seems to indicate use only by a special branch of the militia. This type of pistol was used during a period covering several years, as others bearing later dates are also known.

United States.

No. 4. (N711). FLINT-LOCK PISTOL. Caliber .58. Nine-and-one-half-inch, smooth-bore barrel. All mountings are of brass. There are no sights. The lock-plate is stamped with the United States eagle above the letters "U.S.", and with the words "Harpers Ferry 1807." Stamped upon the barrel are the figures "U.S.P. 1452." This is a model 1806, Harpers Ferry pistol, the second model ever made by a United States government armory, the first being the model 1804, also of Harpers Ferry. The model 1806 was made at Harpers Ferry during the years 1806, 1807 and 1808 and as far as is known was not made at any other time and at any other place than at Harpers Ferry.

United States.

No. 5. (N4057). FLINT-LOCK PISTOL. Caliber .58. Nine-and-one-half-inch, round, smooth-bore, iron barrel. All mountings are of brass. There are no sights. The lock-plate is stamped with the United States eagle above the letters "U.S." and with the words "Harpers Ferry 1807." The barrel is stamped with the figures "U.S.P. 1312." This is another example of the model 1806 Harpers Ferry pistol, but the one here illustrated is of the 1807 issue. See No. 4 described above.

United States.

No. 6. (N4059). FLINT-LOCK PISTOL. Caliber .55. Nine-and-one-half-inch, round, smooth-bore, iron barrel. All mountings are of brass. Has a sight in front but none in the rear. Stamped upon the barrel are the letters "M.P." and upon the left side of the body "H.M." The lock-plate bears the words "H. Deringer Phila." Henry Deringer was the son of Henry Deringer, the latter being well known in Colonial times as a maker of Kentucky rifles. The younger Deringer was born October 26th, 1786, at Easton, Pennsylvania, and died at Philadelphia, in February, 1868. He was famous as the maker of a peculiar percussion-cap pocket pistol of elegant design and workmanship popularly called a "Deringer." With one of these pistols President Lincoln was assassinated. Sawyer thinks that probably the first pistols made by Deringer were those of the model of 1808, a specimen of which is here shown. He further says that no proof is available that this militia pistol, marked merely as this one is, was not made by his father, nor has record of its service been found. It is probable that pistols of this type did duty with the Pennsylvania militia up to the end of the War of 1812.

United States.

No. 7. (N2759). FLINT-LOCK PISTOL. Caliber .78. Seven-and-one-half-inch, half-octagonal, smooth-bore, iron barrel. All mountings are of brass. There are no sights upon the pistol. It formerly had a swivel ramrod and a brass band, the latter being around the fore-stock. The top of the barrel is engraved with a floral design. Engraved upon the lock-plate are the letters "U. S." No proof-marks are given. As this is an English model of about 1798, and having come from a sale of United States government stores, it is likely that this piece was captured during the War of 1812.

United States.

No. 8. (N4067). FLINT-LOCK PISTOL. Caliber .78. Seven-and-one-half-inch, half-octagonal, smooth-bore, iron barrel. All mountings are of brass. There are no sights. This pistol is equipped with a swivel-ramrod and carries a brass band around the fore-stock. Engraved upon the top of the barrel is a floral design and upon the lock-plate are the letters "U.S." No proof-marks are given. This piece, like No. 7, appears to be an English model of about 1798. It also came from a sale of United States government stores, which fact makes it probable that this pistol was captured during the War of 1812.

United States.

No. 9. (N4060). FLINT-LOCK PISTOL. Caliber .69. Eight-and-one-quarter-inch, round, smooth-bore, iron barrel. All mountings are of brass. This pistol is not equipped with sights. The brass band on the fore-stock is held in place by a spring. Stamped upon the lock-plate is the name "Evans." This is a "Valley Forge" pistol. Mount Joy Forge, the first name of this armory, was established in 1742 by Stephen Evans, Daniel Walker and Joseph Williams. With the same name and the same partners it operated as an arms manufacturing plant until about 1752 when it came under the control of Joseph Potts who changed its name to Valley Forge. When the American army under Washington evacuated Valley Forge, the armory was destroyed so as not to fall into the hands of the British. In 1785 it was rebuilt by Isaac and David Potts. In 1808 the firm was known as O. and E. Evans. Later, the arms made by the forge were marked either "Valley Forge", or "W. L. Evans" or with both names on the same lock-plate. The pistol here shown is a copy of the French army (Charleville) pistol of the "Year 13" (1805) and probably dates from about 1808.

United States.

No. 10. (N4058). FLINT-LOCK PISTOL. Caliber .69. Nine-and-one-half-inch, round, smooth-bore, iron barrel. Fitted with brass mountings. Has a steel belt-hook upon the left side. There are no sights. The lock-plate bears the figure of an eagle over the words "U. States" and "S. North, Berlin, Con." This is a model of 1812 pistol and was made in North's new factory at Middletown, Conn., although almost all pistols of this model bear the stamp of the old factory at

Berlin. This pistol and all others made by Simeon North for the government thereafter were regulation pistols which were issued by the government to the regular army and also sold to the states to equip their militia. Pistols of the type here shown rendered valuable service on land and sea during the War of 1812.

United States.

No. 11. (N4062). FLINT-LOCK PISTOL. Caliber .69. Eight-and-one-half-inch, round, smooth-bore, iron barrel. All parts are made of iron except the flash-pan which is of brass. The pistol is not provided with sights. Stamped upon the left side of the barrel are the letters "U.S.P." The right side, above the vent, bears the letters "H.H.P." Upon the lock-plate are the words "S.North, U.S.", but the usual figure of the eagle is missing. This pistol fires an ounce ball and is of the model of 1814. It was also made in North's new factory at Middletown, Conn. Pistols of this kind were issued by the government to the regular army and also sold to the states to equip the militia.

United States.

No. 12. (N4061). FLINT-LOCK PISTOL. Caliber .54. Eight-and-one-half-inch, round, smooth-bore, iron barrel. All metal parts are made of iron except the flash-pan which is of brass. The pistol is not equipped with sights. The left side of the barrel is stamped "U.S.P." The right side, above the vent, bears the letters "J.W." The lock-plate is marked "S. North U.S. Midln. Con." and bears the usual eagle. This pistol is an example of the model of 1816. The model of 1814 of North's pistols were of .69 caliber (see specimen No. 11 above), and took an ounce ball. While thoroughly serviceable, their recoil was excessive. It was decided that a new issue should use a half-ounce ball of caliber .54. S. North was commissioned to make about 20,000 of these which were identical with the 1814 model except in caliber. Models 1814 and 1816 are said to have been issued for service in the Seminole War.

United States.

No. 13. (N4063). FLINT-LOCK PISTOL. Caliber .69. Ten-and-one-half-inch, round, smooth-bore, iron barrel. All metal parts are made of iron. This pistol has a fore-sight but no rear sight. The left side of the barrel is stamped with the letters "P.V.", together with the head of an eagle and the date "1818." The lock-plate is stamped "Springfield, 1818," and "U.S." surmounted by an eagle. One thousand of these pistols are said to have been made at the Springfield armory. The cock is of the goose-neck variety, which reversion seems strange at this late date, as the cock which was reenforced under the jaw and which had a flat or rounded face, had long been used for regulation arms because it was less apt to break than the goose-neck variety. Some collectors claim that the locks of these model 1818 Springfield pistols were purchased ready-made in Europe, in 1815, due to a shortage of pistol-locks in the United States.

United States.

No. 14. (N4064). FLINT-LOCK PISTOL. Caliber .54. Nine-and-one-half-inch, round, smooth-bore, iron barrel. All metal parts are of iron. Has front and rear sights. Stamped upon the left side of the barrel are the letters "U.S.P." This is an example of the model 1819 army pistol. The lock-plate is marked "S. North U.S. Midln. Con." and bears the usual eagle. Upon the lock-plate is also stamped the date of issue, 1821. The model of 1819 army pistol had a safety-catch to the rear of the cock, as is here shown, to hold the piece at half-cock; the navy pistol did not have this device. It was thought that the safety-catch would permit a cavalryman to snatch the pistol from the holster or thrust it in with less chance of accidental discharge. However, this safety-device proved to be more of a hindrance than a help and was discontinued in all other models. Twenty thousand of these model 1819 pistols were made by North for which he received $8.00 each. These pistols were used in the Black Hawk, Seminole and Mexican Wars.

United States.

No. 15. (N4065). FLINT-LOCK PISTOL. Caliber .54. Eight-inch, round, smooth-bore, iron barrel. All metal parts are made of iron except the flash-pan which is of brass. This pistol is equipped with front and rear sights. The left side of the barrel is stamped "U.S.P." J.H." The lock-plate is stamped "U.S. S.North, 1827." The usual eagle is missing. A belt-hook is attached to the left side of the stock. This pistol is a shortened reproduction of the model of 1819. Some pistols of the 1827 model have a different curve to the grip than those of the model 1819.

United States.

No. 16. (N4054). FLINT-LOCK PISTOL. Caliber .54. Eight-inch, round, smooth-bore, iron barrel. The flash-pan is of brass, all of the other metal parts are of iron. This pistol is equipped with front and rear sights. No marks of identification are given. This is a model of 1827 army pistol, and was probably made by W. L. Evans. Usually this model made by Evans is stamped either W. L. Evans, or Valley Forge, or both, but does not have the U. S. mark. This pistol corresponds with the regulation model of 1827 in all respects, and was made for militia and for private sale as well as for the government.

United States.

No. 17. (N4066). FLINT-LOCK PISTOL. Caliber .54. Eight-inch, round, smooth-bore, iron barrel. All iron parts of this navy pistol are coated with tin to prevent or minimize corrosion by salt water. The top of the barrel is stamped with the letters "U.S.", and the lock-plate "W.L. Evans V. Forge, 1831, U.S.N." This is either an 1819 or an 1827 model. Pistols of this type were made by North, Evans, Henry and by the United States government at Harpers Ferry. They are marked with the name of the maker, the date of their manufacture and "U.S.N." and are not provided with safety-catches. Repetitions of this model were issued for about twenty years, some of the pistols of later date being plated with tin as the one illustrated here.

United States.

No. 18. (N709). FLINT-LOCK PISTOL. Caliber .54. Eight-and-one-quarter-inch, round, smooth-bore, iron barrel. This is a model 1836 United States army pistol. For a more detailed description of this model see No. 19 below. The known contractors who made the model 1836 pistol for the government were: R. Johnson, Middletown, Connecticut; and A. Waters, Millbury, Massachusetts. The lock-plate of the specimen here shown is stamped "U.S. Johnson, Middn. Conn., 1837." Upon the top of the barrel are the figures "U.S.P. JCS."

United States.

No. 19. (N4072). FLINT-LOCK PISTOL. Caliber .54. Eight-and-one-quarter-inch, round, smooth-bore, iron barrel. The fore-sight and pan are made of brass; other metallic parts are of iron. This is a model 1836, United States army pistol. In this model the lock-plate, frizzen and cock were case-hardened in mottled colors; the frizzen-spring, trigger, barrel-tang and all screws were blued; the other steel parts were left in the bright condition and polished. The black walnut stock was smoothed, finished with linseed oil, rubbed off, and left dull. Pistols of this model were elegantly shaped, conspicuous by their colors and nearly the equals of the old-time dueling pistols in the matter of excellent workmanship. The brass pan of this model was placed horizontally and had a shield to protect the eyes of the user. The charge for pistols of this model consisted of fifty grains of rifle powder and a half-ounce spherical bullet. The lock-plate of the specimen here shown is stamped "U.S. Johnson, Middn., Conn., 1837." Upon the barrel are the figures "U.S.P. JCB."

United States.

No. 20. (N4073). FLINT-LOCK PISTOL. Caliber .54. Eight-and-one-quarter-inch, round, smooth-bore, iron barrel. This is another example of the model 1836 United States army pistol similar to No. 19 described above, but of the 1838 issue, which date is stamped upon the lock-plate. The lock-plate is also stamped with the words "U.S. Johnson, Middn., Conn." Stamped upon the top of the barrel are the figures "U.S. L.S." The trigger-guard bears the letters "I.K."

United States.

No. 21. (N7040). FLINT-LOCK PISTOL. Caliber .54. Eight-and-one-quarter-inch, round, smooth-bore, iron barrel. This is another example of the model of 1836 United States army pistol but of the issue of 1839. For a more detailed description of this model see No. 19. The lock-plate is stamped "U.S. R.Johnson, Middn., Conn., 1839," and the barrel "U.S.P. E.B."

United States.

No. 22. (N4074). FLINT-LOCK PISTOL. Caliber .54. Eight-and-one-quarter-inch, round, smooth-bore, iron barrel. This is still another example of the model 1836 United States army pistol but of the 1840 issue. See No. 19 for a more detailed description of this model. The lock-plate is stamped "U.S. R.Johnson, Middn., Conn., 1840," and the barrel "U.S.P. JCS MS." The left side of the stock also bears various government inspectors' stamps.

United States.

No. 23. (N4075). FLINT-LOCK PISTOL. Caliber .54. Eight-and-one-eighth-inch, round, smooth-bore, iron barrel. This is another pistol belonging to the 1836 model. It is of the 1841 issue. See No. 19 for more information regarding this model. Upon the lock-plate of the specimen illustrated here is stamped "U.S. R.Johnson, Middn., Conn., 1841," and upon the barrel "U.S.P. NWP." The left side of the stock also bears various government inspectors' stamps.

United States

No. 24. (N4076). Eight-and-one-eighth-inch, round, smooth-bore, iron barrel. This also is an example of the model 1836, United States army pistol, but of the 1842 issue. See No. 19 for further information regarding this model. The lock-plate is stamped "U.S. R.Johnson, Middn., Conn., 1842," and the barrel "U.S.P. JCB." Upon the left side of the stock are various government inspectors' stamps.

United States.

No. 25. (N4077). FLINT-LOCK PISTOL. Caliber .54. Eight-and-one-eighth-inch, round, smooth-bore, iron barrel. This is a model 1836, United States army pistol of the 1843 issue. See No. 19. The lock-plate is stamped "U.S. R.Johnson, Middn., Conn., 1843," and the barrel "U.S.P. J.H." Upon the left side of the stock are the usual government inspectors' marks.

United States.

No. 26. (N4078). FLINT-LOCK PISTOL. Caliber .54. Eight-and-one-eighth-inch, round, smooth-bore, iron barrel. This is a model 1836, United States army pistol of the 1844 issue. See No. 19. The lock-plate is stamped "U.S. R.Johnson, Middn., Conn., 1844," and the barrel "U.S.P. JH." Government inspectors' stamps are upon the left side of the stock.

United States.

No. 27. (N4079). FLINT-LOCK PISTOL. Caliber .54. Eight-and-one-eighth-inch, round, smooth-bore, iron barrel. The fore-sight and flash-pan are of brass; other metallic parts being of steel. This is a model 1836, United States army pistol. These pistols were made under contract for the government by Robert Johnson of Middletown, Conn., and by A. H. Waters and Company of Millbury, Mass. The Waters family had been makers of firearms since Richard Waters came from England in 1832. The lock-plate of the specimen shown here is stamped "A.H. Waters & Co., Milbury, Mass.", (only one "1" in Millbury) together with the head of an eagle. The top of the barrel is marked "U.S.P. JH." There is no date upon the lock-plate.

United States.

No. 28. (N4080). FLINT-LOCK PISTOL. Caliber .54. Eight-and-one-eighth-inch, round, smooth-bore, iron barrel. This is another model 1836, United States army pistol, but of the 1837 issue. See No. 19. The lock-plate is stamped "A. Waters, Milbury, Ms., 1837," together with the head of an eagle. Stamped upon the barrel are the figures "U. S. P. LF."

United States.

No. 29. (N4081). FLINT-LOCK PISTOL. Caliber .54. Eight-and-one-eighth-inch, round, smooth-bore, iron barrel. This is another model 1836, United States army pistol of the issue of 1838. See No. 19. The lock-plate is stamped "A. Waters, Milbury, Ms., 1838," together with the head of an eagle. The barrel is stamped "U. S. P. EB."

United States.

No. 30. (N4082). FLINT-LOCK Pistol. Caliber .54. Eight-and-one-eighth-inch, round, smooth-bore, iron barrel. This is another model 1836, United States army pistol but of 1839 issue. See No. 19. The lock-plate is stamped "A. Waters, Milbury, Ms., 1839," and with the head of an eagle. The barrel is stamped "U. S. P. LF 42." There are various government inspectors' stamps upon the left side of the stock.

United States.

No. 31. (N4083). FLINT-LOCK PISTOL. Caliber .54. Eight-and-one-eighth-inch, round, smooth-bore, iron barrel. This model 1836, United States army pistol is of the 1840 issue and was made by A. Waters. See No. 19. The lock-plate is stamped "A. Waters, Milbury, Ms., 1840," and with the head of an eagle. Stamped upon the barrel are the figures "U. S. P. JH." Inspectors' stamps are upon the left side of the stock.

United States.

No. 32. (N4084). FLINT-LOCK PISTOL. Caliber .54. Eight-and-one-eighth-inch, round, smooth-bore, iron barrel. Another model of 1836 army pistol but of the 1841 issue. See No. 19. The lock-plate is stamped "A. Waters, Milbury, Ms., 1841," together with the head of an eagle. The barrel is stamped "U. S. P. JH." The left side of the stock bears the usual government inspectors' stamps.

United States.

No. 33. (N4085). FLINT-LOCK PISTOL. Caliber .54. Eight-and-one-quarter-inch, round, smooth-bore, iron barrel. This is an 1842 issue of the 1836 model, army pistol. See No. 19. The lock-plate is stamped "A. Waters, Milbury, Ms., 1842," and also bears the head of an eagle. The barrel is marked "U. S. P. JCS." The bottom of the stock bears the name of a former owner, "P. L. Vermint."

United States.

No. 34. (N4086). FLINT-LOCK PISTOL. Caliber .54. Eight-and-one-eighth-inch, round, smooth-bore, iron barrel. This is an 1843 issue of the 1836 model, army pistol. See No. 19. The lock-plate is stamped "A. Waters, Milbury, Ms., 1843," and also bears the head of an eagle. Various government inspectors' stamps appear upon the left side of the stock. The barrel is marked "U. S. P. JH."

United States.

No. 35. (N4087). FLINT-LOCK PISTOL. Caliber .54. Eight-and-one-eighth-inch, round, smooth-bore, iron barrel. This is an 1836 model, army pistol of the 1844 issue. See No. 19. The lock-plate is stamped "A. H. Waters & Co., Milbury, Mass., 1844," and with the head of an eagle. The barrel is marked "U. S. P. EB." No government inspectors' stamps are upon the stock.

United States.

PLATE 77.

UNITED STATES SINGLE-SHOT MILITARY PISTOLS, REMODELED FLINT-LOCK PISTOLS AND PERCUSSION PISTOLS

No. 1. (N716). PERCUSSION PISTOL. (Altered from Flint-Lock.) Caliber .54. Nine-and-three-quarters-inch, round, smooth-bore, steel barrel. Equipped with front and rear sights. The butt is inlaid with an oval, silver escutcheon bearing the initials "C.W.". This is a model 1806, Harpers Ferry, army pistol, which was altered from the flint-lock system by enlarging the touch-hole so as to screw a cylindrical lug into it and then screwing the cone into the lug. This pistol fired an half-ounce spherical bullet. The lock-plate is stamped "Harpers Ferry 1808" together with an eagle over the letters "U. S." The original models of this pistol were made only at Harpers Ferry during the years 1806, 1807 and 1808.

United States.

No. 2. (N4069). PERCUSSION PISTOL. (Altered from Flint-Lock.) Caliber .54. Eight-and-one-half-inch, round, smooth-bore, iron barrel. Equipped with a front sight but none in the rear. This is a model 1816 army pistol, made by Simeon North. He made 20,000 of these for the United States government. The specimen shown here was altered from the flint-lock system in the same way as the pistol described under No. 1. The lock-plate is stamped with the figure of an eagle and with the words "S. North, Midltn, Conn., U.S." No date is given.

United States.

No. 3. (N3284). PERCUSSION PISTOL. (Altered from Flint-Lock.) Nine-and-one-half-inch, round, smooth-bore, iron barrel. Fitted with front and rear sights. This is a model 1819, army pistol, made by Simeon North. The army models of this date were provided with a safety-catch to hold the piece at half-cock as is shown in this pistol and the one following. This safety device was abandoned in subsequent models as it proved to be a hindrance. North made 20,000 pistols of this model for the United States government at $8.00 a piece. The specimen shown here was altered from the flint-lock system in the same manner as No. 1. The lock-plate is stamped "S. North, Midltn., Conn., U.S.," together with the figure of an eagle and the date of issue, "1821."

United States.

No. 4. (N4070). PERCUSSION PISTOL. (Altered from Flint-Lock.) Caliber .54. Nine-and-three-eighths-inch, round, smooth-bore, iron barrel. Fitted with front and rear sights. This is a model 1819, army pistol, made by Simeon North. This army model is also provided with a safety-catch behind the hammer. This pistol was altered from the flint-lock system in the same fashion as No. 1. The lock-plate is stamped "S. North, Middtn., Conn., U.S.," and also bears the figure of an eagle and the date of issue "1822."

United States.

No. 5. (N6783). PERCUSSION PISTOL. (Altered from Flint-lock.) Caliber .54. Eight-inch, round, smooth-bore, iron barrel. Equipped with front and rear sights. This is a model 1836, army pistol, made by Robert Johnson. It takes a charge of fifty grains of rifle powder and a half-ounce spherical ball. The specimen shown here was altered from the flint-lock system in the same way as No. 1. The lock-plate is stamped "U.S. R. Johnson, Middn., Conn., 1840." Upon the barrel are the figures "U. S. P.   JH.   45." Pistols of this model were noted for their handsome appearance.

United States.

No. 6. (N4089). PERCUSSION PISTOL. (Altered from Flint-Lock.) Caliber .54. Eight-inch, round, smooth-bore, iron barrel. Equipped with front and rear sights. This is a model 1836, army pistol, made by Robert Johnson. Its charge consisted of fifty grains of powder and an half-ounce spherical ball. This pistol was changed from the flint-lock system in the same way as No. 1. The lock-plate is stamped "U.S. R. Johnson, Middn., Conn., 1844," and the barrel, "U. S. P. JCB."

United States.

No. 7. (N4091). PERCUSSION PISTOL. (Altered from Flint-Lock.) Caliber .54. Eight-inch, round, smooth-bore, iron barrel. Equipped with front and rear sights. This is another example of the model of 1836, United States army pistol, as made by Robert Johnson. The charge is the same as for No. 6. The pistol illustrated here was altered from the flint-lock system by brazing a lump to the right barrel flat so as to cover the touch-hole, after which the cone-seat was placed on top and the cone screwed into the lump. The lock-plate is stamped "U.S. R. Johnson, Middn., Conn., 1837," and the barrel, "U. S. P. PW."

United States.

No. 8. (N723). PERCUSSION PISTOL. (Altered from Flint-Lock.) Caliber .54. Eight-inch, round, smooth-bore, iron barrel, equipped with front and rear sights. This specimen illustrates another type of conversion to the percussion system. This was accomplished by plugging up the vent, or touch-hole, and then screwing a cone into the barrel. The lock-plate is stamped "U.S. R. Johnson, Middn., Conn., 1842," and the barrel, "U. S. P. JH." Upon the left side of the stock are two government inspectors' stamps.

United States.

No. 9. (N6782). PERCUSSION PISTOL. (Altered from Flint-Lock.) Caliber .54. Eight-inch, round, smooth-bore, iron barrel. Equipped with front and rear sights. This is another example of the 1836 model pistol which was converted to the percussion system in the same manner as described under No. 8. The lock-plate is stamped

"U.S. R. Johnson, Mddn., Conn., 1843," and the barrel, "U. S. P. JH." There are two government inspectors' stamps upon the left side of the stock.

United States.

No. 10. (N4068). PERCUSSION PISTOL. (Altered from Flint-Lock.) Caliber .54. Nine-and-one-half-inch, round, smooth-bore, iron barrel. This is a model 1819, army pistol, as made by Simeon North. The army pistols of this model were usually provided with a safety-device to hold the piece at half-cock. This device is missing in the specimen here shown which seems to indicate that the pistol was used by the navy. This pistol was converted into the per-cussion system in the same way as No. 8. There is neither name nor date upon the lock-plate. The left side of the barrel is stamped with the Liége proof-mark which seems to indicate that the pistol got into Belgian hands, very likely after being discarded by the United States government, and was then proved in Belgium before being again disposed of.

United States.

No. 11. (N4071). PERCUSSION PISTOL. (Altered from Flint-Lock.) Caliber .54. Nine-and-three-eighths-inch, round, smooth-bore, iron barrel. Equipped with a front sight but no rear sight. This is a model 1819, army pistol, as made by Simeon North. It is provided with a safety-catch to the rear of the hammer to hold the pistol at half-cock. The specimen shown here was converted to the percussion system by brazing a lump to the right of the barrel flat so as to cover the touch-hole, after which the cone-seat was placed on top and the cone screwed into the lump. The lock-plate is stamped "S. North, Middtn., Conn., U.S." and also bears the figure of an eagle and the date of issue, "1821."

United States.

No. 12. (N4090). PERCUSSION PISTOL. (Altered from Flint-Lock.) Caliber .54. Eight-inch, round, smooth-bore, iron barrel. Equipped with front and rear sights. This is a model 1836 army pistol made by A. Waters. It fires a charge of fifty grains of rifle powder and a half-ounce ball. This specimen was altered to the percussion system by the same method as was employed for No. 11. The lock-plate is stamped with the head of an eagle and "A. Waters, Milbury, Ms., 1837." The barrel is stamped "U. S. P. JCS" and "N. Carolina."

United States.

No. 13. (N4088). PERCUSSION PISTOL. (Altered from Flint-Lock.) Caliber .54. Eight-inch, round, smooth-bore, steel barrel, equipped with front and rear sights. This is another example of the model 1836, army pistol, as made by A. Waters. This pistol was altered to the percussion system by screwing a cone into the barrel and plugging up the touch-hole. The lock-plate is stamped with the head of an eagle and "A. Waters, Milbury, Ms., 1842." Upon the barrel are the figures "U.S.P. JH." The average weight of the pistols of this model was two pounds, eleven ounces.

United States.

No. 14. (N6786). PERCUSSION PISTOL. (Altered from Flint-Lock.) Caliber .54. Eight-inch, round, smooth-bore, iron barrel which is equipped with front and rear sights. This pistol was converted to the percussion system in the same manner as No. 13. The lock-plate is stamped with the head of an eagle and with "A. Waters, Milbury, Ms., 1843." The barrel is stamped "U. S. P. NWP." Two government inspectors' stamps appear upon the stock.

United States.

No. 15. (N4092). PERCUSSION PISTOL. (Altered from Flint-Lock.) Caliber .54. Eight-inch, round, smooth-bore, iron barrel, equipped with front and rear sights. This specimen was converted into the percussion system in the same manner as No. 13. The lock-plate is flush with the wood of the stock. Stamped upon the lock-plate is the head of an eagle, and the words "A. Waters & Co., Milbury, Ms.";　but no date. There are no stamps upon either barrel or stock.

United States.

No. 16. (N4145). PERCUSSION PISTOL. Caliber .54. Eight-inch, round, smooth-bore, iron barrel which is fitted with a front sight but has none in the rear. This is a model 1842, United States, army pistol, the first model to use the percussion-cap. The lock-plate of the specimen here shown is stamped "Palmetto Armory, S.C." It also bears the figure of the palmetto palm, and "Columbia, S.C., 1852." The barrel bears the date "1853" and the letters "P.V." together with the figure of a palm. The left side of the barrel is stamped "Wm.

Glaze & Co." The Palmetto Armory made arms for the government until the outbreak of the Civil War, and from 1861 to 1865 for the Confederacy.

United States.

No. 17. (N6785). PERCUSSION PISTOL. Caliber .54. Eight-inch, round, smooth-bore, iron barrel, provided with a front sight but no rear sight. This is another example of the 1842 model, army pistol, the first model to use the percussion-cap. The lock-plate is stamped "U.S. H. Aston, Middtn., Conn., 1847," and the barrel, "U.S.P. SK 1848." Henry Aston was born in London, in 1803. He arrived in the United States with his father on July 14, 1819, and soon thereafter found employment in the establishment of Simeon North as a workman of high skill, making United States presentation pistols for commodores Hull and McDonough. In 1845 he received a government contract for model 1842 pistols at $14.00 per pair, which included spare cone, screw driver, ball screw, bullet mold and spring vise.

United States.

No. 18. (N4093). PERCUSSION PISTOL. Caliber .54. Eight-inch, round, smooth-bore, iron barrel which has a sight in front but none in the rear. This is another example of the 1842 model pistol, as made by Henry Aston. Upon the lock-plate are the words "U.S. H. Aston, Middtn., Conn., 1849." The barrel is stamped "U.S.P. GW" and with the date "1850."

United States.

No. 19. (N721). PERCUSSION PISTOL. Caliber .54. Eight-inch, round, smooth-bore, iron barrel which has a sight in front but none in the rear. The mountings are of brass. This is a model 1842, army pistol. The lock-plate is stamped "U.S. H. Aston, Middtn., Conn., 1850" and the barrel "U.S.P. JH 1850."

United States.

No. 20. (N6784). PERCUSSION PISTOL. Caliber, length of barrel, sights and mountings are the same as those of No. 19. Stamped upon the lock-plate are the words "U.S. H. Aston, Middtn., Conn., 1850," and upon the barrel, "U.S.P. WN 1850." The left side of the stock is stamped with two government inspectors' stamps.

United States.

No. 21. (N4094). PERCUSSION PISTOL. Caliber, length of barrel, sights and mountings are the same as those of No. 19. The lock-plate is stamped "U.S. H. Aston, Middtn., Conn., 1852," and the barrel "U.S.P. NWP."

United States.

No. 22. (N3285). PERCUSSION PISTOL. Caliber .54. Eight-inch, round, smooth-bore, steel barrel, equipped with a front, but no rear sight. All mountings are of brass. This model 1842 pistol was made by Ira N. Johnson. In November, 1852, Johnson bought up the interests of Henry Aston, Peter Ashton, Sylvester C. Bailey and John North, and finished by himself a contract which he had obtained for 10,000 of these pistols. The lock-plate is stamped "U.S. I. N. Johnson, Middtn., Conn., 1854." The barrel is stamped "U.S.P. JOB" and also with the word "Steel." This pistol was used by the navy as the figure of an anchor is stamped upon the barrel.

United States.

No. 23. (N4095). PERCUSSION PISTOL. Caliber, length of barrel, sights and other features are the same as those of No. 22. The lock-plate and barrel are stamped with the same words and figures as No. 22. This is another example of the navy pistol, model 1842.

United States.

PLATE 78.

UNITED STATES PISTOL-CARBINES AND BOX-LOCK PISTOLS
(SINGLE-SHOT PERCUSSION)

No. 1. (N4100). PERCUSSION PISTOL. Caliber .54. Eight-inch, round, smooth-bore, iron barrel, equipped with a front sight but no rear sight. This is a model 1842, army pistol, fitted with a carbine stock. To remove the stock, the button back of the trigger-guard is pressed. This releases a catch and frees the pistol from the stock. Some stocks of the kind illustrated here were submitted to the government for trial but did not meet with approval. The lock-plate is stamped "U.S. H. Aston & Co., Middtn., Conn., 1851," and the barrel, "U.S.P. WN." Upon the tang is the date "1851."

United States.

No. 2.   (N4101).   PERCUSSION PISTOL.   Caliber .58.

Eleven-and-one-half-inch, rifled, steel barrel, equipped with a front sight and a three-leaved rear sight. This is a model 1855 pistol-carbine. The form of stock shown here met with government approval. In the specimen illustrated the door of the lock is opened to show the Maynard priming magazine. By the movement of the hammer a strip of tape about ten inches long containing 50 pellets of detonating powder was fed to the cone, only one pellet appearing above the cone with each movement. Instead of this tape primer, the ordinary copper cap could be used. The lock-plate is stamped "U.S. Springfield, 1855." The barrel is stamped with the letters "V.P." and the head of an eagle facing to the left, which is the government proof-mark. Upon the tang is the date "1855."

United States.

No. 3. (N3283). PERCUSSION PISTOL. Caliber .58. Eleven-and-one-half-inch, rifled, steel barrel. This pistol-carbine is identical with specimen No. 2 in all respects except the hammer-screw. This screw as here shown, is probably not of the regulation type.

United States.

No. 4. (N4102). PERCUSSION PISTOL. Caliber .58. Eleven-and-one-half-inch, rifled, steel barrel. This model 1855 pistol-carbine differs from No. 2 in having a steel back-strap instead of a brass one, and in having the lock plate dated 1856 instead of 1855. In both carbines the rear leaf sights are numbered 1, 2, 3 and 4, and in other respects they are similar. The figures upon the leaves of the sights represent hundreds of yards.

United States.

No. 5. (N4103). PERCUSSION PISTOL. Caliber .58. Eleven-and-one-half-inch, rifled, steel barrel. This model 1855 carbine is a reproduction of a Confederate pistol-carbine which is marked "Fayetteville, C.S.A., 1862." The Confederates made some of these arms but omitted the Maynard tape primer and used the type of lock here shown.

United States.

No. 6. (N4096). PERCUSSION PISTOL. Caliber .54. Five-and-one-half-inch, smooth-bore, steel barrel. All mountings are of brass. There are no sights. This is an 1843 "box-lock" army pistol. Box-lock is a survival of a term applied by Henry Nock, noted London gunsmith to a flint-lock of his design, having the cock on the inner side of the lock-plate. The lock-plate is stamped "N. P. Ames, Springfield, Mass., U.S.R. 1843." The letters U.S.R. signify United States Mounted Riflemen. The barrel is stamped "U.S.R. 1843, JCB P." Pistols of this model were made at the Springfield Armory, and by N. P. Ames and by H. Deringer. It is estimated that each of these contractors produced only 2,000 pistols of this model. Some of these pistols were smooth-bore and others were rifled.

United States.

No. 7. (N4097). PERCUSSION PISTOL. Caliber .54. Five-and-one-half-inch, round, smooth-bore, steel barrel. All mountings are of brass but there are no sights. This specimen is an 1843 "box-lock" pistol which is almost identical with No. 6 except that it is stamped with the letters "U.S.N." which signify United States Navy. The lock-plate is stamped "N.P. Ames, Springfield, Mass., U.S.N. 1845." The barrel was formerly stamped but the letters have become undecipherable. The lock-plate is removed to show the mechanism.

United States.

No. 8. (N4098). PERCUSSION PISTOL. Caliber .54. Five-and-one-half-inch, round, smooth-bore, steel barrel. There are no sights. All mountings are of brass. This is another example of the model 1843 "box-lock" pistol. See No. 6. The lock-plate is stamped with the words "U.S. Deringer, Philadel." Stamped upon the barrel are the letters "R.P." No date is given. This is probably a United States Mounted Rifleman's pistol.

United States.

No. 9. (N4099). PERCUSSION PISTOL. Caliber .54. Five-and-one-half-inch, rifled, steel barrel which is equipped with front and rear sights. There are no stamps upon the barrel. In other respects this pistol is the same as No. 8 described above. The lock-plate is stamped "U.S. Deringer, Philadel."

United States.

No. 10a. (N2921). PERCUSSION PISTOL. Caliber .67. Eight-and-one-quarter-inch, round, smooth-bore, steel barrel. Has brass mountings and carries an iron ring in the butt. This pistol is not equipped with sights. The lock-plate is stamped with the letters "U.S." together with the dates "1851" and "1852." The top of the barrel is decorated with painted designs. This is certainly not an United States army pistol as the top of the barrel is stamped with the Liége proof-mark. Stamped upon the left side of the stock are the figures "HN4" and "ED".

Belgium?

No. 10b. (N2726). PERCUSSION PISTOL. This pistol is the same in all respects to No. 10a described above, except that the left side of the stock is stamped with the figures "F. F. 1852."

Belgium?

PLATE 79.

CHIEFLY EUROPEAN PERCUSSION MILITARY PISTOLS

No. 1. (N2722). PERCUSSION PISTOL. Caliber .70. Seven-and-one-quarter-inch, damaskeened, octagonal, steel barrel. The grip is checkered and has a German silver cap-box in the butt. This pistol is equipped with front and rear sights and has a wooden ramrod. The lock-plate and hammer are engraved with floral designs, while the trigger-guard and ferrule are made of engraved German silver. Stamped upon the left side of the barrel is the Liége proof-mark. No maker's name is given.

Belgium.

No. 2. (N4264). PERCUSSION PISTOL. Caliber .55. Eight-inch, octagonal, brass barrel. This pistol has a smooth, round grip covered with a brass butt-plate. It also is fitted with a fore-sight and a concave rear sight but has no ramrod. The trigger-guard is of brass and unornamented. The barrel is stamped "P M J R 1838" followed by the word "Lancers." Upon the lock-plate is the name "Lane & Reed, Boston." The name of the former owner, "L. Dennis," is engraved upon a silver escutcheon upon the left side of the stock. This pistol was remodeled from the flint-lock to the percussion system.

United States.

No. 3. (N4263). PERCUSSION PISTOL. Caliber .60. Eight-and-one-half-inch, round, steel barrel. This pistol has a smooth, round grip and a wooden ramrod but neither front nor rear sights. The trigger-guard is of brass and is engraved with a flower. Upon the left side of the barrel are two British proof-marks and the lock-plate is stamped with the name "Joseph Golcher." Joseph Golcher was a member of the famous Golcher (also written Goulcher) family of Pennsylvania gunsmiths who operated in Philadelphia, Lancaster and elsewhere from the 1820's through the 1870's. Joseph, who had a shop in Philadelphia, but later removed to the Pacific Coast, specialized in locks, and one of his locks has obviously at some time been placed on this imported English pistol, probably when it was remodelled from a flint-lock to the percussion system.

England.

No. 4. (N4259). PERCUSSION PISTOL. Caliber .56. Seven-and-one-quarter-inch, round, steel barrel having a bell-shaped muzzle. The grip is carved and pitted with small marks. Trigger-guard and butt-plate are made of brass. There are no sights and no marks of identification upon this pistol. It is probably of English manufacture and was remodeled from the flint-lock system.

England?

No. 5. (N719). PERCUSSION PISTOL. Caliber .52. Six-inch, damaskeened, octagonal, steel barrel having a slightly bell-shaped muzzle. The checkered grip is covered with an iron butt-plate. This pistol is equipped with front and rear sights and formerly carried a wooden ramrod. Stamped upon the left side of the barrel is an unknown proof-mark. All parts of the pistol are unornamented and no maker's name is given.

Germany.

No. 6. (N4250a, b). PERCUSSION PISTOL. Caliber .65. Nine-inch, round, steel barrel, bound with a brass band near the muzzle. This piece has a smooth sloping grip to which the gunstock shown below may be attached thus converting the pistol into a carbine. There are front and rear sights upon the barrel. Upon the outside of the lock-plate is a safety-device which fits into the hammer at half-cock thus preventing the cap from being struck. The left side of the barrel is stamped with the number "963" and "V. 18," while the lock-plate is stamped with two proof-marks in the form of a crown over the letter "K." This is an Austrian military pistol-carbine of the model of 1855 and was loaded with a paper cartridge through the muzzle.

Austria.

No. 7. (N722). PERCUSSION PISTOL. Caliber .64. Seven-and-one-half-inch, round, steel barrel. This pistol has a checkered grip

having a flat, steel butt-plate which is stamped with the number "4423."
There are front and rear sights upon this pistol. Inscribed upon the
top of the barrel are the words "Lorenz Bossel in Suhl." No proof-
marks are given.

Saxony, Germany.

No. 8. (N4256). PERCUSSION PISTOL. Caliber .64. Seven-
inch, damaskeened, octagonal, steel barrel which is slightly flared at the
muzzle. The interior of the barrel is finely rifled. The grip is checkered and has an iron butt-plate. All steel mountings are finely engraved.
The pistol is engraved with front and rear sights. Upon the lock-plate
are the words "Deprez à Heerstal près de Liége." No proof-marks
are visible.

Belgium.

No. 9. (N731). PERCUSSION PISTOL. Caliber .65. Eight-
and-one-half-inch, round, ribbed, steel barrel. This pistol has a smooth,
sloping grip the butt of which is covered with a brass knob in the form
of a face. There are neither front nor rear sights. The trigger-guard
is of brass and ornamented with a raised human figure in relief, while
the rear portion of the barrel was formerly chiseled with floral designs
also in relief. No proof-marks or maker's name are given.

Germany.

No. 10. (N4247). PERCUSSION PISTOL. Caliber .62.
Eight-and-one-quarter-inch, round, tapering, steel barrel bound with a
brass band. The grip is large and smooth and has a brass butt with a
ring. This pistol is equipped with front and rear sights but has no
ramrod. The trigger-guard is made of heavy brass and is provided
with a spur to secure greater steadiness in aiming. Stamped upon the
guard are the figures "16. U. 3. 70." The percussion-cap may be protected by a unique safety-device which fits around the cap and prevents
the hammer from hitting it. Stamped upon the left side of the barrel
is the date "1852" and upon the right, "1866." The lock-plate is
stamped with a crown over the words "Suhl G. H." This is a cavalry
pistol.

Germany.

No. 11. (N4251). PERCUSSION PISTOL. Caliber .67. Nine-
and-one-quarter-inch, round, steel barrel, fitted with a swivel ramrod.
There are no sights. This pistol has a smooth, heavy grip having a

brass butt-plate fitted with an iron ring. All mountings are massive and unornamented. Upon the lock-plate is an unknown proof-mark and the number "853." This is said to be an Austrian army pistol, model of 1845.

Austria.

No. 12. (N4255). PERCUSSION PISTOL. Caliber .57. Nine-and-one-half-inch, round, tapering, steel barrel, having a rear-sight in the form of a groove. An iron ramrod is carried beneath the barrel. The grip is smooth and has an iron butt-plate to which a ring is attached. The lock-plate is stamped with the Russian coat-of-arms and the number "864." The plate opposite the lock is stamped with the number "799." This is a Russian army pistol.

Russia.

No. 13. (N4254). PERCUSSION PISTOL. Caliber .80. Eight-and-three-quarters-inch, round, steel barrel. The grip is smooth, sloping and provided with a brass butt-plate. There are no sights. Stamped upon the lock-plate is the British crown and the word "Tower." Upon the barrel are two English proof-marks. This pistol is not remodeled from the flint-lock system, although at first sight it appears to be.

England.

No. 14. (N6841). PERCUSSION PISTOL. Caliber .75. Eight-and-three-quarters-inch, round, steel barrel. The smooth, sloping grip is covered with a brass butt-plate. The pistol has no sights. The trigger-guard is made of brass. Stamped upon the barrel are two British proof-marks. This pistol is no doubt, a "Tower" piece, but is not so marked.

England.

No. 15. (N724). PERCUSSION PISTOL. Caliber .72. Seven-and-one-half-inch, round, steel barrel. The heavy, smooth grip has a brass butt-plate provided with an iron ring. There are front and rear sights. The ramroad is hollow at the ramming end which thus also serves as a measure for powder. Stamped upon the left side of the barrel are the figures "C De 17-6 NA MF," and upon the right side, "S. 1852." The stock is stamped with the number "31." All mountings are of heavy brass. This is a French cavalry pistol, model

1822, which was remodeled to the percussion system. It was made in St. Etienne.

France.

No. 16. (N5904). PERCUSSION PISTOL. Caliber .72. Seven-and-one-quarter-inch, round, steel barrel. This pistol has a thick, round grip covered with a heavy brass butt. There is a front sight but none in the rear. The ramrod is missing. The trigger-guard is of brass and is stamped with a crown over the letter "P". Other brass mountings are similarly stamped. The lock-plate is marked with a crown over the letter "J". This pistol is an 1806 model and was used either by the French cavalry or navy. It was remodeled to the percussion system and was originally made at St. Etienne.
Donor: Frank Warth.

France.

No. 17. (N4241). PERCUSSION PISTOL. Caliber .70. Seven-and-one-half-inch, round, steel barrel. The grip is heavy and smooth and has an iron ring in the brass butt-plate. Fitted with front and rear sights. Stamped upon the lock-plate are the words "Mre. Rte. (Rle.)? de Mutzig." The stock is marked "Del Forge 710." Upon the top of the back-strap are the figures "Mle. 1822." All mountings are of brass. The ramrod serves also as a measure for powder. This is a French cavalry pistol of the model 1822 which was remodeled to the percussion system in 1841.

France.

No. 18. (N4243). PERCUSSION PISTOL. Caliber .68. Seven-and-one-half-inch, round, steel barrel. The grip is finely checkered as is also the inner back-strap. There are front and rear sights and formerly there was also a ramrod. All mountings are made of unornamented red brass. Marked upon the top of the barrel is the letter "B" in relief. This pistol is a French officer's model of 1816 which was remodeled to the percussion system in 1841. The gun was made in Maubeuge.

France.

No. 19. (N4242). PERCUSSION PISTOL. Caliber .68. Eight-and-one-quarter-inch, damaskeened, octagonal barrel, which is finely rifled. The muzzle is bell-shaped. The checkered grip is fitted with an iron butt-plate and ring. All steel mountings were originally blued. The right side of the barrel is stamped with the figures "1844 M.R." This 1844 model, French officer's pistol was made in Chatellerault.

France.

No. 20. (N2733). PERCUSSION PISTOL. Caliber .60. Four-and-one-half-inch, round, steel barrel. The smooth grip has a steel butt-plate. There are no sights. All mountings are of steel and unornamented. Stamped upon the left side of the barrel are the letters "A G D." This pistol is an 1842 model and was made for the gendarmerie or French armed police. It was made in Tulle.

France.

No. 21. (N2734). PERCUSSION PISTOL. Caliber .58. Four-and-three-quarters-inch, round, steel barrel. The grip is smooth and is covered with a steel butt-plate. There are no sights. A ramrod is carried beneath the barrel. All mountings are of steel and are unornamented. Stamped upon the left side of the barrel are two proof-marks. This is an 1842 model pistol and is so stamped behind the breech. It was made in the royal arsenal at St. Etienne for the French gendarmes or armed police.

France.

No. 22. (N4244). PERCUSSION PISTOL. Caliber .67. Four-and-one-half-inch, round, steel barrel. There is a ramrod but there are no sights. Engraved upon the lock-plate are the words "Mre. Rle. St. Etienne." This pistol belongs to the 1855 model and was used by the Italian and French gendarmes. It has been remodeled from the flint-lock to the percussion system. Its number is "274," and it was made in St. Etienne.

France.

No. 23. (N3976). PERCUSSION PISTOL. Caliber .62. Four-and-three-quarters-inch, round, steel barrel. The grip is smooth and round, having the butt covered with a brass butt-plate. All mountings are of heavy brass. There are no sights. The trigger-guard is missing. Stamped upon the left side of the stock are the numbers "13" and "168". The place of manufacture of this pistol is not given. It is an 1855 gendarme model and is so dated behind the breech. It was probably made in St. Etienne.

France.

No. 24. (N2735). PERCUSSION PISTOL. Caliber .58. Four-and-three-quarters-inch, round, steel barrel. The smooth grip is covered with a steel butt-plate. There are no sights. All mountings are of steel and unornamented. Two French proof-marks are stamped upon the left side of the barrel. This is an 1842 model gendarme pistol and is so dated behind the breech. It was made in the royal arsenal in St. Etienne.

France.

No. 25. (N4245). PERCUSSION PISTOL. Caliber .58. Four-and-one-half-inch, round, steel barrel. There is a steel plate upon the butt. There are no sights. A small ramrod is carried beneath the barrel. Two French proof-marks are stamped upon the left side of the barrel. The date "1842" is stamped behind the breech. This is another example of the French pistol used by the gendarmerie of that country. It was made in Chatellerault.

France.

No. 26. (N5848). PERCUSSION PISTOL. Caliber .65. Four-and-three-quarters-inch, round, steel barrel. Grip and butt-plate are the same as those of No. 25. Stamped upon the left side of the barrel are the figures "B 1812." The lock-plate is engraved with the name "Maubeuge." This is an 1802 model pistol which was later re-modeled to the percussion system.

Donor: Anton Beer.

France.

No. 27. (N4246). PERCUSSION PISTOL. Caliber .58. Four-and-three-quarters-inch, round, steel barrel which is finely rifled. The smooth, round grip has a cap-box in the butt, the lid of which is engraved with the letters "G. R." This pistol is equipped with front and rear sights and a ramrod. All mountings are of steel and are un-ornamented. Stamped upon the left side of the barrel are the letters "M.L." The trigger-guard is stamped with the letter "P" with a shield. The place of manufacture upon the lock-plate is effaced. This pistol is an 1855 model and was made for the officers of the French gendarmerie.

France.

No. 28. (N4269). PERCUSSION PISTOL. Caliber .50. Three-and-five-eighths-inch barrel, browned to imitate the spirals of damaskeened steel. The checkered, sloping grip has a steel butt-plate. There are no sights. The lock-plate is of great length and it and the hammer are engraved with scroll designs. Trigger-guard and ferrule are similarly engraved. Stamped upon the top of the barrel is the Liége proof-mark. No other marks of identification are given.

Belgium.

PLATE 80.

DERINGERS AND SIMILAR PISTOLS

No. 1. (4288). PERCUSSION PISTOL. Caliber .44. Four-and-one-half-inch, rifled barrel, which has browned spirals on the exterior to imitate damaskeened steel. The cross-hatched grip has no butt-plate. This pistol is fitted with front and rear sights. Trigger-guard and ferrule are made of engraved brass. The ramrod is made of wood. Hammer and lock-plate are engraved. Stamped upon the lock-plate are the words "Deringer, Phila."

United States.

No. 2. (N760). PERCUSSION PISTOL. Caliber .45. Four-and-three-eighths-inch, rifled, iron barrel. The grip is checkered and its butt is partly covered by a triangular brass strip. The end of the stock is tipped with brass. There are front and rear sights. Trigger-guard and hammer are of iron and ornamented with engravings of scroll designs. Barrel and lock-plate are stamped "Deringer, Philadel'a."

United States.

No. 3. (N4286). PERCUSSION PISTOL. Caliber .41. Four-inch, rifled, octagonal, steel barrel. The checkered grip is inlaid with an oval silver escutcheon, and the end of the stock is tipped with brass. There is no butt-plate. The pistol is equipped with front and rear sights. The hammer is apparently new. The trigger-guard is of engraved steel. Lock-plate and barrel bear the words "Deringer, Phila."

United States.

No. 4. (N6776). PERCUSSION PISTOL. Caliber .45. Three-and-one-half-inch, rifled, steel barrel. This pistol has a small checkered grip with a triangular white metal butt-plate. The stock is also tipped with white metal and there are several escutcheons of the same metal about the slots in the fore-stock. There are front and rear sights but the ramrod is missing. Hammer and lock-plate are engraved with scrolls and the latter is inscribed with the name "Deringer." The top of the barrel is stamped with the same name.

Donor: Frank Shrosbree.

United States.

No. 5. (N4287). PERCUSSION PISTOL. Caliber .41. Three-and-one-half-inch, rifled, steel barrel. The grip is partly checkered and is inlaid with a white metal escutcheon in the form of a shield. The stock extends to the muzzle where it ends in a bulb-like projection. There are front and rear sights. Hammer and lock-plate are engraved. The barrel and lock-plate are stamped "Deringer, Philadel." Stamped upon the left side of the barrel is the letter "P." This pistol, like the other Deringers, was made by Henry Deringer in Philadelphia.

United States.

No. 6. (N4289). PERCUSSION PISTOL. Caliber .45. Three-and-one-eighth-inch, rifled, steel barrel. The checkered grip is inlaid with a brass shield on top and has a brass cap-box in the butt. Sights and ramrod are of the usual kind. Trigger-guard is of engraved brass. Stamped upon the left side of the barrel is the letter "P," while the top of the barrel is stamped "Deringer, Philadel." The lock-plate is engraved and bears the same words as those stamped upon the top of the barrel.

United States.

No. 7. (N4290). PERCUSSION PISTOL. Caliber .45. Two-and-three-quarters-inch, rifled, steel barrel. The grip is checkered in three sections and has a brass butt-plate. Equipped with the usual form of sights. The stock extends almost to the muzzle and is tipped with brass. The trigger-guard is also of brass and is engraved. Barrel and lock-plate are stamped with the name "Deringer."

United States.

No. 8. (N4293). PERCUSSION PISTOL. Caliber .44. Two-and-three-quarters-inch, rifled, steel barrel. The small, checkered grip has a brass butt-plate and shield, the latter being inlaid on top of the grip. This pistol is provided with front and rear sights. No provision is made for carrying a ramrod upon this weapon. Hammer and lock-plate are engraved. The barrel and lock-plate are stamped with the words "Deringer, Philadel." Stamped upon the left side of the barrel is the letter "P."

United States.

No. 9. (N754). PERCUSSION PISTOL. Caliber .40. Two-and-one-half-inch, rifled barrel, which is browned to resemble damaskeened steel. The checkered grip is fitted with a small, triangular butt-plate. This pistol has front and rear sights but no ramrod. The trigger-guard is made of engraved brass. Hammer and lock-plate are engraved with scroll designs. The name "Deringer" appears nowhere upon this pistol. On the under-side of the barrel is the Liége proof-mark, which may indicate that the pistol was made in the United States and later exported to Belgium, or that it was made in Belgium and, therefore, is an imitation Deringer.

United States?

No. 10. (N4294). PERCUSSION PISTOL. Caliber .45. Two-and-one-quarter-inch, rifled barrel, with spiral bands made to resemble damaskeened steel. The grip is partly cross-hatched and is fitted with a triangular butt-plate. This piece is equipped with front and rear sights but has no ramrod. The trigger-guard is of engraved brass. Hammer and lock-plate are also engraved. This pistol is probably a Deringer but is stamped with no special marks of identification.

United States?

No. 11. (N2634). PERCUSSION PISTOL. Caliber .50. Two-and-one-quarter-inch, rifled, steel barrel. The small, cross-hatched grip is inlaid with a brass butt-plate and a shield. Trigger-guard is of engraver brass. Both barrel and lock-plate are stamped with the words "Deringer, Philadel."

United States.

No. 12. (N4296). PERCUSSION PISTOL. Caliber .40. Two-and-one-quarter-inch, rifled, steel barrel. The grip is partly cross-hatched and is fitted with a cap-box in the butt. Has the usual form of sights, and the engraved brass trigger-guard. The hammer and lock-plate are also engraved. Stamped upon the top of the barrel are the words "N. Curry & Bros., San Franco, Cala. Agents" and "Deringer, Philadel." The left side of the barrel is stamped with a "P."

United States.

No. 13. (N4295). PERCUSSION PISTOL. Caliber .36. Two-and-one-quarter-inch, rifled barrel, with spiral bands resembling those of damaskeened steel. The grip is partly cross-hatched, having the butt covered with a triangular plate of brass. There is no ramrod. The trigger-guard is of engraved brass. Hammer and lock-plate are also engraved. This pistol is probably a Deringer but bears no special marks of identification.

United States.

No. 14. (N4297). PERCUSSION PISTOL. Caliber .40. One-and-three-quarters-inch, rifled barrel, which has browned spirals on the exterior to resemble damaskeened steel. The grip is partly cross-hatched and the butt formerly had a butt-piece. There are front and rear sights. The trigger-guard is of engraved brass. Hammer and lock-plate are engraved with scroll designs. Upon the top of the barrel are the words "A. Wurfflein, Phila." The pistol number is "1313."

United States.

No. 15. (N4298). PERCUSSION PISTOL. Caliber .40. One-and-three-quarters-inch, rifled, steel barrel. Fitted with front and rear sights. The trigger-guard is of engraved brass. Hammer and lock-plate are also engraved. Upon the top of the barrel are the words "Made for F.H. Clarke & Co., Memphis, Tenn." and "Deringer, Philadel." The left side of the barrel is stamped with the letter "P."

United States.

No. 16. (N4299). PERCUSSION PISTOL. Caliber .45. Two-and-one-eighth-inch, rifled, steel barrel. The grip, which is partly cross-hatched, has the butt inlaid with a brass strip and the top with a brass shield. Fitted with front and rear sights. Hammer and lock-plate are engraved. The trigger-guard is of engraved brass. Stamped upon the lock-plate and barrel are the words "Slotter & Co., Phila." The right side of the barrel is stamped with the words "Wart Steel." This pistol was patterned after the Deringer.

United States.

No. 17. (N4285). PERCUSSION PISTOL. Caliber .22. Four-and-one-eighth-inch, brass barrel, which has been nickel-plated. The grip is partly cross-hatched and fitted with a wire shoulder-stock. The pistol is equipped with front and rear sights. There is neither ramrod nor trigger-guard upon this piece. Stamped upon the lock-plate are the words "Wurfflein, Phila." All parts of the pistol are unornamented. It was made up of parts left in the Deringer factory by using the Deringer lock and grip and attaching a brass barrel. It was loaded with a ball, no powder being used. A heavily charged cap served to expel the bullet from the barrel. This pistol was designed for indoor target practice.

United States.

No. 18. (N4291). PERCUSSION PISTOL. Caliber .44. Two-and-five-eighths-inch, rifled, steel barrel. Has a small, checkered grip, with a triangular butt-plate. There are front and rear sights, and an ivory or bone ramrod. The trigger-guard is of steel and slightly engraved. The hammer is also engraved. Barrel and lock-plate are marked with the words "Deringer, Philadel'a."

United States.

No. 19. (N4292). PERCUSSION PISTOL. Caliber .42. Two-and-three-quarters-inch, octagonal, damaskeened, steel barrel. The pistol has a small grip which was formerly checkered. The butt is covered with a triangular piece of brass. There is a front sight but none in the rear. The trigger-guard is of brass, as are also the other mountings on the pistol. The lock-plate is engraved with scroll designs, and the barrel is stamped with the name "Deringer."

United States.

No. 20. (N3289). PERCUSSION PISTOL. Caliber .57. Three-and-three-quarters-inch, damaskeened, octagonal barrel. The grip is checkered but has no butt-plate. Equipped with front and rear sights. The lock-plate and hammer are engraved with floral designs, as are also the trigger-guard and ferrules. The ramrod is of wood. The top of the barrel is engraved "Cork, CB-431." Upon the lock-plate is the name "G. Richardson." This type of weapon was known as a pocket pistol.

Ireland.

No. 21. (N4281). PERCUSSION PISTOL. Caliber .45. Four-and-three-eighths-inch, rifled, octagonal, steel barrel. The checkered grip is fitted with a triangular butt-plate. The stock is tipped with silver, and the trigger-guard is made of the same metal and is engraved. Equipped with front and rear sights, and a wooden ramrod. The engraved lock-plate is inscribed with the name "T. Perkins" and the barrel is stamped "London." This is an English pistol which is patterned after the famous Deringer.

England.

No. 22. (N4283). PERCUSSION PISTOL. Caliber .53. Three-and-three-quarters-inch, damaskeened, octagonal, steel barrel. The grip is checkered but has no butt-plate. The trigger-guard is made of engraved German silver. The ramrod is of wood. Lock-plate and hammer are elegantly engraved. Engraved upon the barrel are the words "Richardson, Kilkenny." This is another foreign pistol modeled after the American Deringer.

Ireland.

No. 23. (N4271). PERCUSSION PISTOL. Caliber .60. Five-and-one-half-inch, octagonal, steel barrel. The grip is checkered and fitted with a buttplate in the shape of a flower. There are front and rear sights. Has an engraved, steel trigger-guard. Hammer and lock-plate are also engraved and the latter is inscribed with the name "J.P. Moore." Upon the top of the barrel are the words "J.P. Moore, New York."

United States.

No. 24. (N4282). PERCUSSION PISTOL. Caliber .38. Four-inch, octagonal, steel barrel. The checkered grip is provided with a small cap-box in the butt. The pistol is fitted with front and rear sights. The trigger-guard forms part of the inner back-strap and is stamped "German Silver." The tip of the stock and all other mountings are of the same metal. The lock-plate is engraved and bears the name "T. Atwood." Stamped upon the barrel is the word "London."

England.

No. 25. (N4284). PERCUSSION PISTOL. Caliber .36. Four-and-one-quarter-inch, rifled, octagonal barrel made of damascus steel. The curved, cross-hatched grip is fitted with a percussion-cap box in the butt. Equipped with front and rear sights, and a wooden ramrod. The trigger-guard and one ferrule are made of engraved silver. The hammer and lock-plate are engraved, and the latter bears the name "Richardson," who was a London gunsmith.

England.

No. 26a. (N4279). PERCUSSION PISTOL. Caliber .45. Four-and-one-half-inch, heavy, octagonal barrel, made of damaskeened steel. There is no butt-plate upon the checkered grip. However, the top of the grip is inlaid with a small silver escutcheon. There is a sight in front but none in the rear. None of the mountings are engraved. The lock is of the back-action type, and was made by Winchester of London. Stamped upon the barrel are the words "London Fine Twist."

England.

No. 26b. (N4280). PERCUSSION PISTOL. Caliber .45. Four-

and-one-half-inch, heavy, octagonal barrel, made of damaskeened steel. There is no butt-plate upon the checkered grip. A small, silver escutcheon is inlaid upon the top of the grip. There is a sight in front but none in the rear. Lock-plate and barrel are marked in the same manner as No. 26a.

England.

PLATE 81.

PERCUSSION PISTOLS
TOP OR CENTER-HAMMER TYPES

No. 1. (N4339). PERCUSSION PISTOL. Caliber .44. Three-and-one-half-inch, octagonal, steel barrel. This pistol has a smooth, flat grip and a top-hammer but no sights or ramrod. All parts are unornamented. Upon the left side of the lock-plate is the Liége proof-mark and another in the form of a crown over the letter "T." The numeral "1" also appears on the same side. The letter "T" is also engraved on the body where the trigger-guard is fastened to the pistol. No maker's name is given. The pistols illustrated under No. 1 to No. 11 are almost all of the same kind. The workmanship is coarse and frequently inferior, which may account for the fact that in many instances neither maker's name nor proof-marks are given. Only the most salient features of each of these weaponss will be given below.

Belgium.

No. 2. (N4341). PERCUSSION PISTOL. Caliber .40. Three-and-one-half-inch, octagonal, steel barrel. The under-side of the barrel and breech are stamped with the number "14." No proof-marks or maker's name are given.

Belgium?

No. 3. (N6757). PERCUSSION PISTOL. Caliber .38. Three-and-one-quarter-inch, octagonal, steel barrel. No proof-marks or other marks of identification are given.

Belgium?

No. 4. (N6764). PERCUSSION PISTOL. Caliber .44. Three-and-three-eighths-inch, octagonal, steel barrel. The left side of the barrel is stamped with the letters "PU." No other marks are given.

Belgium?

No. 5. (N6774). PERCUSSION PISTOL. Caliber .47. Three-

and-three-quarters-inch, octagonal, steel barrel, which may be unscrewed from the breech by inserting a special wrench into the muzzle. The left side of the barrel is stamped with the Liége proof-mark.

Donor: Edward R. Ryan.

Belgium.

No. 6. (N4348). PERCUSSION PISTOL. Caliber .47. Two-and-five-eighths-in, octagonal, steel barrel. The left side of the barrel is stamped with the Liége proof-mark. No maker's name is given.

Belgium.

No. 7. (N741). PERCUSSION PISTOL. Caliber .48. Three-and-three-quarters-inch, octagonal, steel barrel. All parts are unornamented except the lock-plate, which is crudely engraved. The Liége proof-mark is stamped upon the left side of the barrel, while the under-side is stamped with the letter "L."

Belgium.

No. 8. (N742). PERCUSSION PISTOL. Caliber .48. Three-and-one-half-inch, round, steel barrel. The trigger-guard and lock-plate are engraved with scrolls. English and Belgium proof-marks are stamped upon the left side of the barrel, while the under-side of the breech is marked with the number "934."

England.

No. 9. (N4342). PERCUSSION PISTOL. Caliber .71. Three-and-one-quarter-inch, octagonal, steel barrel, which can be unscrewed by inserting a special wrench into the muzzle. Lock-plate and folding trigger are engraved with scrolls. The left side of the barrel bears the Liége proof-mark. No maker's name is given.

Belgium.

No. 10. (N6766). PERCUSSION PISTOL. Caliber .48. Two-and-one-half-inch, octagonal, steel barrel. All mountings and other parts of this pistol are unornamented and no marks of identification are given.

Belgium?

No. 11. (N5906). PERCUSSION PISTOL. Caliber .35. Two-and-three-quarters-inch, round, steel barrel. All mountings and other parts of this pistol are unornamented. No proof-marks or maker's name are visible.

Donor: Christ Warth.

Belgium?

No. 12. (N4350). PERCUSSION PISTOL. Caliber .42. Two-and-three-quarters-inch, round, steel barrel. This pistol has a smooth, flat grip, a superposed hammer but no sights, ramrod or safety-catch. The trigger-guard is engraved with a flower. Two English proof-marks are stamped beneath the barrel. The lock-plate is partly engraved and is stamped with the words "C. Balton & Co., London."

England.

No. 13. (N744). PERCUSSION PISTOL. Caliber .50. Three-and-one-half-inch, octagonal, steel barrel. This pistol has an oval, cross-hatched grip having the top and butt inlaid with silver plates. A super-posed hammer is provided with a safety-catch and a folding trigger. There are no sights, ramrod, or trigger-guard. Body and lock-plate of this pistol are finely engraved with scroll designs and the bottom of the barrel is stamped with two English proof-marks. Engraved upon the lock-plate are the words "R.S. Clark, London."

England.

No. 14. (N4345). PERCUSSION PISTOL. Caliber .45. Three-and-one-quarter-inch, round, steel barrel. All parts are unornamented. The trigger-guard is missing. Stamped upon the left side of the barrel is the Liége proof-mark and the under-side is marked "DR." No maker's name is given.

Belgium.

No. 15. (N4346). PERCUSSION PISTOL. Caliber .44. Three-and-one-quarter-inch, round, steel barrel. Has a smooth, flat grip which is decorated with a floral scroll. The trigger-guard is slightly engraved and the bottom of the barrel is stamped with two English proof-marks. The lock-plate is engraved with floral garlands and bears the name "Ketland & Co., London." This piece was altered from the flint-lock system.

England.

No. 16. (N4352). PERCUSSION PISTOL. Caliber .40. Three-and-three-eighths-inch, round, steel barrel. The trigger-guard and lock-plate are slightly engraved with scrolls and flowers. Upon the bottom of the barrel are three English proof-marks which probably are of the Birmingham Proof-House. However, the lock-plate is engraved with the words "Thomps, London."

England.

No. 17. (N4351). PERCUSSION PISTOL. Caliber .36. Three-and-three-eighths-inch, round, steel barrel, having a brass breech. Has a small, oval, checkered grip with a cap-box in the butt and a silver plate inlaid on the top. The trigger is of the folding variety. Stamped beneath the barrel are two Birmingham proof-marks. Upon the lock-plate are the words "E.W. Bailey, London." Bailey probably was the agent or dealer of a Birmingham gunsmith and not himself a maker of firearms.

England.

No. 18. (N2770). PERCUSSION PISTOL. Caliber .48. Two-and-seven-eighths-inch, round, steel barrel. The grip is smooth and oval and is covered with black enamel. The trigger-guard is engraved with a flower and the lock-plate is ornamented with scrolls. Upon the left side of the barrel is the Liége proof-mark. No other marks of identification are given.

Belgium.

No. 19. (N4347). PERCUSSION PISTOL. Caliber .44. Three-inch, round barrel, probably made of damaskeened steel. The smooth grip is made of briar root. There is a cap-box in the butt. The only ornamentation is on the lock-plate which is slightly engraved. The right side of the barrel is marked with a crown over the letters "F.M." while the left side bears the Liége proof-mark. No maker's name is given.

England?

No. 20. (N7041). PERCUSSION PISTOL. Caliber .42. Two-and-one-quarter-inch, round, steel barrel, which can be unscrewed by means of a special wrench inserted into the muzzle. The oval grip is partly checkered. The lock-plate is partly engraved with flowers. No marks of identification are given.

Belgium?

No. 21. (N4349). PERCUSSION PISTOL. Caliber .45. Two-and-one-half-inch, round, steel barrel. The grip is partly checkered and is fitted with two cheek-pieces which are carved with flowers. The superposed hammer is provided with a safety-catch. A diamond-shaped design is engraved upon the lock-plate. The Liége proof-mark is at its usual place, upon the left side of the barrel. This pistol was probably altered from the flint-lock system.

Belgium.

No. 22. (N4359). PERCUSSION PISTOL. Caliber .42. Two-and-one-quarter-inch, round, steel barrel, which can be unscrewed from the brass breech. The pistol is provided with a checkered grip, the checking being only on top where the grip is also inlaid with a silver escutcheon in the form of a shield. The trigger is of the folding kind. Beneath the barrel are two Birmingham proof-marks; the lock-plate is engraved with scrolls and the name "H. Smith, London."

England.

No. 23. (N740). PERCUSSION PISTOL. Caliber .40. Two-and-three-quarters-inch, octagonal, steel barrel. The smooth, oval grip is slightly fluted on the sides. The hammer and under-side of the barrel are stamped with the number "52." No marks of identification are given.

England.

No. 24. (N4357). PERCUSSION PISTOL. Caliber .48. Two-and-one-half-inch, octagonal, brass barrel. This piece was altered from the flint-lock to the percussion-cap system. The top-hammer formerly had a safety-catch. A star is engraved upon the iron trigger-guard and the barrel is stamped with the Liége proof-mark.

Belgium.

No. 25. (N4360). PERCUSSION PISTOL. Caliber .48. Two-and-one-half-inch, round, steel barrel. The lock-plate is engraved with crude floral designs. The left side of the barrel is stamped with the Liége proof-mark and another in the shape of a crown over the letter "S." No maker's name is given.

Belgium.

No. 26. (N739). PERCUSSION PISTOL. Caliber .40. Two-and-one-half-inch, brass barrel, which may be unscrewed. The iron trigger-guard is ornamented with an engraved star. The left side of the barrel is stamped with the Liége proof-mark; the right side is marked with the letters "S.P."

Belgium?

No. 27. (N4358). PERCUSSION PISTOL. Caliber .45. Two-and-one-quarter-inch, round, steel barrel, which may be unscrewed from the breech. The under-side of the barrel bears the letter "R.H." followed by three Birmingham proof-marks. This pistol was remodeled from the flint-lock system. It bears no maker's name.

England.

No. 28. (N4361). PERCUSSION PISTOL. Caliber .44. Two-and-one-eighth-inch, round, steel barrel. Has a checkered grip with a cap-box in the butt. The lock-plate is engraved with flowers and the bottom of the barrel is stamped with the Liége proof-mark.

Belgium.

No. 29. (N4362-3). PAIR OF PERCUSSION PISTOLS. Caliber .38. Two-and-one-quarter-inch, round, blued steel barrels, which can be unscrewed from the brass breeches. The grips are finely cross-hatched and each is inlaid with a silver disc on top. The folding-triggers act as safety-devices. Each lock-plate is engraved with the words "Richd. Hollis, London," and each barrel is stamped with three proof-marks, and the letters "R.H."

England.

No. 30. (N4375). PERCUSSION PISTOL. Caliber .32. One-and-three-quarters-inch, tapering, steel barrel, which may be unscrewed from the German silver breech. Has a closely checkered grip, fitted with a cap-box in the butt and inlaid with a silver shield on top. The top-hammer is finely engraved. Upon the bottom of the barrel are three English proof-marks. The lock-plate is finely engraved, made of German silver, and is inscribed with the words "Richards, London."

England.

No. 31. (N4367). PERCUSSION PISTOL. Caliber .35. One-and-one-half-inch, octagonal, rifled, steel barrel. Has a solid ivory grip. The hammer is finely engraved and fitted with a safety-catch; as a further precaution the trigger may be folded up, provided that the pistol is but half-cocked. The lock-plate is elegantly engraved with scrolls, while the bottom of the barrel is stamped with two English proof-marks and twice with the numerals "2." Upon the top of the barrel are the words "Beattie, London."

England.

No. 32. (N4364). PERCUSSION PISTOL. Caliber .35. One-and-five-eighths-inch, round, damaskeened barrel. Has a smooth, oval grip which is enameled in black and inlaid with a small silver disc on top. Lock-plate and folding-trigger are engraved. No maker's name or proof-marks are given.

France?

No. 33. (N4365). PERCUSSION PISTOL. Caliber .28. One-

and-one-half-inch, tapering, rifled, steel barrel. The smooth, oval grip is inlaid with a silver shield on top. The lock-plate is engraved with scrolls. Stamped upon the bottom of the barrel is the number "59." No marks of identification are given.

France?

PLATE 82.

PERCUSSION PISTOLS,
CANNON-BARRELS, LARGE PISTOLS, AND
TOP OR CENTER-HAMMERS.

No. 1. (N4258). PERCUSSION PISTOL. Caliber .70. Six-and-three-quarters-inch, half-octagonal, iron barrel. The grip is checkered and the engraved back-strap extends to the butt. In the butt is a cap-box with an engraved cover. The rear sight is concave; there is no front sight. The lock is missing. This is a Spanish presentation piece, the barrel of which is made from horse-shoes worn by the Queen's horse. Iron from horse-shoes and carriage rims was frequently used for gun barrels, it being considered to be of superior quality, probably because it had been subjected to numerous stresses and strains and, therefore, was deemed to be tougher than ordinary iron. This pistol was presented to an officer whose name, unfortunately now partly obliterated, was inlaid with gold on top of the barrel. A gift of this kind was considered to be one of the greatest favors which could be bestowed upon an officer. The presentation was made in 1842 and the piece was made in Eibar.

Spain.

No. 2. (N4257). PERCUSSION PISTOL. Caliber .66. Six-and-three-quarters-in, half-octagonal, iron barrel. This piece has a finely checkered grip fitted with an engraved steel butt-plate which has a ring. The rear sight is concave; there is no front sight. Lock-plate and all mountings are delicately engraved. The hammer is shaped to resemble the head of an animal. The following inscription is inlaid with gold in the barrel: "Construido De Herraduras Por Gabriel Oqunoureno En Eybar Ano 1843," which translated means "Made of horse-shoes by Gabriel Oqunoureno in Eybar, in the year 1843."

Spain.

No. 3. (N5634). PERCUSSION PISTOL. Caliber .75. Six-and-one-quarter-inch, octagonal, iron barrel. The grip is partly checkered and has a back-strap which extends to the butt where it ends in a plate fitted with a ring. There is a belt-hook upon the left side. The pistol has neither front nor rear sights. The mountings are but slightly engraved. Stamped upon the grip is the number "120" while the barrel is engraved with the words "En Eibar Por Gastelu Ano 1866." This piece was made in Eibar.

Spain.

No. 4. (N2727). PERCUSSION PISTOL. Caliber .68. Six-inch, octagonal, damaskeened barrel. The grip is cross-hatched and is inlaid with a silver plate. There is a cap-box in the butt. The pistol is provided with front and rear sights. Lock-plate, hammer and trigger-guard are engraved with scrolls. No maker's name or proof-marks are given.

?

No. 5. (N2730). PERCUSSION PISTOL. Caliber .65. Six-and-three-quarters-inch, rifled damaskeened barrel which is fitted with front and rear sights. The fluted grip has a brass cap-box in the butt. Lock-plate and hammer are engraved with designs in the form of leaves. Trigger-guard and ferrules are of brass and similarly engraved. The breech is stamped with the letter "H" and two other marks which are probably proof-marks.

?

No. 6. (N2723). PERCUSSION PISTOL. Caliber .58. Seven-inch, octagonal, steel barrel. Has a finely cross-hatched grip. A steel cap-box with an engraved cover is in the butt. The stock extends almost to the muzzle where it is tipped with horn. Fitted with front and rear sights and a swivel ramrod. Lock-plate, hammer and trigger-guard are finely engraved. Upon the top of the barrel is the inscription "James Beattie, 223 Regent St., London." The left side of the barrel bears the Liége proof-mark.

England.

No. 7. (N4270). PERCUSSION PISTOL. Caliber .58. Seven-and-seven-eighths-inch, octagonal barrel. Has a rather small, smooth grip, fitted with a cap-box. Equipped with front and rear sights and a swivel ramrod. The trigger-guard is of engraved brass. Hammer and lock-plate are engraved with fine scroll designs. The barrel is stamped "London" and the lock-plate "Hopkins."

England.

No. 8. (N4274). PERCUSSION PISTOL. Caliber .65. Five-and-five-eighths-inch, octagonal, damaskeened barrel. Has a cross-hatched grip with a brass capped butt. Front and rear sights. Trigger-guard and ferrules are of engraved brass. The lock-plate is crudely engraved. Stamped upon the barrel are the words "E.&W. Bond, 45 Cornhill, London."

England.

No. 9. (N4273). PERCUSSION PISTOL. Caliber .48. Five-and-one-half-inch, octagonal barrel, having a front sight but none in the rear. The checkered grip has a silver butt-plate and the stock is also fitted with a silver tip near the muzzle. The lock-plate, trigger-guard and ferrules are of steel and finely engraved. This piece was made by Goddard and Company of London and is so marked.

England.

No. 10. (N4276). PERCUSSION PISTOL. Caliber .55. Four-and-one-half-inch, octagonal, steel barrel. The smooth, oval grip is not provided with a butt-plate. Fitted with front and rear sights. The trigger-guard and ferrules are of silver and are engraved. Lock-plate and hammer are engraved with scroll designs. The barrel is stamped "London" and the lock-plate bears the name "Baker."

England.

No. 11. (N2732). PERCUSSION PISTOL. Caliber .50. Five-and-one-half-inch, damaskeened barrel. The left side of the oval, check-ered grip is inlaid with a silver plate in the form of two birds sitting on a branch of a tree. There are front and rear sights. Hammer and lock-plate are engraved. Upon the lock-plate is the name "Richd. Hollis" and the barrel is stamped "London."

England.

No. 12. (N4278). PERCUSSION PISTOL. Caliber .55. Four-and-three-eighths-inch, octagonal, damaskeened barrel. Has a fine, cross-hatched grip having a silver butt-plate. Another silver plate is inlaid on top of the grip. The pistol is fitted with front and rear sights and with a swivel ramrod. Lock-plate, back-strap, trigger-guard and hammer are engraved with fine scrolls. No maker's name or proof-marks are given.

England.

No. 13. (N4272). PERCUSSION PISTOL. Caliber .55. Four-

and-one-quarter-inch, octagonal, damaskeened barrel which is fitted with front and rear sights. The end of the stock is tipped with lead. All mountings are of steel and show signs of former engraving. The lock-plate is stamped with the name "Sharpe" and the barrel, "London." In shape and some other ways this pistol resembles a large Deringer.

England.

No. 14. (N4277). PERCUSSION PISTOL. Caliber .42. Four-and-one-quarter-inch, rifled, octagonal barrel. Has a curved, checkered grip with a cap-box in the butt. The trigger-guard is of engraved brass while the ferrules are of the same alloy but plain. Lock-plate and hammer are finely engraved with scroll designs. Stamped upon the barrel is the word "London" while the lock-plate is inscribed with the name "Ino. (John) Wadsworth."

England.

No. 15. (N4275). PERCUSSION PISTOL. Caliber .50. Five-and-one-quarter-inch, octagonal barrel. Has a checkered grip fitted with a brass cap-box. Equipped with front and rear sights. Hammer, lock-plate and trigger-guard are engraved. The barrel is stamped with the word "London" and the lock-plate with the name "J.Bishop."

England.

No. 16. (N2995). PERCUSSION PISTOL. Caliber .52. Four-and-one-half-inch, bronze barrel which is ornamented on top with incised figures. Has a smooth, oval grip of dark wood. The pistol is fitted with three sights which are useless since they are in the same line with the superposed hammer. There is no trigger-guard or ramrod. A small thong of leather fastened below the hind sight serves as a carrying strap. There is no half-cock position for the hammer which must be brought to full-cock at once whenever the pistol is to be fired. Three undecipherable marks are on the bottom of the barrel.

China?

No. 17. (N735). PERCUSSION PISTOL. Caliber .38. Four-and-one-quarter-inch, cannon-shaped, octagonal, brass barrel. There are no sights or ramrod. A spring-bayonet was formerly attached to the under-side of the barrel near the muzzle. Trigger-guard and barrel are engraved with designs resembling lotus leaves. No marks of identification are to be found on this pistol.

Japan.

No. 18. (N4335). PERCUSSION PISTOL. Caliber .50. Six-and-one-half-inch, brass, cannon-muzzled barrel which is octagonal at the breech. Has a smooth, oval grip but no sights or ramrod. The trigger-guard is of iron. The left side of the lock-plate is engraved with wreaths and inscribed with the name "P. Larkin."

England?

No. 19. (N734). PERCUSSION PISTOL. Caliber .40. Four-and-one-quarter-inch, octagonal, cannon-shaped, brass barrel with bronzed finish. The grip is partly checkered and is fluted down the back. There are no sights, ramrod or butt-plate. The trigger-guard is of iron. On the left side of the barrel are some undecipherable marks; otherwise there are no marks of identification.

England?

No. 20. (N2898). PERCUSSION PISTOL. Caliber .55. Four-and-three-quarters-inch, cannon-shaped brass barrel to which there is attached a three-inch spring bayonet which may be folded up beneath the barrel when the blade is not needed. The grip is smooth and inlaid with wire in the form of scrolls. The superposed hammer has a safety-device. The lock-plate is engraved but bears no maker's name.

England.

No. 21. (N3292). PERCUSSION PISTOL. Caliber .50. Three-and-one-quarter-inch, cannon-shaped, brass barrel. Has a smooth, oval grip, but no sights, ramrod, or butt-plate. The lock-plate is engraved with a star and the left side of the barrel is stamped with the Liége proof-mark. No maker's name is given.

Belgium.

No. 22. (N737). PERCUSSION PISTOL. Caliber .45. Three-and-one-quarter-inch, octagonal, brass barrel. The top-hammer is missing. There are no sights or ramrod. The trigger-guard is of iron. All parts are unornamented. Stamped upon the under-side of the barrel is the number "94" and upon the left side the Liége proof-mark.

Belgium.

No. 23. (N4266). PERCUSSION PISTOL. Caliber .66. Six-and-one-half-inch, half-octagonal, cannon-shaped, iron barrel. The smooth grip is fitted with an iron butt-plate which is rather heavy. There are no sights. Trigger-guard, hammer and lock-plate are incised, and were formerly inlaid with gold. This pistol probably was remodeled from the flint-lock system. It is of Spanish manufacture but bears no special marks of identification.

Spain.

No. 24. (N738). PERCUSSION PISTOL. Caliber .45. Two-and-five-eighths-inch, octagonal, cannon-shaped, brass barrel. There are no sights, ramrod or safety-catch. Has a plain, iron trigger-guard. Upon the left side of the lock-plate are English and Belgium proof-marks and the word "London." The right side of the lock-plate bears the name "Ketland."

England.

No. 25. (N4355). PERCUSSION PISTOL. Caliber .40. Two-inch, octagonal, cannon-shaped, brass barrel. Has a thick, wooden grip, bearing carvings of flowers, and a plain, iron trigger-guard. The left side of the barrel is stamped with the Liége proof-mark; no maker's name is given.

Belgium.

No. 26. (N4353). PERCUSSION PISTOL. Caliber .40. Two-and-five-eighths-inch, cannon-shaped, brass barrel. The trigger-guard is of plain iron. All parts of the pistol are unornamented. The under-side of the barrel is stamped with the number "1." No other marks of identification are given.

England?

No. 27. (N4354). PERCUSSION PISTOL. Caliber .44. Two-and-three-eighths-inch, octagonal, brass barrel. There are no sights or ramrod. Has an iron trigger-guard engraved with a star. The lock-plate is engraved with designs of leaves and the left side of the barrel bears the Liége proof-mark. The under-side is stamped with the letters "S.P." This pistol was remodeled from the flint-lock system.

Belgium.

No. 28. (N733). PERCUSSION PISTOL. Caliber .38. One-and-three-quarters-inch, octagonal, brass barrel. Has a plain, iron trigger-guard. Stamped upon the left side of the barrel are English and Belgian proof-marks. No maker's name is given.

England.

No. 29. (N4356). PERCUSSION PISTOL. Caliber .36. One-and-seven-eighths-inch, octagonal, brass barrel. There are no sights.

The trigger-guard is of iron. All parts of the pistol are unornamented. The Liége proof-mark is stamped upon the left side of the lock-plate; the under-side of the barrel is marked with the letters "S.P." No maker's name is given.

Belgium.

No. 30. (N4333). PERCUSSION PISTOL. Caliber .62. Four-and-one-quarter-inch, detachable, round, iron barrel. Has a smooth grip. A bulbous butt covered with an embossed silver plate resembles a human head. The top of the grip bears a plate upon which the letters "W.H." are engraved. The trigger-guard is of iron and engraved with a flower. Two English proof-marks in the form of crowns over the letter "V" are stamped upon the under-side of the barrel. The lock-plate is marked "Griffin, London."

England.

No. 31. (N743). PERCUSSION PISTOL. Caliber .45. Five-inch, smooth, brass barrel. No sights, ramrod or trigger-guard. The pistol has a plain, oval grip. A hole in the butt is for the attachment of a ring or thong. All parts are devoid of ornamentation. No proof-marks or maker's name are given.

England?

No. 32. (N4334). PERCUSSION PISTOL. Caliber .44. Five-and-one-quarter-inch, round, steel barrel which can be unscrewed from the brass breech by inserting a special wrench into the muzzle. The smooth, oval grip is stamped on the left side with the number "689." The lock-plate is engraved with floral designs and bears the Liége proof-mark. No maker's name is given.

Belgium.

No. 33. (N4337). PERCUSSION PISTOL. Caliber .45. Four and-three-quarters-inch, damaskeened barrel which can be unscrewed by means of a special wrench. The checkered grip is inlaid on top with a silver plate which is engraved with the letters "R.S.M." The lock-plate is engraved and bears the words "Clark, London." Stamped upon the left side of the barrel is the Liége proof-mark.

Belgium.

No. 34. (N4368). PERCUSSION PISTOL. Caliber .42. Three-and-five-eighths-inch, octagonal, damaskeened barrel which can be removed by inserting a special wrench into the muzzle. The folding-trigger acts as a safety-device. The lock-plate is engraved and the left side of the barrel bears the Liége proof-mark. Upon the lock-plate are the words "Sharpe, London."

England.

No. 35. (N4344). PERCUSSION PISTOL. Caliber .44. Three-and-one-quarter-inch, octagonal, damaskeened barrel having a detachable muzzle. Has an oval, checkered grip but no sights or ramrod. Trigger-guard and lock-plate are slightly engraved with floral designs. The Liége proof-mark is at its usual place upon the barrel.

Belgium.

No. 36. (N4343). PERCUSSION PISTOL. Caliber .45. Three-and-one-quarter inch, octagonal, steel barrel. Has a smooth, flat grip which is inlaid on top with an oval, silver escutcheon. There are no sights. An iron ramrod is carried beneath the barrel in iron ferrules. The lock-plate and trigger-guard are engraved but bear no maker's name. The barrel bears the Liége proof-mark in the usual place.

Belgium.

No. 37. (N4336). PERCUSSION PISTOL. Caliber .42. Two-and-one-quarter-inch, round, steel barrel, slightly engraved near the muzzle. Has a brass breech and a smooth oval grip. The superposed hammer is provided with a safety-catch. Has no sights or ramrod. The folding-trigger is revealed when the pistol is brought to full cock. This pistol was probably remodeled from the flint-lock system. It was made by W. Jacot, London.

England.

No. 38. (N4338). PERCUSSION PISTOL. Caliber .45. Five-inch, octagonal, steel barrel, to the under-side of which is fastened a spring-bayonet three-and-one-half-inches long. The smooth, oval grip is inlaid on top with a small, silver disc. Has a large top-hammer, but no sights or ramrod. The lock-plate is engraved with flowers. All steel parts were formerly blued. No maker's name is given. Upon the left side of the barrel is the Liége proof-mark.

Belgium.

No. 39. (N2768). PERCUSSION PISTOL. Caliber .50. Six-and-one-quarter-inch, round, silver steel, damascus barrel. The grip is rather broad and checkered; the top-hammer has no safety-catch. There are no sights or ramrod. Lock-plate and trigger-guard are engraved with flowers; the barrel is stamped with the number "1" near the breech. The great length of the barrel seems to indicate that the pistol was intended for target use, although it is difficult to see how it could be used for this purpose without sights.

England?

No. 40. (N745). PERCUSSION PISTOL. Caliber .45. Four-and-one-eighth-inch, octagonal, browned, iron barrel which is fitted with a spring-bayonet having a blade two-and-one-quarter-inches long. Has a flat, checkered grip. The lock-plate is crudely engraved. The Liége proof-mark is stamped upon the left side of the plate. No other marks of identification are given.

Belgium.

No. 41. (N4340). PERCUSSION PISTOL. Caliber .45. Four-and-one-quarter-inch, octagonal, damaskeened barrel. Has an oval, cross-hatched grip, the top and butt of which were formerly inlaid with metal. The trigger is of the folding kind. The lock-plate is engraved with scrolls and bears the words "Sharpe, London." No proof-marks are given.

England.

No. 42. (N746). PERCUSSION PISTOL. Caliber .48. Three-and-three-quarters-inch, octagonal, damaskeened barrel which is fitted with a two-and-one-half-inch spring-bayonet. Has a smooth, mottled, wooden grip having a cap-box in the butt. Top-hammer, trigger-guard and lock-plate are engraved. The left side of the barrel is stamped with the Liége proof-mark. No other marks of identification are given.

Belgium.

PLATE 83

VARIOUS KINDS OF DOUBLE-BARRELED PERCUSSION PISTOLS

No. 1. (N730). PERCUSSION PISTOL. Caliber .62. Nine-and-one-quarter-inch, round steel barrels. Has a checkered grip, inlaid with a silver shield. Equipped with a front sight and a concave rear sight. Ramrod and the two side-hammers are missing. Each lock-plate has a safety-catch and is engraved with the name "Spencer."

The word "Birmingham" appears upon the top of each barrel. This weapon was probably remodeled from the flint-lock system.

England.

No. 2. (N4417). PERCUSSION PISTOL. Caliber .54. Seven-inch, round, steel barrels. The grip is checkered and inlaid with a silver escutcheon. Has a front sight and an iron ramrod, but no rear sight. The lock-plates are provided with safety-catches. All parts are finely engraved with scrolls. The name "J. E. Perry & Co." is engraved upon each lock-plate; the under-side of each barrel bears English proof-marks and the letters "E. S."

England.

No. 3. (N2737). PERCUSSION PISTOL. Caliber .54. Five-and-three-quarters-inch, round, steel barrels. Has a cross-hatched grip with a cap-box in the butt. Hammers and lock-plates are finely engraved with scrolls. The lock-plates bear the words "John Blissett, London." The pistol has front and rear sights and a swivel ramrod. The under-side of each barrel is stamped with English proof-marks and the number "2," while the top of the barrels bear the words "John Blissett, 321 High Holborn, London, Maker to H. R. H. Duke of Essex."

England.

No. 4. (N2738). PERCUSSION PISTOL. Caliber .68. Seven-and-one-half-inch, round, steel barrels. Equipped with a checkered grip. Although now missing, it formerly had a cap-box in the butt. The ramrod is also gone. There is a sight in front, but none in the rear. Trigger-guard, hammers and lock-plates are engraved with scrolls. The left barrel is stamped with the Liége proof-mark and the right with the number "40." These figures can be seen only by removing the barrels from the stock.

Belgium.

No. 5. (N4416). PERCUSSION PISTOL. Caliber .69. Six-and-three-quarters-inch, damaskeened barrels. Has a fluted grip with a silver cap-box in the butt, the cover of which is missing. The only sight is in front. The trigger-guard is of engraved silver. The left barrel is stamped with the Liége proof-mark and the letter "N"; the right barrel is marked with the number "166." No maker's name is given.

Belgium.

No. 6. (N2739). PERCUSSION PISTOL. Caliber .74. Seven-inch, steel barrels which are equipped with a front sight and a concave rear sight. The grip is checkered and has a German silver cap-box in the butt. The German silver trigger-guard is engraved. Lock-plates and hammers are engraved with scrolls. The left barrel is stamped with the Liége proof-mark and the number "3458"; the right is marked with the number "36."

Belgium.

No. 7. (N4427). PERCUSSION PISTOL. Caliber .34. Five-and-one-quarter-inch, round, steel barrels which are cast in one piece. Equipped with a front sight and ramrod, but no rear sight. The round body is engraved with a design representing a cannon draped by an American flag. The two superposed hammers are operated by one trigger. Upon the top of the barrels are the figures "W.P. 1835" (?) and also the letters "P.M." The bottom of the pistol bears the number "1131." It was made by Bacon and Company of Norwich, Connecticut.

United States.

No. 8. (N749). PERCUSSION PISTOL. Caliber .35. Five-and-one-half-inch, round, rifled, steel barrels cast in one piece. Equipped with a front sight, but has no rear sight or ramrod. Fitted with two superposed hammers which are operated by one trigger. All parts of the pistol are unornamented. The top is stamped "Allen & Thurber," and the bottom bears the number "713." It was made in Worcester, Massachusetts.

United States.

No. 9. (N4428). PERCUSSION PISTOL. Caliber .34. Four-inch, round, rifled, steel barrels which are cast in one piece. The body is round and engraved with a design of leaves. Has a fore-sight, but no rear sight. Is equipped with a wooden ramrod. Two hammers are operated by one trigger; upon cocking and pulling the trigger, first the right hammer falls, and by pulling a second time the left hammer is made to fall. Stamped upon the top of the barrels is the name "Bruce & Davis" and the words "Cast Steel." The under-side of the pistol is marked with the number "13."

United States.

No. 10. (N752). PERCUSSION PISTOL. Caliber .36. Two-and-seven-eighths-inch, steel barrels which are rifled, and cast in one piece. Action is the same as No. 9. In pistols of this type the hammers are brought to full-cock at once, there being no notch to hold the hammers at half-cock. The top is stamped "Allen & Wheelock" and the bottom of the barrels bear the number "316."

United States.

No. 11. (N757). PERCUSSION PISTOL. Caliber .43. Three-and-three-eighths-inch, octagonal, steel barrels. Has two hammers and two triggers. There are no sights. All parts of the pistol are unornamented. No marks of identification are given.

United States?

No. 12. (N4419). PERCUSSION PISTOL. Caliber .41. Four-and-five-eighths-inch, octagonal, steel barrels, joined by a rib. Has a fluted grip and is equipped with front and rear sights and a ramrod. The body is engraved with scrolls. There are no proof-marks upon the barrels, neither is the maker's name given, although the pistol was made by C. Kemlein.

Germany.

No. 13. (N6761). PERCUSSION PISTOL. Caliber .44. Three-and-one-half-inch, octagonal, steel barrels. Has a smooth, oval grip, two superposed hammers, but no sights or ramrod. All parts of the pistol are unornamented. The under-side of both barrels is stamped with the number "31." No maker's name is given.

England?

No. 14. (N6778). PERCUSSION PISTOL. Caliber .38. Three-and-one-half-inch, octagonal, steel barrels. Formerly had two superposed hammers which are now missing. The body is of polished steel and ornamented with scrolls. Has neither sights nor ramrod. Stamped upon the under-side of the barrel are the letters "EO." No other marks of identification are given.

England?

No. 15. (N751). PERCUSSION PISTOL. Caliber .38. Four-and-three-quarters-inch, round, rifled, steel barrels. Has a round grip inlaid with a silver shield, but no sights or ramrod. The body is slightly engraved, but all other parts of the pistol are unornamented. Each barrel is stamped with an English proof-mark; in addition the left one is marked with the proof-mark of Liége. No maker's name is given.

England.

No. 16. (N2994). PERCUSSION PISTOL. Caliber .40. Three-and-one-half-inch, round, steel barrels. All parts of the pistol are unornamented. The left barrel is stamped with the Liége proof-mark and both barrels bear a proof-mark in the form of a crown over the letter "C." In addition the right barrel is stamped with the number "19." No maker's name is given.

England.

No. 17. (N6772). PERCUSSION PISTOL. Caliber .43. Three-and-one-eighth-inch, round, steel barrels. Has a smooth, round grip, two superposed hammers, but no sights or ramrod. The former trigger-guard is missing. The left barrel is stamped with the Liége proof-mark and with an English mark in the form of a crown over the letter "V." The right barrel is unstamped.
Donor: H. A. Kirchner.

England.

No. 18. (N4422). PERCUSSION PISTOL. Caliber .41. Three-and-one-quarter-inch, octagonal, brass barrels, cast in one piece. Has an unfinished, home-made grip. The trigger-guard is of iron and unornamented. The left barrel is stamped with an English proof-mark in the form of a crown over the letter "V" and is also marked with the Liége proof-mark. No maker's name is given.

England.

No. 19. (N750). PERCUSSION PISTOL. Caliber .40. Three-and-one-quarter-inch, octagonal, brass barrels, cast in one piece. Has a round, smooth, lacquered grip. The trigger-guard is of iron and engraved with scrolls. The body and sides are of brass, and are also engraved. The left barrel bears both the Liége proof-mark and another in the form of a crown over the letter "R." The under-side of the right barrel is stamped with the letter "G."

England.

No. 20. (N4426). PERCUSSION PISTOL. Caliber .40. Three-and-one-half-inch, round, steel barrels. The checkered grip is inlaid with a silver oval. Trigger-guard and body are ornamented with engraving. The left barrel is stamped both with the Liége proof-mark and another in the form of a crown over the letter "R." No maker's name is given.

England.

No. 21. (N2779). PERCUSSION PISTOL. Caliber .42. Three-and-three-eighths-inch, round, steel barrels. The grip is inlaid with a silver escutcheon in the form of an oval. Trigger-guard and body are engraved with floral and scroll designs. The left barrel is stamped with the Liége proof-mark and the under-side of the right barrel bears an English proof-mark. No maker's name is given.

England.

No. 22. (N753). PERCUSSION PISTOL. Caliber .45. Four-and-three-eighths-inch, round, steel barrels. Has a smooth, flat grip which is rather thick. The body is slightly engraved, but all the other parts of the pistol are unornamented. No maker's name or other marks of identification are given.

England?

No. 23. (N4423). PERCUSSION PISTOL. Caliber .40. Three-and-one-quarter-inch, round, steel barrels. Has a coarsely checkered grip. The left barrel is stamped with the Liége proof-mark, and each barrel is marked with proof-marks in the form of a crown over the letter "R." No maker's name is given.

England.

No. 24. (N5907). PERCUSSION PISTOL. Caliber .42. Three-and-one-eighth-inch, round, steel barrels. Trigger-guard and body are engraved. The left barrel is stamped with the Liége proof-mark and the right one is marked with the number "3." No maker's name is given.
Donor: Christ Warth.

Belgium?

No. 25. (N4425). PERCUSSION PISTOL. Caliber .36. Three-and-one-quarter-inch, round, steel barrels. The body is engraved with scroll designs, and the trigger-guard is scratched with an eight-pointed star. The left barrel bears the usual Liége proof-mark and in addition another proof-mark in the form of a crown over the letter "R." The right barrel also bears the latter proof-mark. No maker's name is given.

England?

No. 26. (N4420). PERCUSSION PISTOL. Caliber .41. Three-and-one-quarter-inch, octagonal, steel barrel. Has a smooth, oval grip. The left trigger is more curved than the right one. Each of the barrels is stamped upon the bottom with an undecipherable mark. No other marks of identification are given.

Belgium?

No. 27. (N6763). PERCUSSION PISTOL. Caliber .45. Three-and-one-quarter-inch, round, steel barrels which may be unscrewed from the breech. The body is slightly engraved, but the rest of the pistol is unornamented. The left barrel is stamped with the Liége proof-mark and both barrels are stamped with English proof-marks in the form of a crown over the letter "V."

England.

No. 28. (N4430). PERCUSSION PISTOL. Caliber .36. Two-and-three-quarters-inch, round, steel barrels. All steel parts are blued. Upon the bottom of each barrel is scratched the numeral "XXI." There are no other marks of identification.

United States.

No. 29. (N2780). PERCUSSION PISTOL. Caliber .50. Two-and-five-eighths-inch, round, damaskeened, steel barrels. The left trigger is more curved than the right one. All parts of the pistol are un-ornamented. It has neither sights nor ramrod. The body, right trigger, left barrel and the rib joining the barrels are stamped with the number "1." No proof-marks or maker's name are given.

France?

No. 30. (N4424). PERCUSSION PISTOL. Caliber .40. Three-and-one-eighth-inch, round, steel barrels. The grip is coarsely checkered, and the body bears rude engravings of flowers. The rest of the pistol is unornamented. The left barrel is stamped with the Liége proof-mark, and both barrels are stamped with proof-marks in the form of a crown over the letter "U." The letter "H" appears beneath the body near the trigger-guard.

Germany?

No. 31. (N6773). PERCUSSION PISTOL. Caliber .40. Three-and-one-half-inch, round, steel barrels. All parts are unornamented. The under-side of each barrel is stamped with the Liége proof-mark. No other marks of identification are given.

Donor: Dr. Edwin W. Bartlett.

Belgium.

No. 32. (N4421). PERCUSSION PISTOL. Caliber .45. Three-and-three-quarters-inch, octagonal, steel barrels, made in one piece. The grip is finely checkered, and is inlaid with a silver plate. Equipped with front and rear sights and a swivel ramrod. Hammers, lock-plates, body and trigger-guards are finely engraved with scrolls. Each barrel is stamped with two proof-marks. Engraved upon the top of the pistol are the words "J. Wilson, London."

England.

No. 33. (N7064). PERCUSSION PISTOL. Caliber .43. Three-and-five-eighths-inch, octagonal, smooth-bore, damaskeened barrels. The body is engraved with scroll-like floral designs, and the trigger-guard filed with an eight-pointed star. An oval, silver plate is inlaid on top of the smooth grip. The left barrel bears the Liége proof-mark.
Donor: Emil Peters.

Belgium.

No. 34. (N4418). PERCUSSION PISTOL. Caliber .36. Four-and-one-quarter-inch, round, rifled, steel barrels. The grip is L-shaped, made of briar wood, and has an engraved steel butt-plate. Equipped with front and rear sights. All metal parts of this pistol are finely engraved. The left barrel is stamped with the Liége proof-mark and the number "3," each breech being also stamped with the same number. The right barrel is marked "14000" and "Invention Delvigne."

France.

No. 35. (N4432). PERCUSSION PISTOL. Caliber .28. One-and-five-eighths-inch, round, steel barrels which may be unscrewed from the breeches by the aid of a special wrench. The grip is finely checkered, and is also slightly carved. Trigger-guard and body are engraved with flowers. There are no sights or ramrod. The left barrel is stamped with the number "25," and the right one with the letters "AF" sur-mounted by a crown. No maker's name is given. The total weight of this pistol is nine ounces.

England.

No. 36. (N4431). PERCUSSION PISTOL. Caliber .32. Two-and-one-quarter-inch, round, steel barrels. The grip is polished, and made of mottled wood. There are no sights or ramrod. Equipped with two folding-triggers. Body and triggers are engraved with scrolls.

The left barrel is stamped with the Liége proof-mark, and both barrels bear other marks in the form of a crown over the letter "M."

Belgium.

No. 37. (N4414). PERCUSSION PISTOL. Caliber .60. Six-and-three-quarters-inch, superposed, octagonal, steel barrels. Has an oval grip with a brass ring in the butt. There are front and rear sights and a ramrod, the latter being carried at the right of the barrels. Has two top-hammers. Each barrel is stamped with German proof-marks and the figures "5.5 gr. N.G.P.N/71, 20 gr.Bl.," indicating the kind of powder and the weight of the lead ball. Each barrel is also stamped with the number "10." No maker's name is given.

Germany.

No. 38. (N2740). PERCUSSION PISTOL. Caliber .55. Six-and-one-quarter-inch, superposed, octagonal, steel barrels. Has an oval, checkered grip having a steel cap-box in the butt. Equipped with a fore-sight and a ramrod at the left side, but has no rear sight. Trigger-guard and lock-plates are engraved with scrolls. There are two top-hammers. The upper barrel is marked with the Liége proof-mark, and each barrel bears the figures "12.8 A.G." The lower barrel is also stamped with the number "13.0." No maker's name is given.

Belgium.

No. 39. (N2772). PERCUSSION PISTOL. Caliber .47. Three-and-one-half-inch, round, superposed, steel barrels. Has a flat, fluted grip with a cap-box in the butt. There are no sights or ramrod. One of the top-hammers is partly broken away. Trigger-guard and body are slightly engraved. The lower barrel is stamped with a crown over the letter "D."

England?

No. 40. (N2773). PERCUSSION PISTOL. Caliber .48. Three-and-one-eighth-inch, octagonal, superposed, steel barrels. Has a smooth, unornamented grip. The barrels are capable of being rotated upon a horizontal axis, thus bringing each nipple under the hammer which is superposed and serves both barrels. The pistol has no sights, but is equipped with a ramrod and folding-trigger. Each barrel is stamped with two proof-marks which are probably English. No maker's name is given.

England?

No. 41. (N4415). PERCUSSION PISTOL. Caliber .36. Three-and-one-quarter-inch, round, superposed, steel barrels. Has a checkered grip inlaid with a silver plate. The barrels are capable of being rotated as those described under No. 40. There are no sights or ramrod. The trigger is of the folding kind. Both barrels are stamped with two proof-marks each. No maker's name is given.

England.

No. 42. (N6777). PERCUSSION PISTOL. Caliber .38. Two-and-seven-eighths-inch, round, superposed, steel barrels. Has a checkered grip inlaid with a silver oval. The barrels are capable of being rotated upon a horizontal axis, as has been described under No. 40. There are no sights or ramrod. The trigger is of the concealed or folding kind. Each barrel is stamped with two proof-marks which are probably English. No maker's name is given.
Donor: Frank Shrosbree.

England?

No. 43. (N6343). PERCUSSION PISTOL. Caliber .40. One-and-seven-eighths-inch, round, superposed, iron barrels, each rifled with four grooves. The grip is made of black wood, and is fluted. An engraved, iron plate covers the butt. Body, trigger and a part of the breech are engraved with floral designs. The number "29" is stamped upon the lower barrel. No other marks of identification are given.

France?

No. 44. (N6779). PERCUSSION PISTOL. Caliber .40. Two-inch, superposed, steel barrels. Has a smooth grip, inlaid with a silver shield. There is but one hammer. The barrels can be rotated as those of No. 40. The pistol is provided with a safety-catch. Trigger-guard, hammer and body are engraved. One barrel is stamped with the number "4," and the left side of the body bears the words "Wheeler & Son."

England?

PLATE 84.

UNDER-HAMMERS, SELF-COCKING TOP-HAMMERS, AND OTHER PERCUSSION PISTOLS

No. 1. (N4398). PERCUSSION PISTOL. Caliber .37. Eleven-and-one-quarter-inch, half-octagonal, steel barrel. Has a smooth, flat grip, two rear sights and a fore-sight. The hammer is placed beneath the barrel. One of the reasons for placing hammers beneath barrels in percussion weapons was to avoid injuring the eyes and face by flying particles of the metallic cap. This pistol has neither trigger-guard nor ramrod. All parts are unornamented. The sides of the barrel are stamped "A.D. Laws." No other information is given.

United States.

No. 2. (N748). PERCUSSION PISTOL. Caliber .45. Ten-inch, steel barrel, the rear portion of which is octagonal. Has a flat, unpolished, L-shaped grip. The pistol is equipped with front and rear sights, but has no ramrod. Because of the great sweep of the under-hammer and its proximity to the trigger, it usually was impractical to place trigger-guards upon this variety of weapon. No marks of identification are given upon the pistol shown here.

United States.

No. 3. (N4399). PERCUSSION PISTOL. Caliber .35. Six-and-three-quarters-inch, round, tapering, steel barrel. Has a thick, smooth, clumsy grip. The rear sight may be elevated or lowered by means of a screw. This under-hammer pistol has a German silver trigger-guard. Stamped upon the top of the barrel are the words "M.S. Sanderson, Geo.V. Seaver, 1858," followed by the pistol number "132." The place of manufacture is not given.

United States.

No. 4. (N4401). PERCUSSION PISTOL. Caliber .34. Seven-and-one-half-inch, half-octagonal, rifled, steel barrel. Has a smooth, polished grip; the butt is bound with a German silver band. The front and rear sights are carefully made, and indicate that the pistol was used either for target practice or for dueling. The pistol cannot be brought to half-cock, but is brought to full-cock at once when the under-hammer is pulled down. Upon the top of the pistol are the words "Gibbs Tiffany & Co., Sturbridge, Mass.," and "Made for Rogers Bros. & Co., 52 Market St., Philad." Upon the under-side of the barrel is the pistol number "152."

United States.

No. 5. (N4402). PERCUSSION PISTOL. Caliber .34. Seven-and-three-quarters-inch, half-octagonal, rifled, steel barrel. Has a smooth grip bound with a brass strip. There are front and rear sights, but no ramrod. Upon the top of the barrel are the words "Gibbs, Tiffany & Co., Sturbridge, Mass." Another part of the barrel is marked

"E. Hutchings & Co., Agents, Balto." Upon the under-side of the barrel is the number "152."

United States.

No. 6. (N2790). PERCUSSION PISTOL. Caliber .48. Five-and-seven-eighths-inch, octagonal, rifled, steel barrel. Has a smooth, oval grip, the top of which is cast with an iron spur. Upon the body of the pistol are signs of former engravings. The right side of the barrel is stamped with the number "21." The maker's name is not given.

United States.

No. 7. (N727). PERCUSSION PISTOL. Caliber .32. Five-and-one-eighth-inch, heavy, octagonal, rifled, steel barrel. Has a smooth, flat, polished grip with a flaring butt which is covered with a silver plate. Formerly a ramrod was carried in the ferrules beneath the barrel. All parts of the pistol are unornamented, and no maker's name or pistol number are given.

United States.

No. 8. (N4404). PERCUSSION PISTOL. Caliber .38. Five-and-one-half-inch, half-octagonal, steel barrel. Has a smooth grip of brass which may be opened at the side, exposing a large cavity which serves as a cap-box. The pistol has a front sight, but no rear sight or ramrod. Stamped upon the left side of the body is the name "B.M. Bosworth." No pistol number is given.

United States?

No. 9. (N4400). PERCUSSION PISTOL. Caliber .36. Five-and-seven-eighths-inch, half-octagonal, rifled, steel barrel. Has a smooth, round grip, and front and rear sights. All steel parts were formerly blued but unornamented. The place of manufacture and the maker's name are not given upon this piece. The left side of the barrel is stamped with the number "4."

United States.

No. 10. (N6755). PERCUSSION PISTOL. Caliber .36. Five-inch, steel barrel which is fluted near the breech. Has a smooth, flat grip, carved with a design of a heart at the butt. Equipped with front and rear sights, but no ramrod. The top of the barrel is stamped "W. Raymond." No pistol number is given.

United States.

Nos. 11 and 12. (N4405, N4406). PAIR OF PERCUSSION PISTOLS. Caliber .28. Four-and-three-eighths-inch, rifled, steel barrels, the rear portions of which are octagonal and inlaid with bands of silver and gold. The fore-end of each stock is tipped with an engraved silver band. Stamped upon the top of each barrel are the words "Cast Steel." No maker's name or pistol number are given.

United States?

No. 13. (N5908). PERCUSSION PISTOL. Caliber .32. Five-and-one-quarter-inch, half-octagonal, rifled, steel barrel. Has a smooth, flat grip, the sides of which are inlaid with small ovals of silver. The butt was formerly covered with a plate. There is a sight in front, but none in the rear. The body of the pistol is slightly engraved. Stamped upon the top of the barrel are the words "Mead & Adriance, Pocket Rifle, Cast Steel Warranted." The number "3" appears upon the underside of the barrel.

Donor: Christ Warth.

United States?

No. 14. (N3293). PERCUSSION PISTOL. Caliber .38. Five-and-one-half-inch, octagonal, rifled, steel barrel. Has an oval checkered grip, the top inlaid with a plate engraved with the initials "C.R." The body is of engraved German silver. There are front and rear sights. Only a slight pressure upon the trigger is necessary to set off the under-hammer. The top of the body is engraved "Redferry, London," and the barrel is stamped with British proof-marks and the number "3."

England.

No. 15. (N4403). PERCUSSION PISTOL. Caliber .45. Six-inch, octagonal, rifled, steel barrel. Has an oval grip which is coarsely checkered. The body is of brass, engraved with scrolls. There are front and rear sights. The head of the under-hammer fits flush into the body when the pistol is discharged. Engraved upon the top of the pistol are the words "Bentley, Patentee, London." The left side of the barrel is stamped with two English proof-marks, and the right side with the number "11."

England.

No. 16. (N4409). PERCUSSION PISTOL. Caliber .50. Nine-and-three-quarters-inch, round, steel barrel. Has a round, smooth, ebony grip, fitted with a cap-box in the butt. There are neither front nor rear sights. The hammer is placed upon the side, and by pulling the ring the pistol is cocked, and at the same time the folding-trigger is pushed outward ready for use. The proof-mark is in the form of a British crown over the letters "AF." Stamped upon the under-side of the barrel is the number "18." No maker's name is given.

England.

No. 17. (N732). PERCUSSION PISTOL. Caliber .32. Four-and-one-half-inch, octagonal, rifled, steel barrel. Has a smooth, round, grip having the butt covered by a silver plate and band. The side-plates upon the body are of brass, and are finely engraved. Upon the top of the barrel is the name "H. Sheets." No pistol number is given.

United States.

No. 18. (N6765). PERCUSSION PISTOL. Caliber .30. Five-and-one-quarter-inch, half-octagonal, steel barrel. All parts of the pistol are unornamented. Neither maker's name nor pistol number are given upon this piece.

United States.

No. 19. (N4407). PERCUSSION PISTOL. Caliber .36. Five-and-three-quarters-inch, half-octagonal, rifled, steel barrel. Has a smooth, oval constricted grip which is bound with a brass strip. There are front and rear sights. The top of the barrel is stamped with the words "H.J. Hale, Warranted Cast Steel," and the under-side with the pistol number "120." This piece was made in Southbridge, Connecticut.

United States.

No. 20. (N4408). PERCUSSION PISTOL. Caliber .32. Four-and-one-half-inch, half-octagonal, rifled, steel barrel. Has an L-shaped grip which is half bound with brass bands. Equipped with front and rear sights. No pistol number or maker's name is given.

United States?

No. 21. (N4411). PERCUSSION PISTOL. Caliber .31. Three-and-three-eighths-inch, half-octagonal, rifled, steel barrel. Has a smooth, sloping grip which has a brass back-strap. Has a fore-sight and a concave rear sight. The pistol number "398" is stamped upon the back-strap beneath the grip. No maker's name is given, although this pistol was made by the Quinebog Manufacturing Company.

United States.

No. 22. (N4410). PERCUSSION PISTOL. Caliber .34. Four-and-one-quarter-inch, half-octagonal, rifled steel barrel. The grip is smooth and polished. There are front and rear sights. There is no half-cock notch. The body is engraved with scrolls, and all steel parts were originally blued. The left side of the barrel is stamped with the words "Bacon & Co., Norwich, C-T" followed by the words "Cast Steel." Upon the under-side of the barrel is the number "46."

United States.

No. 23. (N6758). PERCUSSION PISTOL. Caliber .46. Five-and-one-eighth-inch, half-octagonal, steel barrel. The grip is smooth and round, and screwed to an iron back-strap. Pulling the trigger raises the hammer to a given height, and then allows it to fall upon the cap. The hammer can be raised in no other way. There are no sights. The left side of the barrel is stamped "Patented 1837," and the hammer bears the words "Allen's Patent." Upon the left side of the barrel is the pistol number "206." This pistol was probably made in Grafton, Massachusetts.

Donor: H. P. Fischer.

United States.

No. 24. (N4390). PERCUSSION PISTOL. Caliber .61. Five-inch, tapering, brass barrel. The body is engraved with scrolls. This pistol was formerly a pepper-box from which the barrels have been removed and the brass barrel shown substituted. The brass ring in back of the breech is engraved "J.B.R. 1861," followed by the number "3." The self-cocking top-hammer is stamped "Allen's Patent," and the back-strap under the grip is marked with the number "303." This piece was probably made in Worcester, Massachusetts.

United States.

No. 25. (N728). PERCUSSION PISTOL. Caliber .35. Five-and-one-half-inch, half-octagonal, steel barrel which may be unscrewed from the breech. Has a smooth, round grip which is screwed to an iron back-strap. This pistol cannot be cocked by merely pulling the trigger as is the case with some of the pistols described above; it is brought to full-cock by depressing the ear of the hammer by means of the thumb; it is then discharged in the usual way. This piece was made by Stocking and Company of Worcester, Massachusetts. The pistol number is "35."

United States.

No. 26. (N2625). PERCUSSION PISTOL. Caliber .45. Four-and-one-half-inch, octagonal, steel barrel. The hammer is double-acting, that is, pulling the trigger raises the hammer and then allows it to fall upon the cap. There are no sights. Stamped upon the hammer are the words "Patented April 16, 1845," and upon the left side of the barrel a proof-mark in the form of a crown over the letter "F." This pistol was made by Allen and Wheelock of Worcester, Massachusetts, and exported from the United States.

United States.

No. 27. (N2767). PERCUSSION PISTOL. Caliber .36. Four-and-one-half-inch, half-octagonal, steel barrel. Has a round, smooth grip, attached to a brass back-strap. Pulling the ring trigger raises the top-hammer, and then allows it to fall upon the cap. The force necessary to raise the hammer, by pulling the trigger, is so great that it renders the pistol unfit as a weapon of precision. There is no ear upon the hammer, hence it can only be raised by pulling the trigger. No maker's name or pistol number is given, however, the pistol probably was made by Allen and Thurber of Norwich, Connecticut.

United States.

No. 28. (N6759). PERCUSSION PISTOL. Caliber .32. Four-and-three-quarters-inch, round, steel barrel. Has a flat, top-hammer whose action is the same as that described under No. 27. The grip is smooth and round, and is inlaid with oval silver plates. There are no sights. Stamped upon the barrel are the words "S. Sutherland, Richmond, Va.," and the pistol number "163."

United States.

No. 29. (N2877). PERCUSSION PISTOL. Caliber .35. Four-and-one-quarter-inch, half-octagonal, steel barrel. The smooth, round grip is screwed to an iron back-strap. Trigger action is the same as that of No. 27. Pistols of this class cannot be brought to half-cock, and because of the top-hammer, have no sights. The body of the pistol is engraved, the mountings being unornamented. Stamped upon the hammer are the words "Manhattan Mfg. Co., New York," while barrel and trigger bear the number "647."

United States.

No. 30. (N4391). PERCUSSION PISTOL. Caliber .36. Three-and-one-quarter-inch, half-octagonal, rifled, steel barrel. Trigger action is the same as that of No. 27. The body is the only part of the pistol which is ornamented. The hammer is stamped "Patented April 16,

1845," and the bottom of the barrel is marked with the number "103." This piece was made by Allen and Wheelock of Worcester, Massachusetts.

United States.

No. 31. (N4392). PERCUSSION PISTOL. Caliber .36. Three-and-one-quarter-inch, half-octagonal, steel barrel. Trigger action is the same as described under No. 27. The hammer is stamped "Allen's Patent" and the under-side of the barrel "Cast Steel." This pistol is badly rusted, but is still in working order. Its number is "42," and it was made by Allen and Wheelock.

United States.

No. 32. (N4393). PERCUSSION PISTOL. Caliber .32. Three-and-one-quarter-inch, half-octagonal, steel barrel. Action is the same as that of No. 27. The body was formerly engraved. No maker's name appears upon the pistol, but it was probably made under Allen's Patent. number "21" appears upon the under-side of the barrel.

United States.

No. 33. (N4395). PERCUSSION PISTOL. Caliber .30. Two-and-one-quarter-inch, half-octagonal, steel barrel which may be unscrewed from the breech. The action is the same as that of No. 27. All steel parts were originally blued. The hammer is stamped with the words "Allen's Patent," and the under-side of the barrel bears the shop number "998."

United States.

No. 34. (N4394). PERCUSSION PISTOL. Caliber .30. Two-and-one-quarter-inch, half-octagonal, steel barrel which may be unscrewed from the breech. The action is the same as that of No. 27. All steel parts were originally blued. The hammer is stamped "Allen's Patent" and the under-side of the barrel with the number "56." This pistol was probably made in Worcester, Massachusetts.

United States.

No. 35. (N4306). PERCUSSION PISTOL. Caliber .54. Seven-and-one-half-inch, round, steel barrel. This pistol is fitted with a front sight and a swivel ramrod, but has no back sight. The side-hammer is engraved with floral designs. All other parts of the pistol are unornamented. Upon the inside of the back-strap are the figures "2W" and also the pistol number "43." The maker's name is not given, but this pistol was made by Allen and Wheelock of Worcester, Massachusetts.

United States.

No. 36. (N4304). PERCUSSION PISTOL. Caliber .40. Eight-and-three-eighths-inch, half-octagonal, rifled, steel barrel. The rear sight can be elevated or lowered by means of a thumb screw in back. Ramrod and extractor are combined. All parts of the pistol are unornamented. The left side of the barrel is stamped "Allen & Thurber, Worcester," followed by the words "Cast Steel." The under-side of the barrel is marked with the pistol number "231." This weapon was owned by C. H. Longstreet of Syracuse, New York, and carried by him while paymaster on ship during the Civil War. It was presented to William A. Lawrence in 1872.

United States.

No. 37. (N4305). PERCUSSION PISTOL. Caliber .44. Seven-and-three-eighths-inch, octagonal, rifled, steel barrel. Has front and rear sights and a wooden ramrod. There is a large, metal fore-end attached to the stock. The lock-plate and hammer bear marks of former engravings. The top of the barrel is stamped "B & S, New York, Cast Steel." This pistol was made by Blunt and Sims of New York.

United States.

No. 38. (N4328). PERCUSSION PISTOL. Caliber .35. Seven-and-one-quarter-inch, half-octagonal, rifled, steel barrel. All parts of the pistol are unornamented. Cocking the hammer brings the pistol at once to full-cock. The top of the pistol is marked "Allen & Thurber, Worcester." Upon the bottom of the barrel is the pistol number "76."

United States.

No. 39. (N4330). PERCUSSION PISTOL. Caliber .35. Six-and-one-half-inch, half-octagonal, rifled, steel barrel. The top-hammer is set slightly to the right of the center. All parts of the pistol are unornamented. The top of the barrel is stamped "Stocking & Co., Worcester," followed by the words "Warranted Cast Steel." The pistol number "A12" appears upon the under-side of the barrel.

United States.

No. 40. (N4316). PERCUSSION PISTOL. Caliber .38. Six-and-three-eighths-inch, octagonal, rifled, steel barrel. Trigger-guard, hammer and back-strap are engraved. The under-side of the barrel is stamped with the letters "R-C" and the shop number "91." The top of the barrel is marked "B & S, New York," and "Cast Steel." This pistol was made by Blunt and Sims.

United States.

No. 41. (N4331). PERCUSSION PISTOL. Caliber .36. Five-and-one-half-inch, half-octagonal, steel barrel. The top-hammer is placed a little to the right of the center to facilitate aiming. All parts of this pistol are unornamented. The top of the barrel is stamped "V. W. Marston & Knox, New York 1854." Upon the under-side of the barrel is the pistol number "593."

United States.

No. 42. (N4329). PERCUSSION PISTOL. Caliber .35. Four-and-one-half-inch, half-octagonal, rifled, steel barrel. Equipped with front and rear sights. The top-hammer is placed a little to the right of the center as in No. 41. All parts of the weapon are unornamented. Marked upon the top of the barrel are the words "Allen & Wheelock," while upon the under-side is the number "114." Pistols of this type were made during the period of 1856 to 1865.

United States.

No. 43. (N6258). PERCUSSION PISTOL. Caliber .31. Four-and-one-quarter-inch, half-octagonal, smooth-bore, steel barrel. All steel parts were formerly blued, and the pistol is devoid of ornamentation. There are no sights. No maker's name is given. The pistol number "91" is stamped below the barrel, just in front of the trigger-guard.

Donor: E. C. Morawetz.

United States.

PLATE 85.

MISCELLANEOUS PERCUSSION PISTOLS

No. 1. (N572 a, b). PERCUSSION PISTOL. Caliber .64. Twelve-and-one-quarter-inch, steel barrel. Has a checkered grip to which the accompanying gunstock can be attached, thus converting the pistol into a sort of carbine. The pistol is fitted with front and rear sights and a swivel ramrod. The left side of the barrel carries a spring bayonet ten-and-one-half inches long. The hammer is engraved and has a safety-catch. Trigger-guard, lock-plate and the top of the grip

are also carefully engraved. The lock-plate is inscribed with the name
"Pattison," and the top of the barrel "Dublin."

Ireland.

No. 2. (N2741). PERCUSSION PISTOL. Caliber .55. Six-
and-one-half-inch, octagonal, steel barrel which is exceedingly heavy
and thick. The smooth, L-shaped grip is rather thick, and clumsy.
Equipped with back and front sights. Trigger-guard and top of lock-
plate are slightly engraved. Upon the top of the lock-plate is also the
name "H. Estes." No other marks of identification are given. This
pistol is probably home-made.

France?

No. 3. (N4314). PERCUSSION PISTOL. Caliber .50. Six-
and-one-quarter-inch, round, tapering, steel barrel which is octagonal at
the breech. Has a smooth, thick grip enclosed in an iron back-strap.
There are neither front nor rear sights. To cock this pistol, the hammer
is pulled back; the trigger is pushed forward simultaneously. The top
of the barrel is stamped with the name "C. Tissier."

France?

No. 4. (N2775). PERCUSSION PISTOL. Caliber .60. Five-
and-one-quarter-inch, octagonal, steel barrel having a crude, cross-
hatched grip, fitted with an engraved butt-piece, and inlaid with a silver
plate on top. Equipped with front and rear sights. Trigger-guard,
hammer and lock-plate are finely engraved with scroll designs. The
under-side of the barrel formerly carried ferrules for a ramrod. This
piece was made by James Beattie, London, and is so marked.

England.

No. 5. (N4315). PERCUSSION PISTOL. Caliber .69. Five-
and-one-half-inch, octagonal, steel barrel. Has a checkered grip, fitted
with a cap-box in the butt, and inlaid with a silver escutcheon in the
form of a shield on top. Equipped with front and rear sights. Trigger-
guard and lock-plate are engraved with scrolls. Formerly there was a
belt-hook upon the left side of the weapon. The under-side of the
barrel bears two English proof-marks. No maker's name is given.

England.

No. 6. (N4308). PERCUSSION PISTOL. Caliber .68. Five-
and-one-half-inch, octagonal, steel barrel which is slightly engraved
near the muzzle. Has a finely checkered grip, fitted with a cap-box in

the butt, and inlaid with a silver plate on top. The pistol is equipped with front and rear sights and a swivel ramrod. A belt-hook is carried upon the left side. Hammer, trigger-guard and top of the pistol are finely engraved. The safety-catch for the hammer has been lost. Stamped upon the top of the barrel are the words "Bridle, Barnstaple"; the left side is marked with two English proof-marks.

England.

No. 7. (N4310). PERCUSSION PISTOL. Caliber .50. Five-and-one-half-inch, octagonal, steel barrel. Has a finely checkered grip having a silver butt with a cap-box. Equipped with front and rear sights and a swivel ramrod. The left side of the pistol is fitted with a belt-hook. Hammer, lock-plate and trigger-guard are finely engraved. All steel parts were formerly blued. The bottom of the barrel is stamped with two English proof-marks; the top is engraved with the words "Beckwith, London."

England.

No. 8. (N3291). PERCUSSION PISTOL. Caliber .38. Five-and-one-half-inch, octagonal, silver-steel, damascus barrel. The grip is finely checkered and has a cap-box in the butt. Equipped with front and rear sights, a swivel ramrod and a folding-trigger. Hammer and lock-plate are engraved with floral designs. The left side of the barrel is stamped with the Liége proof-mark, and inscribed with the words "Purdy, London." The word "London" is also engraved on top of the barrel.

England.

No. 9. (4307). PERCUSSION PISTOL. Caliber .38. Six-inch, silver-steel, damascus barrel. The grip is finely checkered and has a cap-box in the butt. Hammer, lock-plate and trigger-guard are engraved with floral and scroll designs. The left side of the barrel is stamped with the Liége proof-mark, and the lock-plate is engraved with the name "Tarratt, London." This piece was probably used as a target or dueling pistol.

England.

No. 10. (N4311). PERCUSSION PISTOL. Caliber .41. Four-and-one-half-inch, octagonal, steel barrel. The grip is finely checkered, and has a box in the butt for percussion-caps. Lock-plate and body are made of engraved German silver. The left side of the pistol carries a belt-hook. The top of the breech is engraved "V. Libeau, London."

England.

No. 11. (N4317). PERCUSSION PISTOL. Caliber .46. Four-and-one-eighth-inch, octagonal, steel barrel. Has a finely checkered grip inlaid with a German silver disc on top. Has front and rear sights, but no ramrod. Hammer and trigger-guard are of engraved steel. The left side of the barrel is stamped with two English proof-marks, but no maker's name is given.

England.

No. 12. (N4312). PERCUSSION PISTOL. Caliber .38. Three-and-three-quarters-inch, octagonal, damaskeened barrel. Has a smooth, oval grip. The cap-box in the butt has an embossed cover in the form of a lion's head. Equipped with front and rear sights and a swivel ramrod. The lock-plate and ferrules are made of engraved, German silver. Hammer and trigger-guard are also engraved. The under-side of the barrel is stamped with the number "8", and the left side with the Liége proof-mark. The former owner's name "I. Brown" is engraved upon the inner back-strap. No maker's name is given.

England?.

No. 13. (N4325). PERCUSSION PISTOL. Caliber .48. Two-and-seven-eighths-inch, steel barrel which is very finely rifled. The grip is made of finely checkered ebony, and has a cap-box in the butt with an embossed cover in the form of a woman's head. Equipped with front and rear sights, but no ramrod. Hammer, lock-plate and folding-trigger are carefully engraved with floral designs. All steel parts were formerly blued. The left side of the barrel is stamped with the Liége proof-mark. No maker's name is given. This pistol is one of a pair, the mate of which is No. 24 shown on this plate.

England?.

No. 14. (N759). PERCUSSION PISTOL. Caliber .41. Three-and-one-half-inch, octagonal, steel barrel which is inlaid with gold and silver poppies. The grip is profusely inlaid with scrolls and flowers. Equipped with front and rear sights, and folding-trigger, but no ramrod. The bottom of the barrel is stamped with two English proof-marks and the letters "R. H." Engraved upon the top of the breech are the words "Lewis & Tomes, London."

England.

No. 15. (N4318). PERCUSSION PISTOL. Caliber .47. Three-and-one-half-inch, octagonal, steel barrel. The checkered grip is inlaid with a diamond-shaped plate of silver on top. This pistol is equipped with a rear sight, but has no front sight or ramrod. Trigger-guard, hammer and lock-plate are engraved with floral designs. The left side of the breech is stamped with the Liége proof-mark. No other marks of identification are given.

England?.

No. 16. (N4319). PERCUSSION PISTOL. Caliber .44. Three-and-one-quarter-inch, octagonal, steel barrel. Has a rear sight, but neither fore-sight nor ramrod. The folding-trigger is missing. The lock-plate is engraved with broad scroll designs, the rest of the pistol being unornamented. This weapon is of French manufacture, but has neither maker's name nor proof-marks.

France.

No. 17. (N4320). PERCUSSION PISTOL. Caliber .47. Three-and-one-quarter-inch, octagonal, steel barrel which may be un-screwed from the breech. The crudely checkered grip is partly lacquered. Has only one sight, and that is in the rear. Lock-plate, hammer and trigger-guard are engraved with broad floral designs. The left side of the barrel bears the Liége proof-mark. No maker's name is given.

Belgium.

No. 18. (N4321). PERCUSSION PISTOL. Caliber .45. Two-and-one-half-inch, octagonal barrel which may be unscrewed from the breech. This pistol is equipped with a smooth, round grip, a folding-trigger and a rear sight. The lock-plate and hammer are engraved with broad scroll designs. Upon the left side of the breech is the Liége proof-mark. No other marks of identification are given.

Belgium.

No. 19. (N2771). PERCUSSION PISTOL. Caliber .38. Four-and-one-quarter-inch, round, steel barrel. This pistol has an L-shaped, or "saw-handle" grip which is checkered, and has a cap-box in the butt. The gun is equipped with front and rear sights. The breech is of German silver. Two English proof-marks are stamped upon the bottom of the barrel while the top bears the name "E. W. Bailey."

England.

No. 20. (N3290). PERCUSSION PISTOL. Caliber .50. Five-and-seven-eighths-inch, octagonal, steel barrel. Has an engraved German silver grip which is cast with a spur. Equipped with front and rear sights. The trigger-guard is also made of engraved German silver, and is fitted with an auxiliary grip to give greater steadiness in aiming. This pistol was probably used in target work. Its left side is stamped with two English proof-marks, but no maker's name is visible.

England.

No. 21. (N4309). PERCUSSION PISTOL. Caliber .43. Five-and-one-half-inch, octagonal, steel barrel which was formerly blued. The smooth grip is fitted with a spur which indicates that the pistol was intended for target practice. Equipped with front and rear sights and a swivel ramrod. Trigger-guard, hammer and the back of the pistol are engraved with scrolls. The top of the barrel is stamped with the words "Cast Steel," and the inside of the back-strap is stamped "VIII." No other marks of identification are given.

United States.

No. 22. (N6760). PERCUSSION PISTOL. Caliber .48. Seven-inch, round, steel barrel. Has a fluted, ebony grip with a steel button at the butt. There are sights in front and rear. The back portion of the barrel is octagonal, and finely engraved with flowers. The folding-trigger is similarly ornamented. Engraved upon the top of the barrel is an undecipherable name, apparently French. No proof-marks are given.

France?.

No. 23. (N4322). PERCUSSION PISTOL. Caliber .38. Three-and-one-quarter-inch, round, steel barrel. Has a smooth, oval grip, but neither ramrod nor sights. Side-hammer, folding-trigger and lock-plate are engraved. The under-side of the barrel bears an English proof-mark and the number "553," while the left side is stamped with the Liége proof-mark. No maker's name is given.

England.

No. 24. (N4323). PERCUSSION PISTOL. This pistol is one of a pair and is the mate to No. 13, already described.

No. 25. (N5636 a, b). PAIR OF PERCUSSION PISTOLS. Caliber .48. Three-and-one-quarter-inch, damaskeened, fluted, steel barrels. The grips are of fluted ivory, covered with steel butt-plates.

Each pistol is provided with a rear sight, but there is none in front. Hammers, lock-plates and folding-triggers are finely engraved. The top of each breech is engraved with a bird, while one side of the lock-plate is engraved with a running retriever. Each barrel is stamped with a proof-mark in the form of a crown over the letter "V", and also with the Liége proof-mark. No maker's name is given.

England.

No. 26. (N4327). PERCUSSION PISTOL. Caliber .35. Two-inch, fluted, damaskeened barrel. The grip is of ivory, and has an engraved steel cap-box in the butt. The lock-plate and folding-trigger bear scroll designs in relief. There is only one sight which is in the rear. The left side of the barrel is stamped with a proof-mark in the form of a crown over the letter "T," and bears also the Liége proof-mark. The under-side of the barrel bears the numeral "1." No maker's name is given.

Belgium.

No. 27. (N4324). PERCUSSION PISTOL. Caliber .48. Three-and-one-quarter-inch, round, steel barrel. The smooth, bulbous grip is fitted with an embossed butt-plate. The top-hammer is placed a little to one side of the barrel. There are neither sights nor ramrod. The folding-trigger is engraved with flowers, and the left side of the barrel is stamped with the Liége proof-mark. Upon the right lock-plate is the name "Louis Malherbe."

France?.

No. 28. (N4326). PERCUSSION PISTOL. Caliber .44. Two-and-one-quarter-inch, round, steel barrel. Has a fluted, ebony grip which formerly had a butt-plate. There is only one sight which is in the rear. The hammer is on top, and is set to one side of the center. The folding-trigger and the lock-plate are elegantly engraved with scrolls. Upon the left side of the barrel is the usual Liége proof-mark. No maker's name is given.

No. 29. (N4313). PERCUSSION PISTOL. Caliber .40. Three-and-one-quarter-inch, octagonal, steel barrel. The grip is smooth, and oval, and is ornamented with five brass nails. Has front and rear sights and a swivel ramrod. Lock-plate, hammer and trigger-guard are engraved with scroll designs. Upon the left side of the lock-plate are two British proof-marks; engraved on top of the barrel are the words "Colgan, Limerick."

Ireland.

No. 30. (N4332). PERCUSSION PISTOL. Caliber .32. Four-and-one-quarter-inch, half-octagonal, steel barrel, the bore of which is smooth. Has a small, round, wooden grip, a superposed hammer, but no sights or ramrod. All parts of the pistol are unornamented. The under-side of the barrel near the breech is cracked. The pistol number "36" appears on the back-strap under the grip. No maker's name is given but this pistol was made by Edward Deacon.

United States.

No. 31. (N4369). PERCUSSION PISTOL. Caliber .36. Three-and-one-quarter-inch, half-octagonal, smooth-bore, steel barrel. The number "321" appears on the back-strap near the grip. The rest of the features of this pistol are the same as those of No. 30. No maker's name is given, but the pistol was made by Edward Deacon.

United States.

No. 32. (N4370). PERCUSSION PISTOL. Caliber .34. Two-and-three-eighths-inch, half-octagonal, smooth-bore, steel barrel. The pistol number "176" appears upon the under-side of the barrel. The other characteristics of this pistol are similar to those of No. 30. Stamped upon the left side of the barrel is the name "Edward Deacon."

United States.

No. 33. (N4371). PERCUSSION PISTOL. Caliber .31. Two-and-one-quarter-inch, half-octagonal, smooth-bore, steel barrel. The pistol number "95" appears upon the under-side of the barrel. The other features of the pistol are the same as those of No. 30. The pistol here illustrated was made by Edward Deacon, but his name does not appear upon the weapon.

United States.

No. 34. (N4372). PERCUSSION PISTOL. Caliber .34. Two-and-one-quarter-inch, half-octagonal, steel barrel which may be unscrewed from the breech; it is shown detached in the illustration. There are no sights. All steel parts are blued but unornamented. The pistol number "312" is stamped upon the under-side of the barrel. The maker's name is not given, but this pistol was made by Allen and Thurber of Worcester, Massachusetts.

United States.

No. 35. (N4387). PERCUSSION PISTOL. Caliber .36. Five-inch, octagonal, ,rifled, steel barrel. Has a smooth, round grip, and front and rear sights. The nipple is placed in the rear of the breech. This weapon is known as a "Gallery Pistol" and is stamped with the number "87" upon the under-side of the barrel. No maker's name is given although this piece was made by Allen and Thurber.

United States.

No. 36. (N4389). PERCUSSION PISTOL. Caliber .33. Three-and-three-quarters-inch, round, steel barrel. Equipped with front and rear sights. The nipple is placed in the rear of the breech. This weapon was used in target practice and is therefore sometimes known as a "Gallery Pistol." The number "53" is stamped upon the back-strap beneath the grip. No maker's name is given.

United States?.

No. 37. (N769). PERCUSSION PISTOL. Caliber .34. Two-and-three-quarters-inch, round, tapering, steel barrel. Has neither sights nor ramrod. All parts are unornamented. Neither factory number nor maker's name are given.

United States?.

No. 38. (N2876). PERCUSSION PISTOL. Caliber .30. Four-and-one-quarter-inch, tapering, steel barrel. There are neither sights nor ramrod. All parts are unornamented. The maker's name is effaced. The pistol number "94" appears upon the left side of the breech.

United States?.

No. 39. (N6775). PERCUSSION PISTOL. Caliber .35. Four-inch, round, tapering, steel barrel. Has no sights, ramrod, or trigger-guard. The pistol number "91" appears upon the under-side of the barrel. No other marks of identification are given.
Donor: George Baxter.

United States?.

No. 40. (N6762). PERCUSSION PISTOL. Caliber .36. Two-and-one-eighth-inch, round, tapering barrel. There are no sights, ramrod, or trigger-guard. Barrel and body are made in one piece. The under-side of the barrel bears the number "78". No maker's name is given.

United States?.

No. 41.   (N4379).   PERCUSSION PISTOL.   Caliber .38. Three-and-one-quarter-inch, round, steel barrel.   Body and barrel are cast in one piece.   There are no sights, ramrod, or trigger-guard. This pistol has no ornamentation, maker's name or shop number.

?

No. 42. (N4382). PERCUSSION PISTOL. Caliber .32. Two-and-three-quarters-inch, round, tapering, steel barrel which is attached to a bronze body. All parts are unornamented. No maker's name or shop number are given.

United States?.

No. 43.   (N4373).   PERCUSSION PISTOL.   Caliber .30. Two-and-one-quarter-inch, round, tapering, steel barrel. Stamped upon the top of the barrel is the word "Hornet"; the left side is marked with the number "320." No maker's name is given.

United States.

No. 44.   (N4374).   PERCUSSION PISTOL.   Caliber .32. Two-and-one-quarter-inch, round, tapering, steel barrel. All parts of the pistol are unornamented, except the trigger-guard which is slightly engraved. No pistol number or maker's name is given.

United States?.

No. 45.   (N4376).   PERCUSSION PISTOL.   Caliber .32. Two-inch, round, tapering, steel barrel. Has a sheath trigger. The other features are the same as those of No. 44. No marks of identification are given.

United States?.

No. 46.   (N4383).   PERCUSSION PISTOL.   Caliber .34. Two-and-one-quarter-inch, round, tapering, steel barrel which is attached to a bronze body. The left side of the body is stamped "Hero, M.F.A. Co." The initials stand for Manhattan Fire Arms Company, the makers of the pistol. This company was located in New York City. No pistol number is given.

United States.

No. 47. (N4377). PERCUSSION PISTOL. Caliber .30. Two-and-one-quarter-inch, round, tapering, steel barrel. The left side of the barrel is stamped with the numbers "14" and "67". No maker's name is given.

United States?.

No. 48. (N4380). PERCUSSION PISTOL. Caliber .35. Three-and-one-quarter-inch, round, tapering, steel barrel, attached to a bronze body. The sides of the body are stamped with the name "Faro"; no other marks of identification are given.

United States?.

No. 49. (N4396). PERCUSSION PISTOL. Wooden model or pattern of Ethan Allen's single shot pistol. This model represents only a portion of the grip and body, and was made in Worcester, Massachusetts.

United States.

No. 50. (N4366). PERCUSSION PISTOL. Caliber .45. Two-and-one-quarter-inch, round, steel barrel. Has an engraved, brass body and a smooth, oval grip. The left side of the body is engraved with the word "Tryon," while the right side bears the word "Philadelphia." The pistol number is "11".

United States.

PLATE 86.

PERCUSSION PEPPER-BOXES, FOREIGN AND UNITED STATES

No. 1. (N2786). PEPPER-BOX REVOLVER. Caliber .50. Four, three-and-one-half-inch, smooth-bore damanskeened barrels which unscrew from the breech to load. The barrels are numbered from 1 to 4. The under-hammer is operated by a ring-trigger which cocks, revolves and fires a barrel at a single action. The revolver is stamped with the Liége proof-mark and "Mariette Breveté," the last two words meaning "patented by Mariette."

France.

No. 2. (N4439). PEPPER-BOX REVOLVER. Caliber .36. Four, two-and-three-quarters-inch, damaskeened, smooth-bore barrels, which unscrew from the breech when loading. The body is engraved with scrolls. The grip is of smooth, polished wood. The action and the marks upon this revolver are the same as those of No. 1.

France.

No. 3. (N2785). PEPPER-BOX REVOLVER. Caliber .40. Six three-inch, damaskeened, smooth-bore barrels which unscrew from the breech to load.   The barrels are numbered from 1 to 6.   Grip, action and marks upon this pepper-box are the same as No. 1.

France.

No. 4.   (N4441).   PEPPER-BOX REVOLVER.   Caliber .38. Six, three-inch, damaskeened, smooth-bore barrels which may be unscrewed from the breech-block.   The under-hammer is operated by the ring trigger which cocks, revolves, and fires each barrel at a single action. The revolver is stamped with the Liége proof-mark and with the words "Mariette Breveté."

France.

No. 5.   (N788).   PEPPER-BOX REVOLVER.   Caliber .30. Eight, two-and-three-quarters-inch, damaskeened barrels which unscrew from the breech to load.   The barrels are numbered from 1 to 8.   The grip-plates are made of smooth ivory, and the body is ornamented with elaborate inlaid designs of gold and silver.   This is an under-hammer double-action pepper-box.   It is stamped with the Liége proof-mark and with the words "Mariette Breveté."   It is also engraved with the words "Le Page Mautier Arq. du Roi, Paris."   The abbreviation Arq. stands for arquebusier, and the whole phrase means "Le Page Mautier Gunsmith to the King, Paris."

France.

No. 6.   (N2784).   PEPPER-BOX REVOLVER.   Caliber .35. Eight, six-and-one-quarter-inch, damaskeened, smooth-bore barrels, which unscrew from the breech, to load.   The barrels are numbered from 1 to 8.   The length and large number of barrels make this weapon difficult to aim.   It is another example of the double-action, under-hammer pepper-box.   It is marked with the Liége proof-mark and the words "Mariette Breveté."

France.

No. 7.   (N4440).   PEPPER-BOX REVOLVER.   Caliber .36. Four, six-inch, damaskeened, smooth-bore barrels which unscrew from the breech, to load.   This is still another example of the double-action, under-hammer revolver.   It is marked in the same way as No. 6.

France.

No. 8.   (N3295).   PEPPER-BOX REVOLVER.   Caliber .36. Two-and-three-quarters-inch, fluted and ribbed cylinder containing six smooth bores.   Each bore of this double-action pepper-box is marked with an English proof-mark. This revolver is provided with a flat top-hammer and a nipple-shield. Upon the left side of the body is the word "Southall."

England.

No. 9. (N4467). PEPPER-BOX REVOLVER. Caliber .36. Two-and-three-quarters-inch, fluted cylinder having six smooth bores. This revolver is equipped with a flat top-hammer and a nipple-shield. The groove between each barrel is stamped with a British proof-mark. No maker's name is given.

England.

No. 10. (N3294). PEPPER-BOX REVOLVER. Caliber .35. Two-and-three-quarters-inch, fluted and ribbed cylinder containing six smooth bores. This double-action pepper-box is provided with a flat top-hammer and a nipple-shield. The grip is very finely checkered. The inner back-strap and butt-plate are cast in one piece. Engraved upon the left side of the German silver body are the words "Redfern & Bourne."

England.

No. 11. (N2623). PEPPER-BOX REVOLVER. Caliber .34. Two-inch, fluted and ribbed cylinder containing six smooth bores. Has a nipple-shield and a flat top-hammer. Trigger cocks, revolves, and fires at a single action. The groove between each bore is stamped with an English proof-mark. Upon the left side of the body are the words "Chas. Osborne, London."

England.

No. 12. (N4463). PEPPER-BOX REVOLVER. Caliber .36. Two-inch, fluted and ribbed cylinder containing six smooth bores. This is a double-action pepper-box having a flat top-hammer and a nipple-shield. The groove between each bore is stamped with British proof-marks, but no maker's name is given.

England.

No. 13. (N4464). PEPPER-BOX REVOLVER. Caliber .36. Two-and-one-quarter-inch, fluted and ribbed cylinder containing six smooth bores. There is no nipple-shield, but the action is the same as No. 12. Engraved upon the left side of the body are the words "J. R. Cooper's Self-Acting Revolving Pistol" and upon the right side "John

Graveley Importer." The groove between each bore is marked with what appear to be Birmingham proof-marks.

England.

No. 14. (N3298). PEPPER-BOX REVOLVER. Caliber .38. Three-inch, fluted and ribbed cylinder containing six smooth bores. This is a double-action, under-hammer pepper-box having no nipple-shield. The groove between each bore bears English proof-marks. Upon the top of the body is the name "Boyd" and on the left side of the frame, the words "J. R. Cooper's Patent."

England.

No. 15. (N4462). PEPPER-BOX REVOLVER. Caliber .36. Three-inch, fluted cylinder containing six smooth bores. This is another example of the under-hammer double-action pepper-box. The triangular groove in the right side of the frame is to facilitate the placing of percussion-caps upon the nipples. The proof-marks are of the same kind and are placed in the same position upon the cylinder as those of No. 14. This piece was also made under the J. R. Cooper patent.

England.

No. 16. (N4447). PEPPER-BOX REVOLVER. Caliber .26. Two-inch, fluted and ribbed cylinder containing five smooth bores. This double-action, under-hammer pepper-box is equipped with a ring trigger. All steel parts are blued. The end of the cylinder at the muzzle is marked with the figures "R-C 3." No maker's name is given.

England.

No. 17. (N4461). PEPPER-BOX REVOLVER. Caliber .31. Three-and-one-half-inch, fluted and ribbed cylinder containing six smooth bores. The action of this pepper-box is the same as that of No. 16. Has a nipple-shield and an under-hammer, the latter being operated by a ring trigger. The muzzle end of the cylinder is stamped with the marks "R-C 37." The maker's name is not given.

England.

No. 18. (N3297). PEPPER-BOX REVOLVER. Caliber .38. Three-inch, fluted and ribbed cylinder containing six smooth bores. Hammer, action, and trigger are the same as those of No. 17. Upon the left side of the frame are the words "J. R. Cooper's Patent." This piece was probably made in Birmingham.

England.

No. 19. (N5782). PEPPER-BOX REVOLVER. Caliber .30. Four-and-one-half-inch, grooved hexagonal, brass cylinder containing six smooth bores. Each bore has to be brought into line with the hammer by revolving the cylinder by hand. The piece is equipped with a nipple-shield, and is said to have been made in Pennsylvania. It ranks among the rarest of pepper-boxes. Stamped upon the left lock are the letters "A I S." No other marks are given.

Donor: Mrs. F. B. Leidersdorf.

United States.

No. 20. (N2068). PEPPER-BOX REVOLVER. Caliber .30. Four-inch, grooved, octagonal, brass cylinder containing four smooth bores. Each bore has to be brought into line with the hammer by revolving the cylinder by hand. The pepper-box has a nipple-shield similar to No. 19 which piece it greatly resembles. The specimen here illustrated is one of the earliest types of pepper-boxes and is, therefore, very rare. It is supposed to have been made in Nazareth, Pennsylvania. The letters "I E H" appear upon the left lock-plate.

United States.

No. 21. (N4438). PEPPER-BOX REVOLVER. Caliber .34. Four, three-and-three-quarters-inch barrels which are brazed together. The nipples are set into the rear ends of the breeches. The barrels have to be revolved by hand and are held in position by means of a spring on the left side of the frame. When not in use, the trigger may be folded up. No marks of any kind are given upon this weapon.

?

No. 22. (N3296, N4442). PAIR OF PEPPER-BOX REVOLV-ERS. Caliber .26. Eight, five-and-one-quarter-inch, damaskeened barrels, which unscrew from the breech-block to load. The front ends of the barrels of one of these pistols are marked with two punched dots; those of the other with three. The barrels of each pistol are numbered from 1 to 8. Each pepper-box is double-acting and is operated by the ring trigger. The accessories in the case include some bullets, a powder-flask and a bullet-mold, the latter also serving as a wrench for unscrewing the barrels. Each revolver bears the Liége proof-mark, and is also stamped with the words "Mariette Breveté."

France.

No. 23. (N4446). PEPPER-BOX REVOLVER. Caliber .32. Two-and-one-half-inch, steel cylinder containing four smooth bores.

The cylinder has to be revolved by hand in order to fire the barrels in succession. The revolver is equipped with a nipple-shield and a top-hammer having an offset head to make cocking easier. Stamped upon the cylinder are the words "Pecare and Smith's Patent 1849 New York Cast Steel."

United States.

No. 24. (N4444). PEPPER-BOX REVOLVER. Caliber .31. Two-and-one-half-inch, cylinder containing four smooth bores. The cylinder has to be revolved by hand. The pepper-box is provided with a nipple-shield and a top-hammer. Stamped upon the cylinder are the words "Pecare and Smith's Patent 1849 New York Cast Steel."

United States.

No. 25. (N4437). PEPPER-BOX REVOLVER. Caliber .26. Three-and-one-quarter-inch, fluted, steel cylinder containing three smooth bores. The cylinder is revolved by hand but the hammer is cocked by pulling the trigger. The cylinder is stamped with the words "Cast Steel" and the left side of the hammer, "Manhattan Mfg. Co., New York." The revolver bears the number "49."

United States.

No. 26. (N3353). PEPPER-BOX REVOLVER. Caliber .32. Three-inch, fluted cylinder containing three smooth bores. The cylinder is revolved by hand, but the hammer is cocked by pulling the trigger. The cylinder is stamped with the words "Cast Steel" and the left side of the hammer, "Manhattan Mfg. Co., New York." The revolver bears the number "284."

United States.

No. 27. (N4488). PEPPER-BOX REVOLVER. Caliber .31. Three-inch, fluted and ribbed cylinder containing six smooth bores. The triangular groove in the right of the body is to facilitate the placing of the percussion-caps upon the nipples. This is an example of an American, double-action, under-hammer pepper-box having a ring-trigger. Upon the top of the body appear to be the words "Lunt & Sym New York."

United States.

No. 28. (N4454). PEPPER-BOX REVOLVER. Caliber .31. Three-and-one-half-inch, fluted cylinder having six smooth bores. This piece has a special top-hammer having a ring by which the pepper-box may be cocked, thus making the arm single-action. It can also be used as a double-action revolver by pulling the trigger. Stamped upon the cylinder are the words "Cast Steel," and also upon the cylinder, but near the nipples, are the figures "1" and "53." The maker's name is not given but the weapon was probably made by the Manhattan Fire Arms Manufacturing Company of New York.

United States.

No. 29. (N5910). PEPPER-BOX REVOLVER. Caliber .32. Two-and-one-half-inch, fluted cylinder having six smooth bores. This is a double-action pepper-box having a flat top-hammer and a nipple-shield. The barrel is stamped with the words "Cast Steel" and the left side of the hammer, "Manhattan F.A. Mfg. Co., New York." Upon the inside of the trigger-guard is the number "292."
Donor: Fred Bues.

United States.

No. 30a. (N2993). PEPPER-BOX REVOLVER. Caliber .31. Two-and-one-quarter-inch, fluted cylinder having five smooth bores. This double-action pepper-box has flat top-hammer and a nipple-shield. The barrel is marked "Cast Steel" and the left side of the hammer, "Manhattan F.A. Mfg. Co., New York." Stamped upon the cylinder near the nipples is the number "595."

United States.

No. 30b. (N4450). PEPPER-BOX REVOLVER. Same as No. 30a except that the cylinder is not marked "Cast Steel" and the revolver number is "870."

United States.

No. 31. (N4453). PEPPER-BOX REVOLVER. Caliber .31. Three-and-one-quarter-inch, fluted cylinder with six smooth bores. Action and markings are the same as those of No. 30, except that the inner side of the trigger-guard is marked with the number "46."

United States.

No. 32. (N4451). PEPPER-BOX REVOLVER. Caliber .31. Two-inch, fluted cylinder having six smooth bores. The action is the same as that of No. 30. Stamped upon the cylinder are the words "Manhattan F.A. Co., New York Cast Steel." The number "20" is marked upon the inside of the trigger-guard and upon the cylinder near the nipples. There are no marks upon the hammer.

United States.

No. 33. (N4452). PEPPER-BOX REVOLVER. Caliber .31. Two-and-one-half-inch, fluted cylinder, having six smooth bores. This is another double-action pepper-box equipped with a top-hammer and a nipple-shield. There are no marks upon the cylinder except for a "W" near the nipples. The revolver number is "2." The piece was probably made by Sprague and Marston of New York.

United States.

No. 34. (N4470). PEPPER-BOX REVOLVER. Caliber .34. Four-inch, fluted and ribbed cylinder having six smooth bores. This is still another example of the double-action pepper-box equipped with top-hammer and nipple-shield. Upon the cylinder are the words "Sprague & Marston, New York, Cast Steel Warranted," while the hammer is marked "W.W. Marston Patented 1849 New York." The inside of the trigger-guard is stamped with the figures "K 6."

United States.

No. 35. (N4473). PEPPER-BOX REVOLVER. Caliber .35. Three-and-one-quarter-inch, ribbed cylinder with six smooth bores. The action is the same as that of No. 34. Stamped upon the cylinder are the words "Sprague & Marston New York," and upon the hammer "W.W. Marston, Patented 1849 New York." The figures "K 35" are marked upon the inside of the trigger-guard.

United States.

No. 36. (N4483). PEPPER-BOX REVOLVER. Caliber .32. Two-and-one-quarter-inch, fluted and ribbed cylinder containing six smooth bores. The action is the same as that of No. 34. The cylinder is stamped with the words "W.W. Marston & Knox New York," and the hammer, "W.W. Marston New York 1854." Upon the inside of the trigger-guard is the number "3."

United States.

No. 37. (N4475). PEPPER-BOX REVOLVER. Caliber .32. Three-and-one-quarter-inch, fluted and ribbed cylinder containing six smooth bores. The action is the same as that of No. 34. The cylinder is stamped "W.W. Marston & Knox New York," the hammer, "W.W. Marston New York 1854," and the shank of the trigger bears the number "368."

United States.

No. 38. (N3837). PEPPER-BOX REVOLVER. Caliber .31. Three-and-one-half-inch, fluted and ribbed cylinder containing six smooth bores. The nipple-shield is missing. The pepper-box is cocked by pressure of the thumb upon the ear-piece of the top-hammer. Cocking the piece in this manner also revolves the cylinder. The cylinder is marked with the words "Stocking & Co., Worcester," and the inside of the trigger-guard, "A 29."

United States.

No. 39. (N4487). PEPPER-BOX REVOLVER. Caliber .31. Three-and-one-half-inch, fluted and ribbed cylinder containing six smooth bores. The operation of the top-hammer is as described under No. 38. Stamped upon the cylinder are the words "Stocking & Co., Worcester," and upon the left side of the hammer, "Patent Secured 1848." Upon the cylinder near the nipples is the number "58," and upon the inside of the trigger-guard the figures, "B 58."

United States.

No. 40. (N4486). PEPPER-BOX REVOLVER. Caliber .26. Four-and-one-quarter-inch, fluted and ribbed cylinder with six smooth bores. The operation of the top-hammer is as described under No. 38. The cylinder is stamped "Stocking & Co., Worcester," and the left side of the hammer, "Patent Secured 1848." The number "26" appears upon the cylinder near the nipples.

United States.

PLATE 87.

PERCUSSION PEPPER-BOXES
OF UNITED STATES MANUFACTURE

No. 1. (N4465). PEPPER-BOX REVOLVER. Caliber .32. Three-and-one-half-inch, fluted and ribbed cylinder containing six bores. Has a smooth, flat grip. The body is flat, and slightly engraved. This is one of the earliest examples of the Allen patent double-action pepper-boxes. It was made during the period of 1837-1842. Stamped upon the left side of the flat top-hammer are the words "Allen & Thurber, Grafton, Mass.," and upon the left side of the body "Allen's Patent."

United States.

No. 2. (N2782). PEPPER-BOX REVOLVER. Caliber .31.

Three-and-one-quarter-inch, ribbed and fluted cylinder containing six chambers. There is an oval, silver plate upon both sides of the plain, wooden grip. The action is the same as that described under No. 1. The cylinder is marked "Allen & Thurber, Norwich, C-T, Patented 1837, Cast Steel," and the left side of the hammer, "Allen's Patent." The revolver number is "79."

United States.

No. 3. (N4449). PEPPER-BOX REVOLVER. Caliber .32. Two-and-one-half-inch, fluted cylinder containing five smooth bores. Has plain, smooth grip plates of bone. All iron parts are somewhat corroded, but the body shows evidences of former engraving. This is another example of the Allen double-action pepper-box. The left side of the hammer is marked "Allen's Patent 1845," and the inside of the trigger-guard bears the number "40."

United States.

No. 4. (N4455). PEPPER-BOX REVOLVER. Caliber .34. Three-and-one-half-inch, fluted cylinder containing six smooth-bore chambers. The action is the same as that of No. 1. The left side of the hammer is marked "Allen's Patent 1845" and the inside of the trigger-guard bears the number "443 2," the figure "2" is separated from the rest of the number and probably denotes some special class or style of pepper-box.

United States.

No. 5. (N4457). PEPPER-BOX REVOLVER. Caliber .32. Three-and-one-quarter-inch, fluted cylinder containing six smooth bores. Has the usual flat top-hammer and the double-action peculiar to these pepper-boxes. The cylinder is marked "Allen & Thurber Worcester," the left side of the hammer, "Allen's Patent 1845," and the inside of the trigger-guard, "97 2." This piece was made in Worcester, Massachusetts.

United States.

No. 6. (N4474). PEPPER-BOX REVOLVER. Caliber .32. Three-and-one-quarter-inch, ribbed cylinder containing six smooth-bore chambers for loose ammunition just as the other pepper-boxes already described. The cylinder is marked "Allen & Thurber Worcester Patented 1837 Cast Steel," and the left side of the hammer, "Allen's

Patent." Stamped upon the inside of the trigger-guard is the number "404."

United States.

No. 7. (N4471). PEPPER-BOX REVOLVER. Caliber .34. Four-and-one-quarter-inch, ribbed cylinder containing six smooth-bore barrels. The action is the same as that of No. 1. The cylinder is stamped "Allen & Thurber Worcester Patented 1837 Cast-Steel," and the inside of the trigger-guard is marked with the number "23."

United States.

No. 8. (N4459). PEPPER-BOX REVOLVER. Caliber .34. Five-inch, fluted cylinder with six smooth-bore chambers. The trigger-guard of this double-action pepper-box is provided with a spur to give greater steadiness in aiming. The cylinder is stamped with the words "Allen & Thurber Worcester," and with the number "207," while the left side of the hammer is marked "Allen's Patent 1845." The inside of the trigger-guard is marked with the number "403."

United States.

No. 9. (N4468). PEPPER-BOX REVOLVER. Caliber .36. Five-and-one-quarter-inch, fluted and ribbed cylinder containing six smooth-bore barrels. This double-action pepper-box is also provided with a spur trigger-guard similar to that of No. 8. The cylinder is marked "Patented 1837 Cast Steel," and the left side of the hammer, "Allens Patent." Upon the inside of the trigger-guard is the number "13".

United States.

No. 10. (N4466). PEPPER-BOX REVOLVER. Caliber .31. Three-and-one-half-inch, ribbed cylinder containing six smooth-bore chambers. This is not a double-action pepper-box; it is cocked by pulling back the eared hammer which at the same time revolves the cylinder. The cylinder is stamped "Allen & Thurber Worcester Cast-Steel." Both the cylinder and the inside of the trigger-guard are marked with the number "113."

United States.

No. 11. (N3354). PEPPER-BOX REVOLVER. Caliber .32. Three-and-one-quarter-inch, fluted and ribbed cylinder containing six smooth-bore chambers. The action is the same as that of No. 1.

Stamped upon the cylinder are the words "Allen & Thurber Worcester Patented 1837 Cast Steel," and upon the left side of the hammer, "Allens Patent." The number "6" is stamped upon the inside of the trigger-guard.

United States.

No. 12. (N4472). PEPPER-BOX REVOLVER. Caliber .36. Four-and-one-quarter-inch, fluted and ribbed cylinder containing six smooth-bore chambers. This is another double-action pepper-box. The cylinder is marked "Allen & Thurber Worcester" over which are stamped in larger type the words "J. G. Bolen N. Y." The cylinder is also stamped "Patented 1837 Cast-Steel" and with the numbers "122" and "5". No number appears upon the inside of the trigger-guard.

United States.

No. 13. (N784). PEPPER-BOX REVOLVER. Caliber .32. Three-and-one-quarter-inch fluted cylinder containing six smooth-bore chambers. The revolver is double-action. Stamped upon the cylinder are the words "Allen & Thurber Worcester," and upon the left side of the hammer, "Allen's Patent 1845." The number "233" is stamped on the inside of the trigger-guard.

United States.

No. 14. (N787). PEPPER-BOX REVOLVER. Caliber .32. Four-and-one-quarter-inch, fluted cylinder having six smooth-bore chambers. This is a double-action pepper-box. The cylinder is stamped with the words "Cast Steel," and the left side of the hammer, "Manhattan Mfg. Co., New York." Upon the inside of the trigger-guard is the number "14".

United States.

No. 15. (N4445). PEPPER-BOX REVOLVER. Caliber .32. Three-inch, fluted cylinder having four smooth-bore chambers. This is also a double-action revolver. The cylinder is marked "Patented 1845 Allen & Wheelock" and "1621." Marked upon the left side of the hammer are the words "Allen's Patent Jan. 13, 1857," and upon the inside of the trigger-guard, the number "1621."

United States.

No. 16. (N4456). PEPPER-BOX REVOLVER. Caliber .32. Three-and-one-quarter-inch fluted cylinder containing six smooth-bore chambers. The cylinder of this double-action pepper-box is marked "Allen & Wheelock Worcester Patented April 16," and the left side of the hammer, "Allen's Patent 1845." Upon the inside of the trigger-guard are the numbers "117 1."

United States.

No. 17. (N6804). PEPPER-BOX REVOLVER. Caliber .32. Two-and-one-half-inch, fluted cylinder with five smooth-bore chambers. The cylinder of this double-action pepper-box is stamped "Allen & Wheelock Worcester Patented April 16," and the left side of the hammer, "Allen's Patent 1845." The trigger-guard is missing.

United States.

No. 18. (N4448). PEPPER-BOX REVOLVER. Caliber .31. Two-and-one-half-inch, fluted cylinder containing five smooth-bore chambers. The nipple-shield is missing. Below the pepper-box is a bullet-mold for casting balls for this kind of a revolver. The cylinder and the inside of the trigger-guard are stamped with the number "219," and the left side of the hammer is marked "Patented April 16, 1845."

United States.

No. 19. (N4485). PEPPER-BOX REVOLVER. Caliber .31. Two-and-one-quarter-inch fluted and ribbed cylinder containing six smooth-bore chambers. The grip plates are smooth, and inlaid upon each side with an oval silver plate. The pepper-box is double-action. Stamped upon the cylinder are the words "Patented 1837 Cast Steel," and upon the left side of the hammer, "Allen's Patent." The number "33" appears upon the inside of the trigger-guard.

United States.

No. 20. (N4482). PEPPER-BOX REVOLVER. Caliber .32. Three-and-one-quarter-inch, fluted and ribbed cylinder containing six smooth-bore chambers. This double-action pepper-box is equipped with a ring trigger. Each of the smooth grip plates is inlaid with an oval silver plate. The cylinder is marked "Patented 1837 Cast-Steel," and the left side of the hammer, "Allen's Patent." The revolver number is "7".

United States.

No. 21. (N4476). PEPPER-BOX REVOLVER. Caliber .31. Three-and-one-quarter-inch, fluted and ribbed cylinder containing six smooth-bore chambers. The grip plates are made of smooth bone. Stamped upon the cylinder of this double-action pepper-box are the words "Allen's Patent Worcester Mass., Patented 1837 Cast Steel," and upon the left side of the hammer, "Allen's Patent." The inside of the trigger-guard is marked "3 258."

United States.

No. 22. (N4478). PEPPER-BOX REVOLVER. Caliber .32. Three-and-one-quarter-inch, fluted and ribbed cylinder containing six smooth-bore chambers. The smooth grip is inlaid upon both sides with small, oval, silver plates. Upon the cylinder are the words "Patented 1837 Cast Steel," and upon the left side of the hammer, "Allen's Patent." The cylinder near the nipples and the inside of the trigger-guard is stamped with the number "17."

No. 23. (N4469). PEPPER-BOX REVOLVER. Caliber .36. Five-and-one-quarter-inch, fluted and ribbed cylinder containing six smooth-bore chambers. The cylinder of this double-action pepper-box is marked "Allen & Thurber Worcester Patented 1837 Cast Steel." Upon the inside of the trigger-guard is the number "23," and upon the left side of the hammer, "Allen's Patent."

United States.

No. 24. (N4479-N4481). PEPPER-BOX REVOLVER. Caliber .31. Three-and-one-quarter-inch, fluted and ribbed cylinder containing six smooth-bore chambers. This double-action pepper-box is shown in its original case, along with the accessories which came with an arm of this kind. The left side of the hammer is stamped with the words "J. G. Bolen, N. Y.," and the top of the same, "Allen's Patent." There are no marks upon the cylinder. Stamped upon the inside of the trigger-guard is the number "92." The accessories consist of a bullet-mold, an extractor and a powder-flask, the latter being marked "Dixon & Sons Sheffield." The lower portion of the advertisement given upon the inside cover of the case reads: "For Sale, Wholesale and Retail by J. G. Bolen, 104 Broadway, Between Wall and Pine Streets, New York."

United States.

No. 25. (N4458). PEPPER-BOX REVOLVER. Caliber .31. Three-and-one-quarter-inch, fluted and ribbed cylinder containing six smooth-bore chambers.   The only mark upon the cylinder of this pepper-box is the number "35", stamped near the nipples.   All metal parts are brightly polished.   Cocking the under-hammer by hand also revolves the cylinder.   No maker's name is given although this piece was made by Bacon and Company of Norwich, Connecticut.

United States.

No. 26.   (N4489).   PEPPER-BOX REVOLVER.   Caliber .32. Three-and-one-quarter-inch, fluted and ribbed cylinder with six smooth-bore chambers.   This is not a double-action revolver; cocking the under-hammer by hand revolves the cylinder.   Stamped upon the cylinder are the words "Bacon & Co., Norwich, C-T, Cast Steel" and the number "15," the latter being placed near the nipples.

United States.

No. 27.   (N789).   PEPPER-BOX REVOLVER.   Caliber .36. Two-and-one-quarter-inch, fluted and ribbed cylinder with six smooth-bore chambers.   This is a double-action pepper-box.   The cylinder is marked "Cast Steel," and the top of the hammer is stamped with the letter "c" enclosed in a circle, and the words "The Washington"; the rest of the die made no impression.   This piece was made by the Washington Arms Company.

United States.

No. 28.   (N4484).   PEPPER-BOX REVOLVER.   Caliber .32. Two-and-one-quarter-inch, fluted and ribbed cylinder having six smooth-bore chambers.   The cylinder of this double-action pepper-box is marked "Cast Steel," and the top-hammer is stamped with the letter "C" enclosed in a circle, and with the words "The Washington Arms Co."   Upon the shank of the trigger is the number "36."

United States.

No. 29.   (N6803).   PEPPER-BOX REVOLVER.   Caliber .32. Three-and-one-quarter-inch cylinder of the usual type containing six smooth-bore chambers.   The words "Cast Steel" appear upon the cylinder of this double-action pepper-box.   The shank of the trigger is marked with the number "2".   No maker's name is given, but this revolver was made by the Washington Arms Company.

United States.

No. 30.   (N4494).   PEPPER-BOX REVOLVER.   Caliber .26. Three-and-one-half-inch, fluted and ribbed cylinder containing ten smooth-bore chambers. The cylinder is completely covered by an en-graved cylinder of iron, shown partly withdrawn in the illustration. This is a double-action pepper-box having a folding-trigger. The end of the short retracting hammer strikes the percussion-cap placed upon the nipple, the latter being set in line with the chamber. The only marks upon this revolver consist of the figures "16 A. H" (no period after the "H"), stamped upon the cylinder. The maker and locality are unknown.

?

No. 31. (N785). PEPPER-BOX REVOLVER. Caliber .26. Three-and-one-half-inch, fluted and ribbed cylinder containing ten smooth-bore chambers. There formerly was a metal casing similar to that shown in No. 30. It completely enclosed the cylinder. Action, trigger, and position of the nipples are the same as those of No. 30. Stamped upon the cylinder are the figures "A. H 69" (no period after the "H"). No other marks of identification are given.

?

No. 32. (N4433). PEPPER-BOX REVOLVER. Caliber .31. Three-and-one-half-inch, ribbed cylinder containing four smooth-bore chambers. This revolver has a revolving hammer-head which is con-cealed within the breech. The cylinder unscrews for capping, as shown in the illustration. The cocking is done by means of the ring-trigger, and the firing by the outer trigger. Stamped upon the cylinder are the words "G. Leonard Jr. Charlestown Patented 1849 Cast Steel" and also the number "17."

United States.

No. 33. (N4435). PEPPER-BOX REVOLVER. Caliber .26. Two-and-one-quarter-inch, fluted cylinder containing five bores each of which is rifled with five grooves. The large cylinder is attached to a smaller one, one-and-three-eighths-inches in length, by means of a shaft which permits the long cylinder to unscrew and travel freely forward for loading. The cylinder breaks downward for capping (See No. 34). Triggers, action and hammer are the same as No. 32. Stamped upon the cylinder are the words "Robbins & Lawrence Co., Windsor, Vt., Patent 1849," and upon the butt the number "4122."

United States.

No. 34. (N4434). PEPPER-BOX REVOLVER. Caliber .26. Two-and-one-quarter-inch, fluted cylinder containing five bores each of which is rifled with five grooves. The action and other features of this revolver are the same as those described under No. 33. The cylinder is marked in a like manner. Stamped upon the butt is the number "1967."

United States.

No. 35. (N4436). PEPPER-BOX REVOLVER. Caliber .30. Three-and-one-eighth-inch, fluted and ribbed cylinder containing five bores each of which is rifled with five grooves. The cylinder is heavily browned, and the body is polished and brightly blued. The other features, including the marking, are the same as those of No. 33. Butt and breech are marked with the number "3057."

United States.

No. 36. (N4490). PEPPER-BOX CYLINDER. Caliber .34. Unfinished cylinder containing seven smooth-bore chambers. The cylinder is simply bored and fluted and illustrates one step in the process of manufacturing Leonard's patent pepper-box revolver.

United States.

No. 37. (N4491). PEPPER-BOX CYLINDER. Caliber .36. Unfinished cylinder containing six smooth-bore chambers. This cylinder is fluted and bored, and has a breech-plate with nipples. This specimen, like No. 36, shows a step in the manufacturing of Leonard's patent pepper-box.

United States.

No. 38. (N4492). PEPPER-BOX BUTT. A casting of the butt, and a portion of the body of a pepper-box revolver as it came from the mold before any finishing was done upon it. This specimen was obtained from the Robbins and Lawrence factory at Windsor, Vermont, after it was discontinued.

United States.

No. 39. (N4493). WOODEN PATTERN. This pattern or model is intended to illustrate the action of a pepper-box. The pattern was made by E. Allen in 1837 in Canton, Massachusetts.

United States.

PLATE 88.

PERCUSSION REVOLVERS, FOREIGN AND UNITED STATES

No. 1. (N4498). PERCUSSION REVOLVER. Caliber .40. Five-and-five-eighths-inch, octagonal, smooth-bore, steel barrel. There are six chambers in the cylinder. The flat top-hammer is equipped with a safety-catch which allows the cylinder to rotate freely in order to load and cap the chambers. A nipple-shield is in the rear of the cylinder. All mountings are engraved. No maker's name is given, although the revolver was made by Adams and Company.

England.

No. 2. (N4497). PERCUSSION REVOLVER. Caliber .38. Five-and-seven-eighths-inch, octagonal, steel barrel which is fastened to the axis of the cylinder by means of a set screw. There are six chambers in the cylinder. The revolver is equipped with a flat top-hammer, a nipple-shield and a fore-sight. All mountings are engraved, and each chamber is stamped with a proof-mark. This revolver, like No. 1, is double-action. Engraved upon the top of the barrel are the words "Adams & Co."

England.

No. 3. (N4495). PERCUSSION REVOLVER. Caliber .48. Five-and-five-eighths-inch, octagonal, rifled, steel barrel which is fastened to the pistol frame and to the axis of the cylinder. This double-action revolver has six chambers, and a flat top-hammer which has a slot through which to sight. Besides these, the gun also has a nipple-shield, fore-sight and a folding ramrod. All mountings are engraved. Upon the top of the revolver are the words "C.H. Gilks, Tower Hill, London."

England.

No. 4. (N2314). PERCUSSION REVOLVER. Caliber .40. Five-and-one-quarter-inch, octagonal, rifled, steel barrel which is fastened to the axis of the cylinder by means of a set screw. This is another form of double-action revolver. It is equipped with six chambers, a flat top-hammer, a nipple-shield and a fore-sight. All mountings are engraved. The barrel and each chamber are stamped with proof-marks. Engraved upon the top of the barrel are the words "T. W.

Boyd, Montreal." This gunsmith made flat top-hammer revolvers and bodies for pepper-boxes.

Canada.

No. 5. (N4496). PERCUSSION REVOLVER. Caliber .44. Five-and-three-quarters-inch, octagonal, rifled, steel barrel which is fastened to the axis of the cylinder. This double-action revolver has a cylinder containing six chambers, a flat top-hammer with a slot to sight through, a nipple-shield and a fore-sight. All mountings are of steel and engraved. Barrel and chambers are stamped with proof-marks. Engraved upon the top of the barrel are the words "Boyd, Montreal."

Canada.

No. 6. PERCUSSION REVOLVER. Caliber .36. Six-inch, octagonal, steel barrel. This is also a double-action revolver. It has a cylinder containing six chambers, a flat top-hammer operated by a ring trigger and a safety on one side in the form of a button. There is no nipple-shield or rear sight. All mountings are engraved. The cylinder is stamped with the Liége proof-mark and the inner side of the back-strap "H. Colleye, Breveté."

Belgium?

No. 7. (N794). PERCUSSION REVOLVER. Caliber .34. Three-and-one-quarter-inch, octagonal, steel barrel. Has six chambers in a plain cylinder, a flat top-hammer provided with a V-shaped groove through which to sight. There is no nipple-shield or safety-catch. The ring-trigger, body, and butt-plate of this double-action revolver are engraved. The cylinder revolves to the right instead of to the left, as is the case in some types. The inner side of the back-strap is stamped with the word "Breveté" and a partially effaced name.

France?

No. 8. (N2746). PERCUSSION REVOLVER. Caliber .32. Four-inch, octagonal, rifled, steel barrel. This revolver has front and rear sights and six chambers. All parts are unornamented. The cylinder is revolved by means of the lever on top of the revolver, and the firing is done by means of a concealed hammer. The barrel may be removed by means of the lever on the left side of the weapon. In the butt is a cap-box, stamped with the name "Devisme," while the barrel is marked with the number "1267" and a proof-mark. This revolver was made in Paris.

France.

No. 9. (N2751). PERCUSSION REVOLVER. Caliber .41. Six-and-one-quarter-inch, octagonal, rifled, steel barrel. There are six chambers for loose ammunition, and also front and rear sights. The cylinder bears scrolls in relief, and is revolved by means of the lever on the right side of the revolver. The firing hammer is concealed. By means of another lever on the left side of the weapon, the barrel may be removed. A short ramrod is carried in the butt. Upon the top of the barrel are the words "Devisme À Paris." The barrel is stamped with the number "472" and the grip with "3192."

France.

No. 10. (N2875). PERCUSSION REVOLVER. Caliber .44. Six-and-one-quarter-inch, octagonal, rifled barrel. This double-action revolver has five chambers, and front and rear sights. A lever ramrod formerly was carried on the left side of the barrel. The cylinder may be removed by withdrawing the cylinder pin. The body is slightly ornamented and stamped "Adam's Patent," followed by the number "5560 CD." The cylinder bears the same number and, in addition, the Liége proof-mark.

England.

No. 11. (N2871). PERCUSSION REVOLVER. Caliber .45. Six-and-one-quarter-inch, octagonal, rifled barrel. There are five chambers in the cylinder, and the revolver is equipped with front and rear sights, but no ramrod. The safety-catch is missing. There are signs of former engraving upon the frame. Barrel and cylinder are stamped with an English proof-mark in the form of a crown over the letter "V." No maker's name or revolver number are given, although this piece was most likely made by Adams.

England.

No. 12. (N4518). PERCUSSION REVOLVER. Caliber .45. Six-and-one-eighth-inch, octagonal, rifled barrel. There are five chambers, and front and rear sights. The loading lever works through a hole in the frame. All steel parts except the hammer are blued, but otherwise unornamented. The frame of this double-action revolver is stamped "Patent No. 2529." Barrel and cylinder bear English proof-marks. Neither name nor number are given upon this revolver, which probably was made under Adams' patent.

England.

No. 13. (N4521). PERCUSSION REVOLVER. Caliber .38. Four-and-one-quarter-inch, octagonal, rifled barrel. This double-action revolver has five chambers, front and rear sights, and a lever ramrod which works through a hole in the frame. All steel parts were formerly blued. The frame is stamped "Patent, No. 3955," while barrel and cylinder are marked with English proof-marks. No maker's name is given, although this revolver was probably made by Adams.

England.

No. 14. (N4523). PERCUSSION REVOLVER. Caliber .38. Four-and-one-half-inch, octagonal, rifled barrel. Has six chambers, front and rear sights, and a safety-catch in the form of a spring behind the cylinder. The hammer is earless. All parts are unornamented. The barrel and frame of this double-action revolver are stamped with English proof-marks. No maker's name or revolver number are given. Probably made by Adams.

England.

No. 15. (N4519). PERCUSSION REVOLVER. Caliber .45. Six-inch, octagonal, rifled barrel. Has front and rear sights, and a lever ramrod. A safety-catch on the left side of the frame allows the cylinder to be revolved for loading and capping. All parts are unornamented. The frame and cylinder of this double-action revolver are stamped with numerous proof-marks, including that of Liége. The piece is also marked "Adams' Patent 1856." Upon the top of the barrel are the words "A. Francotte à Liége." The revolver number is "N A 302."

Belgium.

No. 16. (N2743). PERCUSSION REVOLVER. Caliber .38. Four-inch, octagonal, rifled barrel. Has five chambers, and front and rear sights, but no ramrod. There is a safety-catch, in the form of a spring behind the cylinder. The hammer is earless, as this is a double-action revolver, and the cocking was done by the trigger. The body is engraved with scrolls, and the frame is stamped "Adam's Patent 7678." Upon the top of the barrel are the words "Manfd. by A. Francotte Licensed by Deane Adams & Deane, London." There is also an English proof-mark upon the barrel.

England.

No. 17. (N4522). PERCUSSION REVOLVER. Caliber .41. Four-and-seven-eighths-inch, octagonal, rifled barrel. This double-

action revolver has five chambers, front and rear sights, a ramrod
working through a groove on the left side of the frame, and a safety-
catch in the form of a spring behind the cylinder. The hammer is
earless. All parts of the weapon are unornamented. The cylinder is
stamped with proof-marks. The number "9369," and the word "London" appear upon the frame. No maker's name is given.

England.

No. 18.   (N3306).   PERCUSSION REVOLVER.   Caliber .48.
Five-and-three-quarters-inch, octagonal, rifled barrel. Has five chambers, and a rear sight; the front sight is missing. A lever ramrod is
carried beneath the barrel. The revolver is cocked and the cylinder
revolved by pulling back the side-hammer. Upon the frame are the
words "Kerr's Patent, 1301." The number "1301" also appears upon
the cylinder, and, in addition, there are some English proof-marks.
Engraved upon the lock-plate are the words "London Armoury."

England.

No. 19.   (N4517).   PERCUSSION REVOLVER.   Caliber .48.
Five-and-three-quarters-inch, octagonal, rifled barrel. This revolver has
front and rear sights, and a cylinder containing five chambers. The
action and other features of this piece are the same as those of No. 18.
Upon the frame are the words "Kerr's Patent, 8865." The cylinder
is stamped with English proof-marks and the steel plate on the grip is
engraved "London Armoury Co."

England.

No. 20.   (N4135).   PERCUSSION REVOLVER.   Caliber .38.
Six-inch, octagonal, rifled barrel. The cylinder has five chambers.
Equipped with a lever ramrod, and a safety-catch, the latter located
back of the cylinder. Cylinder and body are stamped with the number
"222." The body is marked "Adams Patent, May 3, 1858," and the
ramrod, "Kerr's Patent, April 14, 1857," together with the number
"1820." Upon the top of the barrel are the words "Manufactured by
Mass. Arms Co., Chicopee Falls."

United States.

No. 21.   (N2742).   PERCUSSION REVOLVER.   Caliber .50.
Seven-and-one-half-inch, octagonal, rifled barrel. Has five chambers,
and a front sight, but none in the rear. A lever ramrod is carried
beneath the barrel. The earless hammer is provided with a spring
safety-catch. There is a cap-box in the butt. All parts are unorna-

mented, but were originally blued. The cylinder is stamped with English proof-marks in the form of a crown over the letter "V." Upon the body are the words "Sheath's Patent No. 5."

England.

No. 22. (N3307). PERCUSSION REVOLVER. Caliber .40. Four-and-one-half-inch, octagonal, rifled barrel. Has six chambers, front and rear sights, and a lever-rod. A groove along the top of the barrel facilitates aiming. The earless hammer is provided with a spring safety-catch. Body and butt-plate are engraved, and all steel parts are blued. The cylinder and barrel are stamped with English proof-marks, and the barrel is marked "Tardrew & Son Bideford." No revolver number is given.

England.

No. 23. (N4520). PERCUSSION REVOLVER. Caliber .38. Four-and-one-half-inch, octagonal, rifled barrel. Has five chambers, front and rear sights, a lever ramrod which fits on a pin on the left side of the frame, and an earless hammer. This revolver has a double trigger; pulling the lower one brings the hammer back, and a slight pressure upon the upper trigger discharges the revolver. All mountings are engraved with scrolls, and all steel parts are blued and polished. The cylinder is stamped with English proof-marks. Upon the frame, the figures "No. 6152 T." are engraved. This revolver was made by William Tranter, in Birmingham.

England.

No. 24. (N5618-9). PERCUSSION REVOLVER IN CASE. Caliber .38. Four-and-one-quarter-inch, octagonal, rifled barrel. Has five chambers, front and rear sights, a lever ramrod which fits on a pin on the left side of the frame, and an earless hammer. This revolver also has a double trigger; pulling the lower one brings the hammer back and revolves the cylinder, while a slight pressure upon upper trigger discharges the revolver. Body, butt-plate and barrel are finely engraved with scrolls. All steel parts are polished and blued. English proof-marks are stamped upon the cylinder, and the frame bears the figures "3757 T." The accessories in the case consist of a box of conical bullets, a box of lubricant, a box of caps, an oil-can, a loading lever, a powder-flask, a nipple wrench, a bullet mold, and a charge extractor. The revolver was made by William Tranter of Birmingham.

England.

No. 25. (N4525). PERCUSSION REVOLVER. Caliber .36. Six-inch, octagonal, rifled barrel. Has six chambers in the cylinder. The front and rear sights are missing. This is a single-action revolver, that is, the cylinder can only be revolved by pulling back the hammer. The barrel is easily removed by releasing the lever beneath it. Body, trigger-guard and butt are engraved with scrolls. The cylinder is stamped with the Liége proof-mark, and the barrel, with the number, "951," and the name "Deprez."

France?

No. 26. (N4524). PERCUSSION REVOLVER. Caliber .28. Three-and-three-quarters-inch, octagonal, rifled barrel. Has five chambers, and a back sight; the front sight is missing. A spring on the left side prevents the hammer from striking the barrel, and enables the cylinder to be rotated for loading and capping. The cylinder itself is easily removed by withdrawing the pin. The body of this double-action revolver is slightly ornamented. No maker's name or revolver number are given; the piece is probably of French manufacture.

France?

No. 27. (N4502). PERCUSSION REVOLVER. Caliber .32. Four-inch, round, rifled barrel. Has six chambers. The top-hammer is placed slightly to the right of the center. Pulling back the hammer revolves the cylinder and cocks the revolver. Body and cylinder are engraved. Stamped upon the body are the words "Warner's Patent, Jan. 1851," while the top of the barrel is marked "Springfield Arms Co." The revolver number is "190."

United States.

No. 28. (N4506). PERCUSSION REVOLVER. Caliber .28. Two-and-five-eighths-inch, round, smoth-bore barrel. Has six chambers, and a front sight, but no back sight or ramrod. The hammer is placed slightly to the right of the center. Pulling back the hammer cocks the revolver and revolves the cylinder. The body is slightly engraved. No maker's name is given although the piece was made under the James Warner patent probably by the Springfield Arms Company. The patent was granted to Warner January 7, 1851. The number on the specimen illustrated here is "53."

United States.

No. 29. (N4507). PERCUSSION REVOLVER. Caliber .28. Two and-one-half-inch, round, smooth-bore, steel barrel. Has six chambers, and a front sight, but no back sight or ramrod. The top-hammer is placed slightly to the right of the center. Pulling back the hammer revolves the cylinder and cocks the revolver. Barrel, body, and cylinder are engraved. No maker's name is given. It was made under Warner's patent, and is stamped with the numbers "11" and "75."

United States.

No. 30. (N4508). PERCUSSION REVOLVER. Caliber .28. Two-and-one-half-inch, round, smooth-bore barrel. Has six chambers and a front sight, but no back sight or ramrod. There is a set-trigger in back of the regular one. The action is the same as that of No. 29. No maker's name is given. This revolver was also made under the Warner patent. The revolver number is "24."

United States.

No. 31. (N793). PERCUSSION REVOLVER. Caliber .30. Two-and-five-eighths-inch, round, rifled barrel. Has six chambers, and front and back sights, but no ramrod. This revolver has a ring-trigger in the rear of which is a set-trigger. This, like all Warner revolvers of this type, is single-action. The cylinder is engraved, and the numbers "688" and "25" appear on the inside of the back-strap. No maker's name is given. The revolver was probably made by the Springfield Arms Company of Springfield, Massachusetts.

United States.

No. 32. (N4512). PERCUSSION REVOLVER. Caliber .28. Three-inch, round, rifled barrel. Has a cylinder with six chambers, a front sight, and a lever ramrod. A notch in the hammer serves as a back sight. The barrel and cylinder are blued, but all parts are unornamented. Pulling the hammer back revolves the cylinder and cocks the revolver. Stamped upon the top of the frame are the words "James Warner, Springfield, Mass., U. S. A." The revolver number is "78."

United States.

No. 33. (N4501). PERCUSSION REVOLVER. Caliber .40. Six-and-one-quarter-inch, round, rifled, steel barrel. Has six chambers, and front and rear sights, but no ramrod. The cylinder may be removed, by turning the catch in front of the cylinder around the barrel. The cylinder has to be turned by hand after each shot. This revolver is extremely heavy. It is stamped "Wesson & Leavitt's Patent" and

"Patented No. 26, 1850." Upon the top of the barrel are the words "Mass. Arms Co., Chicopee Falls."

United States.

No. 34. (N4503). PERCUSSION REVOLVER. Caliber .32. Six-inch, round, rifled barrel. Has six chambers, and front and rear sights, but no ramrod. The cylinder may be removed by releasing the catch in front of it and lifting up the barrel as illustrated in No. 33. Frame and hammer are slightly engraved. This is another Wesson and Leavitt revolver, patented November 26th, 1850. It was made by the Massachusetts Arms Company and bears the number "206."

United States.

No. 35. (N796). PERCUSSION REVOLVER. Caliber .32. Six-inch, round, rifled barrel. Has six chambers in the cylinder. This is a single-action revolver; that is, pulling back the hammer revolves the cylinder and cocks the revolver. All parts except the trigger-guard and back-strap are inlaid with gold. This revolver was also made under Wesson and Leavitt's patent by the Massachusetts Arms Company of Chicopee Falls. The revolver number is "40."

United States.

No. 36. (N2749). PERCUSSION REVOLVER. Caliber .32. Six-inch, round, rifled, steel barrel. Has six chambers, and front and rear sights, but no ramrod. Pulling back the hammer revolves the cylinder and cocks the revolver. This revolver is fitted with the Maynard primer, in which the percussion-caps are mounted upon a strip of paper coiled in a magazine near the lock of the weapon. Cocking the revolver brings one of the caps over the nipple of the chamber, and pulling the trigger explodes the cap and at the same time cuts it off the coil. This weapon has but one nipple which directs the fire from the percussion-cap into the pin-hole of the chamber. The revolver was made under Wesson's and Leavitt's patent by the Massachusetts Arms Company of Chicopee Falls. The revolver number is undecipherable.

United States.

No. 37. (N4560). PERCUSSION REVOLVER. Caliber .31. Five-inch, round, rifled barrel. Has six chambers. This revolver is also provided with the Maynard primer (See No. 36). There is but one nipple, from which the fire of the percussion-cap is directed into the various pin-holes of the chambers. Barrel and body are engraved with floral designs. The cylinder has to be revolved by hand. Pressing the small button in the trigger-guard releases the cylinder to effect rotation. This revolver was also made under the patent of Wesson and Leavitt (See No. 36).

United States.

No. 38. (N4564). PERCUSSION REVOLVER. Caliber .26. Two-and-three-quarters-inch, octagonal, rifled barrel. Has six chambers, and is provided with the Maynard primer (See No. 36). The nipple is as described under No. 37. The body is slightly engraved, and all steel parts were formerly blued. The piece was made by the Massachusetts Arms Company, and bears the numbers "208" and "660."

United States.

No. 39. (N4563). PERCUSSION REVOLVER. Caliber .28. Two-and-one-half-inch, round, rifled barrel. Has six chambers, and front and rear sights, but no ramrod. Fitted with the Maynard primer, part of which is still in the magazine (shown open in the illustration). The cylinder has to be revolved by hand. The nipple is as described under No. 37. Both body and cylinder are engraved with floral designs. The revolver was made by the Massachusetts Arms Company and bears the numbers "436" and "569."

United States.

No. 40. (N4562). PERCUSSION REVOLVER. Caliber .28. Three-and-one-half-inch, round, rifled barrel, fitted with front and rear sights. This revolver is also provided with the Maynard primer. (See No. 36.) Pulling back the hammer revolves the cylinder and cocks the revolver. Cylinder and body are engraved, and the barrel is blued. This piece was made by the Massachusetts Arms Company of Chicopee Falls, and bears the numbers "319" and "101."

United States.

No. 41. (N4500). PERCUSSION REVOLVER. Caliber .32. Three-inch, octagonal, rifled, steel barrel. Has five chambers, and a front sight, but no rear sight or ramrod. This double-action revolver has a top-hammer after the manner of the pepper-boxes. The body and cylinder are slightly engraved. This piece, which was made by Josiah Ells of Pittsburgh, Pa., bears the number "50."

United States.

No. 42. (N4499). PERCUSSION REVOLVER. Caliber .28. Three-and-one-half-inch, octagonal, rifled barrel. Has five chambers, a front sight, and a rear sight, the latter being in the form of a groove. This double-action revolver also has a flat top-hammer just as No. 41, and like it, was made by Josiah Ells of Pittsburgh. The patent was granted to Ells, August 1st, 1854. The revolver number is "22."

United States.

No. 43. (N4565). PERCUSSION REVOLVER. Caliber .44. Seven-inch, octagonal, rifled barrel. Front and rear sights. Has five chambers and a lever ramrod. Pulling the hammer back revolves the cylinder and cocks the revolver. By removing the thumb screw in front of the trigger guard, a tube of disc-primers can be inserted. The cocking of the revolver feeds the discs automatically to the nipples. Stamped upon the top of the frame are the words "Butterfield's Patent, Dec. 11, 1855, Philada." The revolver number is "26," and the piece was made by Jesse S. Butterfield in Philadelphia, Pa.

United States.

No. 44. (N4556). PERCUSSION REVOLVER. Caliber .32. Three-and-one-half-inch, octagonal, rifled barrel. There are five chambers in the cylinder, each of which carries two charges, one in front of the other. The two hammers are operated by the sheath trigger; one hammer fires the forward charge, the other the rear one. There is a deep groove in the right side of the frame for loading the cylinder. All steel parts are blued. Stamped upon the top of the barrel are the words "Walch Fire-Arms Co., New York. Pat'd Feb. 8, 1859." No revolver number is given.

United States.

No. 45. (N4509). PERCUSSION REVOLVER. Caliber .31. Four-inch, octagonal, rifled barrel. Has six chambers, a lever ramrod, a front sight, and a rear sight, the latter being in the form of a groove. Pulling the hammer back revolves the cylinder and cocks the revolver. All parts of the weapon are unornamented. Upon the top of the barrel are the words "C.R. Alsop, Middletown, Conn., Patented July 17th, August 7th, 1860, May 14, 1861." The peculiarity of the Alsop revolver is the use of a cam of odd shape for moving the cylinder backwards and forwards. The revolver is stamped with the number "612."

United States.

No. 46. (N6770). PERCUSSION REVOLVER. Caliber .38. Three-and-one-half-inch, octagonal, rifled barrel. Has five chambers, a front sight, a lever ramrod which is carried beneath the barrel, and a rear sight which is in the form of a groove. The action and the markings upon the top of the barrel of this Alsop revolver are the same as those of No. 45. The revolver number is "121."

United States.

No. 47. (N4526). PERCUSSION REVOLVER. Caliber .28. Three-inch, octagonal, rifled barrel, having a rear sight in the form of a groove. Has five chambers in a cylinder which is engraved with floral designs. There is no ramrod or trigger-guard. Pulling back the hammer revolves the cylinder and cocks the revolver. The weapon is stamped with the number "48." No maker's name is given, but the revolver is said to have been made by Bliss and Goodyear of New Haven, Connecticut.

United States.

No. 48. (N4535). PERCUSSION REVOLVER. Caliber .31. Four-inch, octagonal, rifled, steel barrel which is equipped with front and rear sights. A lever ramrod is carried beneath the barrel. The cylinder is engraved with scenes of the chase, the body, with scrolls. All steel parts were formerly blued. The revolver is single-action. The top of the barrel is marked "Bacon Mf'g. Co., Norwich, Conn.," and the bottom is stamped with the number "161."

United States.

No. 49. (N4533). PERCUSSION REVOLVER. Caliber .31. Four-inch, round, rifled barrel having a front sight. The rear sight is in the form of a notch cut into the hammer. A lever ramrod is carried beneath the barrel. Stamped upon the top of the barrel of this single-action revolver are the words "Bacon Mf'g. Co., Norwich, Conn." The revolver number is "1210."

United States.

No. 50. (N4532). PERCUSSION REVOLVER. Caliber .32. Five-and-one-half-inch, octagonal, rifled barrel. Has five chambers, a front sight, a rear sight in the form of a groove, and a lever ramrod. All parts of this single-action revolver are unornamented. The top of the barrel is stamped "W.W. Marston Phoenix Armory, New York City." The revolver number is "833."

United States.

PLATE 89.

PERCUSSION REVOLVERS,
CHIEFLY UNITED STATES

No. 1. (N4513). PERCUSSION REVOLVER. Caliber .31. Three-and-one-half-inch, octagonal, rifled barrel having no sights or ramrod. There are five chambers in the cylinder. This is a double-action revolver having a top-hammer and having the nipples placed in the sides of the cylinder as in some pepper-boxes. The cylinder is engraved with scenes of the chase, but the rest of the weapon is unornamented. The barrel is stamped "Allen & Wheelock," and the hammer "Patented April 16, 1845." The revolver number is "141." This piece was made in Worcester, Massachusetts.

United States.

No. 2. (N4514). PERCUSSION REVOLVER. Caliber .30. Two-and-three-eighths-inch, octagonal, rifled barrel. Has five chambers in a cylinder which is engraved with scenes of the chase. There are no sights or ramrod. This double-action revolver has a top-hammer, and has the nipples placed in the sides of the cylinder and not in the end. Stamped upon the barrel are the words "Allen & Wheelock, Worcester, Mass., Allen's Patent April 16, 1865." The revolver number is "566."

United States.

No. 3. (N4527). PERCUSSION REVOLVER. Caliber .38. Three-inch, octagonal, rifled barrel. Has five chambers in the cylinder. The rear sight is in the form of a groove; the trigger is sheathed. Pulling back the hammer revolves the cylinder and cocks the revolver. No maker's name is given but the weapon is said to have been made under Allen's patent. The revolver bears the number "500."

United States.

No. 4. (N4505). PERCUSSION REVOLVER. Caliber .27. Three-inch, octagonal, rifled barrel. The cylinder has five chambers. All metal parts are silver-plated, and the cylinder is engraved with scenes of the chase. The trigger-guard and ramrod are combined. Releasing a catch causes the trigger-guard to fall downward which, by means of geared teeth, then acts on the rammer and forces the ball home. This single-action revolver is equipped with a side-hammer, and a cylinder-pin, the latter withdrawing from the rear. The left side of the barrel is stamped "Allen & Wheelock, Worcester, Mass., U.S., Allen's Pt's. Jan. 13, Dec. 15, 1857, Sept. 7." (Year not given after last date.) The revolver number is "836."

United States.

No. 5. (N4504). PERCUSSION REVOLVER. Caliber .32. Four-inch, octagonal, rifled barrel. The cylinder has five chambers, and is engraved with a forest scene. Trigger-guard and ramrod are combined. Releasing a catch in the rear of the trigger-guard causes the latter to fall downward, a toothed wheel at the other end, then engages the grooves in the ramrod and forces the ball home. The cylinder pin may be unscrewed and withdrawn from the rear, thus freeing the cylinder. The left side of the barrel is marked "Allen & Wheelock, Worcester, Mass., U.S., Allen's Pt's. Jan. 13, Dec. 15, 1857." The number upon the revolver is "330."

United States.

No. 6. (N4119). PERCUSSION REVOLVER. Caliber .36. Five-inch, octagonal, rifled barrel. There are six chambers in the cylinder. Trigger-guard and ramrod are combined, as in Nos. 4 and 5. Has a center-hammer and a cylinder-pin, the latter withdraws from the front. The left side of this single-action revolver is stamped "Allen & Wheelock, Worcester, Mass., U.S., Allen's Pt's. Jan. 13, Dec. 15, 1857, Sept. 7, 1858." The revolver number is "409."

United States.

No. 7. (N4120). PERCUSSION REVOLVER. Caliber .36. Eight-inch, octagonal, rifled barrel. Has six chambers in a cylinder which is engraved with a forest scene showing wild animals. Trigger-guard and ramrod are combined, as in Nos. 4 and 5. The hammer is on the right side. The cylinder-pin withdraws from the rear. Stamped upon the left side of the barrel are the words "Allen & Wheelock, Worcester, Mass., U.S., Allen's Pt's. Jan. 13, Dec. 15, 1857, Sept. 7, 1858." The revolver number is "520."

United States.

No. 8. (N4117). PERCUSSION REVOLVER. Caliber .45. Seven-and-one-half-inch, half-octagonal, rifled barrel. This is a single-action revolver, and has six chambers in the cylinder. Trigger-guard and ramrod are combined, as in Nos. 4 and 5. Has a center-hammer. By pressing a spring beneath the cylinder-pin, it may be withdrawn from the front.   The left side of the barrel is stamped "Allen & Wheelock, Worcester, Mass., U.S., Allen's Pt's. Jan. 13, Dec. 15, 1857, Sept."   (The day of the month and the year of the last date are not given, but September 7, 1858, is probably intended.)   The revolver is marked with the number "97."

United States.

No. 9.   (N4118).   PERCUSSION REVOLVER.   Caliber .44. Seven-and-one-half-inch, half-octagonal, rifled barrel.   This single-action revolver has six chambers in the cylinder.   Trigger-guard and rammer are combined.   (The guard is shown down in the picture to illustrate the loading action.   (See Nos. 4 and 5.)   The hammer is of the center variety.   By pressing a spring beneath it, the cylinder-pin may be withdrawn from the front.   Stamped upon the left side of the barrel are the words "Allen & Wheelock, Worcester, Mass., U.S., Allen's Pt's. Jan. 13, Dec. 15, 1857, Sept."   (The day of the month and the year are not given.   See No. 8.)   The revolver number is "129."

United States.

No. 10   (N795).   PERCUSSION REVOLVER.   Caliber .32. Four-inch, round, rifled barrel.   Has five chambers in a cylinder which is engraved with various scenes of the chase.   This is a single-action revolver, equipped with a lever ramrod and steel trigger-guard.   Stamped upon the top of the barrel are the words "Hopkins & Allen Mfg. Co., Norwich, Ct."   Upon the bottom of the barrel is the number "850."

United States.

No. 11.   (N4539).   PERCUSSION REVOLVER.   Caliber .31. Four-inch, octagonal, rifled barrel.   Chambers, ramrod, trigger-guard, and cylinder are the same as those of No. 10.   The top of the barrel is stamped "Hopkins & Allen Mfg. Co., Norwich, Ct.," and the bottom of the barrel and the ramrod bear the number "642."

United States.

No. 12.   (N5708).   PERCUSSION REVOLVER.   Caliber .40. Four-inch, blued barrel.   Has a smooth, oval grip, a brass body, a front sight, and a rear sight in the form of a groove.   The sheath trigger makes a trigger-guard unnecessary.   There are two hammers which are operated by the single trigger.   This is a Lindsay pistol which was patterned after the rifle of the same name.   In each weapon there were two charges in the barrel, one on top of the other.   Upon discharge, the charge in front was fired first by one hammer, and the second charge then fired by the other hammer. Body and barrel of the pistol are engraved. The pistol is in its original card-board case which also contains the accessories, consisting of bullets, powder-flask and bullet-mold. Stamped upon the pistol are the words "Lindsay's Young America, Man'f'd. by J.P. Lindsay-Man'fg. Co., New York. Patented Feb. 8, 1859, and Oct. 9, 1860." The pistol number is "698."

Donor: Salvation Army.

United States.

No. 13. (N6756). PERCUSSION REVOLVER. Caliber .40. Four-inch, blued barrel. This is another Lindsay pistol identical in all respects to No. 12, except for the pistol number, which in this specimen is "1091."

United States.

No. 14. (N3308). PERCUSSION REVOLVER. Caliber .38. Four-and-one-quarter-inch, octagonal, rifled barrel. Has front and rear sights, five chambers in the cylinder, and an earless hammer provided with a safety-catch in the form of a spring. The body of this double-action revolver is engraved with scrolls. No maker's name or shop number are given, but the back-strap is stamped with the initials "W.H."

United States.

No. 15. (N4530). PERCUSSION REVOLVER. Caliber .32. Three-inch, octagonal, rifled barrel. This weapon is sometimes known as the "Whitney-Beals ring trigger pocket revolver." A trigger operated fork, for revolving the cylinder, is pivoted within a shield on the left side. There is a row of notches near each edge of the cylinder, and the tips of the fork engage them alternately. Stamped upon the tip of the barrel are the words "Address E. Whitney, Whitneyville, Ct." The revolver number is "26."

United States.

No. 16. (N4529). PERCUSSION REVOLVER. Caliber .32. Four-inch, octagonal, rifled, steel barrel. There are seven chambers in the cylinder. All metal parts, except the barrel, are made of brass and are nickel-plated. This is another example of the "Whitney-Beals ring trigger pocket revolver" having the same action as that described under No. 15. The barrel is also marked in the same way, but the revolver number is "B 47."

United States.

No. 17. (N4553). PERCUSSION REVOLVER. Caliber .31. Four-inch, octagonal, rifled barrel. Has five chambers in a cylinder which is engraved with floral designs. A small slit in the hammer serves as a rear sight. Barrel and hammer are blued and the body is made of brass. In order to revolve the cylinder, the hammer is brought to half-cock; the forward trigger is then pressed, which, releasing a catch, permits the cylinder to be turned by hand. The chamber is then fired by pulling the main trigger. The top of the barrel is stamped "E. Whitney, N. Haven, Ct.," and the revolver number is "V76."

United States.

No. 18. (N4511). PERCUSSION REVOLVER. Caliber .28. Three-and-one-half-inch, octagonal, rifled barrel. Has six chambers in an engraved cylinder, front and rear sights, and a lever ramrod. The trigger is of the mortise or sheath variety. Pulling back the hammer revolves the cylinder and cocks the revolver. This arm resembles the Colt revolver made during the same period. Stamped upon the top of the barrel are the words "E. Whitney, N. Haven." The revolver number is "983."

United States.

No. 19. (N4538). PERCUSSION REVOLVER. Caliber .32. Three-and-one-half-inch, octagonal, rifled barrel. There are five chambers in the cylinder. All parts of this single-action revolver are unornamented. The top of the barrel is stamped "E. Whitney, N. Haven," and the revolver number is "L 9."

United States.

No. 20. (N6771). PERCUSSION REVOLVER. Caliber .32. Three-and-one-half-inch, octagonal, rifled barrel. Has five chambers in the cylinder, a front sight, a concave rear sight, a lever ramrod, a mortise hammer, and a brass trigger-guard. All parts of this single-action revolver are unornamented. The top of the barrel is stamped "E. Whitney, N. Haven," and the revolver number is "19955 F."

United States.

No. 21. (N4543). PERCUSSION REVOLVER. Caliber .32. Six-inch, octagonal, rifled barrel. This single-action revolver has five chambers in the cylinder, a brass fore-sight, a lever ramrod, a brass trigger-guard, and a rear sight which consists of a groove cut into the hammer. The grip is made of polished wood and is oval in form. This revolver was intended to be carried in a holster suspended from a belt. The top of the barrel is stamped "E. Whitney, N. Haven," and the bottom bears the number "16835."

United States.

No. 22. (N4541). PERCUSSION REVOLVER. Caliber .36. Seven-and-five-eighths-inch, octagonal, rifled barrel. Has six chambers in the cylinder. This is a single-action, navy, Whitney revolver, having a lever ramrod, brass trigger-guard, a brass front sight, and a rear sight which consists of a groove in the frame above the cylinder, together with a notch in the hammer. The grip is smooth and oval. Stamped upon the top of the barrel are the words "E. Whitney, N. Haven," while the ramrod is marked with the number "24312 F."

United States.

No. 23. (N4544). PERCUSSION REVOLVER. Caliber .32. Five-inch, octagonal, rifled barrel. Has a lever ramrod and five chambers. Frame and trigger-guard are made of steel; all parts are unornamented. This is a Cooper revolver. See No. 25 for a more detailed description of this weapon. The barrel is stamped with the words "Manf'd by J.M. Cooper & Co., Pittsburgh, Pa., Patd. Apr. 25, 1854. Patd. Jan. 7, 1851. Reissd. July 26, 1859." The original patentee of this revolver was James Maslin Cooper of Pittsburgh, Pa. Various parts of the weapon are stamped with the number "131."

United States.

No. 24. (N4546). PERCUSSION REVOLVER. Caliber .31. Four-inch, octagonal, rifled barrel. There are six chambers in the cylinder. Trigger-guard and back-strap are made of brass. This is another example of the Cooper double-action revolver. See No. 25 for further details. The barrel is marked "Cooper Firearms Mfg. Co., Frankford, Phila., Pa. Pat. Jan. 7 1851, Sept. 4 1860, Sept. 1 1863, Sept. 22 1863." The revolver number is "11128."

United States.

No. 25. (N4549). PERCUSSION REVOLVER. Caliber .32. Four-inch, octagonal, rifled barrel. Has five chambers in the cylinder. Trigger-guard and back-strap are made of brass. This is a Cooper double-action revolver, the peculiarities of which are the improvements: for locking the cylinder, except when being rotated, for retaining the exploded cap on the tube until rotated past the hammer, and longitudinal grooves in the cylinder-pin for holding oil and collecting dirt. The top of the barrel is stamped "Cooper Firearms Mfg. Co., Frankford, Phila., Pa.," followed by patent dates precisely as in specimen No. 24. The revolver number is "2971."

United States.

No. 26. (N4547). PERCUSSION REVOLVER. Caliber .36. Three-and-seven-eighths-inch, octagonal, rifled barrel. Has five chambers, a lever ramrod, a brass trigger-guard and a back-strap of the same metal. This is a double-action Cooper revolver which possesses all the peculiarities of that weapon. (See No. 25.) The top of the barrel is marked "Cooper Firearms Mfg. Co., Frankford, Phila., Pa.", followed by patent dates precisely as in specimen No. 24. The revolver numbers are "14885" and "326227."

United States.

No. 27. (N4536). PERCUSSION REVOLVER. Caliber .31. Four-inch, octagonal, rifled barrel. There are five chambers in the cylinder. The rear sight is in the form of a notch cut into the hammer. The cylinder is engraved with a scene depicting a stage-coach holdup; this scene was also represented upon contemporaneous Colt revolvers. As a matter of fact, the Manhattan Fire Arms Company, who made the revolver here shown, manufactured imitation Colts. The body is engraved with scrolls, and all steel parts were formerly blued. Stamped upon the top of the barrel of this single-action revolver are the words "Manhattan Fire Arms Mf'g. Co., New York." The bottom of the frame bears the words "Patented Dec. 27, 1859," and the number "2596."

United States.

No. 28. (N4534). PERCUSSION REVOLVER. Caliber .31. Five-inch, octagonal, rifled, steel barrel. Has five chambers in a partly fluted cylinder. The rear sight is in the form of a notch cut into the hammer. The body is engraved with scrolls. Stamped upon the top of this single-action revolver are the words "Manhattan Fire Arms Manuf'g. Co., New York." The number upon the weapon is "994."

United States.

No. 29. (N4542). PERCUSSION REVOLVER. Caliber .38. Six-and-one-half-inch, octagonal, rifled barrel. There are five chambers in the cylinder. Equipped with a brass trigger-guard and a brass bound grip. The cylinder is engraved with scenes from the Civil War, and the barrel was formerly blued. The revolver is single-action and the top of the barrel is marked "Manhattan Fire Arms Co., Newark, N.J. Patented March 8, 1864." The revolver number is "54412."

United States.

No. 30. (N4545). PERCUSSION REVOLVER. Caliber .38. Five-inch, octagonal, rifled barrel. There are five chambers. The cylinder is engraved with scenes from the Civil War. Equipped with a brass trigger-guard and a brass bound grip. Pulling back the hammer revolves the cylinder and cocks the revolver. The top of the barrel is stamped "Manhattan Fire Arms Co., Newark, N.J. Patented March 8, 1864." Upon the cylinder are the words "Patented Dec. 27, 1859." The revolver number is "67048."

United States.

No. 31. (N4123). PERCUSSION REVOLVER. Caliber .45. Eight-inch, round, rifled barrel. Has six chambers in the cylinder. By removing the thumb screw in the frame behind the cylinder, the revolver can be "broken" by grasping the grip in the left hand and forcing the barrel down with the right. This causes the cylinder to be released from the frame, and renders the examination of the inside of the barrel easy. The frame of the revolver is stamped "Starr Arms Co., New York, Starr's Patent Jan. 15, 1856," and the number is "33436."

United States.

No. 32. (N4124). PERCUSSION REVOLVER. Caliber .41. Six-inch, round, rifled barrel. Has six chambers in a plain cylinder. This revolver is opened or "broken" in the same way as No. 31 described above. However, unlike the preceding Starr revolver, the one shown here is semi-double-acting. Upon pulling the large trigger, the cylinder revolves and brings the arm to full-cock. To fire, the small secondary trigger behind the first, is then pulled. The frame of the revolver is stamped "Starr Arms Co., New York, Starr's Patent Jan. 15, 1856," and the revolver numbers are "2207" and "2180."

United States.

No. 33. (N4125). PERCUSSION REVOLVER. Caliber .45. Six-inch, round, rifled barrel. The cylinder has six chambers. All parts are of steel, and were formerly blued. The revolver is opened or "broken" as No. 31. The revolver may be used as a double-action weapon by pulling the large trigger which cocks, revolves and fires at a single action; or the arm can be cocked by pulling back the hammer and then firing by means of the small trigger which is almost concealed in the frame behind the large one. The frame of the revolver is stamped "Starr Arms Co., New York, Starr's Patent Jan. 15, 1856," and the number is "14308."

United States.

No. 34. (N4122). PERCUSSION REVOLVER. Caliber .44. Seven-and-one-half-inch, octagonal, rifled barrel. There are six chambers in the cylinder. All parts are of steel, and the barrel and frame are heavily blued. This is a United States army revolver and bears the government condemnation stamp. The arm is single-action. Upon the top of the barrel are the words "Rogers & Spencer, Utica, N.Y." The left side of the frame, the under-side of the barrel, and also the cylinder and butt-plate, are stamped with the number "3279."

United States.

No. 35. (N4557). PERCUSSION REVOLVER. Caliber .44. Seven-and-one-half-inch, octagonal, rifled barrel. Has six chambers. All metal parts are made of steel and are unornamented. This arm is known as a "Pettengill Army Revolver." (See No. 36.) The United States bought 5,000 of these at $20.00 apiece. Stamped upon the top of the frame of this double-action revolver are the words "Pettengill's Patent 1856" and "Raymond & Robitaille Patented 1858." The revolver number is "3213." Many, if not all, of the Pettengill revolvers were made by Rogers Spencer and Company, in Willow Vale, New York.

United States.

No. 36. (N4558). PERCUSSION REVOLVER. Caliber .44. Five-and-one-half-inch, octagonal, rifled barrel. There are six chambers in the cylinder. All metal parts are of steel, made rust-proof with a white metal. The cylinder is partly fluted, and the chambers are numbered from one to six. The larger varieties of this revolver, of which this is an example, were known as "Pettengill's army revolvers." Navy and pocket sizes were also made. The hammer is concealed, and the whole mechanism has been termed complicated and frail. The trigger cocks, revolves, and fires the weapon at a single action. The top of the frame is marked "Pettengill's Patent 1856" and "Raymond & Robitaille Patented 1858." In addition to this, the top of the barrel is stamped with the word "Queen." The revolver number is "3438."

United States.

No. 37. (N4559). PERCUSSION REVOLVER. Caliber .31. Four-and-one-half-inch, octagonal, rifled barrel. Has six chambers. All metal parts are of steel, and the barrel is blued. The handle of the lever ramrod is split, and serves as a spring to retain it under the barrel. This, another example of the "Pettengill revolver," was probably intended for pocket or household use. (See Nos. 35 and 36.) The top of the frame is stamped "Pettengill's Patent 1856," and the bottom near the trigger-guard is marked "Raymond & Robitaille Patented 1858." The revolver number is "179."

United States.

No. 38. (N4548). PERCUSSION REVOLVER. Caliber .30. Four-and-one-quarter-inch, round, rifled barrel. Has six chambers. A groove on top of the frame serves as the rear sight. Trigger-guard is made of brass. The top of the barrel of this single-action revolver is stamped "Western Arms Co., New York." The revolver numbers are "17" and "11717."

United States.

No. 39. (N4552). PERCUSSION REVOLVER. Caliber .32. Four-and-one-eighth-inch, round, rifled barrel. Has six chambers, a lever ramrod, a front sight, and a rear sight in the form of a groove on top of the frame. The cylinder is blued. Stamped upon the top of this single-action revolver are the words "Address W. Irving, 20 Cliff St., N. Y." The revolver numbers are "19" and "3049."

United States.

No. 40. (N4537). PERCUSSION REVOLVER. Caliber .31. Four-and-one-quarter-inch, octagonal, rifled barrel. Has five chambers in a fluted cylinder. All steel parts were formerly blued. The arm is single-action. The top of the barrel bears the words "The Union Arms Co.," and the serial number is "6307."

United States.

No. 41. (N6769). PERCUSSION REVOLVER. Caliber .30. Three-and-three-quarters-inch, octagonal, rifled barrel. The revolver is double-action, and the hammer is placed slightly to the right of the center. Body and cylinder are engraved with floral designs. Upon the hammer are the words "Patented Aug. 1 and April 28." Year not given. There are no other marks of identification.

United States.

No. 42. (N4126). PERCUSSION REVOLVER. Caliber .44. Eight-inch, octagonal, rifled barrel. There are five chambers in the cylinder. All metal parts are made of steel, and were formerly blued. Has a curved side-hammer which, however, strikes through the center of the frame. The grip is oval and checkered. The cylinder-pin withdraws from the rear. This type of revolver was invented by Benjamin F. Joslyn. Its peculiarities are described "as a series of parts in such relation to the hammer-shaft as to enable it to operate upon them to cause revolution of the cylinder the desired distance and to lock it while in communication with the barrel." These revolvers were used in the United States army and navy, although this one is not designated with any marks of these services. The top of the barrel bears the words "B.F. Joslyn, Pat'd. May 4th, 1858." The revolver is stamped with the number "1047" and the ramrod with "1001." This piece was probably made in Worcester, Mass.

United States.

No. 43. (N4127). PERCUSSION REVOLVER. Caliber .36. Seven-and-one-eighth-inch, octagonal, rifled barrel. There are six chambers in the cylinder. Has a lever ramrod and an enormous trigger-guard. The hammer is operated through the frame. In this type of revolver an attempt was made to get a gas-tight joint between barrel and cylinder, so each chamber is reamed at the mouth and fits, in its turn, over the rear end of the barrel. Pulling with the second finger on the ringed trigger draws back the cylinder from the barrel, rotates, and also cocks the hammer. Releasing the lever lets the cylinder move forward to fit over the barrel, and also leaves the weapon cocked. The top of the barrel is stamped "Savage R.F.A. Co., Middletown, Ct. H.S. North Patented June 17, 1856, January 18, 1859, May 15, 1860." This piece was probably used in the navy and is marked with the numbers "580" and "140." It was made by the Savage Revolving Fire Arms Company.

United States.

No. 44. (N3505). PERCUSSION REVOLVER. This revolver is similar to No. 46, except for the revolver number, which in the specimen here illustrated is "1127."

United States.

No. 45. (N4540). PERCUSSION REVOLVER. Caliber .36. Seven-and-one-half-inch, octagonal, rifled barrel. Has six chambers.

Equipped with a lever ramrod, and a brass trigger-guard and back-strap. All steel parts were formerly blued. The grip is made of ivory. Engraved upon the cylinder is a picture representing a naval engagement, the illustration being signed "W.L. Ormsby, SC, New Orleans, April 1862." The same illustration and designer's name appears upon the similar Colt revolvers of the same period. Stamped upon the top of the barrel of this single-action revolver are the words "Metropolitan Arms Co., New York." This company was established in 1859 and manufactured imitation Colt revolvers. Various parts of the revolver are marked with the number "6211."

United States.

No. 46. (N4128). PERCUSSION REVOLVER. Caliber .44. Seven-and-one-half-inch, round, rifled barrel. There are six chambers in the cylinder. Has a center-hammer and a lever ramrod. This is a Freeman revolver, invented by A. T. Freeman of Binghamton, N. Y. The chief feature of this weapon is a two-part cylinder pin, so made that neither part can drop out. The revolver is single-action and the top of the barrel is marked "Freeman's Pat. Dec. 9, 1862. Hoard's Armory, Watertown, N. Y." The revolver number is "1432."

United States.

No. 47. (N4238). PERCUSSION REVOLVER. Caliber of upper barrel, .41, of lower barrel, .65. Six-and-one-half-inch, octagonal, rifled upper barrel. The lower barrel is unrifled and is five inches in length. All metal parts are of steel and are unornamented. The grip is made of checkered wood. The butt is fitted with an eye which serves as an attachment for carrying the revolver at the belt. This type of weapon was invented by Dr. (Colonel) Alexandre François Le Mat of New Orleans, by whom it was patented in 1856. The cylinder has nine chambers which fire their bullets through the rifled upper barrel. The lower barrel serves as an axis on which the cylinder turns, and is intended to use a charge of buckshot. The nose of the hammer is hinged, adjusting to discharge either barrel. These revolvers were made in France and England, and found considerable use during the Civil War. Later they were made to take metallic cartridges. The top of the barrel is engraved "Col. LeMat Br' e s. g. dg. Paris," and the revolver number is "1121."

France.

No. 48. (N4237). PERCUSSION REVOLVER. Caliber of upper barrel, .44; of lower barrel, .65. Seven-inch, half-octagonal, rifled, steel upper barrel. The lower barrel is unrifled and is five-and-one-eighth-inches in length. All metal parts are made of steel, and finely blued, with the exception of the ramrod and hammer. Has an oval, checkered grip. The butt is fitted with a ring which serves as an attachment for carrying the revolver at the belt. This is another LeMat revolver. (See No. 47.) Stamped upon the top of the barrel are the words "Col. Le Mat's Patent." This specimen was made for Slidell & Buregard of Charleston, S. C., for the Southern Confederacy. The revolver number is "10."

France.

No. 49. (N3300). PERCUSSION REVOLVER. Caliber of upper barrel, .44; of lower barrel, .65. Six-and-five-eighths-inch, octagonal, rifled upper barrel. The lower barrel is unrifled and is five inches in length. All metal parts are of steel, and were formerly browned. The cylinder has nine chambers, and fires bullets through the upper rifled barrel. The lower barrel serves as an axis on which the cylinder turns and fires buckshot. This is another LeMat revolver. (See No. 47.) The specimen here shown was used by the Confederates in the Civil War but neither maker's name nor revolver number are to be found upon this piece.

France?

No. 50. (N4240). PERCUSSION REVOLVER. Caliber of upper barrel, .36; of lower barrel, .50. Four-and-one-quarter-inch, octagonal, rifled upper barrel. The lower barrel is unrifled and is two-and-three-quarters-inches in length. There is a cylinder in the butt for carrying percussion-caps. The cylinder has nine chambers for bullets while the central barrel, which at the same time serves as an axis on which the cylinder revolves, fires buckshot. This is another LeMat revolver. (See No. 47.) There is no maker's name given upon the specimen shown here. The revolver number is "72."

France?

PLATE 90.

RIM-FIRE BREECH-LOADING PISTOLS

No. 1. (N4879). RIM-FIRE PISTOL. Caliber .22. Three-and-one-half-inch, half-octagonal, rifled barrel. This pistol fires a metallic, rim-fire cartridge; has a brass body, a smooth, oval grip, but no ramrod or trigger-guard. Pulling the small, steel projection beneath the frame toward the trigger, causes the barrel to tip downward so that it may be loaded or the empty shell removed. The pistol is not equipped with a self-ejector. The top of the barrel is stamped "Frank Wesson, Worcester, Mass., Pat'd. Oct. 25, 1859 & Nov. 11, 1862." The pistol number is "9336."

United States.

No. 2. (N3311). RIM-FIRE PISTOL. Caliber .22. Three-and-one-half-inch, half-octagonal, rifled barrel. This pistol, which fires a rim-fire, metallic cartridge, has a nickel-plated barrel and body. Pressing a small, steel button on the left side of the frame causes the barrel to tip downward so that it may be loaded or the empty shell removed. This pistol was probably formerly equipped with a self-ejector. Stamped upon the top of the barrel are the words "J. Stevens & Co., Chicopee Falls, Pat. Sept. 6, 1864." The pistol number is "3767."

United States.

No. 3. (N5789). RIM-FIRE PISTOL. Caliber .22. Three-and-one-half-inch, half-octagonal, rifled barrel. Barrel and body are nickel-plated. Has a smooth, oval grip. Formerly there was a small button on the left side of the frame which when pressed caused the barrel to tip downward so that it could be loaded. (See No. 2.) The pistol is provided with a self-ejector which removes the cartridge from the barrel when the latter is tipped downward. The top of the barrel is marked "J. Stevens & Co., Chicopee Falls, Pat. Sept. 6, 1864," and the pistol number is "17247."

United States.

No. 4. (N4596). RIM-FIRE PISTOL. Caliber .22. Three-and-one-half-inch, half-octagonal, rifled barrel. The body is of brass but was probably once nickel-plated. The action and markings upon the Stevens pistol shown here are the same as those of No. 3. The pistol number is "18070."

United States.

No. 5. (N4597). RIM-FIRE PISTOL. Caliber .22. Three-and-one-half-inch, half-octagonal, rifled barrel. The body is of brass and is nickel-plated. This is another Stevens pistol similar in action and markings to No. 3. The pistol number is "24351."

United States.

No. 6. (N6851). RIM-FIRE PISTOL. Caliber .28. Three-inch, half-octagonal, rifled barrel, which was formerly blued. Equipped with a small, crescent-shaped trigger. There formerly was a button on the left side of the frame which, when pressed, caused the barrel to tip downward so that it could be loaded or so that the empty shell could be removed. There is no self-ejector. The grip is small and flat and the top of the barrel is marked "Stevens & Co., Vest Pocket Pistol, Chicopee Falls, Mass." The pistol number is "125."

United States.

No. 7. (N4880). RIM-FIRE PISTOL. Caliber .22. Three-and-one-half-inch, octagonal, rifled barrel, which was formerly blued, and which is equipped with front and rear sights. The body is of engraved brass. Pressing down upon a catch, immediately in front of the hammer, causes the barrel to tip downward so that a cartridge may be inserted in the breech. There is no self-ejector. The top of the barrel is marked "T.J. Stafford, New Haven, Ct.", and the butt-plate is stamped "Patented March 19, 1860." The pistol number is "1963."

United States.

No. 8. (N4878). RIM-FIRE PISTOL. Caliber .32. Three-and-one-half-inch, octagonal, rifled barrel, which is provided with a front sight. A slit in the hammer serves as a back sight. The frame is of brass and silver-plated. Pressing the catch on the frame beneath the barrel enables the latter to slide forward for loading, as is shown in the illustration. The cartridge fits into the groove immediately behind the hammer. This groove also acts as an extractor when the barrel is drawn forward. Stamped upon the top of the barrel are the words "L.B. Taylor & Co., Chicopee, Mass." Barrel and butt are marked with the number "1153."

United States.

No. 9. (N4608). RIM-FIRE PISTOL. Caliber .32. Three-and-three-quarters-inch, octagonal, rifled barrel which is fitted with a front sight. There is no rear sight. This pistol has a brass body and a smooth, oval grip. Pressing the small button beneath the frame enables the barrel to be swung to the side for the removal of the empty cartridge and the insertion of a new one. There is no self-ejector. No maker's name is given on any part of the pistol which probably was made by Frank Wesson. The inside of the back-strap bears the number "499."

United States.

No. 10. (N4609). RIM-FIRE PISTOL. Caliber .30. Three-and-five-eighths-inch, octagonal, rifled barrel which is fitted with a front sight. A groove in the ear of the hammer serves as a rear sight. Has a brass body which is nickel-plated; the barrel being also plated with the same metal. Pressing the small button beneath the frame enables the barrel to be swung to the side for the removal of the empty cartridge and the insertion of a new one. The pistol has no self-ejector. Stamped upon the top of the barrel are the words "Merwin & Bray, New York," and on the inside of the back-strap is the number "744."

United States.

No. 11. (N4610). RIM-FIRE PISTOL. Caliber .22. Three-and-one-quarter-inch, octagonal, rifled barrel. This is another rim-fire pistol as made by Merwin and Bray, and is similar in action and markings to No. 10. The breech is shown closed. Stamped upon the inside of the back-strap is the number "403."

United States.

No. 12. (N4598). RIM-FIRE PISTOL. Caliber .40. Two-and-seven-eighths-inch, half-octagonal, rifled barrel. There is no rear sight, ramrod, or trigger-guard. Pulling back the roughened catch in the frame toward the muzzle, causes the barrel to tip downward, as shown, so that a cartridge may be inserted in the breech. Equipped with a self-ejector. The top of the barrel is marked "Ballard's" and the left side is stamped "Ballard's Worcester, Mass., Pat'd. June 22, 1867." The pistol number is "899."

United States.

No. 13. (N4621). RIM-FIRE PISTOL. Caliber .32. Three-inch, half-octagonal, rifled barrel. Fitted with front and rear sights. The barrel swings to the right, and a lever arrangement beneath it operates the ejector. Stamped upon the left side of the barrel are the words "Allen & Wheelock, Worcester, Mass." The pistol number is "59."

United States.

No. 14. (N4619). RIM-FIRE PISTOL. Caliber .32. Five-inch, octagonal, rifled barrel. The barrel swings to the right and a lever arrangement beneath it operates the ejector. The action is the same as that of No. 13. The left side of the barrel is marked "Allen & Wheelock, Worcester, Mass.," and the pistol number is "626."

United States.

No. 15. (N4620). RIM-FIRE PISTOL. Caliber .32. Three-and-three-quarters-inch, half-octagonal, rifled barrel. Fitted with front and rear sights. The barrel swings to the right, and a lever arrangement beneath it operates the ejector. Stamped upon the left side of the barrel are the words "E. Allen & Co., Worcester, Mass., Allen's Pat. Mch. 7, 1865," and on the bottom is the pistol number "1099."

United States.

No. 16. (N4623). RIM-FIRE PISTOL. Caliber .22. Three-and-one-eighth-inch, half-octagonal, rifled barrel. Fitted with front and rear sights. The action and markings on this pistol are the same as those of No. 15. The bottom of the barrel bears the pistol number "13282." Made by E. Allen and Company of Worcester, Mass.

United States.

No. 17. (N4622). RIM-FIRE PISTOL. Caliber .22. Three-inch, half-octagonal, rifled barrel. Has front and rear sights. All metal parts are made of steel, and are rust-proofed. The barrel swings to the right, as shown, and a lever arrangement beneath it operates the ejector. Fitted with a smooth, wooden grip. The left side of the barrel bears the words "Forehand & Wadsworth, Worcester, Mass., Pat. Mch. 7, 1865." The pistol number is "2029."

United States.

No. 18. (N4615). RIM-FIRE PISTOL. Caliber .22. Two-and-one-half-inch, round, smooth-bore barrel. There are no sights. The grip is oval in shape. All metal parts were formerly nickel-plated. When the hammer is brought to full-cock the barrel may be swung to the right in order to load or to extract the empty cartridge. There is no self-ejector. This type of pistol was sometimes called a "vest-pocket pistol." No maker's name is given upon this specimen but it bears the number "279."

United States.

No. 19. (N6787). RIM-FIRE PISTOL. Caliber .22. Two-and-three-eighths-inch, round, smooth-bore barrel. Has no sights. Fitted with an oval, home-made grip. When the hammer is brought to full-cock, the barrel may be swung to the right in order to load or to extract the empty cartridge. There is no self-ejector. No maker's name is given, and the pistol number is missing.

United States.

No. 20. (N2776). RIM-FIRE PISTOL. Caliber .41. Two-and-one-half-inch, octagonal, rifled barrel. There is a sight in front, but none in the rear. Has a brass body, and a smooth, oval grip. Pressing a small button beneath the frame enables the barrel to be swung to the right for the removal of the empty cartridge and the insertion of a new one. This pistol has no self-ejector. The top of the barrel bears the word "Southerner" while the left side is stamped "Brown Mf'g. Co., Newburyport, Mass., Pat. Apr. 9, 1867." The pistol number is "2236."

United States.

No. 21. (N4611). RIM-FIRE PISTOL. Caliber .41. Two-and-one-half-inch, octagonal, rifled barrel. The body is of brass while the barrel is of blued steel. Action, sights and other characteristics of this Brown pistol are the same as those of No. 20. Shown open. The markings upon the barrel are the same as those of No. 20, but the pistol number is "9977."

United States.

No. 22. (N4612). RIM-FIRE PISTOL. Caliber .28. Three-and-one-quarter-inch, half-octagonal, rifled barrel. This pistol is provided with a front sight while a groove in the hammer serves as a rear sight. The body is of brass, and the grip is made of smooth rosewood. Pressing a small button beneath the frame enables the barrel to be swung to the right for the removal of the empty cartridge and the insertion of a new one. Not fitted with a self-ejector. The top of the barrel is stamped "Cowles & Son, Chicopee, Mass.," and the pistol number is "1206."

United States.

No. 23. (N4617). RIM-FIRE PISTOL. Caliber .22. Two-and-three-quarters-inch, half-octagonal, rifled barrel. Has a front sight, but no rear sight, ramrod, or trigger-guard. The body is made of brass, and the grip is smooth and flat. Pressing the small button beneath the frame enables the barrel to be swung to the side for the removal of the empty cartridge and the insertion of a new one. There is no self-ejector. The right side of the barrel bears the words "J.M. Marlin, New Haven, Ct.", while the top is stamped with the letters "O.K." The pistol number is "2389."

United States.

No. 24. (N4616). RIM-FIRE PISTOL. Caliber .32. Two-and-three-quarters-inch, round, smooth-bore barrel. Has no ramrod, trigger-guard or sights. All metal parts are nickel-plated. When the hammer is brought to full-cock the barrel may be swung to the right in order to load, or to extract the shell. There is no self-ejector. No maker's name is given, but the inside of the grip is stamped with the number "179."

United States.

No. 25. (N4613). RIM-FIRE PISTOL. Caliber .22. Two-and-three-quarters-inch, round, smooth-bore barrel. There is a sight in front, but none in the rear. The body is of brass and, like the barrel, was formerly nickel-plated. Pressing a small button beneath the frame enables the barrel to be "broken" to the side. There is no self-ejector. The top of the barrel bears the name "Pointer" but no maker's name is given. The pistol number is "12."

United States.

No. 26. (N4618). RIM-FIRE PISTOL. Caliber .22. Two-and-one-half-inch, round, smooth-bore, brass barrel. There are no sights. All metal parts are made of brass, except the hammer, which is of steel. Has a flat, oval grip. When the hammer is brought to full-cock, the barrel may be swung to the right for loading or extraction. Not equipped with a self-ejector. The left side of the barrel is stamped with the word "King Pin," but no maker's name is given. Stamped upon the inside of the grip is the number "601."

United States.

No. 27. (N4614). RIM-FIRE PISTOL. Caliber .22. Two-and-one-quarter-inch, round, smooth-bore barrel. Has a flat, unpolished, oval grip. There are no sights. The body is of brass and was formerly nickel-plated. When the hammer is brought to full-cock, the barrel may be swung to the right for loading and extraction. Not equipped with a self-ejector. This pistol is stamped with the number "84" but no maker's name is given.

United States.

No. 28. (N4877). RIM-FIRE PISTOL. Caliber .32. Three-and-seven-eighths-inch, octagonal, rifled barrel. Has a smooth, oval grip, and a brass body, the latter was formerly nickel-plated. Pushing the small screw, on the left side of the barrel, in front of the hammer, toward the muzzle, releases a catch which enables the barrel to be swung to the right in a vertical plane. The pistol may now be loaded or the empty shell may be removed by means of the rack and pinion ejector operated by a lever on the under-side of the barrel. No maker's name is given on this pistol but it was patented by R. L. and J. Eichinson of Springfield, Mass., and made by Wesson and Harrington, probably in Hartford, Conn. The pistol number is "225."

United States.

No. 29. (N4602). RIM-FIRE PISTOL. Caliber .32. Four-inch, round, rifled barrel. All metal parts are made of steel, and the barrel was formerly blued. Has a smooth, oval grip. Turning the barrel to the left and drawing it forward ejects the empty shell, and allows a new one to be inserted. The top of the barrel bears the words "Rupertus Pat'd. Pistol Mfe. Co., Philadelphia." The number "6" is stamped on the inside of the grip.

United States.

No. 30. (N4605). RIM-FIRE PISTOL. Caliber .38. Two-and-one-quarter-inch, rifled barrel having a V-shaped ridge on top. The barrel is slightly engraved and decorated with an arrow on each side pointing toward the muzzle. The body is of brass and is ornamented with scrolls at the sides, and checkering at the butt. Has a front sight; a slit in the hammer serves as a rear sight. Bringing the hammer to half-cock, and drawing the small stud on the right side of the body to the rear, releases a catch which enables the barrel to be turned to the left side for loading or unloading. See No. 31. There is no self-ejector. The top of the barrel is marked "Moore's Pat. F. A. Co., Brooklyn, N.Y.", and the under-side, "D.Moore's Pat. Feb. 24, 1863," together with the number "1133."

United States.

No. 31. (N6794). RIM-FIRE PISTOL. Caliber .38. Two-and-one-quarter-inch, rifled barrel having a V-shaped ridge on top. This is another example of the Moore pistol. It is shown "broken" and is identical in all respects to No. 30, except with regard to the pistol number, which is "3424" in the specimen here shown.

United States.

No. 32. (N4606). RIM-FIRE PISTOL. Caliber .38. Two-and-one-quarter-inch, rifled barrel. The action and ornamentation of this Moore pistol are the same as those of No. 30. The top of the barrel is stamped "D. Moore Patent Feb. 19, 1861." No pistol number is given.

United States.

No. 33. (N777). RIM-FIRE PISTOL. Caliber .38. One-and-three-quarters-inch, rifled barrel having a V-shaped ridge on top. The barrel is slightly engraved. The body is of brass, and is ornamented with scrolls at the sides, and checkering at the butt. Sights are the same as in No. 30. The action is identical to that specimen also. Stamped upon the top of the barrel are the words "National Arms Co." while the bottom is marked with the number "1798." This pistol was made in Brooklyn, N. Y.

United States.

No. 34. (N4603). RIM-FIRE PISTOL. Caliber .38. Two-and-one-quarter-inch, rifled barrel. The barrel is slightly engraved, and was formerly covered with another metal, probably silver. The grip is made of wood and not of brass as in similar pistols of this kind. The action is the same as that of No. 30. The top of the barrel is stamped "National Arms Co., Brooklyn, N.Y.", and the bottom bears the number "1138."

United States.

No. 35. (N3836). RIM-FIRE PISTOL. Caliber .38. Two-and-one-quarter-inch, rifled barrel. The body is of steel and slightly engraved; the grip is of checkered wood. There is no ornamentation upon the barrel. The action and sights are the same as those of No. 30. The top of the barrel is stamped "National Arms Co., Brooklyn, N.Y.", while the bottom bears the number "10253."

United States.

No. 36. (N4604). RIM-FIRE PISTOL. Caliber .38. Two-and-one-quarter-inch, rifled barrel which is slightly engraved with scrolls. The body is of brass, and is also slightly engraved. Has a front sight; a deep groove in the hammer serves as a back sight. The action is the same as that of No. 30. The top of the barrel is stamped "National Arms Co., Brooklyn, N.Y.", and the bottom with the number "2712."

United States.

No. 37. (4592). RIM-FIRE PISTOL. Caliber .41. Four-inch, octagonal, rifled, steel barrel having a V-shaped ridge on top.

The grip is made of checkered ebony, and the barrel is heavily blued. Pressing the rough, steel button upon the upper part of the frame releases a catch whereby a portion of the breech may be turned to the left side exposing the breech of the barrel. A cartridge can then be inserted. The movable portion of the breech also operates the ejector, and contains the firing-pin which transmits the motion of the hammer to the rim of the cartridge. This pistol was invented by Henry Hammond and is sometimes called the "Bull Dozer." The top of the barrel is stamped "Connecticut Arms & Manf'g. Co., Naubuc, Conn.", and the breech "Patented Oct. 25, 1864." The inside of the grip bears the number "2639."

United States.

No. 38. (N4888). RIM-FIRE PISTOL. Caliber .22. Three, three-and-one-eighth-inch, superposed, rifled, steel barrels. The left side of the barrels is equipped with a sliding knife-bayonet which is shown exposed in the illustration. Has a small, smooth, wooden grip. Turning the flat, steel catch, in front of the hammer, to the right, causes the barrels to drop downward so that they can be loaded or the empty shells removed. See No. 40. There is no self-ejector. An indicator on the right side of the frame shows the number of barrels which have been fired. Stamped upon the left side of the frame are the words "Wm. W. Marston, Patented. May 26, 1857, New York City." The left side of the barrels is stamped with the number "773."

United States.

No. 39. (N4886). RIM-FIRE PISTOL. Caliber .26. Three, four-inch, superposed, rifled barrels. The barrels are blued, and the body is made of brass. Has a small, smooth, wooden grip. Each barrel is fired singly by a firing-pin, inside of the frame, which advances behind the breech of the next higher barrel after each shot. The device on the right side of the frame numbered 0 1 2 3 indicates which barrels have been fired. The barrels are opened or "broken" as those of No. 38. No maker's name is given upon the pistol but it was made by Wm. W. Marston. No pistol number is given.

United States.

No. 40. (N4887). RIM-FIRE PISTOL. Caliber .32. Three, three-and-one-eighth-inch, superposed, rifled barrels which were formerly blued. Action and method of opening or "breaking" are the same as in No. 39. The three-pronged device on the right side of the frame acts as an extractor.   Stamped upon the left side of the frame are the words "Wm. W. Marston, Patented May 26, 1857, New York City, Improved 1864."   The pistol number is "2614."

United States.

No. 41.   (N4885).   RIM-FIRE PISTOL.   Caliber .32.   Three, four-inch, superposed, rifled barrels.   All metal parts are nickel-plated. The action and the method of opening or "breaking" are the same as those of No. 39.   A three-pronged device on the right side of the frame acts as an extractor.   The markings are the same as those on No. 40, and pistol number is "275."

United States.

No. 42.   (N3832).   RIM-FIRE PISTOL.   Caliber .32.   Two, three-inch, superposed, octagonal, rifled barrels.   Has a front sight, but none in the rear.   The barrels are blued, while the body is of brass, silver-plated.   Has a polished, rosewood grip.   Releasing a small catch in front of the trigger makes it possible to rotate the barrels to the left or right to load or unload, or to bring the loaded one in front of the hammer.   Has no self-ejector.   The barrels are stamped "American Arms Co., Boston, Mass., Wheeler's Pat., Oct. 31, 1865-June 19, 1866." The butt-plate is marked with the number "4964."

United States.

No. 43.   (N4883).   RIM-FIRE PISTOL.   Caliber of one barrel .22; of the other .32.   Two, three-inch, superposed, octagonal, rifled, steel barrels.   This is another Wheeler pistol.   It is shown open.   In all other respects it is similar to No. 42.   The butt-plate is stamped with the number "1888."

United States.

No. 44.   (N779).   RIM-FIRE PISTOL.   Caliber .28.   Two, two-and-one-eighth-inch, round, smooth-bore barrels.   Has front and back sights.   The grip is rather large and is checkered.   The barrels unscrew from the frame for loading.   There is no self-ejector.   Each barrel is marked with two English proof-marks.   Upon the left side of the frame are the words "Woodward's No. 490 Reg. Feb. 4, 1863."   The rib of the barrels and the butt-plate are stamped with the number "490."

England.

No. 45.   (N4893).   RIM-FIRE PISTOL.   Caliber .32.   Four, two-and-three-quarters-inch, rifled barrels.   The trigger is in the form of a roughened, steel button unprotected by a guard.   The body is of brass.   At each pull of the hammer the striker on the face revolves and strikes the rim of a cartridge.   This is continued until all four cartridges have been fired in succession.   Pressing the catch on the left side of the frame allows the barrels to swing downward in order to load.   There is no self-ejector.   The left side of the frame is stamped with the words "Starr's Pat's. May 10, 1864."   Barrel and frame are marked with the number "12."   This pistol was made in New York City.

United States.

No. 46.   (N4891).   RIM-FIRE PISTOL.   Caliber .41.   One, two-and-three-quarters-inch, round, rifled barrel.   Trigger is the same as that of No. 45.   The body is of brass, and was formerly plated; the grip is round, checkered and made of wood.   Pressing the catch on the left side of the frame allows the barrels to swing downward, as shown in the illustration.   The shell is extracted by pulling back the ejector which constitutes part of the breech.   The hammer strikes a firing-pin which in turn fires the cartridge.   The left side of the frame is marked "Starr's Pats. May 10, 1864."   Barrel and frame are stamped with the number "5."   This pistol was made in New York City.

United States.

No. 47.   (N4892).   RIM-FIRE PISTOL.   Caliber .32.   Four, three-inch, rifled, steel barrels.   Trigger is the same as that of No. 45. The body is of brass, plated with silver.   Has a smooth, polished, wooden grip.   The action on firing is the same as that of No. 45.   There is no self-ejector.   The left side of the frame is stamped with the words "Starr's Pat's. May 10, 1864."   Barrel and frame are stamped with the pistol number "1."   Like other pistols of this kind, the one here shown was made in New York City.

United States.

No. 48.   (N4899).   RIM-FIRE PISTOL.   Caliber .28.   Four, three-inch, rifled, steel barrels.   Has front and rear sights.   The barrels were formerly blued.   The body is of brass and unornamented.   At each pull of the hammer the striker on the face revolves and strikes the rim of a new cartridge.   This is continued until all four cartridges have been fired in succession.   When the hammer is brought to half-cock and the small button beneath the frame is pressed, the barrels may be slid forward for loading or for the removal of the empty shells.   Right and left sides of the barrels are stamped with English proof-marks. This is an English imitation of the Sharp's pistol. No maker's name or pistol number are given.

England.

No. 49. (N2781). RIM-FIRE PISTOL. Caliber .30. Four, three-and-one-eighth-inch, rifled, steel barrels. There are no sights. All metal parts are of steel, and decorated with line-designs. The hammer has a revolving striker as that described under No. 48. There is no firing-pin intervening between the striker and the cartridge. Pressing the catch on the left side of the frame allows the barrels to swing downward in order to load. There is no ejector. This pistol, which is probably of French manufacture, bears neither maker's name or pistol number.

France?

No. 50. (N6999). RIM-FIRE PISTOL. Caliber .26. Four, three-inch, rifled barrels. All metal parts are made of polished steel which is engraved with floral designs. Has a smooth, polished, ebony grip. The striker is of the revolving variety as described under No. 48. A lever on the left side of the frame acts as a safety-device, and also releases the barrels so they may be tipped downward. See illustration. There is no self-ejector. The muzzle of the barrels is stamped with the Liége proof-mark and with another in the form of a crown over the letter "N." Frame and barrels bear the number "885." No maker's name is given.

France?

PLATE 91.

RIM-FIRE PISTOLS AND REVOLVERS

No. 1. (N4900). RIM-FIRE PISTOL. Caliber .32. Four, three-inch, rifled barrels. Has front and rear sights. The barrels are blued, and the frame is made of brass which is silver-plated. The grip is of polished rosewood. At each pull of the hammer the striker on the face revolves and strikes the rim of a new cartridge. This is continued until all four cartridges are fired in succession. When the hammer is brought to half-cock, and the small screw beneath the frame is pressed inward, the barrels may be slid forward for loading. There is no self-ejector. The right side of the barrel is stamped "C. Sharp's & Co., Philada. Pa.", while barrel and butt bear the number "4076."

United States.

No. 2. (N4898). RIM-FIRE PISTOL. Caliber .28. Four, three-inch, rifled barrels. Has front and rear sights, a brass body and checkered ebonite grip plates. The barrels were formerly blued. The action is the same as that of No. 1. The right side of the barrel is stamped "C. Sharp's & Co., Philada. Pa.", and the left side "C. Sharp's Patent, 1859." Barrel and butt bear the number "13192."

United States.

No. 3. (N3835). RIM-FIRE PISTOL. Caliber .22. Four, two-and-one-half-inch, rifled barrels. Has front and rear sights, a brass frame which was formerly silver-plated, blued barrels, and a round, rosewood grip. The action is the same as that of No. 1. The right side of the barrel is stamped "C. Sharp's & Co., Philada. Pa.", and the left side "C. Sharp's." Barrel and butt bear the number "13649."

United States.

No. 4. (N4903). RIM-FIRE PISTOL. Caliber .22. Four, two-and-one-half-inch, rifled barrels. Has front and rear sights, barrels which were formerly blued, a brass frame, and ebonite grip plates which are decorated with floral designs in relief. The action is the same as that of No. 1. The right side of the barrel is stamped "C. Sharp's & Co., Philada. Pa.", and the left side "C. Sharp's Patent 1859." Barrel and butt bear the number "811."

United States.

No. 5. (N4901). RIM-FIRE PISTOL. Caliber .32. Four, two-and-one-half-inch, rifled barrels. Has front and rear sights, steel barrels which were formerly blued, and an unornamented steel frame. The grip is of smooth, round, polished wood. The action is the same as that of No. 1. However, the stud for releasing the barrels is on the left side of the frame, and not beneath it as in No. 1. Stamped upon the right side of the frame are the words "C. Sharp's Patent Jan. 25, 1859." The pistol number is "2420."

United States.

No. 6. (N4902). RIM-FIRE PISTOL. Caliber .32. Four, three-inch rifled, steel barrels. Has front and rear sights, blued barrels, and a silver-plated brass body. The action is the same as that of No. 1. When the hammer is brought to half-cock and the small stud on the left side of the frame is pressed down, the barrels may be slid forward to load. There is no self-ejector. The right side of the barrel is marked "C. Sharp's Patent, Jan. 25, 1859," and the pistol number is "13979."

United States.

No. 7. (N4896). RIM-FIRE PISTOL. Caliber .31. Four, three-and-one-half-inch, rifled barrels. Has front and rear sights, a steel frame, and large, checkered ebonite grip plates. At each pull of the hammer the striker on the face revolves and strikes the rim of a new cartridge. This is continued until all four cartridges have been fired in succession. When the hammer is brought to half-cock and the small stud on the left side of the frame is pressed down, the barrels may be slid forward to load. Not fitted with a self-ejector. The top of the barrel is marked "Address Sharp's & Hankins, Philadelphia, Penn." The pistol number is "3374."

United States.

No. 8. (N4895). RIM-FIRE PISTOL. Caliber .32. Four, three-and-one-half-inch, rifled barrels. Has front and rear sights, blued barrels, an unornamented steel frame, and an oval polished grip. The action is the same as that of No. 7. The top of the barrel is stamped "Address Sharp's & Hankins, Philadelphia, Penn.", and the right side "C. Sharp's Patent Jan. 25, 1859." The pistol number is "12206."

United States.

No. 9. (N780). RIM-FIRE PISTOL. Caliber .32. Four, three-and-one-half-inch, rifled barrels. The grip plates are of checkered ebonite. The action is the same as that of No. 7. The markings on the barrel are the same as those of No. 8. The pistol number is "4161."

United States.

No. 10. (N4897). RIM-FIRE PISTOL. Caliber .32. Four, three-and-one-half-inch, rifled barrels. All metal parts are of steel; the barrels were formerly blued. The grip is rather large, and is made of checkered ebonite. The action is the same as that of No. 7. The markings upon the barrel are the same as those on No. 8. The pistol number is "7187."

United States.

No. 11. (N4890). RIM-FIRE PISTOL. Caliber .22. Two, two-inch, superposed, octagonal, rifled barrels. Has a front sight, nickel-plated body and barrels, and a small, oval grip, but no rear sight.

Releasing a small catch in front of the trigger makes it possible to rotate the barrels to the left or right in order to load or unload, or to bring the loaded one in front of the hammer. There is no self-ejector. The barrels are stamped "Frank Wesson, Worcester, Mass., Pt. Dec. 15/68." The pistol number is "1690."

United States.

No. 12. (N4889). RIM-FIRE PISTOL. Caliber .32. Two, two-and-one-half-inch, superposed, rifled, octagonal barrels. Has a front sight; a groove in the hammer serves as a rear sight. The barrels were formerly blued. The body is of brass; the grip is of polished rosewood. The barrels are released for loading as in No. 11. Stamped upon the barrels are the words "Frank Wesson, Worcester, Mass., Pat. Dec. 15, 1868." The pistol number "452" is stamped upon the inside of the grip.

United States.

No. 13. (N4884). RIM-FIRE PISTOL. Caliber .38. Two, three-inch, superposed, octagonal, rifled barrels. Has a polished, rosewood grip. A brass body. Has a front sight; a groove in the hammer serves as a rear sight. A knife-bayonet, one-and-one-half-inches in lenghth is carried beneath the barrels and is shown exposed in the illustration. The barrels are released for loading and extraction as in No. 11. Stamped upon the barrels are the words "Frank Wesson, Worcester, Mass., Pat. Dec. 15, 1868," and also "Pt. July 29, 1869." The number "302" is stamped on the inside of the grip.

United States.

No. 14. (N4910). RIM-FIRE REVOLVER. Caliber .22. Eight, three-inch, smooth-bore, steel barrels. Has a groove in the hammer for a rear sight, but has no front sight. Opening a gate in the rear of the barrels, to the right of the hammer, allows the insertion of cartridges or the removal of the empty ones. There is no self-ejector. This pistol is really a pepper-box adapted to the rim-fire system. It is single-action. The body is of silver-plated brass, and the grip of polished rosewood. Stamped upon the right side of the body are the words "Rupertus' Pat. Pistol Mfg. Co., Philada.", and the left side "Rupertus' Patent July 19th, 1864." Upon the butt is the number "2125."

United States.

No. 15. (N4911). RIM-FIRE REVOLVER. Caliber .22. Six, two-and-one-half-inch, rifled barrels. There are no sights. All metal parts are made of blued steel. The cylinder is revolved by pulling back the hammer. Loosening the left-handed bolt in front of the muzzles, allows the barrels, or rather the cylinder, to be removed for the insertion of cartridges or the removal of the empty ones. There is no self-ejector. This revolver, like the preceding one, is really a pepper-box adapted to the rim-fire system. It bears no maker's name, but is a Rupertus Patent Pistol, made by Tryon Brothers of Philadelphia. The pistol number is "674."

United States.

No. 16. (N4833). RIM-FIRE REVOLVER. Caliber .28. Six, one-and-three-quarters-inch, rifled barrels in a fluted cylinder. There are no sights. All metal parts are nickel-plated. A gate in the rear of the cylinder allows cartridges to be introduced into the various chambers and permits the removal of the empty ones. The folding-trigger, cocks, revolves and fires the piece at a single action. This is really a pepper-box adapted to the rim-fire system. No maker's name is given but the weapon is probably of French manufacture. The inside of the grip bears the number "4" and the letters "AD" and "ML." No proof-marks are to be found upon the revolver.

France?

No. 17. (N6845). RIM-FIRE REVOLVER. Caliber .22. Seven, one-and-one-half-inch, smooth-bore chambers in a steel cylinder. Has no sights. Body and grip are made of engraved brass. Removing the right-handed bolt in front of the muzzles, allows the cylinder to be removed for the insertion of the cartridges or the removal of the empty ones. Has no self-ejector. If the muzzle is placed in the palm of the hand, and the third finger placed through the hole in the grip, then when the weapon is enclosed by the fist the revolver serves as a "brass knuckle." The left side of the frame is stamped "My Friend." The butt bears the number "3842." No maker's name is given but this revolver was patented Dec. 26, 1865, by George C. Bunsen of Belleville, Illinois.

Donor: Frank Shrosbree.

United States.

No. 18. (N4917). RIM-FIRE REVOLVER. Caliber .22. Seven, one-and-one-half-inch, smooth-bore chambers in a steel cylinder. The body is made of engraved brass which is silver-plated. The cylinder is of blued steel. The action and method of use of this weapon is the same as No. 17. The left side of the frame is stamped "My Friend, Pat'd. Dec. 26, 1865." No maker's name is given, although this piece was made by George C. Bunsen. See No. 17. The number upon the piece is "5206."

United States.

No. 19. (N4915). RIM-FIRE REVOLVER. Caliber .22. Seven, one-and-one-half-inch, smooth-bore chambers in a steel cylinder. All metal parts are nickel-plated. Cylinder and cylinder-pin are shown removed from the frame. There is no self-ejector; the cylinder-pin being used to push out the empty shells from the chambers. Action and markings upon this weapon are the same as those of No. 18. The revolver number is "13615."

United States.

No. 20. (N5773). RIM-FIRE REVOLVER. Caliber .22. Seven, one-and-one-half-inch, smooth-bore chambers in a steel cylinder. The body is of engraved brass which was formerly plated. This is another combined revolver and "brass knuckle" similar in action to No. 18. The left side of the frame is stamped "My Friend, Pat'd. Dec. 26, 1865," and below the cylinder the words "My Friend" again appear. The revolver number is "4597."

United States.

No. 21. (N4916). RIM-FIRE REVOLVER. Caliber .32. Five, one-and-five-eighths-inch, smooth-bore chambers in a steel cylinder. The body is of engraved brass which is nickel-plated. This is a large-sized combination revolver and "brass knuckle," similar in action to No. 18. The left side of the frame is marked "My Friend, Pat'd. Dec. 26, 1865." The revolver number is "16835."

United States.

No. 22. (N4754). RIM-FIRE REVOLVER. Caliber .22. Three-inch, octagonal, rifled barrel. Has seven chambers for short, rim-fire, metallic cartridges. There is a sight in front, but none in the rear. Barrel and frame are made in one piece. This double-action revolver has a folding-trigger. The cartridges are inserted into the cylinder from the rear and one the right side of the frame. The cylinder-pin is removed by withdrawing from the front. Stamped upon the top of the barrel is name "Bismarck," while the left side bears a proof-mark in the form of a crown over the letter "G." The cylinder is

stamped with a similar proof-mark and, in addition, bears that of Liége.
No maker's name is given. The butt-plate is stamped with the number
"590."

Belgium.

No. 23. (N4818). RIM-FIRE REVOLVER. Caliber .22. Three-
inch, round, rifled barrel. Has six chambers in the cylinder. Has also
a folding-trigger, front and rear sights, and a hand-ejector, the latter
being on the left side. The revolver is double-action and bears an
undecipherable name upon the left side of the frame. The body is
marked with the letter "J," while the cylinder bears the Liége proof-
mark and another in the form of a crown over the letter "V." No
maker's name or revolver number are given.

Belgium.

No. 24. (N4670). RIM-FIRE REVOLVER. Caliber .22. Two-
inch, round, rifled, steel barrel. There are six chambers in the cylinder.
The cartridges for this double-action revolver are inserted into the
cylinder from the rear and on the right side of the frame. There is
no ejector. A safety device upon the right side of the body prevents
the hammer from striking the cartridges while they are being inserted
into the chambers. Each chamber is marked with an English proof-
mark. No maker's name or revolver number are given.

England.

No. 25. (N4836). RIM-FIRE REVOLVER. Caliber .32. Three-
and-one-quarter-inch, round, rifled barrel. There are six chambers in
the cylinder. Has a front sight, while a notch in the hammer serves
as a rear sight. The frame is of engraved brass, while the cylinder is
blued. A hand-ejector is located on the right side of the frame. Pulling back the hammer revolves the cylinder and cocks the revolver. In
order to load, it is necessary to remove the barrel and cylinder. Stamped
upon the barrel are the words "National Arms Co., Brooklyn, N. Y."
and on the cylinder "D. Williamson's Patent June 5-May 17, 1864."
The revolver number is "IC 13."

United States.

No. 26. (N3357). RIM-FIRE REVOLVER. Caliber .32. Three-
and-one-quarter-inch, round, rifled barrel. Has six chambers in the
cylinder. Fitted with a mother-of-pearl grip. All metal parts are
silver-plated. To load, it is necessary to remove the barrel and cylinder.

The barrel of this single-action revolver is stamped "Moore's Pat. Fire Arms Co., Brooklyn, N. Y." The cylinder bears the words "D. Williamson's Patent June 5-May 17, 1864," and the barrel is marked with the number "706."

United States.

No. 27. (N4644). RIM-FIRE REVOLVER. Caliber .25. Two-and-one-eighth-inch, octagonal, rifled barrel. Has six chambers in the cylinder. All metal parts are made of steel, and unornamented. The cartridges are put into the chambers from the rear end, and then covered with a metal disc. The cylinder-pin withdraws from the front. Stamped upon the top of the barrel are the words "F. D. Bliss, New Haven, Ct." Various parts of the weapon bear different numbers, as "879," "793" and "342," which seems to indicate that the revolver is a composite piece.

United States.

No. 28. (N4645). RIM-FIRE REVOLVER. Caliber .28. Three-inch, octagonal, rifled, steel barrel having a rib on top. Has six chambers in the cylinder, a front sight, a groove in the hammer which serves as a rear sight, a silver-plated body, and a blued barrel. To load, barrel and cylinder must be removed. The top of the barrel of this single-action revolver is marked "Conn. Arms Co., Norfolk, Conn." The butt plate bears the number "2574."

United States.

No. 29. (N6797). RIM-FIRE REVOLVER. Caliber .28. Three-inch, octagonal, rifled barrel having a rib on top. There are six chambers in the cylinder. The body is of brass. In order to load this single-action revolver, it is necessary to remove the cylinder from the frame. The top of the barrel is marked "Conn. Arms Co., Norfolk, Conn.", and the cylinder "Patented Mar. 1st, 1864." The revolver number "2628" appears upon the butt-plate.

United States.

No. 30. (N4638). RIM-FIRE REVOLVER. Caliber .28. Three-and-one-half-inch, octagonal, rifled barrel. Has six chambers in the cylinder. A groove in the rear of the frame, on top, serves as a back sight. Has a silver-plated frame and a blued barrel. The cylinder-pin withdraws from the front, releasing the cylinder, which has to be removed in order to load. This single-action revolver was made under

Ellis and White's patent, which was granted July 12, 1859, and July 21, 1863. The top of the barrel bears the words "Eagle Arms Co., New York," and the butt-plate is stamped with the figures "APC 6056."

United States.

No. 31. (N4838). RIM-FIRE REVOLVER. Caliber .28. Three-and-one-half-inch, octagonal, rifled barrel. Has five chambers in the cylinder. A groove in the rear of the frame, on top, serves as a back sight. The frame is nickel-plated, and the barrel was formerly blued. The cylinder is removed from the frame for loading. A hand ejector is located on the right side of the frame in the rear of the cylinder. This is another single-action revolver which was made under the patent granted to Ellis and White. See No. 30. The top of the barrel bears the words "Reynolds Plant's & Hotchkiss, New Haven, Con.", and the side "Merwin & Bray, New York," the agents for whom this revolver was made. The butt bears the number "451."

United States.

No. 32. (N4839). RIM-FIRE REVOLVER. Caliber .28. Three-and-one-half-inch, octagonal, rifled barrel. There are five chambers in the cylinder. Has a front sight, while a groove in the rear of the frame serves as a back sight. The frame is silver-plated. The cylinder is removed from the frame for loading. A hand ejector is located on the right side of the frame behind the cylinder. This single-action revolver was also made under Ellis and White's patent. See No. 30. Upon the top of the barrel are the words "Merwin & Bray Fire Arms Co., N.Y." The number "6144" is stamped upon barrel and butt-plate.

United States.

No. 33. (N4834). RIM-FIRE REVOLVER. Caliber .41. Five-and-one-half-inch, octagonal, rifled barrel. Has six chambers in the cylinder. All steel parts are nickel-plated. The cylinder-pin withdraws from in front to release the cylinder, which must be removed in order to load. Stamped upon the top of the barrel of this single-action revolver are the words "Plant's Mfg. Co. New Haven, Ct.", while the side is marked "Merwin & Bray, New York." The butt-plate bears the number "3632." This revolver, and similar ones, were made for the New York police service.

United States.

No. 34. (N3826). RIM-FIRE REVOLVER. Caliber .40. Six-inch, octagonal, rifled barrel. There are six chambers in the cylinder.

Has a front sight, while a screw on top of the frame serves as a back sight. The cylinder is removed by withdrawing the cylinder-pin from the front. A hand ejector is located on the right side of the frame. The revolver is single-action and was made under the Ellis and White patent. Upon the top of the barrel are the words "Plant's Mfg. Co. New Haven, Ct." The side of the frame bears the letters "M & B N. Y.", standing for Merwin and Bray, New York, the agents for whom these revolvers were made. The revolver number is "268."

United States.

No. 35. (N4837). RIM-FIRE REVOLVER. Caliber .40. Five-and-one-half-inch, octagonal, rifled barrel. Has six chambers in the cylinder. There is a sight in front, while a groove in the frame serves as a back sight. The frame is silver-plated, and the barrel is blued. A hand ejector is located on the right side of the frame behind the cylinder. The cylinder-pin withdraws from the front, releasing the cylinder for loading. This single-action revolver was made under the Ellis and White patent. See No. 30. Stamped upon the top of the barrel are the words "Plant's Mfg. Co. New Haven, Ct.", and on the side, "Merwin & Bray, New York." The butt-plate bears the number "5377."

United States.

No. 36. (N4632). RIM-FIRE REVOLVER. Caliber .36. Seven-and-one-quarter-inch, octagonal, rifled barrel. Has six chambers in the cylinder. Equipped with front and rear sights, and a smooth, polished grip. Trigger-guard and body are made of brass, while the barrel was formerly blued. The cylinder-pin withdraws from in front for the removal of the cylinder in loading. The piece is single-action and the top of the barrel is marked "E.A. Prescott, Worcester, Mass., Pat'd. Oct. 2, 1860." "This revolver invention consists in the peculiar arrangement of a smooth surface in conjunction with a corrugated one rubbing it, fixed obliquely in opposite extremities of a series of chambers mounted on the extremities of revolving arms." Many revolvers of this make were purchased by the United States government for the navy in 1862. The butt-plate is engraved "I. F. Robbins" and the barrel is marked with the number "13."

United States.

No. 37. (N3313). RIM-FIRE REVOLVER. Caliber .36. Seven-inch, octagonal, rifled barrel. Has front and rear sights, a smooth, oval grip, and six chambers. The chambers are loaded from the rear on the right side. This is another Prescott revolver, which has the same action and has the same markings upon the top of the barrel as No. 36. However, the under-side of the barrel is marked with the number "114."

United States.

No. 38. (N4633). RIM-FIRE REVOLVER. Caliber .36. Seven-inch, octagonal, rifled barrel. Has six chambers, front and rear sights, and a smooth, oval grip. The trigger-guard is made of brass, the frame of steel. The barrel has formerly been blued. To load the revolver it is necessary to remove the cylinder from the frame by withdrawing the cylinder-pin. This is an early example of the Prescott revolver, having the same action and markings as No. 36. Cylinder and barrel are stamped with the number "7."

United States.

No. 39. (N4634). RIM-FIRE REVOLVER. Caliber .31. Four-inch, octagonal, rifled barrel. Has six chambers, front and rear sights, and a smooth, oval grip. All metal parts are made of steel, and are unornamented. The cylinder is loaded by withdrawing the cylinder-pin from in front, and removing the cylinder. This is a smaller model of the Prescott revolver; marking and action are the same as No. 36. Cylinder-pin and barrel are stamped with the serial number "563."

United States.

No. 40. (N4635). RIM-FIRE REVOLVER. Caliber .32. Four-inch, octagonal, rifled barrel. Has six chambers, front and rear sights, and a smooth, oval grip. The frame is made of brass, while the barrel is of blued steel. The cylinder is loaded by withdrawing the cylinder-pin from the front, and removing the cylinder from the frame. This specimen, which is single-action, is also a smaller model of the Prescott revolver. It has the same action and the same markings as No. 36. The under-side of the barrel is stamped with the number "132."

United States.

No. 41. (N4666). RIM-FIRE REVOLVER. Caliber .45. Six-and-one-half-inch, round, rifled barrel. Has six chambers. Has a front sight, while a groove in the hammer serves as a rear sight. Equipped with a trigger-guard having a spur to secure greater steadiness in aiming. The revolver is single-action and is rather clumsy and heavy.

The cartridges are inserted into the rear of the cylinder on the right side. Releasing a catch in the cylinder-pin enables the barrel to swing upward and the cylinder to be drawn forward for removing the shells as shown. No maker's name or revolver number are given upon any part of this weapon.

United States.

No. 42. (N4566). RIM-FIRE REVOLVER. Caliber .32. Five-inch, octagonal, rifled barrel. Has six chambers and a front sight, while a groove in the rear of the frame serves as a back sight. The body is brass, while the cylinder and barrel are blued steel. Cylinder-pin and ejector are combined. The piece is single-action and is known as the Pond's Separate-Chambers Revolver. The rear of the cylinder is solid. Each chamber contains a withdrawable sleeve or sheath, and is bored correctly to take a rim-fire, metallic cartridge and is sufficiently thick to allow for the projecting rim of the cartridge. See No. 43. The rear of the cylinder has nicks to admit the nose of the hammer to the ammunition. This weapon was patented by Lucius W. Pond of Worcester, Mass., September 8, 1863. The maker's name is partially erased. The butt-plate bears the number "1060."

United States.

No. 43. (N4567). RIM-FIRE REVOLVER. Caliber .22. Three-and-one-half-inch, octagonal, rifled barrel. Has seven chambers and a front sight, while a groove in the rear of the frame, above the cylinder, serves as a rear sight. Cylinder-pin and ejector are combined. This is another example of the Pond's Separate Chambers Revolver. The action and other characteristics are the same as those of No. 42. The maker's name is not given upon the specimen here shown. Barrel and butt-plate bear the number "374."

United States.

No. 44. (N4639). RIM-FIRE REVOLVER. Caliber .32. Three-inch, round, rifled barrel. Has six chambers, a smooth, rosewood grip and a front sight, but none in the rear. All metal parts are nickel-plated, and the frame is engraved with scrolls. This revolver was invented by Frank P. Slocum of Brooklyn, New York. Its chief peculiarity is a number of independent movable chambers. These are admitted in recesses in a revolving cylinder and are slipped forward against a stationary piston rod to retract the exploded cartridge cases.

They are loaded by placing the cartridge in the lateral recesses of the cylinder, and retracting the movable chambers upon the cartridges. The top of the barrel is stamped "B. A. Co. Patent April 14, 1863." The frame is marked with the number "2382." This piece was made by the Brooklyn Arms Company.

United States.

No. 45. (N4646). RIM-FIRE REVOLVER. Caliber .32. Five-inch, octagonal, rifled barrel. Has six chambers, front and rear sights, and a smooth, oval grip, inlaid with a silver star. All metal parts are made of steel, and are unornamented. The barrel swings upward to load. The cylinder-pin cannot be drawn forward as it is held by a screw inside of the axis of the cylinder. The revolver is single-action and the top of the barrel is marked "L. W. Pond, Worcester, Patd. July 10, 1860," while the under-side bears the number "410."

United States.

No. 46. (N4647). RIM-FIRE REVOLVER. Caliber .32. Three-and-five-eighths-inch, octagonal, rifled barrel. Has six chambers, front and rear sights, and a smooth, oval grip. A small hand ejector is carried in the butt. All metal parts are made of steel, formerly plated. By pressing a button on the left side of the frame, in front of the trigger, the barrel may be swung upward for loading. See illustration. The barrel of this single-action revolver is marked "L. W. Pond, Worcester, Mass., Pat'd. July 10, 1860, Manuf'd for Smith & Wesson Pat'd. April 3, 1865." The revolver number is "3084."

United States.

No. 47. (N6846). RIM-FIRE REVOLVER. Caliber .32. Four-inch, octagonal, rifled barrel. Has six chambers, front and rear sights, and a smooth oval grip. A small hand ejector was formerly carried in the butt. This single-action revolver is opened or "broken" in the same manner as No. 46. The barrel is marked "L. W. Pond, Worcester, Mass., Pat'd. July 10, 1860," and the revolver number is "1401."

United States.

No. 48. (N4637). RIM-FIRE REVOLVER. Caliber .22. Three-inch, octagonal, rifled barrel. Has seven chambers, front and rear sights, and a smooth oval grip. The cylinder-pin withdraws from in front, allowing the cylinder to be removed for loading. All metal parts are made of steel, and are unornamented. Although this piece is not so marked, it was made by L. W. Pond of Worcester, Mass., under the patent of July 10, 1860. The number "14" is stamped upon the under-side of the barrel.

United States.

No. 49. (N4629). RIM-FIRE REVOLVER. Caliber .22. Two-and-one-half-inch, octagonal, rifled barrel. Has seven chambers, front and rear sights, and a smooth, oval grip. All metal parts are made of steel, and were formerly blued. The hammer is fastened to the right side of the frame. Releasing a catch with the thumb enables the cylinder-pin to be withdrawn forward for the removal of the cylinder in order to load or unload. The maker's name and the patent date are almost completely effaced. This piece was made by Lucius W. Pond of Worcester, Mass. It is stamped with the number "899."

United States.

No. 50. (N4784). RIM-FIRE REVOLVER. Caliber .22. Two-and-three-eighths-inch, round, rifled barrel. Has six chambers, front and rear sights, and a round, polished grip. The chambers are charged from the rear end, on the right side of the frame. The cylinder-pin withdraws from the front. The top of the barrel of this single-action revolver is marked "Dead-Shot" and the number "820" is marked upon the under-side. No maker's name is given, but this specimen was made by Lucius W. Pond of Worcester, Mass., under the patent of July 10, 1860.

United States.

PLATE 92.

RIM-FIRE REVOLVERS

No. 1. (N4137). RIM-FIRE REVOLVER. Caliber .41. Seven-and-one-half-inch, half-octagonal, rifled barrel. Has six chambers, front and rear sights, and a smooth, oval grip. The trigger-guard may be released by means of a small catch in the rear, and drawn forward, ejecting the empty shells from the chambers through the rear. The chambers are charged through a small port in the rear of the cylinder on the right side of the frame. The maker's name is almost completely erased from the left side of the barrel. This piece was made by Allen and Wheelock, of Worcester, Mass., and bears the number "165."

United States.

No. 2. (N4660). RIM-FIRE REVOLVER. Caliber .28. Four-inch, octagonal, rifled barrel. There are six chambers in the cylinder. Equipped with a front sight, a notch in the hammer serves as a back sight. The body is slightly engraved with scrolls, and the grip is oval and smooth. The trigger-guard unscrews from the frame, allowing the barrel to slide forward upon the stationary cylinder-pin. The cylinder is then removed for loading. The maker's name is almost completely erased from the top of the barrel of this single-action revolver. This piece was made by Allen and Wheelock, and bears the number "173." Revolvers made by this concern of this type are very rare.

United States.

No. 3. (N4624). RIM-FIRE REVOLVER. Caliber .31. Five-inch, octagonal, rifled barrel. Has the usual sights, and a smooth, polished grip. The cylinder may be loaded by opening the port in the rear, on the left side of the frame. A rack and pinion ejector is fitted in front of the cylinder on the left side. The revolver is single-action. Stamped upon the left side of the barrel are the words "Allen and Wheelock, Worcester, MS. U.S. Allen's Pat's. Sept. 7, Nov. 9, 1858, July 3, 1860." The chief features of the first two patents given above, are described under No. 17. The peculiarities of the 1860 patent provided "the recoil shield of a revolver with a projecting inclined plane so that the cylinder will be free to revolve at the first minute motion of the hammer." The specimen here shown is marked with the numbers "114" and "116."

United States.

No. 4. (N4626). RIM-FIRE REVOLVER. Caliber .28. Four-inch, octagonal, rifled barrel. Has six chambers, front and rear sights, a side-hammer and a smooth, round grip. The revolver is single-action, and the cylinder must be removed from the frame for loading. Stamped upon the left side of the barrel are the words "Allen & Wheelock, Worcester, MS. U.S., Allen's patent Sept. 7, Nov. 9, 1858." The cylinder-pin is stamped with the number "27."

United States.

No. 5. (N4625). RIM-FIRE REVOLVER. Caliber .32. Four-inch, octagonal, rifled barrel. There are six chambers in the cylinder which is engraved with forest scenes. The cylinder-pin is held in place by means of a thumb-screw beneath the frame. To load, the cylinder must be removed from the frame. The revolver is single-action. The

hammer is on the right side of the frame. The left side of the barrel
bears the words "Allen & Wheelock, Worcester, MS. U.S., Allen's
Pat's. Sept. 7, Nov. 9, 1858." Various parts of the revolver are stamped
with the number "74."

United States.

No. 6. (N4627). RIM-FIRE REVOLVER. Caliber .32. Two-
and-three-eighths-inch, octagonal, rifled barrel. Has six chambers in
a cylinder which is engraved with forest scenes. The hammer of this
single-action revolver is on the right side of the frame. To load, the
cylinder must be removed from the frame. This is accomplished by
releasing the thumb-screw which holds the cylinder-pin in place. No
part of the revolver is stamped with any maker's name but the barrel
bears the number "651." This weapon was made by Allen & Wheelock
of Worcester, Mass.

United States.

No. 7. (N4140). RIM-FIRE REVOLVER. Caliber .36. Seven-
and-one-half-inch, octagonal, rifled barrel. All metal parts are of steel
and are unornamented. There are six chambers in the cylinder. The
top of the barrel of this single-action revolver bears the words "Bacon
Mf'g. Co. Norwich, Conn.", and the side, "C. W. Hopkins, Pat. May
27, 1862." This patent was assigned to C. W. Hopkins, T. K. Bacon,
and A. E. Cobb, of Norwich, Connecticut. The chief feature of the
patent consisted of a "means for swinging a cylinder sideways for
loading, etc., without detaching from the frame." When the catch
beneath the ejector is pressed, the cylinder can be swung to the right.
The inside of the grip is marked with the number "12."

United States.

No. 8. (N4667). RIM-FIRE REVOLVER. Caliber .31. Four-
and-seven-eighths-inch, octagonal, rifled barrel. Has six chambers in
a cylinder which is loaded from the rear on the right side. All metal
parts are made of steel, and the frame was originally blued. A hand
ejector is held in place under the barrel by means of a spring at each
end. The cylinder-pin withdraws from the front. The left side of the
frame is stamped "D. D. Cone, Washington, D.C.", and the revolver
number is "709."

United States.

No. 9. (N4664). RIM-FIRE REVOLVER. Caliber .31. Five-
and-seven-eighths-inch, octagonal, rifled, steel barrel. Has seven cham-

bers in the cylinder. The revolver is fitted with a round, curved grip having a ring in the butt-plate. All metal parts are made of steel and are unornamented. A hand ejector is attached to the right side of the frame. No maker's name is given, but this revolver was made by W. Irving, 20 Cliff Street, New York, and manufactured under Reid's patent. This patent, which was granted to James Reid of New York City, on April 28, 1863, had for its chief features "a cylinder which can be used for metallic cartridges or can have nipple plugs screwed into the rear of the chambers for loose ammunition, a swinging plate on the side of the frame behind the cylinder to retain the cartridges in the cylinder." The last, but not the first, feature of this patent is shown in the specimen here illustrated. The revolver number is "53."

United States.

No. 10. (N4663). RIM-FIRE REVOLVER. Caliber .31. Five-inch, octagonal, rifled, steel barrel. Has seven chambers, a smooth, oval grip, a front sight, and a notch in the hammer which serves as a rear sight. The body is of brass and elaborately ornamented with scroll designs. Pressure upon the stud at the right of the hammer allows the barrel and cylinder to swing to the right to load. A small, wire ejector is carried under the barrel. The top of the barrel is marked "D. Moore Patent Sept. 18, 1860," and the bottom is stamped with the number "1846." This piece was probably made by the Moore's Patent Fire Arms Company of Brooklyn, N. Y.

United States.

No. 11. (N4682). RIM-FIRE REVOLVER. Caliber .32. Three-and-one-half-inch, octagonal, rifled, steel barrel. There are six chambers in the cylinder. The frame is made of brass, and is unornamented. The cartridges are inserted into the chambers from the rear, and the cylinder-pin withdraws from in front. Stamped upon the top of the barrel are the words "Whitneyville Armory Ct., U.S.A.", while the butt-plate is marked with the number "1514."

United States.

No. 12. (N4683). RIM-FIRE REVOLVER. Caliber .32. Three-and-one-half-inch, octagonal, rifled barrel. Has six chambers in a fluted cylinder. Equipped with a round, polished grip, a silver-plated brass frame, and a blued, steel barrel. The cartridges are inserted into the chambers from the rear. By releasing a catch in the frame, the cylinder-pin may be withdrawn from in front. The top of the barrel is stamped

"Whitneyville Armory Ct., U.S.A.", and the butt-plate is marked with the number "5412."

United States.

No. 13. (N4685). RIM-FIRE REVOLVER. Caliber .22. Three-and-one-eighth-inch, octagonal, rifled barrel. There are seven chambers in the fluted cylinder. Equipped with a smooth, round grip, and a brass frame. The cartridges are inserted into the cylinder from the rear and on the right side of the frame. The cylinder-pin may be withdrawn from the front. Stamped upon the top of the barrel are the words "Whitneyville, Armory, Ct., U.S.A.", and the butt-plate is marked with the number "26710."

United States.

No. 14. (N4867). RIM-FIRE REVOLVER. Caliber .22. Two-and-one-half-inch, octagonal, rifled, engraved, steel barrel. There are seven chambers in the fluted cylinder. This Whitney revolver is provided with a mother-of-pearl grip, and a finely engraved silver-plated brass body. The remaining features and the markings of this revolver are the same as those of No. 13. The revolver number is "28512" which is stamped upon the butt-plate.

United States.

No. 15. (N4669). RIM-FIRE REVOLVER. Caliber .26. Three-inch, round, rifled barrel. Has five chambers in the cylinder, a smooth, polished grip, a front sight, and a notch in the hammer for the rear sight. The frame is of steel and was formerly silver-plated. Equipped with a lever ramrod, which also serves as the cylinder-pin. Stamped upon the cylinder are the words "Warner's Patent 1857." This patent was granted to James Warner of Springfield, Mass., July 28, 1857, and consisted of a "device for a cylinder which is revolved by a ratchet wheel for improving the certainty of revolution." The butt-plate is marked with the number "2985."

United States.

No. 16. (N4668). RIM-FIRE REVOLVER. Caliber .28. Three-and-one-eighth-inch, octagonal, rifled barrel. The cylinder has five chambers. Has a brass body which was formerly silver-plated. There is a gate on the right side of the frame which, when opened, allows for the insertion of cartridges from the rear of the chambers. The rib on top of the barrel is stamped "Springfield-Arms Co., Mass.", and the butt-plate bears the number "5662."

United States.

No. 17. (N4628). RIM-FIRE REVOLVER. Caliber .22. Two-and-one-half-inch, octagonal, rifled barrel. Has seven chambers for short, rim-fire cartridges. The cylinder is removed from the frame for loading. Stamped upon the left side of the barrel of this single-action revolver are the words "Ethan Allen & Co., Worcester, Mass., Allen's Pats! Sept. 7, Nov. 9, 1858, Sept. 24, 1861." The first patent was "in connection with the cylinder-pin, a surface formed in front of the cylinder to deflect upward the escaping gas." The second patent had "internal lock improvements in a rim-fire metallic cartridge revolver having the hammer on the right side of the frame." The third patent was granted upon the basis of "securing the cylinder-pin by a lever-catch so that both lever and pin can be operated by the thumb at the same time when the cylinder is to be removed." All of these patents are embodied in the above specimen. The under-side of the barrel bears the number "16821."

United States.

No. 18. (N4631). RIM-FIRE REVOLVER. Caliber .22. Two-and-three-eighths-inch, octagonal, rifled barrel. Has seven chambers in the cylinder for short, rim-fire cartridges. Fitted with a smooth, ivory grip. The left side of the barrel of this single-action revolver is stamped "Ethan Allen & Co." Both barrel and frame are plated and engraved. The chief patent features of this type of weapon have been described under No. 17. The cylinder-pin of the specimen here illustrated is marked with the number "204."

United States.

No. 19. (N2903). RIM-FIRE REVOLVER. Caliber .22. Two-and-three-eighths-inch, octagonal, rifled barrel. Has seven chambers in the cylinder for short, rim-fire cartridges. The cylinder is removed from the frame for loading. Fitted with a smooth, ivory grip. Stamped upon the left side of the barrel of this single-action revolver are the words "Ethan Allen & Co., Worcester, Mass., Allen's Pats'. Sept. 7, Nov. 9, 1858, Sept. 24, 1861." For a description of these patents, see No. 17. The cylinder-pin is marked with the number "3905."

United States.

No. 20. (N4630). RIM-FIRE REVOLVER. Caliber .22. Two-and-three-eighths-inch, octagonal, rifled barrel. Has seven chambers for short, rim-fire cartridges. The cylinder is removed from the frame for loading. The barrel and frame are engraved, and the hammer is placed on the right side of the frame. Stamped upon the left side of the barrel are the words "Forehand & Wadsworth, Worcester, Mass., Pats. Sept. 24 & Oct. 22, 1861," while the under-side is marked with the number "38023."

United States.

No. 21. (N4665). RIM-FIRE REVOLVER. Caliber .22. Three-and-one-half-inch, octagonal, rifled barrel. There are seven chambers in the cylinder. The body is made of brass and is silver-plated, while the barrel is blued. This revolver has a swinging plate upon the right side of the frame for the facilitation of inserting and removing cartridges. The top of the barrel is stamped "J. Reid New York." This revolver was made by W. Irving, 20 Cliff Street, New York City. James Reid was the holder of the patent, which was granted to him April 28, 1863. For further details, see No. 9. The under-side of barrel bears the number "420."

United States.

No. 22. (N4656). RIM-FIRE REVOLVER. Caliber .22. Three-and-one-eighth-inch, octagonal, rifled barrel having a rib on top. Has seven chambers. The grip plates are of smooth ivory, while the butt is covered with the bronze head of an animal. The body is of nickel-plated brass, while the barrel is of steel and plated with the same metal. Hunting scenes are engraved upon the cylinder. The hinged barrel tips upward, allowing the cylinder to load and to slide forward on its pin for removal. Stamped upon the top of the barrel are the words "American Standard Tool Co., Newark, N.J." This weapon resembles that made by the Smith and Wesson Company of the same period. The bottom of the barrel is marked with the number "21272."

United States.

No. 23. (N4655). RIM-FIRE REVOLVER. Caliber .22. Three-and-one-eighth-inch, octagonal, rifled barrel having a rib on top. Has seven chambers, front and rear sights, a smooth wooden grip, and a brass body. All metal parts were formerly nickel-plated. The hinged barrel tips upward, allowing the cylinder to load and to slide forward for removal, as shown in the illustration. The markings upon this piece are the same as those of No. 22 and both pieces were made by the American Standard Tool Company. The revolver number is "6471."

United States.

No. 24. (N4909). RIM-FIRE REVOLVER. Caliber .22. Five, one-and-three-quarters-inch, rifled chambers in a steel cylinder. This is a pepper-box of the rim-fire type. All metal parts were formerly nickel-plated. The cartridges are inserted into the chambers from the rear and at the right side of the frame. Pulling back the hammer revolves the cylinder and cocks the revolver. The cylinder is marked "Continental Arms Co., Norwich, Ct., Patented Aug. 28, 1866," and with the number "425."

United States.

No. 25. (N4686). RIM-FIRE REVOLVER. Caliber .22. Two-and-three-quarters-inch, octagonal, rifled barrel. Has seven chambers, front and rear sights, and a polished rosewood grip. Barrel and frame are nickel-plated, and brightly polished. The ejector is fastened to the cylinder-pin, and the latter may be withdrawn forward so as to remove the cylinder. The barrel is marked "Wesson & Harrington, Worcester, Mass., Pat. Feb. 7, June 13, '71." Upon the bottom of the barrel is the number "7373." This revolver has probably never been fired.

United States.

No. 26. (N4658). RIM-FIRE REVOLVER. Caliber .28. Three-inch, octagonal, rifled barrel having a rib on top. Has five chambers, front and rear sights, a brass body, and a smooth, round grip. Barrel and body are nickel-plated. The hinged barrel tips upward, allowing the cylinder to slide forward on its pin for removal. The left side of the barrel is stamped "J. M. Marlin, New Haven, Ct. U.S.A. Pat. July 1, 1878," while the bottom bears the number "3679."

United States.

No. 27. (N4659). RIM-FIRE REVOLVER. Caliber .28. Three-inch, octagonal, rifled barrel having a rib on top. Has five chambers, front and rear sights, a smooth, round grip, and a brass body which was formerly plated with another metal. The hinged barrel tips upward, allowing the cylinder to slide forward on its pin for removal. The barrel is stamped "J. M. Marlin, New Haven, Ct. U.S.A. Pat. July 1, 1878," and "XXX Standard 1872," together with the number "1739."

United States.

No. 28. (N4654). RIM-FIRE REVOLVER. Caliber .22. Three-and-one-eighth-inch, octagonal, rifled, steel barrel having a rib on top. Has seven chambers, front and rear sights, and a polished rosewood grip. The body is of silver-plated steel, while the barrel is elaborately engraved, and was formerly blued. The hinged barrel tips upward, allowing the cylinder to slide forward for removal from its pin. The barrel is stamped "Manhattan Fire Arms Manufg. Co. N. Y.", together with the number "7516."

United States.

No. 29. (N4303). RIM-FIRE REVOLVER. Caliber .22. Three-and-one-eighth-inch, round, rifled barrel having a rib on top. Has seven chambers, front and rear sights, a smooth, round ebony grip, and a nickel-plated, brass body. The barrel is also plated with the same metal. When a spring, in front of the trigger, is pressed the hinged barrel tips up, allowing the cylinder to slide forward on its pin for removal. Stamped upon the top of the barrel are the words "Deringer, Philada.", while the butt-plate is marked with the number "4828."

United States.

No. 30. (N4300). RIM-FIRE REVOLVER. Caliber .32. Three-and-one-half-inch, round, rifled barrel having a rib on top. Has five chambers, front and rear sights, and a round, ivory grip. Barrel, cylinder and frame are elaborately engraved with scroll designs. These parts are also silver-plated. This revolver was very likely a presentation piece. It is opened or "broken" in the same manner as No. 31, which is shown with the barrel tipped upward. The top of the barrel bears the words "Deringer, Philada.", and the butt-plate is marked with the number "316."

United States.

No. 31. (N4302). RIM-FIRE REVOLVER. Caliber .22. Three-inch, round, rifled barrel having a rib on top. Has seven chambers, front and rear sights, and a mother-of-pearl grip. Barrel, cylinder and frame are engraved. The barrel and frame are silver-plated, while the cylinder and hammer are gilded. When a spring in front of the trigger is pressed, the hinged barrel tips up, allowing the cylinder to be loaded and to slide forward on its pin for removal. Upon the top of the barrel are the words "Deringer, Philada. Patd. June 3, 1873," while the butt-plate is marked with the number "1200."

United States.

No. 32. (N4301). RIM-FIRE REVOLVER. Caliber .22. Three-inch, round, rifled, steel barrel with a rib on top. Has seven chambers, front and rear sights, and a round, ivory grip. Barrel, cylinder and frame are engraved, and silver-plated. The action and markings on this revolver are the same as those of No. 31. The butt-plate is marked with the number "671."

United States.

No. 33. (N4865). RIM-FIRE REVOLVER. Caliber .22. Two-and-one-half-inch, octagonal, rifled, steel barrel. Has seven chambers, front and rear sights, and a checkered grip which is made of ebonite. All metal parts were formerly nickel-plated. The hammer is concealed in this double-action revolver. Stamped upon the top of the barrel are the words "Malty Henley & Co., New York, U.S. Pat. Oct. 29, '89. Cal 22 Model 1894." The butt is marked with the number "6942."

United States.

No. 34. (N4704). RIM-FIRE REVOLVER. Caliber .22. Three-inch, round, rifled barrel. Has seven chambers, and front and rear sights. All metal parts were formerly nickel-plated. The chambers are loaded from the rear, and on the right side of the frame. The top of the barrel is stamped "Protector Arms Co., Philada. Patd. Nov. 21, '71," and the butt bears the number "1495."

United States.

No. 35. (N4705). RIM-FIRE REVOLVER. Caliber .22. Three-inch, round, rifled barrel. There are seven chambers in the cylinder. All metal parts are made of steel and are unornamented. Stamped upon the top of the barrel of this single-action revolver are the words "Protector Arms Co., Philada. Patd. Nov. 21, '71," while the butt is marked with the number "989" and the inside of the grip bears the letter "F."

United States.

No. 36. (N4769). RIM-FIRE REVOLVER. Caliber .22. Two-and-one-half-inch, round, rifled barrel. Has seven chambers, a round, ivory grip, a sight in front, but none in the rear. All metal parts are of steel, and were formerly plated. Cylinder, barrel and frame are decorated with scroll designs. The weapon is single-action and is stamped on the top of the frame with the words "Robin Hood No. 1 Long," while the left side of the barrel is marked "Cast Steel, Pat. Feb. 23, '75, Mar. 14, '76." No maker's name is given. The number "1307" is stamped upon the butt.

United States.

No. 37. (N4770). RIM-FIRE REVOLVER. Caliber .22. Two-and-one-quarter-inch, round, rifled barrel. Has seven chambers, a round, wooden grip, and a front sight, but no sight in the rear. All metal parts are made of steel, and are unornamented. The revolver is single-action. The top of the frame is stamped "Robin Hood No. 1 Long," while the barrel is marked "Pat. Feb. 23, '75, Mar. 14, '76." No maker's name is given, but this piece and similar ones were probably made by F. W. Hood. The revolver number is "3406."

United States.

No. 38. (N4771). RIM-FIRE REVOLVER. Caliber .22. Two-and-three-eighths-inch, round, rifled barrel. Has seven chambers, a front sight, and a round, wooden grip. All metal parts are of steel and were formerly plated. Stamped upon the top of the frame are the words "Robin Hood No. 1," while the left side of the barrel is marked "Cast Steel, Pat. Feb. 23, '75, Mar. 14, '76." No maker's name is given, although this piece was probably made by F. W. Hood. The inside of the grip bears the number "9691."

United States.

No. 39. (N4779). RIM-FIRE REVOLVER. Caliber .22. Two-and-one-half-inch, round, rifled barrel. There are seven chambers in the cylinder. All parts are unornamented, except the cylinder, which is chased. The top of the frame is stamped with the word "Alaska," while the left side of the barrel is marked as in No. 38. The number "499" appears upon the butt.

United States.

No. 40. (N4743). RIM-FIRE REVOLVER. Caliber .22. Two-and-one-half-inch, octagonal, rifled barrel. The number of chambers and other characteristics are the same as those of No. 39. The top of the frame is stamped with the word "Brutus," while the left side of the barrel is marked as No. 39. The number "10443" is stamped upon the butt.

United States.

No. 41. (N4790). RIM-FIRE REVOLVER. Caliber .22. Two-and-one-eighth-inch, round, rifled barrel. Has seven chambers in the cylinder. All metal parts are made of steel, and were formerly plated. The top of the frame is stamped "Little John," and the barrel is marked

"Pat. Mar. 14, '76." No maker's name is given. The butt bears the number "3991."

United States.

No. 42. (N4672). RIM-FIRE REVOLVER. Caliber .32. Three-inch, octagonal, rifled barrel. Has seven chambers in the cylinder. Fitted with a front sight, while a groove in the frame serves as a rear sight. All metal parts are made of steel, and are nickel-plated. The grip-plates are missing. Pressing a button upon the left side of the frame releases the cylinder-pin, which may then be drawn forward. The chambers are charged from the right side of the frame, in the rear. Pulling back the hammer revolves the cylinder, and cocks the revolver. The left side of the barrel bears the words "Harrington and Richardson, Worcester, Mass., Pat. May 23, 1876." The revolver number is "3868."

United States.

No. 43. (N4753). RIM-FIRE REVOLVER. Caliber .22. Two-inch, octagonal, rifled barrel. Has seven chambers in the cylinder. All metal parts are made of steel, and are nickel-plated. The grip is made of checkered ebonite. The chambers are charged from the right side of the frame, in the rear. Stamped upon the top of the frame are the words "Young America, Double Action." No maker's name is given, but this piece was probably made by Harrington and Richardson, of Worcester, Mass. The inside of the grip is marked with the number "942."

United States.

No. 44. (N3851). RIM-FIRE REVOLVER. Caliber .22. Two-inch, octagonal, rifled, steel barrel. Has seven chambers for rim-fire cartridges. All metal parts are made of steel and are nickel-plated, except the trigger-guard, which is blued. The grip is made of checkered ebonite. The top of the frame is stamped "Young America, Double Action," and the barrel "H. & R. Arms Co., Worcester, Mass., U.S.A. 22 Rim Fire." This piece was made by the Harrington and Richardson Arms Company and is stamped with the number "240728." The revolver was seized by the Milwaukee police from a person found carrying the same.

Donor: Judge George E. Page.

United States.

No. 45. (N3852). RIM-FIRE REVOLVER. Caliber .22. Two-inch, octagonal, rifled barrel. All metal parts are made of steel and are nickel-plated, except the trigger-guard, which is blued. Has a checkered ebonite grip. Barrel and frame are marked as No. 44. This piece was also made by the Harrington and Richardson Arms Company. The revolver was seized by the Milwaukee police from a person found with the same in his possession.

Donor: Judge George E. Page.

United States.

No. 46. (N4712). RIM-FIRE REVOLVER. Caliber .32. Two-and-five-eighths-inch, octagonal, rifled barrel. Has six chambers, and front and rear sights. All metal parts are of steel, and unornamented. The grip is round, smooth, and is made of wood. The chambers are charged from the right side of the frame in the rear. The left side of the barrel of this single-action revolver is marked "Forehand & Wadsworth, Worcester, Mass., U.S., Pat. Sept. 24, Oct. 22/81-June 27/72," while the top of the frame is stamped "Terror." The revolver number is "3585."

United States.

No. 47. (N5791). RIM-FIRE REVOLVER. Caliber .32. Two-and-one-half-inch, round, rifled barrel. Has six chambers, front and rear sights, and a checkered ebonite grip. All metal parts are made of nickel-plated steel. Releasing a catch in front of the trigger enables the cylinder-pin to be withdrawn forward. The top of the frame bears the words "Forehand & Wadsworth, Double Action No. 32, Worcester, Mass.", and the barrel "Pat'd. July 24, 1877." The butt-plate bears the number "23616."

United States.

No. 48. (N4680). RIM-FIRE REVOLVER. Caliber .32. Two-and-five-eighths-inch, octagonal, rifled barrel. Has five chambers, front and rear sights, and a smooth, ivory grip. All metal parts are of steel, and plated. Pressing a spring on the left side of the frame enables the cylinder-pin to be withdrawn forward for the removal of the cylinder. The piece is single-action and the top of the frame is stamped "Hopkins & Allen Mfg. Co., XL No. 3 N.Y. Pat. Mar. 28, 1871." The under-side of the barrel is marked with the number "Z 422."

United States.

No. 49. (N4778). RIM-FIRE REVOLVER). Caliber .22. Two-and-one-half-inch, round, rifled barrel. Has seven chambers, front and rear sights, and a checkered ebonite grip. Pressing a catch on the left side of frame releases the cylinder-pin so that it can be withdrawn forward for the removal of the cylinder. Stamped upon the top of the frame of this single-action revolver are the words "Ranger 22 Long Pat. Mch. 28, '71, May 27, '79." No maker's name is given, but this piece is said to have been made by Benjamin F. Joslyn of Stonington, Conn. The barrel and the inside of the grip are stamped with the number "5744."

United States.

No. 50. (N4777). RIM-FIRE REVOLVER. Caliber .22. Two-and-one-half-inch, round, rifled barrel. Has seven chambers, front and rear sights, and a polished wooden grip. All metal parts are made of steel, and are nickel-plated. The action and markings upon this revolver are the same as those of No. 49. The barrel and the inside of the grip bear the number "3092."

United States.

PLATE 93.

AMERICAN RIM-FIRE REVOLVERS

No. 1. (N4747). RIM-FIRE REVOLVER. Caliber .22. Three-and-one-half-inch, round barrel. Has seven chambers for short rim-fire cartridges. All metal parts are made of steel, and are unornamented. The cylinder was formerly plated. The grip is fitted with a threaded screw which fits into a shoulder attachment. Stamped upon the top of the frame are the words "Blue Jacket No. 1, Pat. March 28, 1871," while the under-side of the barrel is marked with the number "K 1000." No maker's name is given, but this revolver was probably made by E. S. Laycraft.

United States.

No. 2. (N4748). RIM-FIRE REVOLVER. Caliber .22. Two-and-three-quarters-inch, round, rifled barrel. Has seven chambers for short rim-fire cartridges. There is a sight in front, but none in the rear. All metal parts are of steel, and formerly were plated. The piece is single-action and the chambers are charged from the rear and on the right side of the frame. The top of the frame bears the words "Blue Jacket No. 1," while the bottom of the barrel is marked with the number "3439." No maker's name is given, although this revolver was probably made by E. S. Laycraft.

United States.

No. 3. (N4751). RIM-FIRE REVOLVER. Caliber .22. Two-and-one-half-inch, round, rifled, steel barrel. Has checkered, ebonite grip-plates. The remaining features of this revolver are the same as those of No. 2. The top of the frame is stamped "Blue Jacket No. 1, Pat. May 27, '79," and the under-side of the barrel is marked with the number "304."

United States.

No. 4. (N4750). RIM-FIRE REVOLVER. Caliber .22. Two-and-three-quarters-inch, round, rifled barrel. Has seven chambers for short, rim-fire cartridges. The grip is round and made of wood. Stamped upon the top of the frame are the words "Blue Jacket 1½, Pat. May 27, '79," while the inside of the grip is marked with the number "6703." The remaining features of this revolver are the same as those of No. 2.

United States.

No. 5. (N4749). RIM-FIRE REVOLVER. Caliber .22. Two-and-three-quarters-inch, octagonal, rifled barrel. There are seven chambers in the cylinder. All metal parts are of steel, and nickel-plated. The grip is of checkered ebonite. The barrel and inside of the grip are stamped with the number "1036." The rest of the characteristics are the same as those of No. 2, while the markings are the same as those of No. 3.

United States.

No. 6. (N4692). RIM-FIRE REVOLVER. Caliber .32. Three-inch, octagonal, rifled barrel. Has seven chambers for long **rim-fire** cartridges. It is rather unusual to have seven chambers in a revolver of this caliber as this number makes the chamber walls quite thin. The grip-plates are made of checkered ebonite. Pressing a catch in front of the trigger releases the cylinder-pin so that the fluted cylinder may be removed. Pulling back the hammer revolves the cylinder, and cocks the revolver. The top of the barrel is stamped "America," and the side, "Pat'd. April 23, 1878." Barrel and inside of the grip are marked with the number "1457." No maker's name is given, but this revolver was made by L. Ybarra.

United States.

No. 7. (N4695). RIM-FIRE REVOLVER. Caliber .32. Two-and-three-eighths-inch, octagonal, rifled barrel. Has five chambers for rim-fire cartridges. All metal parts are made of steel, and are plated.

The grip is made of checkered ebonite. The top of the barrel is stamped "Chieftain," while the side is marked "Pat'd. Apr. 23, 1878." The number "595" is stamped upon the under-side of the barrel. No maker's name is given but this revolver was made by L. Ybarra.

United States.

No. 8. (N4728). RIM-FIRE REVOLVER. Caliber .32. Two-and-one-quarter-inch, round, rifled barrel. Has five chambers, a smooth, wooden grip, a front sight, but none in the rear. All metal parts are of nickel-plated steel. The top of the frame is stamped "Conqueror," and the top of the barrel "Pat. Dec. 10, '78." Stamped upon the inside of the grip is the number "815." This is another Ybarra piece having the same action as No. 7. No maker's name is given.

United States.

No. 9. (N4729). RIM-FIRE REVOLVER. Caliber .22. Two-and-three-eighths-inch, round, rifled barrel. There are seven chambers in the cylinder. Action, markings and other characteristics of this specimen are the same as those of No. 8. The number "8759" is stamped upon the inside of the grip. No maker's name is given.

United States.

No. 10. (N4697). RIM-FIRE REVOLVER. Caliber .32. Two-and-one-half-inch, round, rifled barrel. Has five chambers. The top of the barrel is stamped with the word "Crescent," while the side is marked "Pat. Apr. 23, 1878." Action, and other characteristics of this Ybarra revolver, are the same as those of No. 8. The maker's name is not given. Stamped upon the inside of the grip is the number "6502."

United States.

No. 11. (N4698). RIM-FIRE REVOLVER. Caliber .32. Two-and-one-half-inch, octagonal, rifled barrel. The top of the barrel is stamped with the word "Crescent," and the side "Pat. Apr. 23, 1878." Action, and other characteristics of this Ybarra revolver are the same as those of No. 8. No maker's name is given. The under-side of the barrel is stamped with the number "1319."

United States.

No. 12. (N4765). RIM-FIRE REVOLVER. Caliber .22. Two-and-one-eighth-inch, round, rifled barrel. Has seven chambers in the cylinder. Pressing a button on the left side of the frame releases the cylinder-pin so that the cylinder may be removed. The top of the barrel is stamped "Defiance, Pat. April 23, 1878." Action, and other features are the same as those of No. 8. No maker's name is given upon this Ybarra revolver. The inside of the grip and the under-side of the barrel are marked with the number "5929."

United States.

No. 13. (N4760). RIM-FIRE REVOLVER. Caliber .22. Two-and-one-quarter-inch, round, rifled barrel. There are seven chambers in the cylinder. Has a button release for the cylinder-pin as No. 12. Stamped upon the top of the barrel are the words "Defiance, Pat. April 23, 1878." Action and other features of this Ybarra revolver are the same as those of No. 8. The inside of the grip is marked with the number "1700."

United States.

No. 14. (N4763). RIM-FIRE REVOLVER. This revolver is similar in all respects to No. 13, except for the serial number, which in the specimen here shown is "415."

United States.

No. 15. (N4766). RIM-FIRE REVOLVER. Caliber .22. Two-and-one-eighth-inch, round, rifled barrel. Has seven chambers in the cylinder. The top of the barrel is stamped with the word "Defiance," but the patent date is omitted. In other respects this revolver is similar to No. 13. The serial number is "391."

United States.

No. 16. (N4762). RIM-FIRE REVOLVER. Caliber .22. Two-and-one-quarter-inch, round, rifled barrel. Has seven chambers in the cylinder. All metal parts are of steel, nickel-plated, and ornamented with chased designs. Action, markings and other characteristics are the same as those of No. 13. The revolver number is "5332."

United States.

No. 17. (N4764). RIM-FIRE REVOLVER. Caliber .22. Two-and-three-eighths-inch, round, rifled barrel. There are seven chambers in the cylinder. The top of the barrel is marked with the word "Defi-ance," but the patent date is omitted. The remaining features are the same as those of No. 13. The inside of the grip is stamped with the number "340."

United States.

No. 18. (N4735). RIM-FIRE REVOLVER. Caliber .32. Two-and-one-half-inch, round, rifled barrel. Has five chambers in the cylinder. The grip-plates are of checkered ebonite. Pushing forward upon a catch in front of the trigger releases the cylinder-pin so that it can be withdrawn forward. Stamped upon the top of the barrel are the words "Excelsior Pat. Apr. 23, 1878." This revolver was also made by L. Ybarra, although his name is not given upon it. The barrel is stamped with the number "857."

United States.

No. 19. (N4740). RIM-FIRE REVOLVER. Caliber .22. Two-and-one-eighth-inch, round, rifled barrel. There are seven chambers in the cylinder. Has a round, smooth, wooden grip. The cylinder-pin is left-threaded, and, when unscrewed, can be withdrawn forward to remove the cylinder. Stamped upon the top of the barrel are the words "Prairie King, Pat'd. Apr. 23, 1878." The other features are the same as those of No. 13. Marked upon the inside of the grip is the number "942."

United States.

No. 20. (N4742). RIM-FIRE REVOLVER. Caliber .22. Two-and-one-quarter-inch, round, rifled barrel. Has seven chambers in the cylinder. Pressing a catch on the left side of the frame releases the cylinder-pin so that it may be withdrawn forward for the removal of the cylinder. The top of the barrel is stamped with the name "Prairie King," but bears no patent date. Otherwise this piece has the usual characteristics of the Ybarra revolver. Barrel and inside of the grip bear the number "953."

United States.

No. 21. (N4741). RIM-FIRE REVOLVER. Caliber .22. Two-and-one-quarter-inch, round, rifled barrel. Has seven chambers for long rim-fire cartridges. The grip is made of smooth, round ivory. The cylinder-pin is released by means of a button upon the left side of the frame. Barrel, cylinder and frame are decorated with chased designs. The top of the barrel is marked with the name "Prairie King," but no patent date is given. Barrel and inside of the grip are stamped with the number "6618."

United States.

No. 22. (N4702). RIM-FIRE REVOLVER. Caliber .22. Two-and-one-half-inch, round, rifled barrel. Has five chambers for long rim-fire cartridges. Pushing forward upon a catch in front of the trigger releases the cylinder-pin so that it can be withdrawn forward for the removal of the cylinder. The top of the barrel is stamped with the word "Protector," while the side is marked "Pat'd. Apr. 23, 1878." This is another Ybarra revolver, although no maker's name is given. The barrel is stamped with the number "2634."

United States.

No. 23. (N4703). RIM-FIRE REVOLVER. Caliber .32. Two-and-one-half-inch, round, rifled barrel. Has five chambers in the cylinder. Action and markings are the same as those of No. 22. The inside of the grip is stamped with the number "2179."

United States.

No. 24. (N4745). RIM-FIRE REVOLVER. Caliber .22. Two-and-one-quarter-inch, round, rifled barrel. Has seven chambers for long, rim-fire cartridges. All metal parts are made of steel, and were formerly plated. Pushing downward on a catch in front of the trigger releases the cylinder-pin so that it can be withdrawn forward for the removal of the cylinder. The chambers are charged from the rear and on the right side of the frame. Pulling back the hammer revolves the cylinder and cocks the revolver. Stamped upon the top of the frame is the word "Bonanza," while the side is marked "Pat. Dec. 10, '78, Reissued May 10, '81." No maker's name is given, although this revolver was patented by A. L. Sweet. The inside of the grip is stamped with the number "9038."

United States.

No. 25. (N4746). RIM-FIRE REVOLVER. Caliber .22. Two-and-one-eighth-inch, round, rifled barrel. The top of the barrel is stamped "Bonanza" and with the word "Cast Steel." Grip and barrel are marked with the number "2518." In other respects the specimen is similar to No. 24.

United States.

No. 26. (N4781). RIM-FIRE REVOLVER. Caliber .22. Two-and-one-quarter-inch, octagonal, rifled barrel. Has seven chambers in a fluted cylinder, and a checkered ebonite grip. The top of the frame is marked "Express," and the side of the barrel "Pat. Dec. 10th, '78, Reissued May 10, '81." Stamped upon the inside of the grip is the number "8010." In other respects the specimen is similar to No. 24.

United States.

No. 27. (N4780). RIM-FIRE REVOLVER. Caliber .22. Two-and-one-quarter-inch, octagonal, rifled barrel. There are seven chambers in a fluted cylinder. The grip-plates are of checkered ebonite. The top of the barrel is stamped with the word "Guardian," while the barrel bears the words "Pat. Dec. 10th, '78." Stamped upon the inside of the grip is the number "5500." The remaining features are the same as those of No. 24.

United States.

No. 28. (N4788). RIM-FIRE REVOLVER. Caliber .22. Two-and-one-quarter-inch, round, rifled barrel. Has seven chambers in a plain cylinder. The grip is made of polished rosewood. Pressing a button upon the left side of the frame releases the cylinder-pin. The top of the frame is stamped with the words "Little Giant," but no patent date is given. The number "211" is marked upon the barrel. In other respects the specimen is the same as No. 24.

United States.

No. 29. (N4787). RIM-FIRE REVOLVER. Caliber .22. Two-inch, round, rifled barrel. There are seven chambers in the cylinder. The top of the barrel is stamped with the words "Little Giant" and "Pat. Dec. 10, '78." Marked upon the inside of the grip is the number "1035." Otherwise the specimen is the same as No. 24.

United States.

No. 30. (N4662). RIM-FIRE REVOLVER. Caliber .28. Two-and-three-quarters-inch, octagonal, rifled barrel. Has five chambers for rim-fire cartridges. All metal parts are made of steel and are nickel-plated and polished. The grip-plates are made of checkered ebonite; the right plate bears the likeness of Abraham Lincoln, while the left bears that of James A. Garfield. Pressing a catch on the right side of the frame behind the cylinder allows the cylinder and pin to swing out toward the right for loading and ejection, as shown in the illustration. Stamped upon the side of the barrel of this single-action revolver are the words "C. S. Shattuck, Hatfield, Mass., Pat. Nov. 4, 1879." The cylinder-pin is marked with the number "4935."

United States.

No. 31. (N4661). RIM-FIRE REVOLVER. Caliber .28. Two-and-three-quarters-inch, octagonal, rifled barrel. Has five chambers in a fluted cylinder. The grip-plates are made of checkered ebonite. The ear of the hammer has been partly broken off. The cylinder-pin is

stamped with the number "9019."   Action and markings of this revolver
are the same as those of No. 30.

United States.

No. 32. (N4756). RIM-FIRE REVOLVER. Caliber .22. Two-
and-one-quarter-inch, round, rifled barrel. Has seven chambers, front
and rear sights, and a checkered ebonite grip. All metal parts are of
steel and were formerly plated. The cylinder-pin withdraws from the
front. Stamped upon the top of the barrel is the name "Iroquois," while
the left side is marked "E. Remington & Sons, Ilion, N.Y." No revolver
number is given.

United States.

No. 33. (N4706). RIM-FIRE REVOLVER. Caliber .32. Two-
and-three-quarters-inch, round, rifled barrel. Has five chambers in a
fluted cylinder. All metal parts are made of steel and were formerly
plated. Pressing a catch in front of the trigger releases the cylinder-
pin. The top of the barrel is stamped with the name "Marquis of
Lorne." No maker's name is given, although this revolver was made
by E. Remington & Sons. The inside of the grip bears the number
"245."

United States.

No. 34. (N4708). RIM-FIRE REVOLVER. Caliber .22. Two-
and-one-half-inch, round, rifled barrel. Has seven chambers, front and
rear sights, and a smooth, round, wooden grip. The top of the frame
bears the name "Marquis of Lorne." No maker's name is given,
although this revolver was made by E. Remington and Sons. The
inside of the grip bears the characters "125ADL." The action is the
same as that of No. 33.

United States.

No. 35. (N4707). RIM-FIRE REVOLVER. Caliber .32. Two-
and-one-quarter-inch, round, rifled barrel. Has five chambers in a plain
cylinder. The top of the frame is stamped with the name "Marquis
of Lorne," but no maker's name is given. Stamped upon the inside of
the grip is the number "588." The action is the same as that of No. 33,
and like it, this specimen was made by E. Remington and Sons.

United States.

No. 36. (N4866). RIM-FIRE REVOLVER. Caliber .22. Two-
and-three-eighths-inch, half-octagonal, rifled barrel. Has seven cham-

bers, front and rear sights, and a polished rosewood grip. Pressing a catch in the cylinder-pin allows the latter to be drawn forward for the removal of the cylinder. Stamped upon the cylinder are the words "Hopkins & Allen Mfg. Co., Norwich, Ct." Various parts of the weapon are marked with the number "101."

United States.

No. 37. (N4681). RIM-FIRE REVOLVER. Caliber .22. Two-and-one-quarter-inch, octagonal, rifled barrel. Has seven chambers, front and rear sights, and a mother-of-pearl grip. All metal parts are made of steel and are silver-plated, except the trigger-guard, which is blued. Frame and barrel are decorated with floral designs. The hammer is fitted with a patented, hinged ear. Pressing a catch upon the left side of the frame releases the cylinder-pin so that it may be drawn forward for the removal of the cylinder. The top of the frame of this double-action revolver is marked "Hopkins & Allen M'f'g. Co. Pat. Mar. 28, '71, Jan. 5, '86, X.L.I. Double Action." The under-side of the barrel and the grip are stamped with the number "5794."

United States.

No. 38. RIM-FIRE REVOLVER. Caliber .22. Two-and-one-quarter-inch, octagonal, rifled barrel. Has seven chambers, front and rear sights, and checkered ebonite grip-plates. All metal parts are made of steel and are nickel-plated and polished, except the hammer and trigger-guard, which are blued. The cylinder-pin is released by a catch below it and may be withdrawn forward for the removal of the cylinder. The top of the frame is stamped "XL Double Action, Hopk's & Allen Arms Co., Norwich, Ct. U.S.A." The inside of the back-strap bears the number "1500." This revolver was seized by the Milwaukee police from a person found with the same in his possession.

Donor: Judge George E. Page.

United States.

No. 39. (N4913). RIM-FIRE REVOLVER. Caliber .22. Two-inch, round, rifled barrel having a rib on top. Has six chambers, a front sight, but none in the rear. All metal parts are made of steel, which are nickel-plated and polished, except the folding-trigger, which is blued. The grip is made of checkered ebonite and bears the letter "K" upon each side. The hammer is concealed. Pressing a small screw on the upper part of the frame, behind the cylinder, allows the barrel and cylinder to "break" downward, ejecting all cartridges at the same time, as shown in the illustration. When in this position the chambers also can be loaded. The revolver is double-action and is marked upon the top of the barrel with the words "New Baby Mod. 4-5-10-6-13-11." No maker's name is given, although this specimen was made by H. Kolb of Philadelphia, Pa. The inside of the grip is stamped with the number "140."

United States.

No. 40. (N4757). RIM-FIRE REVOLVER. Caliber .22. Two-and-one-quarter-inch, round, rifled barrel. Has seven chambers, front and rear sights, and a mother-of-pearl grip. All metal parts are of steel and were formerly plated. Frame and cylinder are ornamented with chased designs. Stamped upon the top of the barrel are the words "Empire Pat. Nov. 21, '71." This revolver was made by Rupertus of Philadelphia, although his name does not appear upon the specimen. The inside of the grip is stamped with the number "5767."

United States.

No. 41. (N4710). RIM-FIRE REVOLVER. Caliber .32. Two-and-seven-eighths-inch, round, rifled barrel. Has five chambers, front and rear sights, and a round, wooden grip. Pressing a button upon the left side of the frame releases the cylinder-pin, which may then be withdrawn forward for the removal of the cylinder. The top of the barrel bears the word "Scout," and the side "Pat. April 6, 1875." The maker's name is not given. Stamped upon the butt is the number "7137."

United States.

No. 42. (N803). RIM-FIRE REVOLVER. Caliber .32. Two-and-one-half-inch, octagonal, rifled barrel. Has five chambers, front and rear sights, and a round, wooden grip. Pressing a screw upon the left side of the frame releases the cylinder-pin, which may then be withdrawn forward. The cylinder is fluted and its chambers are charged from the rear and on the right side of the frame. Stamped upon the top of the barrel is the name "International No. 3," while the side is marked "Pat. April 6, 1875." The inside of the grip is marked with the number "8400." No maker's name is given.

United States.

No. 43. (N4734). RIM-FIRE REVOLVER. Caliber .22. Two-and-one-quarter-inch, round, rifled barrel. Has seven chambers, a sight in front, but none in the rear. All metal parts are of steel and are nickel-plated, with the exception of the hammer, which is blued. The top of barrel is stamped "Centennial 1876." No maker's name or patent date are given. Barrel and grip are marked with the number "869."

United States.

No. 44. (N4733). RIM-FIRE REVOLVER. Caliber .32. Two-and-one-quarter-inch, round, rifled barrel. Has five chambers in a fluted cylinder. Action and finish are the same as No. 43. The top of the barrel is stamped "Centennial 1876." No maker's name or patent date are given. Various parts of the revolver are stamped with the number "5017."

United States.

No. 45. (N4761). RIM-FIRE REVOLVER. Caliber .22. Two-and-one-eighth-inch, round, rifled barrel. There are five chambers in the cylinder. Has a sight in front, but none in the rear. The left side of the barrel is stamped with the words "Pet Pat. Appl'd. For." The maker's name is not given. Stamped upon the inside of the grip is the number "870."

United States.

No. 46. (N4775). RIM-FIRE REVOLVER. Caliber .22. Two-and-one-half-inch, round, rifled barrel. Has seven chambers and front and rear sights. The cylinder is fluted and probably at one time belonged to another revolver. The grip-plates are made of checkered ebonite. Pushing forward upon a catch in front of the trigger releases the cylinder-pin so that it may be withdrawn for the removal of the cylinder. The top of the barrel bears the word "Spy," while the left side is marked "Pat'd. April 23, 1878." Upon the under-side of the barrel is the number "3613." No maker's name is given.

United States.

No. 47. (N4774). RIM-FIRE REVOLVER. Caliber .22. Two-and-one-eighth-inch, round, rifled barrel. In other respects this revolver is similar to No. 46. The inside of the grip bears the number "2680."

United States.

No. 48. (N4785). RIM-FIRE REVOLVER. Caliber .22. Two-inch, round, rifled, steel barrel. Has seven chambers for short, rim-fire cartridges. Fitted with a front sight, but has none in the rear. The top of the frame is marked "Allen-22." Neither manufacturer's name nor patent date are given. The serial number is "4501."

United States.

No. 49. (N4786). RIM-FIRE REVOLVER. Caliber .22. Two-inch, round, rifled barrel. Has seven chambers and a front sight, but none in the rear. Barrel, cylinder and frame are engraved with scroll designs, and all of these parts were formerly plated. Otherwise the revolver is as has been described under No. 48. The marking is also the same but the serial number is "1219."

United States.

No. 50. (N4693). RIM-FIRE REVOLVER. Caliber .32. Two-and-seven-eighths-inch, round, rifled barrel. Has five chambers for long, rim-fire cartridges. All metal parts are made of steel, and are nickel-plated. The grip-plates are made of smooth ivory. Stamped upon the top of the frame is the word "Earthquake," while the side of the barrel is marked "E. L. Dickinson, Springfield, Mass., Mfg'r." The inside of the grip bears the number "11317."

United States.

PLATE 94.

AMERICAN RIM-FIRE REVOLVERS

No. 1. (N4678). RIM-FIRE REVOLVER. Caliber .22. Two-and-one-eighth-inch, round, rifled barrel. Has seven chambers, front and rear sights, and a small round wooden grip, The top of the barrel is stamped "U.S. Arms Co., N.Y." No patent date is given. The butt is marked with the number "11800."

United States.

No. 2. (N4677). RIM-FIRE REVOLVER. Caliber .22. Two-and-one-quarter-inch, round, rifled barrel. Action and other characteristics are the same as those of No. 1. The top of the frame bears the number "1" and the top of the barrel "U.S. Arms Co., N.Y." The butt is marked with the number "17427." No patent date is given.

United States.

No. 3. (N4674). RIM-FIRE REVOLVER. Caliber .38. Two-and-one-half-inch, octagonal, rifled barrel. Has five chambers, front and rear sights, and a smooth, round, wooden grip. Stamped upon the top of frame are the figures "No. 38" and on top of the barrel "U.S. Arms Co., N.Y." The butt is marked with the number "11064." No patent date is given.

United States.

No. 4. (N4673). RIM-FIRE REVOLVER. Caliber .41. One-and-three-quarters-inch, octagonal, rifled barrel. There are five chambers in the cylinder. Action and other features are the same as those of No. 3. The top of the frame is marked "No. 41" and the top of the barrel "U.S. Arms Co., N.Y." Stamped upon the butt is the number "3692."

United States.

No. 5. (N4789). RIM-FIRE REVOLVER. Caliber .22. Two-and-one-quarter-inch, round, rifled barrel. Has seven chambers in a plain cylinder. Fitted with front and rear sights, and a smooth, round, wooden grip. Stamped upon the top of the frame is the name "Nero," and on the top of the barrel "Patent." The inside of the grip is marked with the number "1150." No maker's name or patent date are given, but this revolver was probably made by C. L. Riker of New York, N. Y.

United States.

No. 6. (N4776). RIM-FIRE REVOLVER. Caliber .22. Two-and-three-eighths-inch, round, rifled barrel. Has seven chambers in a plain cylinder. Other features are as those of No. 5. The top of the barrel is stamped "C.L. Riker, N.Y." No patent date is given. Stamped upon the inside of the grip is the number "49."

United States.

No. 7. (N4684). RIM-FIRE REVOLVER. Caliber .32. Three-inch, round, rifled, brass barrel having a rib on top. Has five chambers in a fluted cylinder. All metal parts are made of brass, except the cylinder, which is of steel. All brass work was formerly plated. The grip-plates are of mother-of-pearl. Stamped upon the top of the frame are the words "Smith's New Model." No patent date is given. The butt is stamped with the number "20507."

United States.

No. 8. (N4725). RIM-FIRE REVOLVER. Caliber .22. Two-and-one-quarter-inch, round, rifled barrel. Has seven chambers for short, rim-fire cartridges. Equipped with a front sight, but has none in the rear. The grip is smooth, round and made of wood. Stamped upon the top of the frame is the word "Defender." No maker's name or patent date are given. The inside of the grip bears the number "3952."

United States.

No. 9. (N804). RIM-FIRE REVOLVER. Caliber .22. Two-and-one-quarter-inch, round, rifled, steel barrel. Action, markings and other features are the same as those of No. 8. The inside of the grip bears the number "581."

United States.

No. 10. (N4726). RIM-FIRE REVOLVER. Caliber .22. Two-and-one-quarter-inch, round, rifled, steel barrel. The cylinder slots are longer than usual in revolvers of this make. Action, markings and other features are the same as those of No. 8. The revolver number is "312."

United States.

No. 11. (N4718). RIM-FIRE REVOLVER. Caliber .32. Two-and-one-half-inch, round, rifled barrel. There are five chambers in a plain cylinder. The top of the frame is marked "Defender." No maker's name or patent date are given. Action and other characteristics are the same as those of No. 8. The inside of the grip is marked with the number "200."

United States.

No. 12. (N4722). RIM-FIRE REVOLVER. Caliber .22. Two-and-one-quarter-inch, round, rifled barrel. Has seven chambers in a fluted cylinder. Otherwise the revolver is the same in action and markings as No. 8. The serial number is "559."

United States.

No. 13. (N4713). RIM-FIRE REVOLVER. Caliber .32. Two-and-one-half-inch, round, rifled barrel. Has five chambers in a fluted cylinder. Action, markings and other characteristics are the same as those of No. 8. The revolver number "25" is marked upon the inside of the grip.

United States.

No. 14. (N4714). RIM-FIRE REVOLVER. Caliber .32. Two-and-one-half-inch, octagonal, rifled barrel. There are five chambers in the fluted cylinder. The slots in the cylinder are longer than usual in this class of revolver. Markings and other features are the same as those of No. 8. The inside of the grip is marked with the number "1052."

United States.

No. 15. (N4719). RIM-FIRE REVOLVER. Caliber .22. Two-and-one-quarter-inch, round, rifled barrel. Has seven chambers for long, rim-fire cartridges. The grip is smooth and made of bone. This is another "Defender" revolver having the same action and features as No. 8. The number "1310" is stamped upon the inside of the grip.

United States.

No. 16. (N4721). RIM-FIRE REVOLVER. Caliber .22. Two-and-one-quarter-inch, round, rifled barrel. Has seven chambers in a fluted cylinder. The front sight is missing. The grip-plates are made of checkered ebonite. Action and markings are the same as No. 8. This revolver is not marked with a serial number.

United States.

No. 17. (N4720). RIM-FIRE REVOLVER. Caliber .22. Two-and-one-quarter-inch, octagonal, rifled barrel. Has seven chambers. The cylinder is fluted, and cut with long notches. The grip is of checkered ebonite. Markings and action are the same as No. 8. The inside of the grip is stamped with the number "753."

United States.

No. 18. (N4716). RIM-FIRE REVOLVER. Caliber .32. Two-and-one-half-inch, octagonal, rifled barrel. There are five chambers in the cylinder. The cylinder is fluted and cut with notches which are in the form of triangular depressions. The grip is larger and of a different form than of the usual "Defender" revolver described above. The grip-plates are of checkered ebonite ornamented with designs representing an eagle carrying four arrows and an olive branch. Stamped upon the top of the barrel is the word "Defender." No maker's name or patent date are given. The inside of the grip is stamped with the number "31."

United States.

No. 19. (N4717). RIM-FIRE REVOLVER. Caliber .22. Two-and-one-quarter-inch, octagonal, rifled barrel. Has seven chambers in the cylinder. The cylinder is fluted and contains longitudinal notches. The ebonite grip is the same as that of No. 18. Marked upon the top of the frame is the word "Defender." The inside of the grip is marked with the number "215."

United States.

No. 20. (N4723). RIM-FIRE REVOLVER. Caliber .22. Two-and-one-quarter-inch, round, rifled barrel. Has seven chambers for long, rim-fire cartridges.   The grip-plates are of checkered ebonite. The top of the frame and both grip-plates are marked "Defender '89." Action and other features are the same as No. 8.   The inside of the grip is stamped with the number "126."

United States.

No. 21. (N4736). RIM-FIRE REVOLVER. Caliber .22. Has seven chambers for short, rim-fire cartridges.   Equipped with front and rear sights and a smooth, round, wooden grip.   Pressing a catch upon the left side of the frame releases the cylinder-pin so that the cylinder can be removed.   The top of the frame is marked "XLCR No. 1."   No maker's name or patent date are given.   The inside of the grip is stamped with the number "5."

United States.

No. 22. (N4758). RIM-FIRE REVOLVER. Caliber .22. Two-and-one-eighth-inch, round, rifled barrel.   Has seven chambers for short, rim-fire cartridges.   Equipped with front and rear sights and a smooth, round, wooden grip.   Pressing a catch on the right side of the frame releases the cylinder-pin so that the cylinder can be removed. Stamped upon the top of the frame is the word "Governor."   No maker's name or patent date are given.   The inside of the grip is stamped with the number "992."   This piece was probably made by Lucius W. Pond, of Worcester, Mass.

United States.

No. 23. (N4759). RIM-FIRE REVOLVER. Caliber .22. Two-and-three-eighths-inch, round, rifled barrel.   Has seven chambers in the fluted cylinder.   The top of the barrel is marked with the word "Governor."   In other respects the revolver is the same as No. 22. Various parts of the piece are stamped with the number "805."

United States.

No. 24. (N4782). RIM-FIRE REVOLVER. Caliber .22. Two-and-one-quarter-inch, round, rifled barrel.   There are seven chambers in the cylinder.   Has a sight in front, but none in the rear.   Stamped upon the top of the barrel is the name "Buffalo Bill."   No maker's name or patent date are given.   The inside of the grip is stamped with the number "707."

United States.

No. 25. (N4772). RIM-FIRE REVOLVER. Caliber .22. Two-and-three-eighths-inch, round, rifled barrel.   Has seven chambers.

There is a sight in front, but none in the rear. The grip is smooth, round, and made of wood. The cylinder-pin unscrews out of the frame so that the cylinder may be removed. Stamped upon the top of the frame is the word "Leader." No maker's name or patent date are given. The inside of the grip bears the number "722."

United States.

No. 26. (N4767). RIM-FIRE REVOLVER. Caliber .22. Two-inch, round, rifled barrel. Has seven chambers. Action and other features are similar to No. 25. The top of the frame is stamped "Creed-moor No. 1," and the inside of the grip is marked with the number "1498." No other marks of identification are given.

United States.

No. 27. (N4755). RIM-FIRE REVOLVER. Caliber .22. Two-and-three-quarters-inch, octagonal, rifled barrel. Has seven chambers. Barrel, cylinder and frame are engraved and were formerly plated. Action and other features are similar to those of No. 25. Stamped upon the top of the frame is the name "Napoleon." Various parts of the revolver are marked with the number "1517." No other information is given upon the specimen.

United States.

No. 28. (N4768). RIM-FIRE REVOLVER. Caliber .22. Two-and-one-quarter-inch, round, rifled barrel. Has seven chambers in a plain cylinder. The cylinder-pin unscrews out of the frame so that the cylinder may be removed. The top of the frame is marked with the words "Tramp's Terror." No maker's name or patent date are given. The inside of the grip is stamped with the number "602."

United States.

No. 29. (N4791). RIM-FIRE REVOLVER. Caliber .22. Two-and-one-eighth-inch, round, rifled barrel. Has seven chambers for long, rim-fire cartridges. All metal parts are of steel, and nickel-plated and polished. The cylinder is decorated with an etched design. Has a smooth, round, bone grip. Action and other features are similar to those of other revolvers of this type. Stamped upon the top of the frame is the name "Rover," while the inside of the grip is marked with the number "747."

United States.

No. 30. (N4744). RIM-FIRE REVOLVER. Caliber .22. Two-

and-one-half-inch, octagonal, rifled barrel. Has seven chambers in a fluted cylinder. The top of the barrel is stamped with the words "Lone Star," and the inside of the grip with the number "256." Action and other features are similar to those of other revolvers of this type.

United States.

No. 31. (N4732). RIM-FIRE REVOLVER. Caliber .22. Two-and-one-half-inch, round, rifled barrel. Has seven chambers for long, rim-fire cartridges. The cylinder is fluted and the grip-plates are made of checkered ebonite. Stamped upon the top of the frame is the word "Sterling." The inside of the grip is marked with the number "108." The remaining features are similar to those of the other revolvers listed above.

United States.

No. 32. (N4731). RIM-FIRE REVOLVER. Caliber .22. Two-and-three-eighths-inch, round, rifled barrel. Has seven chambers for long, rim-fire cartridges in a plain cylinder. There is a sight in front, while a groove in the rear of the frame serves as a back sight. The top of the frame is stamped with the word "Sterling" and the inside of the grip with the number "266." Other features are the same as those of No. 31.

United States.

No. 33. (N4730). RIM-FIRE REVOLVER. Caliber .32. Two-and-five-eighths-inch, round, rifled barrel. Has five chambers for long, rim-fire cartridges. The inside of the grip is stamped with the number "1701." The other features, including the marking, are the same as those of No. 32.

United States.

No. 34. (N4739). RIM-FIRE REVOLVER. Caliber .22. Two-and-one-half-inch, round, rifled barrel. Has seven chambers, a smooth, round, wooden grip, a front sight, but no sight in the rear. Pressing a catch under the cylinder-pin allows the latter to be withdrawn from in front for the removal of the cylinder. The top of the frame is stamped "Hard Pan 1," while the left side of the barrel is marked "Cast Steel Pat. Feb. 23, '75, Mar. 14, '76." The inside of the grip is stamped with the number "2616." No maker's name is given.

United States.

No. 35. (N4738). RIM-FIRE REVOLVER. Caliber .32. Two-and-three-quarters-inch, round, rifled barrel. There are five chambers in the cylinder, which is ornamented with three chased bands. The top of the frame is marked "Hard Pan," but no patent date or maker's name are given. The revolver number is "7874." Other features are the same as those of No. 34.

United States.

No. 36. (N4709). RIM-FIRE REVOLVER. Caliber .32. Two-and-one-quarter-inch, round, rifled barrel. Has five chambers in a fluted cylinder. There is a sight in front, while a groove in the rear of the frame serves as a rear sight. The cylinder-pin is released by pressing a button upon the left side of the frame. Stamped upon the top of the barrel is the name "Peerless." No maker's name or patent date are given. The revolver number is "780," which is stamped both upon the barrel and on the inside of the grip.

United States.

No. 37. (N4737). RIM-FIRE REVOLVER. Caliber .32. Two-and-one-quarter-inch, round, rifled barrel. Has five chambers in a fluted cylinder. The top of the barrel is stamped with the word "Success," while the under-side of the barrel and the inside of the grip are marked with the number "51." No further marks of identification are given. Action and other features are the same as those of No. 36.

United States.

No. 38. (N4701). RIM-FIRE REVOLVER. Caliber .32. Two-and-one-half-inch, round, rifled barrel. Has five chambers in a fluted cylinder. There is a sight in front, while a groove in the rear of the frame serves as a back sight. The top of the barrel is stamped with the word "Southron." The inside of the grip bears the number "474." No other marks of identification are given.

United States.

No. 39. (N4696). RIM-FIRE REVOLVER. Caliber .38. Three-inch, round, rifled barrel. Has five chambers in a fluted cylinder. There is a sight in front but none in the rear. The cylinder-pin is released by pressing a button upon the left side of the frame. Stamped upon the top of the barrel is the word "Bull Dozer" and on the side of the frame "Cala. 38." The inside of the grip bears the number "662." No other marks are given.

United States.

No. 40. (N4700). RIM-FIRE REVOLVER. Caliber .32. Two-and-one-half-inch, round, rifled barrel. Has five chambers for long, rim-fire cartridges. Pressing a catch under the cylinder-pin allows the latter to be withdrawn from in front for the removal of the cylinder. The top of the frame is stamped "32 Avenger N.Y.", and the top of the barrel "Pat. Appld. For." The number "237" is marked upon the inside of the grip.

United States.

No. 41. (N4688). RIM-FIRE REVOLVER. Caliber .38. Two-and-one-half-inch, round, rifled barrel. There are five chambers in the fluted cylinder. Fitted with a curved mother-of-pearl grip. The top of the frame is stamped with the word "Smoker." No maker's name, patent date or revolver number are given.

United States.

No. 42. (N4690). RIM-FIRE REVOLVER. Caliber .32. Two-and-one-half-inch, round, rifled barrel. There are five chambers in the cylinder. Has a small, smooth, wooden grip, the plates of which are too small for the frame. The top of the barrel is stamped with the word "Electric." No maker's name or patent date are given. The inside of the grip is stamped with the number "3794."

United States.

No. 43. (N4691). RIM-FIRE REVOLVER. Caliber .32. Two-and-five-eighths-in, round, rifled barrel. Has five chambers and a front sight. There is no rear sight. The grip-plates are of smooth ivory. Pressing upward on a catch under the cylinder-pin allows the latter to be withdrawn from the front for the removal of the cylinder. The top of the barrel is stamped with the word "Electric," and the inside of the grip is marked with the number "179." No other marks of identification are given.

United States.

No. 44. (N4689). RIM-FIRE REVOLVER. Caliber .32. Two-and-three-eighths-inch, round, rifled barrel. Has five chambers in a fluted cylinder. Equipped with a front sight, while a groove in the rear of the frame serves as a back sight. All metal parts are of steel, and are silver-plated and engraved. The grip-plates are made of mother-of-pearl. Stamped upon the inside of the grip is the number "1135." Action, markings and other characteristics are the same as those of No. 43.

United States.

No. 45. (N4699). RIM-FIRE REVOLVER. Caliber .38. Two-and-one-half-inch, octagonal, rifled barrel. Has five chambers in a fluted cylinder. All metal parts are made of steel and are nickel-plated. Fitted with a checkered ebonite grip. Pressing upon a catch in front of the trigger releases the cylinder-pin so that it may be withdrawn from in front. The top of the frame is stamped "Pioneer," while the inside of the grip bears the number "60." No other marks are given.

United States.

No. 46. (N4711). RIM-FIRE REVOLVER. Caliber .32. Two-and-one-half-inch, round, rifled barrel. Has five chambers in a fluted cylinder. All metal parts are made of steel, which are nickel-plated and polished. The top of the frame is stamped "Aetna," while the inside of the grip is marked with the number "159." No other marks are given.

United States.

No. 47. (N4221). RIM-FIRE REVOLVER. Caliber .38. Two-inch, round, rifled barrel. Has five chambers. The cylinder is partly fluted near one end and has the slots for revolving cut in the middle. Pressing a screw upon the left side of the frame releases the cylinder-pin so that the cylinder may be removed. This piece is not stamped with the name, neither does it bear a maker's name or a patent date. The inside of the grip is marked with the number "296."

United States.

No. 48. (N4752). RIM-FIRE REVOLVER. Caliber .22. Two-and-one-quarter-inch, round, rifled barrel. Has seven chambers for long, rim-fire cartridges. The front sight is missing, and there is no sight in the rear. Pressing a button on the left side of the frame releases the cylinder-pin so that the cylinder may be removed. The top of the barrel is stamped with the word "Liberty," and the inside of the grip bears the number "175." No other marks are given.

United States.

No. 49. (N4792). RIM-FIRE REVOLVER. Caliber .22. Two-and-one-quarter-inch, round, rifled barrel. Has seven chambers for long, rim-fire cartridges in a plain cylinder. Barrel, cylinder and frame are ornamented with chased designs. The cylinder-pin is released by means of a button on the left side of the frame. This piece is not stamped with a name, neither does it bear a maker's name or patent date. The inside of the grip is stamped with the number "15736."

United States.

No. 50. (N4912). RIM-FIRE REVOLVER. Caliber of the upper barrel .22; of the lower barrel .32. Has two, two-and-one-half-inch, round, superposed, smooth-bore barrels. There are eight chambers for short, rim-fire cartridges of .22 caliber, and one chamber in the axis of the cylinder for a rim-fire cartridge of .32 caliber. When all eight chambers of the smaller caliber have been fired, the striking device upon the face of the hammer is lowered and the center chamber discharged on pulling the trigger. The barrels can be tipped downward, as shown, by releasing the catch on top of the frame. There is no self-ejector. Pulling back the hammer revolves the cylinder and cocks the revolver. Stamped upon the top of the piece are the words "Duplex, Pat'd. Dec. 7, 1880," while the left side of the barrels is marked "Osgood Gun Works Norwich, Conn." The revolver number is "4136."

United States.

PLATE 95.

PIN-FIRE PISTOLS AND REVOLVERS

No. 1. (N4796). PIN-FIRE PISTOL. Caliber .40. Four-and-three-quarters-inch, round, rifled, double barrels. The groove, in the rib between the barrels, serves both as a front and rear sight. All metal parts are made of steel, and the frame is slightly ornamented. The fluted, wooden grip is fitted with an iron butt-plate. Pressing a spring upon the left side of the frame releases a catch so that the barrels are tipped downward for loading or removing the empty shells. There is no self-ejector. The pistol is provided with concealed triggers, which are revealed upon bringing the hammers to full cock. The left side of the frame is stamped with the number "16." No maker's name or proof-marks are given.

France.

No. 2. (N2778). PIN-FIRE PISTOL. Caliber .45. Four-and-three-quarters-inch, round, rifled, double barrels. The action is shown open. This pistol has an iron ring attached to the butt-plate. The under-side of the barrels is stamped with the number "2." In other respects the pistol is the same as No. 1.

France.

No. 3. (N2774). PIN-FIRE PISTOL. Caliber .44. Four-and-one-half-inch, octagonal, smooth-bore, double barrels. The fluted, wooden grip is provided with an iron butt-plate which formerly carried a ring. Stamped upon the under-side of the barrels is the letter "A" and the number "69." Action, sights, triggers and other characteristics are the same as those of No. 1.

France.

No. 4. (N4797). PIN-FIRE PISTOL. Caliber .42. Four-and-one-quarter-inch, octagonal, smooth-bore, double barrels. There is an iron ring in the butt-plate. The top of the barrels carries a safety-device which prevents the pins of the cartridges from being struck at the wrong time. Stamped upon the top of the breech-plate is the number "47." In other respects this pistol is similar to No. 1.

France.

No. 5. (N2777). PIN-FIRE PISTOL. Caliber .30. Three-inch, octagonal, rifled, double barrels. Has no ring in the iron butt-plate. A spring beneath the barrels pushes them downward, as shown when the action is open. The number "13" is stamped upon the under-side of the barrels. In other respects this pistol is the same as No. 1.

France.

No. 6. (N4795). PIN-FIRE PISTOL. Caliber .36. Four-and-one-eighth-inch, round, rifled, double barrels. Equipped with front and rear sights. The lock-plate is ornamented with scroll designs, and the hammers are in the form of dolphins' heads. The grip is fluted, and the butt is covered with an iron plate. Pressing forward upon the lever in front of the trigger-guard releases a catch so that the barrels may be tipped downward for loading or removing the empty shells. There is no self-ejector. This type of pistol is sometimes known as the "carriage pistol" due to the fact that some were carried by drivers of stage coaches as a protection against robbers. These weapons were used almost entirely in European countries. No maker's name or proof-marks are given upon this specimen.

France?

No. 7. (N772). PIN-FIRE PISTOL. Caliber .44. Five-inch, round, rifled, double barrels. There is a sight in front, but none in the rear. All metal parts are made of steel and are unornamented. Swinging the lever beneath the frame, to the right, releases a catch so that the barrels may be tipped downward for loading or removing the empty shells. Not fitted with a self-ejector. This is another example of the "carriage pistol," a weapon which was sometimes carried by drivers of stage coaches in European countries. The left side of the lock-plate is stamped with a proof-mark in the form of a crown over the letter "R," while the rib on the bottom between the barrels is marked with the number "2." No maker's name is given.

England.

No. 8. (N758). PIN-FIRE PISTOL. Caliber .36. Three-and-three-quarters-inch, round, rifled barrel. Has a front sight, but the superposed hammer renders this useless. The lock-plate is ornamented with scroll designs, and the hammer is in the form of a dolphin's head. Swinging the lever, beneath the frame, to the right releases a catch so that the barrel may be tipped downward for loading or for removing the empty shells, as shown in the illustration. There is no self-ejector. No proof-mark or maker's name are given upon the pistol, which is probably of French manufacture.

France?

No. 9. (N6793). PIN-FIRE PISTOL. Caliber .44. Four-and-one-eighth-inch, round, rifled, double barrels. Equipped with front and rear sights. Lock-plate and trigger-guard are engraved with scroll designs. The hammers are in the form of dolphins' heads. Swinging the lever beneath the frame to the right releases a catch so that the barrels may be tipped downward for loading or removing the empty shells. This is another carriage pistol. See Nos. 6 and 7. No proof-mark, maker's name or pistol number are given upon this piece.

France?

No. 10. (N4830). PIN-FIRE PEPPER-BOX. Caliber .30. Six rifled chambers in a two-inch cylinder. There are no sights. Has a folding-trigger and a round, ebony grip. The frame is ornamented with engraved floral designs, while the cylinder was formerly blued. To release the cylinder for loading, it is necessary to press down upon the right-angled lever upon the left side of the frame and rotate it through one-hundred-and-eighty degrees; next swing the lever to the right. The cylinder may now be drawn off the cylinder-pin. Stamped upon the bottom of the frame are the words "J.G. Hainfux, Breveté." The inside of the grip bears the number "25," while the rear of the cylinder is marked with the Liége proof-mark.

Belgium.

No. 11. (N3309). PIN-FIRE PEPPER-BOX. Caliber .30. Six rifled chambers in a one-and-seven-eighths-inch cylinder. There are no sights. The trigger on this double-action pepper-box is of the folding variety. The frame is etched with an ornamented leaf design, while the cylinder is blued. To release the cylinder for loading, the handled lever in front of the cylinder is pushed to the right and then the whole is pulled forward. The cylinder-pin is now free so that the cylinder may be drawn off. The breech of the cylinder is stamped with a proof-mark in the form of a crown over the letter "N" and also with the Liége proof-mark. Marked upon the inside of the grip-plates is the number "11." No maker's name is given.

Belgium.

No. 12. (N4832). PIN-FIRE PEPPER-BOX. Caliber .28. Six rifled chambers in a one-and-three-quarters-inch, steel cylinder. Has a folding-trigger and a round, checkered wooden grip. All metal parts are made of blued steel, with the exception of the hammer and trigger, which are unblued. To release the cylinder, the knurled screw in front of the cylinder is unscrewed and then pulled forward. The cylinder may then be removed from its pin. The front of the cylinder is stamped with the Liége proof-mark and with another in the form of a crown over the letter "G." Marked upon the bottom of the frame are the figures "27 1. G.", while the rear of the breech bears the words "A. Francotte à Liége."

Belgium.

No. 13. (N5620). PIN-FIRE PEPPER-BOX. Caliber .28. Six chambers in a one-and-three-quarters-inch cylinder. Has a folding-trigger and a smooth wooden grip which is inlaid with gold. The cylinder-pin unscrews from the front, releasing the cylinder for loading. Stamped upon the trigger are the words "Hers Eibar," while various parts of the pepper-box are marked with the number "23." No proof-marks are given.

Spain.

No. 14. (N4829). PIN-FIRE PEPPER-BOX. Caliber .28. Six smooth-bore chambers in a one-and-three-quarters-inch, steel cylinder. Has a folding-trigger and a smooth brier grip. After opening the gate, the chambers may be loaded from the rear and on the right side of the frame. A hand-ejector is carried in the grip. The cylinder may be removed from its pin by removing the screw in front. Stamped upon the rear of the cylinder is the Liége proof-mark and another in the form of a crown over the letter "G." Upon the inside of the grip are the initials "HNG." No serial number or maker's name are given.

Belgium.

No. 15. (N4831). PIN-FIRE PEPPER-BOX. Caliber .28. Six smooth-bore chambers in a one-and-seven-eighths-inch cylinder. Has a folding-trigger and a smooth, round, wooden grip. A hand-ejector is carried in the grip of this double-action pepper-box. The cylinder-pin unscrews from in front, releasing the cylinder. Stamped upon the rear of the cylinder is the Liége proof-mark and another in the form of a crown over the letter "G." Cylinder and cylinder-pin bear the number "25." No maker's name is given.

Belgium.

No. 16. (N4828). PIN-FIRE PEPPER-BOX. Caliber .36. Six finely rifled chambers in a one-and-seven-eighths-inch, steel cylinder. Has a folding-trigger, a checkered ebony grip and an engraved trigger-guard. All metal parts are made of steel and are finely engraved. A hand-ejector is carried in the grip. The cylinder can be removed by unscrewing a screw in front of the cylinder-pin. A safety-device is carried upon the right side of the frame. The cylinder is stamped with the number "32" and the left side of the frame bears the number "51085." No proof-marks or maker's name are given.

France.

No. 17. (N4827). PIN-FIRE PEPPER-BOX. Caliber .25. Twelve smooth-bore chambers in a one-and-three-quarters-inch cylinder. Has a folding-trigger, and a checkered ebony grip. All metal parts are made of steel. The cylinder is inlaid with copper floral designs, while the body is finely engraved. The cylinder may be removed by unscrewing a screw in front of the cylinder-pin. There is no safety-device. No maker's name, proof-marks or serial number are given.

France?

No. 18. (N4799). PIN-FIRE REVOLVER. Caliber .45. Six-and-one-eighth-inch, round, rifled barrel. There are six chambers in the cylinder. The front sight is missing. A groove in the top of the hammer serves as a back sight. The trigger-guard is provided with an auxiliary grip or spur to give greater steadiness in aiming. The oval wooden grip is fitted with an iron butt-plate which carries a ring. After opening a port the chambers may be charged from the rear on the right side of the frame. A hand-ejector is located on the same side. Stamped upon the left side of the barrel of this single-action revolver are the words "E. Le Faucheux Breveté," while the cylinder bears the Liége proof-mark and another in the form of a crown over the letter "N." The frame is marked with the number "34715." The Le Faucheux Arms Company made pin-fire revolvers both in Paris and in Liége.

France.

No. 19. (N5787). PIN-FIRE REVOLVER. Caliber .48. Six-and-one-eighth-inch, round, rifled barrel. Has six chambers in the cylinder. This is another Le Faucheux revolver having the same characteristics and the same markings upon the barrel as No. 18. The cylinder bears the Liége proof-mark and in addition two others in the form of crowns over the letters "L" and "Z" respectively. The frame is marked with the number "36749."

France.

No. 20. (N4801). PIN-FIRE REVOLVER. Caliber .44. Six-and-one-quarter-inch, round, rifled barrel. Has six chambers in the cylinder. All metal parts are made of steel, and are nickel-plated. This is also an example of a Le Faucheux revolver having the same characteristics and the same marking upon the barrel as No. 18. The cylinder is marked in the same manner as that of No. 19. The frame is stamped with the number "34563."

France.

No. 21. (N799). PIN-FIRE REVOLVER. Caliber .45. Six-and-one-quarter-inch, round, rifled barrel. There are six chambers in the cylinder for pin-fire cartridges. The port on the right side of the frame and the hand-ejector are missing. Action, barrel markings and other features are the same as No. 20. The cylinder bears the Liége proof-mark and in addition two others in the form of crowns over the letters "L" and "Z" respectively. The barrel also bears the last two proof-marks, and the frame is stamped with the number "36797."

France.

No. 22. (N4802). PIN-FIRE REVOLVER. Caliber .45. Four-and-three-quarters-inch, round, rifled barrel. There are six chambers in the cylinder. Has a beaded fore-sight, while a notch in the top of the hammer serves as a rear sight. Barrel and cylinder were formerly blued. Stamped upon the left side of the frame are the words "Invor.

E. Le Faucheux, Breveté, Paris," while the right side is marked "L F 31629." In addition, the rear portion of the barrel bears the letters "S.D."

France.

No. 23. (N4798). PIN-FIRE REVOLVER. Caliber .44. Six-inch, round, rifled barrel, carrying a four-and-three-quarters-inch, folding bayonet beneath the barrel. Has six chambers, a brass fore-sight, while a notch in the top of the hammer serves as a rear sight. The folding-trigger cocks, revolves and fires the piece at a single action. This is another example of the Le Faucheux revolver but it is not so marked, hence it may be an imitation. The left side of the barrel and the rear of the cylinder are stamped with the number "22." No proof-marks are given.

France?

No. 24. (N2750). PIN-FIRE REVOLVER. Caliber .44. Five-and-one-eighth-inch, round, rifled barrel having a rib on top. Has six chambers, a beaded fore-sight, a rear sight and a notch in the top of the hammer which serves as an additional rear sight. The oval, checkered wooden grip is provided with an iron butt-plate which is fitted with a ring. All metal parts of the revolver are ornamented with etched floral designs, with the exception of the sides of the barrel and the hammer. The hand-ejector is missing. Stamped upon the left side of this single-action revolver are the words "E. Le Faucheux, Invor. Breveté." The cylinder bears the Liége proof-mark and another in the form of a crown over the letter "D." The number "60976" appears upon the left side of the frame.

Belgium.

No. 25. (N4806). PIN-FIRE REVOLVER. Caliber .36. Five-and-seven-eighths-inch, round, rifled barrel. Has six chambers, a front sight, and a notch in the top of the hammer which serves as a rear sight. This double-action revolver is provided with a checkered wooden grip having an iron butt-plate which formerly carried a ring. Cylinder and portions of the frame and barrel are inlaid with designs in copper. This is a Le Faucheux revolver but it is not so marked. The cylinder is stamped with the Liége proof-mark but no serial number is given.

Belgium.

No. 26. (N2707). PIN-FIRE REVOLVER. Caliber .44. Five-

and-seven-eighths-inch, octagonal, rifled barrel having a rib on top. The revolver is double-action and has six chambers in the cylinder. Frame and cylinder are engraved with floral designs. This is probably a Le Faucheux revolver but it is not so marked. The cylinder is stamped with the Liége proof-mark and with the number "2."

Belgium.

No. 27. (N2315). PIN-FIRE REVOLVER. Caliber .45. Six-and-one-quarter-inch, round, rifled barrel. Has six chambers for pin-fire cartridges. The trigger-guard is provided with a spur or auxiliary grip to secure greater steadiness in aiming. This is the usual type of single-action Le Faucheux revolver. Marked upon the left side of the frame are the words "Invor. E. Le Faucheux, Breveté, Paris." The right side of the frame is stamped "L F 26539."

France.

No. 28. (N6344). PIN-FIRE REVOLVER. Caliber .45. Five-and-seven-eighths-inch, round barrel, which is rifled with seven grooves. There are seven chambers in the cylinder which is removed by pulling forward upon a device in front of the frame. A bar attached at right angles to the axis of the cylinder-pin serves as a cartridge ejector. In order that the cylinder may not be lost when removed, it is fastened to the barrel by means of a chain, which is illustrated in the picture. Frame, trigger-guard and back-strap are engraved with scroll designs. The side of each chamber and the bottom of the barrel are stamped with undecipherable proof-marks, while the forward end of the cylinder is engraved with a monogram which appears to consist of the letters "GD." The piece is double-acting. Stamped upon the right side of the frame is the number "549." No maker's name is given.

France?

No. 29. (N4800). PIN-FIRE REVOLVER. Caliber .44. Six-and-three-eighths-inch, round, rifled barrel. There are six chambers in the cylinder. Has a triangular fore-sight, while a notch in the top of the hammer serves as a rear sight. All metal parts of this double-action revolver are plated; the cylinder, frame and butt-plate are also engraved. The cylinder is stamped with the Liége proof-mark and with another in the form of a crown over the letter "G." The inside of the grip is marked with the name "H. Malchair." This is a Le Faucheux officer's model, although the maker's name and serial number are not given.

Belgium.

No. 30. (N4808). PIN-FIRE REVOLVER. Caliber .36. Four-and-one-half-inch, octagonal, rifled barrel. Has six chambers for pin-fire cartridges. There is a sight in front, but none in the rear. Cylinder, barrel and frame were formerly blued. Each of these parts is stamped with a proof-mark in the form of a crown over the letter "V." The cylinder and the inside of the grip also bear the number "2." Stamped upon the bottom of the butt are the figures "14455 WG." This is a double-action Le Faucheux revolver which bears no maker's name.

England?

No. 31. (N2708). PIN-FIRE REVOLVER. Caliber .44. Five-and-seven-eighths-inch, octagonal, rifled barrel. Has six chambers and a fore-sight, but no sight in the rear. All metal parts were formerly nickel-plated. The frame is stamped with a proof-mark in the form of serial number are given upon this double-action Le Faucheux revolver. a crown over the letters "L.S.", while the cylinder bears the Liége proof-mark and another in the form of "u̇." No maker's name or serial number are given upon this double-action Le Faucheaux revolver.

Belgium.

No. 32. (N4817). PIN-FIRE REVOLVER. Caliber .28. Five-and-one-half-inch, round, smooth-bore barrel. Has six chambers in a very long cylinder for pin-fire metallic cartridges. This double-action revolver has no sights of any kind. All metal parts are silver-plated, and the smooth ivory grip formerly carried a ring in the butt. The trigger is of the folding variety. Stamped upon the breech of the cylinder is the Liége proof-mark and the number "16," while the port is also stamped with the same number. No maker's name is given.

Belgium.

No. 33. (N4809). PIN-FIRE REVOLVER. Caliber .36. Four-inch, octagonal, rifled barrel. Has six chambers. Equipped with a front sight, while a notch in the top of the hammer serves as a rear sight. The grip-plates are made of carved ebony. All metal parts are heavily blued, with the exception of the hammer. Barrel, cylinder and frame are elaborately engraved, and in addition, the first two are also decorated with gold pins. The piece is double-action, and carries a ring in the butt. A safety-device, in the form of a spring, is located on the right side above the trigger. An ejector is located on the same side. The cylinder is stamped with the Liége proof-mark and another in the

form of a crown over the letter "U." No maker's name or shop number
are given.

Belgium.

No. 34. (N3310). PIN-FIRE REVOLVER. Caliber .30. Four-
and-one-eighth-inch, round, rifled barrel. Has ten chambers for pin-
fire cartridges. Equipped with a beaded fore-sight, while a notch in
the top of the hammer serves as a rear sight. The grip is made of
smooth ivory. All metal parts of the revolver, except the barrel and
hammer, are engraved. This is still another example of the double-
action Le Faucheux revolver, but it bears no maker's name. Upon the
right side of the hammer is the number "19." No proof-marks are
given.

France?

No. 35. (N4812). PIN-FIRE REVOLVER. Caliber .30. Three-
and-five-eighths-inch, round, rifled barrel. Has six chambers and a
front sight, a notch in the top of the hammer serving as a rear sight.
The piece is double-action, and is provided with a folding-trigger. Cylinder and portions of the barrel and frame are engraved with scroll
designs. There is no loading gate, neither is there an ejector. The
cylinder is stamped with the Liége proof-mark and with another in the
form of a crown over the letter "V." Stamped upon the left side of
the barrel is an undecipherable name. No revolver number is given.

Belgium.

No. 36. (N6792). PIN-FIRE REVOLVER. Caliber .28. There
are six chambers in the cylinder. Has a front sight, a folding-trigger
and a notch in the top of the hammer which serves as a rear sight. The
grip is of ebony and the cylinder is blued. This is another example
of the double-action Le Faucheux revolver, but bears no maker's name.
Stamped upon the right side of the barrel is the number "239" while the
cylinder is marked with the Liége proof-mark.

Belgium.

No. 37. (N4825). PIN-FIRE REVOLVER. Caliber .28. Four-
and-seven-eighths-inch, octagonal, rifled barrel. There are twelve
chambers in the cylinder. Has a beaded fore-sight, but no rear sight.
All metal parts of this double-action revolver were formerly plated.
The breech of the cylinder is stamped with the Liége proof-mark and
with another in the form of "R." No serial number is given. This is
another form of the Le Faucheux revolver.

Belgium.

PLATE 96.

PIN-FIRE, NEEDLE AND CENTER-FIRE PISTOLS AND REVOLVERS OF FOREIGN MANUFACTURE

No. 1. (N2560). PIN-FIRE REVOLVER. Caliber .44. Six-and-three-eighths-inch, round, rifled barrel. Has six chambers, a front sight and a notch in the top of the hammer, the notch serving the purpose of a rear sight. After opening a port the chambers may be charged from the rear and on the right side of the frame. An ejector was formerly carried on the same side. Frame and butt-plate are engraved with scroll designs. The grip is smooth and angular, and is made of wood. Barrel and cylinder were formerly blued. The left side of the barrel is stamped "H.B. Solingen," while the cylinder and folding-trigger are marked with the number "396." No proof-marks are given.

Germany.

No. 2. (N4804). PIN-FIRE REVOLVER. Caliber .36. Five-and-three-eighths-inch, octagonal, rifled barrel. There are six chambers in the cylinder. The barrel is removed by releasing the lever on the right side of the frame; after which the cylinder is withdrawn from its pin and charged. There is no ejector. The angular wooden grip is provided with an iron ring in the butt. Cylinder and frame are engraved with floral designs, while the top of the barrel is engraved with the words "R. Drechsler in Baerenstein." Cylinder and hammer are stamped with the number "97" but no proof-marks are visible.

Germany.

No. 3. (N4807). PIN-FIRE REVOLVER. Caliber .36. Five-inch, octagonal, rifled barrel. Has six chambers and a front sight, but no sight in the rear. The revolver is double-acting, and after opening the port the chambers may be charged from the rear on the right side of the frame. An ejector is located upon the same side. The barrel is stamped with the letters "VP $\overset{*}{N}$," while the cylinder bears the Liége proof-mark and also "$\overset{*}{N}$." The port is marked with the number "35." No maker's name is given, although this specimen is probably a Le Faucheux revolver.

Belgium.

No. 4. (N6790). PIN-FIRE REVOLVER. Caliber .36. Four-and-seven-eighths-inch, octagonal, rifled barrel. Has six chambers. Action and other characteristics are the same as those of No. 3. The barrel is marked "$\overset{*}{z}$" and the cylinder bears the Liége proof-mark and also "$\overset{*}{z}$," while the inside of the grip and the cylinder are stamped with the number "32." This is another form of the Le Faucheux revolver. It bears no maker's name.

Belgium.

No. 5. (N4805). PIN-FIRE REVOLVER. Caliber .36. Four-and-five-eighths-inch, round, rifled barrel. There are six chambers in the cylinder. Frame and cylinder are slightly engraved with floral and scroll designs. Action and other features are the same as those of No. 3. The barrel is stamped with the Liége proof-mark and with "$\overset{*}{L}$," while the inside of the grip is marked with the number "15." No maker's name is given.

Belgium.

No. 6. (N4808). PIN-FIRE REVOLVER. Caliber .36. Four-and-one-half-inch, octagonal, rifled barrel. Has six chambers in the cylinder. Cylinder, barrel and frame were formerly blued. All of these parts are stamped with a proof-mark in the form of a crown over the letter "V." Action and other features are the same as those of No. 3. The cylinder and the inside of the grip are marked with the number "2," while the bottom of the butt is stamped with the figures "14455 WG." No maker's name is given.

Belgium.

No. 7. (N6849). PIN-FIRE REVOLVER. Caliber .32. Three-and-one-quarter-inch, octagonal, rifled barrel. There are six chambers in the cylinder. Has a folding-trigger, a smooth oval grip and a front sight, but no sight in the rear. The revolver, which is double-acting, is stamped with the number "4" upon the breech of the cylinder. No maker's name or proof-marks are given.

Belgium.

No. 8. (N4821). PIN-FIRE REVOLVER. Caliber .28. Three-and-one-quarter-inch, octagonal, smooth-bore barrel. Action and other features are similar to those of No. 7. The breech of the cylinder is stamped with the number "4," while the front bears the letter "S" within a circle. Neither maker's name nor proof-marks are given.

Belgium.

No. 9. (N4816). PIN-FIRE REVOLVER. Caliber .28. Three-and-three-eighths-inch, octagonal, rifled barrel. Has six chambers in

the cylinder. Action and other features are the same as those of No. 7.
The grip-plates are made of checkered and carved ivory. All metal
parts were formerly nickel-plated, but are now badly corroded. Cylinder, barrel and grip are stamped with the number "5." No maker's
name or proof-marks are given.

Belgium?

No. 10. (N5767). PIN-FIRE REVOLVER. Caliber .30. Four-
and-one-quarter-inch, octagonal, smooth-bore barrel. Has six chambers
for pin-fire cartridges. Barrel, cylinder and frame are engraved with
line-designs. Action and other characteristics are the same as those of
No. 7. Cylinder, barrel and the base of the hammer are stamped with
the number "4." No maker's name or proof-marks are given.
Donor: August Dellmann.

Belgium?

No. 11. (N4815). PIN-FIRE REVOLVER. Caliber .28. Three-
and-three-eighths-inch, octagonal, smooth-bore barrel. Has the usual
six chambers found in this variety of revolver. All parts of the weapon
are unornamented. The cylinder is stamped with the Liége proof-mark
and with another in the form of "M̊." The barrel also bears the latter
mark. No maker's name or serial number are given.

Belgium.

No. 12. (N4819). PIN-FIRE REVOLVER. Caliber .28. Three-
and-one-quarter-inch, octagonal, rifled barrel. There are six chambers
in the cylinder. All metal parts are plated but otherwise unornamented.
Action and other features are the same as those of No. 7. The cylinder
is stamped with the Liége proof-mark and with another in the form of
"D̊." The barrel also bears the latter mark. No other marks of identification are given.

Belgium.

No. 13. (N6788). PIN-FIRE REVOLVER. Caliber .28. Two-
and-one-half-inch, round, rifled barrel. Has six chambers for pin-fire
cartridges. Equipped with a front sight, while a notch in the top of the
hammer serves as a rear sight. The ejector is missing. Action and
other characteristics are the same as those of No. 7. The cylinder is
stamped with the Liége proof-mark and with another in the form of
"N̊," while the barrel is marked with a crown over the letter "V."
Marked upon the right side of the frame are the letters "GB." No
other marks are given.

Belgium.

No. 14. (N6801). PIN-FIRE REVOLVER. Caliber .28. Three-and-three-eighths-inch, round, rifled barrel. There are six chambers in the cylinder. All metal parts were formerly blued. Action and other features are the same as those of No. 7. The cylinder is stamped with the Liége proof-mark and with "M," while the top of the barrel is marked with the letters "LF" and with a rampant lion within a shield. The side also bears the mark "M." Upon the inside of the grip is the number "8." No maker's name is given.

Donor: Raymond Schulz.

Belgium.

No. 15. (N4820). PIN-FIRE REVOLVER. Caliber .28. Three-and-three-eighths-inch, round, rifled barrel. Has six chambers. All parts of the revolver are unornamented. Action and other features are the same as those of No. 7. The cylinder is stamped with the Liége proof-mark and with another in the form of a crown over the letter "U." There is no proof-mark upon the barrel. No maker's name or serial number are given.

Belgium.

No. 16. (N6848). PIN-FIRE REVOLVER. Caliber .28. Three-and-three-eighths-inch, round, rifled, steel barrel. Has the usual six chambers. Equipped with a round, checkered wooden grip. Barrel, cylinder and frame are engraved with floral designs. The first two were formerly blued. The cylinder is stamped with the Liége proof-mark and with a crown over the letter "U," while the barrel is marked with a crown over the letter "V." The inside of the grip bears the number "14." No maker's name is given.

Donor: Gustav Spankus.

Belgium.

No. 17. (N4813). PIN-FIRE REVOLVER. Caliber .28. Three-and-one-quarter-inch, octagonal, rifled barrel. There are six chambers in the cylinder. The carved grip is decorated with human and animal heads in relief. Action and other features are the same as those of No. 7. All metal parts are nickel-plated. The cylinder is stamped with the Liége proof-mark and with "T," the barrel also bearing the latter mark. Trigger and loading gate are marked with the number "20," but no maker's name is given.

Belgium.

No. 18. (N6850). PIN-FIRE REVOLVER. Caliber .28. Three-and-three-eighths-inch, octagonal, rifled barrel. There are six chambers in the cylinder. Has front and rear sights, a folding-trigger and a smooth, round, wooden grip. The ear of the hammer is broken away. All steel parts were formerly blued. The cylinder-pin may be withdrawn forward by removing a set-screw. Marked upon the top of the barrel are the words "Canon Acier Visse," while the cylinder is stamped "Systeme Le Faucheux Perfre. (Depose)." The cylinder is also stamped with the Liége proof-mark and with another in the form of a crown over "N." Frame and barrel bear the number "3222."

France.

No. 19. (N4823). PIN-FIRE REVOLVER. Caliber .28. Two-and-one-half-inch, round, rifled barrel. Has six chambers in a fluted cylinder for pin-fire cartridges. There is a sight in front, but none in the rear. The ejector is carried in the grip. After removing a set-screw the cylinder-pin may be withdrawn from in front for the removal of the cylinder. The barrel is stamped with the Liége proof-mark and the loading-gate with the number "16." No maker's name is given.

Belgium.

No. 20. (N792). PIN-FIRE REVOLVER. Caliber .30. Three-and-one-half-inch, round, rifled barrel. Has six chambers, a folding-trigger and an earless hammer. There is a sight in front, but none in the rear. Equipped with a checkered wooden grip. Barrel and cylinder are stamped with English proof-marks. Engraved upon the top of the barrel are the words "W. Edwards and Co.", while the left side of the frame is engraved "London, No. 40180."

England.

No. 21. (N4811). PIN-FIRE REVOLVER. Caliber .28. Two-and-seven-eighths-inch, octagonal, rifled barrel. There are six chambers in the cylinder. All metal parts were formerly plated. Equipped with an oval checkered wooden grip. The cylinder bears a partly decipherable inscription: "The Guardian American"—the rest being obliterated. Upon the left side of the frame the number "1" appears two times, while the right side is stamped with the number "10." The breech of the cylinder bears the Liége proof-mark. No maker's name is given.

Belgium?

No. 22. (N4810). PIN-FIRE REVOLVER. Caliber .30. Three-and-one-half-inch, round, rifled barrel. Has six chambers. Frame, cylinder and portions of the barrel are engraved with floral designs. The cylinder is stamped with the Liége proof-mark and with another in the form of a crown over the letter "L." The barrel is also stamped with the latter mark. Stamped upon the right side of the frame is the number "36." No maker's name is given.

Belgium.

No. 23. (N6795). PIN-FIRE REVOLVER. Caliber .28. Three-and-one-half-inch, round, rifled barrel. There are six chambers in the cylinder. The ejector is missing, and the hammer is earless. There are no proof-marks upon cylinder or barrel. Stamped upon the top of the barrel are the figures "7 m/m." and the letter "R" which is within a circle. The left side of the frame formerly bore a name which is now almost completely effaced. Port and trigger are stamped with the number "14." No maker's name is given.

France.

No. 24. (N4822). PIN-FIRE REVOLVER. Caliber .30. Two-and-one-half-inch, octagonal, rifled barrel. Has six chambers in a fluted cylinder. The ejector is missing. All metal parts were formerly blued. Barrel, cylinder and frame are engraved with line-designs. The grip-plates are made of checkered wood. The barrel is stamped with the Liége proof-mark and with "$\overset{*}{s}$"; the cylinder also bearing the latter mark. Stamped upon the loading-gate is the number "6." No maker's name is given.

Belgium.

No. 25. (N4826). PIN-FIRE REVOLVER. Caliber .20. Two-and-three-eighths-inch, octagonal, rifled barrel. Has six chambers, a folding-trigger and a front sight, but none in the rear. The hand-ejector is missing. The grip-plates are made of smooth ivory. All metal parts are nickel-plated. Stamped upon the cylinder is the Liége proof-mark and another in the form of "$\overset{*}{H}$"; the barrel also bears the latter mark. No maker's name or serial number are given.

Belgium.

No. 26. (N4824). PIN-FIRE REVOLVER. Caliber .20. Two-inch, octagonal, rifled barrel. The grip is of carved ebony. All metal parts are nickel-plated. The cylinder is marked with the Liége proof-mark and with "$\overset{*}{G}$"; the barrel also bears the latter mark. Stamped upon the right side of the frame is the number "2778." No maker's name is given.   The other features are similar to those of the pin-fire revolvers listed above.

Belgium.

No. 27. (N4589). NEEDLE-GUN. Caliber .38. Nine-and-three-quarters-inch, octagonal, rifled barrel. Has front and rear sights and a steel trigger-guard, but no hammer. The grip is of wood checkered on the sides but fluted at the butt. This gun fires a lead ball, in the rear of which there is placed a charge of powder followed by a charge of fulminate. The ball is placed in the barrel from on top, after the swinging lever on the right is pressed down to admit the projectile. Stamped upon the left side of the barrel are the figures "Cal: O,34 (German inches) 6 Gran. Pulv.", also "M M M," the letters being in Gothic type, and the number "9156." No maker's name is given. The needle-gun is a breech-loading, center-firing, small arm, exploding the cartridge by a blow from a spring needle. The needle is contained in the breech of this specimen. The early types of these firearms suffered badly from escape of gas and flame at the breech. The needle-gun can justly lay claim to being the parent of the modern breech-loading rifle, inasmuch as it was a bolt-action. In the earlier type, as here shown, the cocking of the action was performed by pulling back on a cocking piece that was concealed with the "needle" or striker situated at the rear of the bolt. In the later models this was performed by the action of drawing back the bolt as is done in all modern bolt-actions. The needle-gun was invented in 1836 by the German gunsmith, Johann Nicholas von Dreyse (1787-1867).

Germany.

No. 28. (N2745). NEEDLE-GUN. Caliber .40. Nine-and-one-eighth-inch, octagonal, rifled barrel having a flared muzzle. Equipped with front and rear sights, but no hammer. The grip is of wood, checkered and studded with German silver nails. The trigger-guard and butt-plate are also made of German silver, and are finely engraved. This is a specimen of the later type of needle-gun, since it is operated by the bolt-action principle. See No. 27. The bolt is drawn back for the insertion of the cartridge and then pushed forward and turned to the right. The cocking piece containing the "needle" is then pushed forward and also turned a little to the right. Upon pulling the trigger the weapon is discharged. The top of the barrel is etched with the words "I. Valt. Funk & Sohn in Suhl." No serial number or proof-marks are given.

Germany.

No. 29. (N4595). CENTER-FIRE PISTOL. Caliber .44. Four-and-one-quarter-inch, rifled, double barrels. Equipped with a front sight, while a groove between the barrels serves as a back sight. Has two rebounding hammers. The pistol is provided with a smooth oval grip having an iron butt-plate fitted with a ring. The frame is engraved with scroll designs. Pressing a catch upon the left side releases the barrels so that they tip downward for loading or for the removal of the empty shells, as shown in the illustration. There is no self-ejector. All metal parts were formerly nickel-plated. Stamped upon the top of the barrels are the words "Armeria De Jose U. Mahuga Malaga." No serial number or proof-marks are given.

Spain.

No. 30. (N4850). CENTER-FIRE REVOLVER. Caliber .45. Three-and-five-eighths-in, round, rifled barrel. Has six chambers in a plain cylinder. Equipped with a front sight, while a groove in the rear of the frame serves as a back sight. The cylinder-pin of this double-acting revolver is hollow and carries the ejector. The revolver has a large checkered wooden grip having a ring in the butt. This type of weapon was issued to the French gendarmie or armed police. The cylinder is stamped with the Liége proof-mark and with a letter "S" inscribed in a circle. No maker's name is given.

France.

No. 31. (N4856). CENTER-FIRE REVOLVER. Caliber .28. Three-and-three-eighths-inch, octagonal, rifled, steel barrel. Has six chambers for center-fire cartridges. Equipped with front and rear sights and a folding-trigger. This is really a Le Faucheaux revolver, using center-fire instead of pin-fire cartridges. Action and other features are the same as those of No. 7. All metal parts were formerly blued. Cylinder, frame and trigger are stamped with the number "5." No maker's name or proof-marks are given.

France.

No. 32. (N3841). CENTER-FIRE REVOLVER. Caliber .28. Three-inch, octagonal, rifled barrel. Has six chambers, front and rear sights and a folding-trigger. The checkered ebonite grip bears the initials of Smith and Wesson. All metal parts are nickel-plated. Cylinder, frame and trigger are stamped with the number "4." The cylinder bears the Liége proof-mark and another in the form of a crown over the letter "Y." No maker's name is given.

Belgium.

No. 33. (N6847). CENTER-FIRE REVOLVER. Caliber .30. Two-and-one-half-inch, octagonal, rifled barrel. Has six chambers, front and rear sights and a folding-trigger. Cylinder and barrel are stamped with proof-marks consisting of two crowns over the letter "U." Trigger and cylinder bear the number "4." No maker's name is given.
Donor: Morris Greenberg.

Germany?

No. 34. (N2709). CENTER-FIRE REVOLVER. Caliber .44. Five-and-three-quarters-inch, octagonal, rifled barrel. Has six chambers, front and rear sights and a checkered wooden grip having a ring in the butt. Barrel, cylinder and frame are engraved with scroll and line designs. The cylinder is stamped with the Liége proof-mark and with a letter "S" inscribed in a circle. The barrel bears the stamp "ż." No maker's name is given.

Belgium.

No. 35. (N4132). CENTER-FIRE REVOLVER. Caliber .44. Five-and-three-quarters-inch, round, rifled barrel. Has six chambers, front and rear sights and an earless hammer. The port is only of sufficient width to retain the cartridges in place, which are inserted into the chambers at the rear and on the right side of the frame. The cylinder-pin is hollow and carries the ejector. Has an angular, smooth, wooden grip. The butt is of iron and carries a ring. Barrel, cylinder and frame bear the name "L. Perrin." In addition, the cylinder is stamped with the number "904." No proof-marks are given upon this piece, which was made by Perrin et Cie of Paris.

France.

No. 36. (N4134). CENTER-FIRE REVOLVER. Caliber .45. Five-and-three-quarters-inch, round, rifled barrel. Barrel, cylinder and frame are stamped with the number "1376," but no proof-marks are given. This specimen was made by Perrin et Cie of Paris, and has the same action and characteristics as No. 35.

France.

No. 37. (N4133). CENTER-FIRE REVOLVER. Caliber .44. Five-and-three-quarters-inch, round, rifled barrel. A spring safety-device is situated upon the left side of the frame. Has an angular checkered wooden grip which is provided with an iron ring in the butt. Barrel, cylinder and frame are stamped "Nr.673." Upon the right side of the frame the words "Perrin & Cie, Bte." appear, while the left side bears the word "Paris."* Action and other features are the same as those of No. 35.

France.

No. 38. (N4138). CENTER-FIRE REVOLVER. Caliber .41. Six-inch, round, rifled barrel. Has six chambers, front and rear sights and a large, smooth, wooden grip which carries an ejector attached to a ring in the butt. All metal parts are blued. A perforated shield is fitted over the breech of the cylinder. This shield is fitted with a port through which the cartridges are inserted into the chambers. The firing-pin of the hammer may be made to rest in an imperforate hole, which then acts as a safety-device and allows the cylinder to be turned when charged. Stamped upon the left side of the frame is the number "403." No proof-marks or maker's name are given.

France.

No. 39. (N3315). CENTER-FIRE REVOLVER. Caliber .41. Six-inch, round, rifled barrel. The left side of the frame is stamped with the number "250," but no proof-marks or maker's name are given. The other features of this specimen are the same as those of No. 38.

France.

PLATE 97.

FOREIGN AND UNITED STATES CENTER-FIRE REVOLVERS

No. 1. (N4841). CENTER-FIRE REVOLVER. Caliber .38. Five-and-one-eighth-inch, octagonal, smooth-bore brass barrel. Has six chambers, front and rear sights and a smooth, round grip which carries a brass ring in the butt. All metal parts of this double-action revolver are made of brass, except the hammer and trigger. Releasing a catch in front of the trigger-guard allows the barrel to tip upward. The empty shells are ejected by pressing the steel ejector upward. Stamped upon the left side of the barrel are the words "AL Spirlt & Cie. 3933," followed by a proof-mark in the form of a crown over the letter "G." The cylinder is also marked with the latter proof-mark, and in addition, bears that of Liége. The frame is stamped with a crowned, rampant lion.

Belgium.

No. 2. (N3314). CENTER-FIRE REVOLVER. Caliber .44.

Four-and-seven-eighths-inch, octagonal, rifled barrel. Has six chambers, front and rear sights and a checkered wooden grip, fitted with an iron ring in the butt. Releasing a catch behind the trigger-guard allows the latter to be lowered, and in so doing forces the barrel and cylinder forward on the cylinder-pin and extracts the shells. See No. 3. The right side of the barrel is stamped "C.F.G." beneath a crown. Upon the same side of the barrel is also the number "2068." This revolver was made by C. F. Galand.

France.

No. 3. (N4840). CENTER-FIRE REVOLVER. Caliber .44. Four-and-seven-eighths-inch, octagonal, rifled barrel. The trigger-guard is provided with an extra grip or spur. The action, which is the same as that of No. 2, is shown open. The cylinder has six chambers and the grip-plates are smooth and made of wood. Stamped upon the top of the barrel are the words "Manufacture Liégeoise D'Armes à Feu," while the right side is marked "C.F. Galand, Invr. Breveté 515." The cylinder is stamped with the Liége proof-mark.

France.

No. 4. (N4842). CENTER-FIRE REVOLVER. Caliber .44. Five-and-one-half-inch, octagonal, rifled barrel. Has six chambers, front and rear sights and a large wooden checkered grip which formerly carried a ring in the butt. Pressing upon a spring on the left side of the frame allows the cylinder and barrel to be swung downward and at the same time automatically ejects the shells. The chambers are also loaded when in this position. This weapon was patented by J. Warvan and is sometimes known as the Waverly Automatic. The ejector is stamped with the number "3." No proof-marks are given.

France.

No. 5. (N4847). CENTER-FIRE REVOLVER. Caliber .36. Five-and-one-quarter-inch, octagonal, rifled barrel. Has six chambers, front and rear sights and a round, wooden grip which carries a ring in the butt. This double-action revolver fires a cartridge which tapers to the front. When the cylinder-pin is removed by withdrawing forward it can be used as an ejector. The right side of the frame is stamped "J.F.J. Bar, Delft," and "W.D.W." Upon the under-side of the barrel is a proof-mark in the form of a crown over the letter "G." Various parts of the specimen are stamped with the serial number "965."

England.

No. 6. (N4858). CENTER-FIRE REVOLVER. Caliber .44. Seven-inch, tapering, rifled barrel. Has six chambers in the cylinder, front and rear sights and a smooth, round, wooden grip fitted with an iron ring in the butt. The piece is single-acting. Pushing down the safety on the left side of the frame prevents the hammer from being cocked in any position. The cylinder-pin can be removed by withdrawing forward, after which it can be used as an ejector. Stamped upon the left side of the frame are the words "Gebr. Mauser & Cie, 1880, Oberndorf." All parts are marked with the number "1701." This is a German military revolver which was made by the Mauser Brothers.

Germany.

No. 7. (N4849). CENTER-FIRE REVOLVER. Caliber .45. Four-and-one-half-inch, round, rifled barrel. Has six chambers, front and rear sights, an auxiliary spur on the guard and a large checkered ebonite grip which formerly carried a ring in the butt. The cylinder-pin is hollow and carries the ejector. Stamped upon the top of the frame are the words "Frontier Double Action," while the right side is marked "E.*" The inside of the grip bears the numbers "50" and "137." No maker's name is given. This is said to be an English army revolver, model of 1890.

England.

No. 8. (N4848). CENTER-FIRE REVOLVER. Caliber .32. Three-and-one-half-inch, octagonal, rifled barrel. Has six chambers, front and rear sights and a folding-trigger. The hammer is earless and the grip is made of checkered wood. After opening a port the chambers may be charged from the rear and on the right side of the frame. All metal parts are made of steel and blued, with the exception of the trigger-guard and hammer, which are left bright. Chambers and barrel are stamped with English proof-marks. The cylinder bears the words "Central Fire Pistol 320 Bore With Self Acting Safety, Patent Dec. 11. '73," while the barrel is marked "J.Lloyd Lewis, Cast Steel, Screwed Barrel." Various parts of the revolver are stamped with the number "570."

England.

No. 9. (N4846). CENTER-FIRE REVOLVER. Caliber .32. Three-and-five-eighths-inch, octagonal, rifled barrel. There are six chambers in the cylinder. Equipped with a front sight, while a groove in the rear of the frame serves as a back sight. The trigger is of the folding variety. An ejector is carried in the butt of the finely checkered wooden grip. All metal parts of this double-acting revolver were formerly blued. The cylinder-pin withdraws from in front. Chambers and barrel are stamped with English proof-marks, and the top of the barrel is engraved with the name "Robert Jones, Liverpool." The number "13" appears upon the cylinder-pin.

England.

No. 10. CENTER-FIRE REVOLVER. Caliber .28. Six-inch, octagonal, rifled barrel with a ridge on top. Has six chambers, triple sights and a smooth ivory grip. Frame and cylinder are engraved with floral designs. The cylinder-pin is hollow and carries an ejector. All metal parts were formerly plated. Stamped upon the top of the barrel of this double-action revolver are the words "British Bull Dog," while the butt bears the number "8099." No maker's name is given. This revolver is patterned after the French type of the same period.

England.

No. 11. (3830). CENTER-FIRE REVOLVER. Caliber .45. Three-and-one-eighth-inch, round, rifled barrel having a ridge on top. Has five chambers in the cylinder. Equipped with a round checkered wooden grip and a front sight, but there is no sight in the rear. All metal parts are nickel-plated. Pressing a screw upon the left side of the frame releases the cylinder-pin so it may be withdrawn forward. The breech of the cylinder is stamped with the Liége proof-mark and with another in the form of "$\overset{*}{\text{K}}$," the barrel also bearing the latter mark. Marked upon the left side of the frame are the figures "14 H.M." The serial number "30" appears upon the cylinder. No maker's name is given.

Belgium.

No. 12. (N4852). CENTER-FIRE REVOLVER. Caliber .45. Two-and-three-quarters-inch, round, rifled barrel with a ridge on top. Has five chambers in the cylinder. The front sight is like that of No. 11, while a groove in the rear of the frame serves as a back sight. Has a checkered wooden grip which formerly had a ring in the butt. The ejector is missing. All metal parts were formerly plated. Stamped upon the top of the barrel are the words "British Bull Dog," while the breech and cylinder bear the Liége proof-mark and another in the form of "$\overset{*}{\text{Z}}$." The left side of the frame is marked with the number "14." No maker's name is given.

England.

No. 13. (N4851). CENTER-FIRE REVOLVER. Caliber .45. Two-and-one-half-inch, round, rifled barrel with a ridge on top. There are five chambers in the fluted cylinder. The ejector is carried in the cylinder-pin which is hollow. Action and other features are the same as those of No. 12. Stamped upon the top of the barrel are the words "British Bull Dog," while the breech of the cylinder bears the Liége proof-mark and "x̽," the right side of the barrel also bearing the latter mark. No maker's name or serial number are given.

England.

No. 14. (N4853). CENTER-FIRE REVOLVER. Caliber .32. Two-and-three-eighths-inch, round, rifled, steel barrel with a ridge on top. Has seven chambers in a fluted cylinder. All metal parts were formerly plated. Action and other features are the same as those of No. 12. The top of the barrel is stamped with the words "British Bull Dog" and "Forehand & Wadsworth." No proof-marks are given. The butt bears the number "116597."

England?

No. 15. (N4854). CENTER-FIRE REVOLVER. Caliber .32. Two-inch, round, rifled barrel having a ridge on top. Has six chambers in a fluted cylinder. Action and other features are the same as those of No. 12. The top of the barrel is stamped with the words "British Bull Dog," while the breech of the cylinder is marked with the Liége proof-mark and with "L̽"; the barrel also bears the latter mark. No maker's name or serial number are given.

No. 16. (N4864). CENTER-FIRE REVOLVER. Caliber .32. Two-inch, octagonal, rifled barrel. Has six chambers in a plain cylinder. Equipped with front and rear sights and a small, round, checkered wooden grip. All metal parts were formerly plated. Stamped upon the top of the barrel is the word "Guardian," while the breech bears the Liége proof-mark. The inside of the grip bears the number "85." No maker's name is given.

England.

No. 17. (N4859). CENTER-FIRE REVOLVER. Caliber .44. Seven-inch, round, rifled barrel. There are six chambers in the fluted cylinder. Equipped with front and rear sights. Pulling back the hammer revolves the cylinder and cocks the revolver. When the loading-gate is pushed down the chambers may be loaded from the rear and on the right side of the frame. To open or "break" the revolver, it is brought to half-cock. The lug beneath the frame is pulled toward the trigger-guard and the catch on the left side of the barrel is pushed in. The barrel is then turned to the left and drawn forward with the cylinder, as shown in No. 18. All metal parts are blued. The grip is made of ivory, and the right plate bears the Mexican emblem of eagle and serpent in high relief. Stamped upon the top of the barrel are the words "Merwin Hulbert & Co., New York, U.S.A. Pat. Jan. 24, April 21, Dec. 15, '74, Aug. 3, '75, July 11, '76, April 17, '77, Pat's Mar. 6, '77." On the side of the barrel are the words "The Hopkins & Allen Manufacturing Co., Norwich, Conn., U.S.A." The butt bears the number "1588." This is apparently a presentation piece.

United States.

No. 18. (N4860). CENTER-FIRE REVOLVER. Caliber .41. Three-and-one-quarter-inch, round, rifled barrel. All metal parts are nickel-plated. The grip is made of checkered ebonite. Action, the method of breaking and other features are the same as those of No. 17. The side of the barrel is stamped "Hopkins & Allen M'f'g. Co. Norwich, Conn., U.S.A. Pat. Jan. 24, April 21, Dec. 15, '74, Aug. 3, '75, July 11, '76, Apr. 17, '77. Pats. Mar. 6, '77." The right side is marked "Calibre 1873 Winchester" and the serial number is "571."

United States.

No. 19. (N4861). CENTER-FIRE REVOLVER. Caliber .38. Three-and-one-half-inch, round, rifled barrel having a rib on top. There are five chambers in the fluted cylinder. Equipped with a front sight, while a groove in the rear of the frame serves as a back sight. Has a sheath trigger, consequently there is no trigger-guard. Action, method of breaking and other features are the same as those of No. 17. All metal parts are silver-plated and elaborately engraved. The grip is smooth and made of ivory. Stamped upon the top of the barrel are the words "Hopkins & Allen Mfg. Co. Norwich, Conn. U.S.A. Pat. Jan. 24, April 21, Dec. 15, '74, Aug. 3, '75, July 11, '76, Apr. 17, '77, Pats. Mar. 6, '77." The left side of the frame is marked "Merwin Hulbert & Co., New York, U.S.A.", while the right side bears the words ".38 Cal." The number "4565" appears upon the butt.

United States.

No. 20. (N3868). CENTER-FIRE REVOLVER. Caliber .36. Three-and-one-quarter-inch, round, rifled barrel having a rib on top.

Has five chambers in a fluted cylinder. Equipped with front and rear sights and an ebonite checkered grip. Pressing upward on the catch on the top of the frame, allows the barrel and cylinder to be tipped downward and automatically ejects the empty shells from the cylinder. The cylinder is loaded when in the same position. All steel parts were formerly blued, but are now gray due to being rubbed with an abrasive. Upon the top of the barrel are the words "Manufactured in U.S." No maker's name is given. The butt is stamped with the number "41654." This revolver was seized by the Milwaukee police from a person found with the same in his possession.
Donor: Judge George E. Page.

United States.

No. 21. (N3867). CENTER-FIRE REVOLVER. Caliber .32. Three-inch, round, rifled barrel having a rib on top. There are five chambers in the fluted cylinder. Pressing upward on the catch on the top of the frame, allows the barrel and cylinder to be tipped downward and automatically ejects the empty shells from the cylinder. The cylinder is loaded when in the same position. All steel parts, except the trigger-guard and hammer, are nickel-plated. Upon the top of the barrel are the words "Eastern Arms Co." The butt is stamped with the number "27382." This revolver was seized by the Milwaukee police from a person found with the same in his possession.
Donor: Judge George E. Page.

United States.

No. 22. (N4862). CENTER-FIRE REVOLVER. Caliber .32. Three-inch, round, rifled barrel having a ridge on top. Has five chambers in a fluted cylinder. Equipped with a front sight, while a groove in the rear of the frame serves as a back sight. The hammer has a hinged ear, which may be folded up so as to prevent accidental discharge. To open or "break" the revolver, it is first brought to half-cock. The lug beneath the frame is pulled toward the trigger-guard and the catch on the left side of the barrel is pushed in. The barrel is then turned to the left and drawn forward with the cylinder, as shown in No. 18. All metal parts were formerly nickel-plated and the grip is made of checkered ebonite. Stamped upon the top of the barrel of this double-action revolver are the words "Merwin Hulbert & Co., New York, U.S.A., Pat. Apr. 17, '77, June 15, '80, Mar. 14, '82, Jan. 9, '83." The butt bears the number "7524."

United States.

No. 23. (N3870). CENTER-FIRE REVOLVER. Caliber .38. Three-inch, octagonal, rifled, steel barrel. Has five chambers in a fluted cylinder for center-fire cartridges. Equipped with a front sight, while a groove in the rear of the frame serves as a back sight. The hammer has a hinged ear, which may be folded up to prevent accidental discharge. The cylinder-pin may be withdrawn forward by pressing upon a catch on the left side of the frame. Marked upon the top of the barrel are the words "Hopkins & Allen M'f'g. Co., Pat. Mar. 28, '71, Jan. 5, '86. X.L. Double Action 38 Cal. Centre Fire." The bottom of the barrel and the inside of the grip are stamped with the number "1289." This revolver was taken by the Milwaukee police from a person found with the same in his possession.

Donor: Judge George E. Page.

United States.

No. 24. (N4844). CENTER-FIRE REVOLVER. Caliber .36. Three-and-one-quarter-inch, rifled barrel having a ridge on top. There are five chambers in the fluted cylinder. Pressing upward on the catch on the top of the frame allows the barrel and cylinder to be tipped downward and then pulling back the movable rim at the breech of the cylinder ejects the empty shells, as shown in the illustration. The cylinder is also loaded when in this position. All steel parts are nickel-plated. The grip is of checkered ebonite. Stamped upon the top of the barrel are the words "American Arms Co., Boston, Mass., U.S.A. Dec. 5, '82, Mar. 27, '83, Reis. Dec. 5, '82." The serial number "11890" appears upon the butt.

United States.

No. 25. (N6060). CENTER-FIRE REVOLVER. Caliber .38. Two-and-one-half-inch, octagonal, rifled barrel. Has five chambers and a front sight; a groove in the rear of the frame serves as a back sight. The grip is of checkered ebonite. The cylinder-pin may be withdrawn forward by releasing the catch in front of the trigger-guard. All metal parts, except the hammer, are nickel-plated and polished. Stamped upon the top of the frame are the words "Double Action No. 6, Hopkins & Allen Arms Co., Norwich, Conn. U.S.A." The butt bears the number "3593." This revolver has apparently never been fired.

Donor: Mrs. W. W. Coleman.

United States.

No. 26. (N3864). CENTER-FIRE REVOLVER. Caliber .32. Two-and-one-quarter-inch, octagonal, rifled barrel. There are five chambers in the cylinder. Has a checkered composition grip and a front sight, while a groove in the rear of the frame serves as a back sight. The cylinder-pin is released as in No. 25. All metal parts are nickel-plated and polished. Stamped upon the top of the frame are the words "X.L. Double Action, Hopkins & Allen Arms Co., Norwich, Ct., U. S. A." The butt is marked with the number "2058." This is another revolver which was seized by the Milwaukee police from a person found with the same in his possession.

Donor: Judge George E. Page.

United States.

No. 27. (N3853). CENTER-FIRE REVOLVER. Caliber .38. Two-and-one-half-inch, octagonal, rifled barrel. There are five chambers in the fluted cylinder. All metal parts, except the trigger and hammer, are nickel-plated and polished. Action and other features are the same as those of No. 26. The top of the barrel is stamped "Forehand Arms Co., Worcester, Mass., Double Action Pat'd. June 2, 1891," and the butt bears the number "36634." This revolver was also seized by the Milwaukee police from a person found with the same in his possession.

Donor: Judge George E. Page.

United States.

No. 28. (N3857). CENTER-FIRE REVOLVER. Caliber .38. Two-and-one-half-inch, octagonal, rifled barrel. Has five chambers in a fluted cylinder. All metal parts, except the trigger-guard, trigger and hammer, are nickel-plated and polished. Stamped upon the top of the frame are the words "The American, Double Action," while the barrel is marked "H. & R. Arms Company, Worcester, Mass., U.S.A., 38 S.&W. Ctge." The butt is stamped with the number "46826." Action and other features are the same as those of No. 26. This specimen, which was made by the Harrington and Richardson Arms Company, was seized by the Milwaukee police from a person having the same in his possession.

Donor: Judge George E. Page.

United States.

No. 29. (N3869). RIM-FIRE REVOLVER. Caliber .32. Two-inch, octagonal, rifled barrel. Has five chambers in a fluted cylinder.

All metal parts, with the exception of the trigger-guard and trigger, are nickel-plated and polished. Action and other characteristics are the same as those of No. 26. Stamped upon the top of the frame are the words "Young America Double Action," while the barrel is marked "H. & R. Arms Company, Worcester, Mass., U.S.A., 32 S. & W. Ctge." The number "177892" appears upon the butt. The revolver was made by the Harrington and Richardson Arms Company and was seized by the Milwaukee police from a person who had the weapon in his possession.

Donor: Judge George E. Page.

United States.

No. 30. (N3871). CENTER-FIRE REVOLVER. Caliber .38. Two-and-one-half-inch, octagonal, rifled barrel. There are five chambers in the fluted cylinder. All metal parts, with the exception of the trigger-guard and trigger, are nickel-plated and polished. The hammer is earless. Action and other features are the same as those of No. 26. Stamped upon the top of the frame are the words "Safety Hammer Double Action," and upon the barrel "H.&R. Arms Company, Worcester, Mass., U.S.A., 38 S.&W. Ctge." The butt bears the serial number "52602." This arm, which was made by the Harrington and Richardson Arms Company, was also taken by the Milwaukee police from a person who had the same in his possession.

Donor: Judge George E. Page.

United States.

No. 31. (N3856). CENTER-FIRE REVOLVER. Caliber .32. Two-and-one-half-inch, octagonal, rifled barrel. Has five chambers in the cylinder. Action, finish and other features are similar to those of No. 26. The top of the frame is stamped "Double Action, Model 1900." No maker's name is given. Stamped upon the inside of the grip is the number "33274." This revolver was also seized by the Milwaukee police from a person having the same in his possession.

Donor: Judge George E. Page.

United States.

No. 32. (N3863). CENTER-FIRE REVOLVER. Caliber .38. Two-and-one-half-inch, octagonal, rifled barrel. Has the usual five chambers in the fluted cylinder. Action, finish and other features are similar to those of No. 26. Stamped upon the top of the frame are the words "H. & R. Double Action Model 1904, 38 Cal.", while the barrel is marked "H.&R. Arms Company, Worcester, Mass., U.S.A., 38 S. & W. Ctge." This specimen was made by the Harrington and Richardson Arms Company and was taken by the Milwaukee police from a person who was carrying the revolver. The serial number is "63121."

Donor: Judge George E. Page.

United States.

No. 33. (N3861). CENTER-FIRE REVOLVER. Caliber .32. Two-and-one-half-inch, octagonal, rifled barrel. There are five chambers in the cylinder. Action, finish and other features are the same as those of No. 26. This Harrington and Richardson revolver is marked upon the frame with the words "H. & R. Double Action Model 1905, 32 Cal.", while the barrel is stamped "H. & R. Arms Company, Worcester, Mass., U.S.A., 32 S. & W. Ctge." The serial number "39464" appears upon the butt. This is also one of the revolvers which was seized by the Milwaukee police.

Donor: Judge George E. Page.

United States.

No. 34. (N3854). CENTER-FIRE REVOLVER. Caliber .32. Two-and-five-eighths-inch, round, rifled barrel. Has five chambers in the cylinder. Action, finish and other characteristics are the same as those of No. 26. The top of the barrel is stamped with the words "U.S. Pistol Co.", but the place of manufacture and the serial number are not given. This revolver was seized by the Milwaukee police from a person found carrying the same.

Donor: Judge George E. Page.

United States.

No. 35. (N4857). CENTER-FIRE REVOLVER. Caliber .32. Two-inch, octagonal, rifled barrel. Has five chambers for long center-fire cartridges. There is a sight in front, but none in the rear. The hammer of this double-acting revolver is concealed. The grip is made of checkered wood, and the ramrod is carried in the cylinder-pin which is hollow. Frame, cylinder and back-strap are elaborately engraved with leaf designs. The right side of the barrel and frame each bear three proof-marks, while the cylinder is stamped with the Liége proof-mark. The top of the frame is engraved with the words "F. Dumoulin & Cie, Liége," while the under-side of the barrel is stamped "Belgium." No serial number is given.

Belgium.

No. 36. (N4239). CENTER-FIRE REVOLVER. Caliber of upper barrel .38; of the lower .50. Four-inch, octagonal, rifled upper barrel; and four-inch, smooth-bore, round lower barrel. This is a breech-loading, center-fire Le Mat revolver. The round, steel barrel serves as an axle upon which the cylinder revolves. The nine chambers of the cylinder fire through the top barrel. The nose of the hammer being adjustable, the single hammer fires both the bullets of the chambers through the top barrel and also the charge of shot from the lower barrel. The revolver is single-acting and is provided with a ring in front of the trigger-guard. An ejector is located on the right side of the frame. All metal parts are of steel and are blued. No maker's name or shop number are given. This revolver is the invention of Colonel Alexandre François Le Mat of New Orleans and was originally made for pin-fire cartridges.

France.

No. 37. (N4928). CENTER-FIRE REVOLVER. Caliber .36. Four-and-one-half-inch, octagonal, rifled barrel. Has six chambers in the cylinder. This revolver is double-acting and is equipped with front and rear sights, and a spring ejector. The frame is made in two parts and may be opened so as to gain access to the internal mechanism, as is shown in the Rast and Gasser revolver illustrated below in No. 38. The various parts may be removed without the use of tools. The right side of the barrel bears Belgium proof-marks, and the breech of the cylinder is stamped with the Liége proof-mark. This revolver was made by Blissett and Sons of London, and bears the serial number "8509."

England.

No. 38. (N3841). CENTER-FIRE REVOLVER. Caliber .315 (8 mm.). Four-and-one-half-inch, round, rifled barrel. There are eight chambers in a plain cylinder. This double-acting revolver fires the Austrian metal-jacketed bullet which is similar to the American .32 long. Has front and rear sights, a checkered wooden grip and a spring ejector. The frame is made in two parts and may be opened in order to gain easy access to the internal mechanism, as shown in the illustration. The various parts may be removed without the aid of tools. This is an Austrian service revolver bearing the number "23422" and the Austrian proof-mark. Stamped upon the frame are the words "Patent Rast &

Gasser Wein." The loading port on the right side of the frame acts also as a safety-device. The revolver was manufactured in Vienna.

Donor: Fred C. Best.

Austria.

PLATE 98.

COLT-PATERSONS AND COLT POCKET-PISTOLS

No. 1. (N4151). PERCUSSION REVOLVER. (Colt-Paterson.) Caliber .36. Four-and-one-half-inch, octagonal, rifled barrel. There is an extra barrel twelve inches in length in the case. Has five chambers for loose ammunition and a small, folding-trigger, which is revealed on bringing the hammer to full-cock. In order to load this revolver, it is necessary to remove the barrel and take the cylinder off the cylinder-pin. There is no ramrod. All steel parts are heavily blued. The grip is common to the early "Texas pistols" of which this is an example. The accessories in the case consist of an extra barrel and cylinder, a cleaning-rod, bullet-mold, cap-box and wrench. The Texas pistol was made at Paterson, New Jersey, about 1838. It was the first form of revolver to come into general use and was popular along the Western frontier. This form of weapon was in great demand in Texas and Mexico during the Mexican War (1846-1848) and one hundred dollars was frequently paid for one even after it had seen considerable service. The barrel of the specimen here shown is marked "Patent Arms M'g. Co., Paterson, N.J. Colt's Pt.", and with the number "31."

U. S. A.

No. 2. (N4153). PERCUSSION REVOLVER. (Colt-Paterson). Caliber .38. Twelve-inch, octagonal, rifled barrel. This target pistol has five chambers and chamfers at the muzzles of the chambers. The piece is single-acting, and has a folding-trigger which is revealed on bringing the hammer to half-cock. Loading is accomplished as in No. 1. All steel parts, with the exception of the frame, are heavily blued. The form of the grip is common to the first types of Colt's revolvers. Engraved upon the top of the barrel are the words "Patent Arms M'g. Co., Paterson, N.J. Colt's Pt." Grip, pin and barrel bear the number "992."

U. S. A.

No. 3. (N4152). PERCUSSION REVOLVER. (Colt-Paterson). Caliber .31. Four-and-one-half-inch, octagonal, rifled barrel. There are five chambers in the cylinder. The grip is smooth and polished, and all metal parts are heavily blued. Action and method of loading are the same as in No. 2. This Texas pistol is without the chamfers at the muzzles of the chambers. Since the great majority of the revolvers of this model were sent to Texas, Colt called them the "Texas pistols." All were made at Paterson, New Jersey. The parts of this model are not readily interchangeable, since most of the work was done by hand. However, the design and precision of the work were considered excellent, and it was done by very good mechanics. The unsatisfactory features of these revolvers were their complexity and fragility. Stamped upon the top of the barrel of the specimen here illustrated are the words "Patent Arms M'g. Co., Paterson, N.J., Colt's Pt." The number "18" also appears upon the barrel.

U. S. A.

No. 4. (N4150a, b). PERCUSSION REVOLVER. (Colt-Paterson). Caliber .26. Three-and-five-eighths-inch, octagonal, rifled, steel barrel. There are five chambers in the cylinder of this single-action revolver. Has a folding-trigger which is revealed on bringing the hammer to half-cock. In order to load this revolver it is necessary to remove the barrel and take the cylinder off the cylinder-pin. Has no ramrod or trigger-guard. All metal parts are heavily blued, and the grip is made of polished wood. Cylinder and frame are engraved with scroll designs. This is a model 1836 "Texas pistol" of pocket size. See No. 3 for more details regarding this model. Stamped upon the top of the barrel are the words "Patent Arms M'g. Co., Paterson, N.J. Colt's Pt." The number "381" also appears upon the barrel. Below the revolver is shown the brass charger, containing powder and bullets.

U. S. A.

No. 5. (N3829). PERCUSSION REVOLVER. (Colt-Paterson). Caliber .28. Two-and-one-half-inch, octagonal, rifled barrel. There are five chambers in the cylinder, and chamfers at the muzzles of the chambers. The revolver is single-action, and the trigger is of the folding variety. This is a model 1839 Texas pistol with a loading lever of the pattern designed for the Walker pistol and corresponding changes in the lug beneath the barrel. This is really a miniature of the Walker except for the trigger and guard of the latter. The top of the barrel is marked "Patent Arms M'g. Co., Paterson, N.J.-Colt's Pt." No serial number is given.

U. S. A.

No. 6. (N4180). PERCUSSION REVOLVER. (Colt-Paterson).
Caliber .31. Six-inch, octagonal, rifled barrel. There are five chambers
in the cylinder, but there are no chamfers at the muzzles of the chambers. In order to load this single-action revolver, it is necessary to
remove the barrel and take the cylinder off the cylinder-pin. There is
no ramrod. This is the second model of Colt's revolver, but it bears
no patent date. It is the first model to bear this peculiar trigger-guard.
Stamped upon the top of the barrel are the words "Address Saml. Colt,
New York City." The cylinder is marked "Colt's Patent, 7987." This
piece was made in Paterson, New Jersey, about 1838.

U. S. A.

No. 7. (16264). PERCUSSION REVOLVER. Caliber .31. Four-
inch, octagonal barrel rifled with seven grooves. There are five chambers in the cylinder, which is engraved with pictures of mounted men
engaged in an Indian fight. The cylinder stops are round. This is an
example of Colt's first model belt revolver, or model of 1848, or more
frequently called the "Wells Fargo," as that express company armed its
messengers with this weapon. There is no loading lever on this model,
which had barrel lengths of three, four, five and six inches. The first
revolvers made by Colt at Hartford, were of this model; they were
made in the three-story building on Pearl Street. Stamped upon the
top of the barrel of the specimen here shown are the words "Address
Saml. Colt New York City," while the left side of the frame is marked
"Colts Patent." Various parts of the piece bear the number "2078."
This revolver was found by the donor on Culp's Hill, after the battle
of Gettysburg, July 3rd, 1863.
Donor: Captain Charles W. Hyde.

U. S. A.

No. 8. (N6684). PERCUSSION REVOLVER. Caliber .31.
Four-inch, octagonal, rifled barrel. This is another example of the
"pocket pistol, model 1848," partly described under No. 7. This model
was made at Hartford, during 1848 and part of 1849, after which it was
superseded by a better arm. The distinguishing features of this model
are a trigger-guard, which is straight at the back; round or elliptical
cylinder stops; no safety-pins or loading lever; and no bearing wheel
on the hammer. Loading was done as in the case of the "Texas pistol."
The wholesale price at the factory was $17.30 for those with four-inch
barrels and a dollar extra for each additional inch increase in barrel

length.   The retail price varied from $20.00 to $25.00 in the New England states.   Various parts of the revolver illustrated here are stamped with the number "3458."

U. S. A.

No. 9. (N6700). PERCUSSION REVOLVER. Caliber .31. Four-inch, round, rifled barrel.   There are five chambers in the cylinder. This is still another example of the "pocket pistol, model of 1848." This piece is provided with an oval trigger-guard.   The specimen here shown is similar to No. 7 and No. 8, with the exception of the barrel and trigger-guard.   The top of the barrel is stamped "Address Saml. Colt, New-York City."   Various parts of the revolver bear the serial number "14835."

U. S. A.

No. 10. (N6683). PERCUSSION REVOLVER. Caliber .31. Five-inch, octagonal, rifled barrel.   This is another example of the "pocket pistol, model 1848."   See Nos. 7 and 8.   It is estimated that 1,500 of this model were made.   The specimen here shown is a presentation piece, as the back-strap is engraved with the words "Prize for Skill at Target Practice Won by Matthew H. Voorhees."   The butt bears the date of presentation, "August 23, 1849."   Stamped upon the top of the barrel are the words "Address Saml. Colt New York-City." The serial number is "1920."

U. S. A.

No. 11. (N4182). PERCUSSION REVOLVER. Caliber .31. Three-inch, octagonal, rifled barrel.   Has five chambers for loose ammunition.   There are no chamfers at the muzzles of the chambers.   In order to load, it is necessary to remove the barrel from the frame. There is no ramrod.   The trigger-guard is straight at the back as in the pocket pistol, model of 1848, made at Hartford.   The revolver here shown, however, was made by Colt at Whitneyville, Connecticut, in 1847; the Patent Arms Company at Paterson, New Jersey, had been discontinued in 1842.   The model illustrated was made with four, five and six-inch barrels and is known as the "pocket pistol," or the "police pistol."   The cylinder, which has rectangular stops, is engraved with a scene representing a stage coach holdup and bears the words "Colts Patent. No. 14012."   Stamped upon the top of the barrel are the words "Address Saml. Colt New York City."   Various other parts of the revolver bear the number "14012."

U. S. A.

No. 12. (N4181). PERCUSSION REVOLVER. Caliber .31. Three-inch, octagonal, rifled barrel. There are five chambers in the cylinder. The trigger-guard is curved at the front and back, otherwise this revolver is the same as No. 11. The top of the barrel is stamped "Address Saml. Colt New York City," while the cylinder is marked "Colts Patent, 55668." Various other parts of the piece are also stamped with the same number. This revolver was made by Colt in Whitneyville, Connecticut.

U. S. A.

No. 13. (N4179). PERCUSSION REVOLVER. Caliber .36. Four-and-one-half-inch, round, rifled barrel. There are five chambers in this single-action revolver. A notch in the top of the hammer serves as a rear sight. Fitted with a lever ramrod beneath the frame. The trigger-guard is curved at the front and back. The cylinder, which is from a different revolver, is turned down at the back and bears the picture of a stage-coach holdup and the words "Colts Patent No. 8420." Other parts are stamped with the number "39954." This revolver is similar to the army pistol, model of 1860, except for the length of barrel, caliber, and the number of shots. Stamped upon the top of the barrel are the words "Address Col. Saml. Colt New-York, U.S. America."

U. S. A.

No. 14. (N2870). PERCUSSION REVOLVER. Caliber .36. Four-and-one-half-inch, octagonal, rifled barrel. Has five chambers in the cylinder which is turned down at the back and which is engraved with a stage-coach holdup. This model was known as "Colt's pocket pistol." The cylinder and various other parts of the revolver are stamped with the number "2317," while the top of the barrel is marked "Address Col. Saml. Colt New-York, U.S. America."

U. S. A.

No. 15. (N4198). PERCUSSION REVOLVER. Caliber .31. Four-inch, octagonal, rifled barrel. There are five chambers in the cylinder. Trigger-guard and back-strap are made of brass, and are silver-plated. The cylinder is straight, and formerly had a stage-coach holdup scene engraved upon it. It bears the words "Colts Patent No. 164227." All other parts of the revolver are also stamped with the same number. This is another example of the "Colt's pocket pistol." The top of the barrel is marked "Address Saml. Colt, New York City."

U. S. A.

No. 16. (N4194).   PERCUSSION REVOLVER.   Caliber .31. Four-inch, octagonal, rifled barrel.   There are five chambers in the cylinder which is straight and which formerly had a stage-coach holdup scene engraved upon it.   Trigger-guard and back-strap are made of steel, and were formerly plated.   This is another example of Colt's pocket pistol.   The cylinder is marked "Colts Patent No. 153484" and the barrel "Address Saml. Colt, New York City."   Various parts of the piece bear the number "153484."

U. S. A.

No. 17. (N4193).   PERCUSSION REVOLVER.   Caliber .31. Four-inch, octagonal, rifled barrel.   Has five chambers in the cylinder. Trigger-guard and back-strap are made of brass, and are silver-plated. Barrel and frame are blued.   The cylinder is engraved with the usual stage-coach holdup, and is stamped "Colts Patent No. 13021."   The barrel and each chamber of the cylinder are marked with English proof-marks.   Each part of the revolver is stamped with the number "213021 E," with the exception of the cylinder.   The top of the barrel is marked "Address Col. Saml. Colt New-York U.S. America."   This revolver was exported from the United States and proved in England.

U. S. A.

No. 18. (N4197).   PERCUSSION REVOLVER.   Caliber .31. Four-inch, octagonal, rifled barrel.   There are five chambers in the cylinder, which is engraved with the usual stage-coach holdup scene. Trigger-guard and back-strap are made of brass, and are silver-plated. All steel parts are blued.   This is another Colt's pocket pistol.   The cylinder is marked "Colts Patent No. 147586," and the barrel "Address Saml. Colt, New York City."   All other parts bear the same serial number as the cylinder.

U. S. A.

No. 19. (N4196).   PERCUSSION REVOLVER.   Caliber .31. Four-inch, octagonal, rifled barrel.   Trigger-guard and back-strap are made of brass, and were formerly silver-plated.   Barrel, cylinder and frame were formerly blued.   Upon the cylinder are the words "Colts Patent No. 4685," and upon the top of the barrel "Address Col. Saml. Colt New-York U.S. America."   All other parts of this pocket pistol, which otherwise is identical with No. 18, bear the number "324685."

U. S. A.

No. 20. (N4195). PERCUSSION REVOLVER. Caliber .31. Four-inch, octagonal, rifled barrel. Has five chambers for loose ammunition. Trigger-guard and back-strap are made of steel, and all steel parts are blued. The usual stage-coach holdup is engraved upon the cylinder, which, in addition, bears the words "Colts Patent No. 3348." All other parts of the revolver bear the same number. The barrel and each chamber are stamped with English proof-marks, and the top of the barrel bears the words "ADDRESS COL. COLT LONDON." Early in January, 1853, Colt opened a foreign plant in London, where arms, nearly identical with those of Hartford, were manufactured.

U. S. A.

No. 21. (N4550). PERCUSSION REVOLVER. Caliber .31. Four-inch, octagonal, rifled barrel. Has five chambers for loose ammunition. All metal parts are nickel-plated, and the frame and cylinder are slightly engraved. Stamped upon the top of the barrel are the words "London Pistol Company," and upon the under-side of the frame "Patented Dec. 27, 1859." This company purchased Colt's London factory and contents. Their pistols were made similar to his, except for a removable side plate and double the number of cylinder stops. The under-side of the barrel bears the number "199."

England.

No. 22. (N4190). PERCUSSION REVOLVER. Caliber .31. Five-inch, octagonal, rifled barrel. There are five chambers in the cylinder. Has small, brass trigger-guard and back-strap. The cylinder is engraved with a scene representing a stage-coach holdup, and is marked "Colts Patent No. 40639." All other parts are stamped with the same number. Stamped upon the top of the barrel are the words "Address Saml. Colt New-York City." This is another form of the pocket pistol.

U. S. A.

No. 23. (N6798). PERCUSSION REVOLVER. Caliber .31. Five-inch, octagonal, rifled barrel. There are five chambers in the cylinder which is stamped with the number "37026." All other parts of the revolver bear the number "39026." In other respects this piece is identical to No. 22.

U. S. A.

No. 24. (N4191). PERCUSSION REVOLVER. Caliber .31. Five-inch, octagonal, rifled barrel. Has five chambers in a cylinder which is engraved with a scene representing a stage-coach holdup. Trigger-guard and back-strap are made of silver-plated brass. Has a smooth, varnished, walnut grip and a blued cylinder and barrel; the frame and hammer are case-hardened in various colors. The cylinder is stamped "Colts Patent No. 7402," and the top of the barrel "Address Col. Saml. Colt New-York U.S. America." Various other parts of the piece are stamped with the number "317402."

U. S. A.

No. 25. (N4187). PERCUSSION REVOLVER. Caliber .31. Five-inch, octagonal, rifled barrel. There are six chambers in the cylinder. Trigger-guard and back-strap are made of brass, and were formerly silver-plated. Frame and hammer are case-hardened in various colors. The number "35866" appears upon the cylinder; all the other parts of this piece are numbered "235866." Other markings and details are the same as those of No. 24.

U. S. A.

No. 26. (N6680). PERCUSSION REVOLVER. Caliber .36. Four-and-one-half-inch, octagonal, rifled barrel. There are five chambers. The cylinder is cut away in the rear, and provided with safety pins. Has a large, brass trigger-guard. Stamped upon the top of the barrel are the words "Address Col. Saml. Colt, New-York, U.S. America." Both cylinder and frame bear the words "Colt's Patent." Various parts bear the serial number "7202."

U. S. A.

No. 27. (N6686). PERCUSSION REVOLVER. Caliber .31. Five-inch, octagonal, rifled barrel. Has six chambers in the cylinder which is engraved with the usual stage-coach holdup scene. This is another representative of the pocket pistol, this one having a large, brass trigger-guard. The top of the barrel is stamped "Address Col. Saml. Colt, New-York U.S. America." Various parts of the revolver bear the number "293139."

U. S. A.

No. 28. (N6708). PERCUSSION REVOLVER. Caliber .31. Five-inch, octagonal, rifled barrel. Has five chambers in the cylinder which is engraved with the usual stage-coach holdup scene. Barrel and each chamber are stamped with Birmingham proof-marks. Back-strap and trigger-guard are made of steel. Upon the top of the barrel

[DDRESS COL. COLT LONDON," indicating that
[e in Colt's London factory some time during 1853-
[s of the revolver bear the shop number "3889."

England.

No. 29.   (N6711).   PERCUSSION REVOLVER.   Caliber .31.
Five-inch, octagonal, rifled barrel.   There are five chambers in the cylinder, which is neither engraved nor provided with safety-pins.   The
slots in the cylinder are square.   Engraved in script upon the top of
the barrel is the name "SAMUEL COLT."   The place of manufacture
is not stated upon this piece, which is a rare form of the pocket pistol.
Various parts of the revolver bear the number "159253."

U. S. A.

No. 30.   (N4192).   PERCUSSION REVOLVER.   Caliber .31.
Four-and-one-half-inch, octagonal, rifled barrel having five chambers
in a semi-fluted cylinder.   The frame and cylinder are engraved; barrel
and hammer are blued; cylinder and back-strap are silver-plated; the
frame is plated with gold.   The grip is made of smooth, polished walnut.
Stamped upon the barrel are the words "Address Col. Saml. Colt New-
York, U.S. America," and on the cylinder "Pat. Sept. 10th 1850."   This
is apparently a composite police pistol, since the frame is stamped with
the number "24243," the trigger-guard with "47134," the butt with
"65267," the barrel-pin with "00343," and the cylinder with "4243."

U. S. A.

No. 31.   (N4173).   PERCUSSION REVOLVER.   Caliber .36.
Four-and-one-half-inch, round, rifled barrel.   There are five chambers
in a semi-fluted cylinder.   Trigger-guard and back-strap are made of
silver-plated brass, while all steel parts are blued.   The grip is made
of smooth, polished steel.   The top of the barrel of this police pistol
is stamped "Address Col. Saml. Colt New York U.S. America," and
the cylinder bears the words "Pat. Sept. 10th 1850."   Various parts
of the revolver are stamped with the number "4131."

U. S. A.

No. 32.   (N6688).   PERCUSSION REVOLVER.   Caliber .31.
Five-inch, octagonal, rifled barrel.   There are five chambers in the
cylinder, which is engraved with a scene representing a stage-coach
holdup.   There are safety-pins in the rear of the cylinder.   The trigger-
guard is rather small.   Back-strap and guard are silver-plated.   Upon

marks, and the left side of the frame is engraved with the words "Colt's Patent." The grip is polished. This model 1851 navy revolver was made in Colt's London factory some time during 1853-1857. All parts bear the number "37325."

England.

No. 13. (N6703). PERCUSSION REVOLVER. Caliber .36. Seven-and-one-half-inch, octagonal, rifled barrel. There are six chambers in the fluted cylinder. The nickel-plating on this specimen is of recent date, as the original arms of this model were blued. The butt is stamped with the letters "U.S.N.", and the top of the barrel bears the words "Address Saml. Colt. New York City." Various parts of the revolver bear the number "59885." The model 1851 navy revolver having the fluted cylinder is very rare.

U. S. A.

No. 14. (N6698). PERCUSSION REVOLVER. Caliber .—. Seven-and-one-half-inch, octagonal, rifled barrel. This variation of the model 1851 navy revolver has a groove in the recoil shield to guide the caps to the nipples. In addition to the usual notch in the top of the hammer, which serves as a rear sight, this piece has also a V-shaped sight placed upon the rear portion of the barrel. The top of the barrel is stamped "Address Col. Saml. Colt, New York U.S. America," while various parts bear the number "164033."

U. S. A.

No. 15. (N6710). PERCUSSION REVOLVER. Caliber .36. Seven-and-one-half-inch, octagonal, rifled barrel. This is a model 1851 navy revolver made in Colt's London factory some time between 1853 and 1857. There is no groove to guide the caps to the nipples. Various parts of the piece bear the number "110708." Stamped upon the top of the barrel are the words "ADDRESS COL. COLT LONDON."

England.

No. 16. (N6697). PERCUSSION REVOLVER. Caliber .36. Seven-and-one-half-inch, octagonal, rifled barrel. Has the usual six chambers in the engraved cylinder. This specimen has a small trigger-guard. There is no groove to guide the caps to the nipples. The top of the barrel is marked "Address Saml. Colt, New York City." Various parts of the revolver bear the number "8357."

U. S. A.

No. 17. (N6707). PERCUSSION REVOLVER. Caliber .36. Seven-and-one-half-inch, octagonal, rifled barrel. This is another London Colt, navy model of 1851. The cylinder is stamped with English proof-marks. There is no groove to guide the caps to the nipples. The top of the barrel bears the words "ADDRESS COL. COLT LONDON." Various parts of the piece are stamped with the number "40369."

England.

No. 18. (N4165). PERCUSSION REVOLVER. Caliber .36. Seven-and-one-half-inch, octagonal, rifled barrel. This piece shows to good advantage the finish that was applied to all regular Colts of this model. Lever, frame and hammer are case-hardened in variegated colors. The barrel, cylinder and trigger-guard are blued by heating in charcoal, while the grip is of oil-finished black walnut. All parts of the piece are stamped with the number "9064." In other respects this revolver is similar to No. 16.

U. S. A.

No. 19. (N6712). PERCUSSION REVOLVER. Caliber .36. Seven-and-one-half-inch, octagonal, rifled barrel. There are six chambers in the cylinder, which is engraved with a scene representing mounted horsemen engaged in a running fight. The figures of this scene are crude and very much out of proportion. There are long stops upon the cylinder, which is also stamped with the Liége proof-mark. Marked upon the top of the barrel are the words "Colt Patent." Various parts of the piece are marked with the number "14248." This is probably a counterfeit Colt made in Belgium for the export trade. See No. 11.

Belgium.

No. 20. (N2872). PERCUSSION REVOLVER. Caliber .36. Seven-and-one-half-inch, octagonal, rifled barrel. Has the usual six chambers in the cylinder. This is a composite model 1851 navy revolver because the barrel bears the number "44273," frame and cylinder are stamped "66717," the brass trigger-guard "44770," and the butt-plate "63678." The top of the barrel is marked "Address Saml. Colt New York City," and the side of the frame "Colt's Patent U.S." There is a notch in the butt-plate which served as a point for the attachment of a carbine stock. All parts of the revolver are unornamented.

U. S. A.

No. 21. (N4162). PERCUSSION REVOLVER. Caliber .36. Seven-and-three-eighths-inch, octagonal, rifled barrel. There are six chambers in the cylinder, which formerly was engraved with the usual naval battle scene peculiar to this model. Trigger-guard and back-strap are made of brass, while the grip-plates are made of smooth ivory. The cylinder is stamped with the number "8891," the butt "150129," while the rest of the parts are stamped with the number "176910." There is a groove in the recoil shield to guide the caps to the nipples. The top of the barrel is stamped "Address Col. Saml. Colt New-York U.S. America."

U. S. A.

No. 22. (N4166). PERCUSSION REVOLVER. Caliber .36. Seven-and-one-half-inch, octagonal, rifled barrel. There are six chambers in the cylinder, which is engraved with the usual naval battle scene. The trigger-guard is made of brass, while the rest of the parts are made of steel. Upon the left side of the frame are the words "Colts Patent U.S." The grip is of checkered ivory, and formerly belonged to a .44 caliber Colt. All parts of this model 1851 revolver are without a serial number. The top of the barrel is marked "Address Saml. Colt New-York U.S. America."

U. S. A.

No. 23. (N4169). PERCUSSION REVOLVER. Caliber .36. Seven-and-one-half-inch, octagonal, rifled barrel. Has six chambers in the cylinder. Trigger-guard, frame and cylinder are gold-plated, and the two latter are also elaborately engraved. Barrel, hammer, ramrod and back-strap are heavily blued, and the grip-plates are made of mother-of-pearl. The top of the barrel is stamped as No. 22. This is either a presentation piece or an officer's revolver, model 1851. The ramrod bears the number "5553," while the frame is stamped "163829."

U. S. A.

No. 24. (N6709). PERCUSSION REVOLVER. Caliber .36. Five-and-three-eighths-inch, octagonal, rifled barrel. Has six chambers in the cylinder. There is a groove in the recoil shield to guide the caps to the nipples. This short form of the model 1851 navy revolver was probably intended for civilian use. The top of the barrel is stamped "Address Saml. Colt, New York City," while cylinder and frame bear the words "Colt's Patent." The barrel bears the number "L 59" and the rest of the parts are stamped with "2186."

U. S. A.

No. 25. (N4170). PERCUSSION REVOLVER. Caliber .36. Four-and-one-half-inch, octagonal, rifled barrel. There are six chambers in the cylinder, which is engraved with the usual naval battle scene peculiar to this model. The trigger-guard is made of steel, and the back-strap of brass. Stamped upon the top of the barrel are the words "Address Saml. Colt Hartford, Ct." This is another example of the short form of the 1851 model navy revolver and was probably intended for civilian use. The butt bears the number "16," trigger-guard "70925," barrel "98136," and the ramrod "8136."

U. S. A.

No. 26. (N6694). CENTER-FIRE REVOLVER. Caliber .45. Eight-inch, round, rifled barrel. Has six chambers. This is a model 1861 army pistol, which has been altered from the percussion-cap system to the center-fire. The hammer passes through the frame in the rear. This revolver has no gate in the rear for loading. The trigger-guard is of brass, while the back-strap is of iron. The cylinder is from an 1851 model navy revolver. Stamped upon the top of the barrel are the words "Address Col. Saml. Colt New-York U.S. America." The cylinder bears the number "14270," while the rest of the parts are stamped "135006."

U. S. A.

No. 27. (N6693). RIM-FIRE REVOLVER. Caliber .44. Eight-inch, round, rifled barrel. There are six chambers in the cylinder. This is a model 1861 army pistol, which has been altered from the percussion-cap system to the rim-fire. The rear of the cylinder containing the nipples has been cut away and a ring containing a firing-pin has been added to replace the removed portion. These features were covered by a patent granted to Alexander Thurer, September 15th, 1868. This revolver fires a tapering, brass cartridge loading into the front of the cylinder. The revolver can be made safe by rotating the ring to the left to the limit of motion and allowing the hammer to rest in the notch. The top of the barrel is stamped "Address Col. Saml. Colt New-York U.S. America." The cylinder is marked with the number "0094," while the rest of the parts are numbered "158793."

U. S. A.

No. 28. (N6749). CENTER-FIRE REVOLVER. Caliber .45. Eight-inch, round, rifled barrel. Has six chambers in the cylinder. This is a model 1861 army revolver, which has been altered from the .44

caliber percussion-cap system to the .45 center-fire. The rear portion of the cylinder containing the nipples has been cut away and a brass ring added to replace the removed part. The cylinder is from an 1851 model navy pistol and has extra stops cut into it. The opening upon the right side of the frame is not equipped with a gate. There is no ejector. The trigger-guard is of brass and the back-strap is of iron. All parts are stamped with the number "143762," except the cylinder, which bears the number "75."

U. S. A.

No. 29. (N6695). CENTER-FIRE REVOLVER. Caliber .45. Seven-and-three-quarters-inch, round, rifled barrel. Has six chambers. This model 1861 army pistol has also been altered from the percussion-cap system to the center-fire. This conversion has a gate upon the right side of the frame for loading. It is also equipped with a spring ejector. The trigger-guard is of brass, and the back-strap is of iron. The cylinder is from an 1851 model navy revolver and has extra stops cut into it. Stamped upon the top of the barrel are the words "Address Col. Saml. Colt New York U.S. America." The cylinder bears the number "5232," while the rest of the parts are stamped "51508" and "372."

U. S. A.

No. 30. (N6696). CENTER-FIRE REVOLVER. Caliber .38. Seven-and-one-half-inch, round, rifled barrel. This is a model 1861 navy pistol, which has been altered from the .36 caliber percussion-cap system. The cylinder is from an 1851 model navy revolver. This piece is provided with a gate and an ejector. The trigger-guard is of brass and the back-strap of iron. The nose of the hammer is fitted with a round firing-pin. Upon the top of the barrel are the words "Address Col. Saml. Colt New-York U.S. America." The butt is stamped "U.S. N." and "89353," the cylinder "89874," while the rest of the parts are marked with the number "10383."

U. S. A.

No. 31. (N6699). RIM-FIRE REVOLVER. Caliber .45. Eight-inch, round, rifled barrel. Has six chambers. This is a model 1861 army pistol, which has been altered from the .44 caliber percussion-cap system to the .45 rim-fire. The frame has been milled out so as to take the straight cylinder. There is no gate or ejector. The butt-plate has a notch for a shoulder stock. There are only three screws in the frame. The nose of the hammer is fitted with a flat firing-pin. The top

of the barrel is marked as is No. 30. Various parts of the revolver bear
the number "61422."

U. S. A.

No. 32. (N4207). RIM-FIRE REVOLVER. Caliber .36. Seven-
and-one-half-inch, round, rifled barrel. There are six chambers in the
cylinder. The chambers are loaded from the rear and on the right side
of the frame. Fitted with a spring ejector. This is a belt pistol which
was altered from loose ammunition. Trigger-guard and back-strap are
made of brass and all steel parts are blued. The top of the barrel is
marked as No. 30, while on the bottom of the barrel are the letters
"R.W.M.", followed by an anchor. All parts are marked with the number "2950."

U. S. A.

No. 33. (N4206). RIM-FIRE REVOLVER. Caliber .38. Seven-
and-one-half-inch, octagonal, rifled barrel. There are six chambers in
the cylinder for rim-fire, metallic cartridges. The chambers are loaded
from the rear and on the right side of the frame, after opening a port.
Provided with a spring ejector. All metal parts are made of steel. The
barrel is marked as that of No. 30. Upon the butt are the letters "U.S.
N." The cylinder bears the number "3700," the barrel "59836," and
the rest of the parts "89358."

U. S. A.

No. 34. (N3317). CENTER-FIRE REVOLVER. Caliber .45.
Seven-and-three-quarters-inch, round, rifled barrel. Has six chambers
for center-fire, metallic cartridges. A rear sight is located upon the
recoil shield. The chambers are loaded from the rear and on the right
side of the frame. A spring ejector is carried on one side. The trigger-
guard is made of brass, while the rest of the parts are of blued steel.
The butt is fitted to take a carbine stock. The cylinder bears the number "5," the barrel "64," and the trigger-guard and butt "46854." The
barrel is marked as that of No. 30.

U. S. A.

No. 35. (N16266). PERCUSSION REVOLVER. Caliber .44.
Seven-and-one-half-inch, octagonal, rifled barrel. This model 1851 navy
revolver, together with the belt, holster, bullet-mold and caps, here
shown, were used by the late Captain Charles W. Hyde of Company B,
First Wisconsin Heavy Artillery, while stationed at Fort Clay, Lexing-

ton, Kentucky, during the Civil War. All parts of the revolver are stamped with the number "110140."

Donor: Captain Charles W. Hyde.

U. S. A.

PLATE 101.

COLTS WITH SIDE-HAMMERS, POLICE AND MISCELLANEOUS COLT REVOLVERS

No. 1. (N4202). PERCUSSION REVOLVER. Caliber .31. Three-and-one-half-inch, round, rifled barrel. Has five chambers in the cylinder for loose ammunition. Pulling back the hammer revolves the cylinder and cocks the revolver. The piece is provided with a front sight, while a groove in the rear of the frame serves as a back sight. All metal parts are made of steel and were formerly blued. This is an example of the "perfected model 1855 pocket pistol," which was patented by Elisha K. Root, December 25, 1855, and purchased by Colt. The chief features of this model are a hammer on the side of the frame, a cylinder inclosed by the frame, and a cylinder-pin which turned with the cylinder and withdrew from the rear. In the specimen here shown there is a set-screw through the cylinder which has to be removed before the cylinder-pin can be withdrawn. The top of the barrel is stamped "Address Col. Colt New York U.S.A." The cylinder is marked "Colts Patent No. 14028," and various other parts are also stamped with the same number.

U. S. A.

No. 2. (N4201). PERCUSSION REVOLVER. Caliber .31. Three-and-one-half-inch, round, rifled barrel. There are five chambers in the cylinder. This is another example of the "perfected model of 1855 pocket pistol." See No. 1. Although this model was not placed on the market until 1857, it is almost always known as the 1855 model. Stamped upon the top of the specimen here shown are the words "Address Col. Colt New-York U.S.A.", and upon the cylinder-pin "May 4th 1858." Various parts of the piece bear the number "10118."

U. S. A.

No. 3. (N4205). PERCUSSION REVOLVER. Caliber .28. Three-and-one-half-inch, octagonal, rifled barrel. There are five chambers in the cylinder which is engraved with a scene representing an Indian fight. The top of the barrel is stamped "Colt's Pt. 1855, Address

Col. Colt Hartford, Ct., U.S.A." Cylinder and other parts are marked with the number "24727." In this specimen there is a small button on the left side of the frame which when pressed releases the cylinder-pin for withdrawal. In all other respects this piece is similar to No. 1.

U. S. A.

No. 4. (N6692). PERCUSSION REVOLVER. Caliber .28. Two-and-one-half-inch, octagonal, rifled barrel. Has the usual five chambers in the cylinder. Stamped upon the top of the barrel are the words "Colt's Pt. 1855. Address Col. Colt, Hartford, Ct. U.S.A." Various parts of the specimen are marked with the number "16064." In all other respects this revolver is similar to No. 1.

U. S. A.

No. 5. (N5771). PERCUSSION REVOLVER. Caliber .28. Three-and-one-half-inch, octagonal, rifled barrel. Has the usual five chambers for loose ammunition. In this specimen there is a small button on the left side of the frame which when pressed releases the cylinder-pin for withdrawal. The top of the barrel is stamped "Colt's Pt. 1855, Address Col. Colt Hartford, Ct., U.S.A." Cylinder and other parts of this piece are marked with the number "20260." In all other respects this revolver is the same as No. 1.

U. S. A.

No. 6. (N4204). PERCUSSION REVOLVER. Caliber .28. Three-and-one-half-inch, octagonal, rifled barrel. In the specimen here shown there is a small catch in the frame which, when pressed, releases the cylinder-pin for withdrawal. The top of the barrel is stamped "Colt's Pt. 1855, Address Col. Colt Hartford, Ct., U.S.A." Cylinder and other parts are marked with the number "3047." In all respects this revolver is similar to No. 1.

U. S. A.

No. 7. (N791). PERCUSSION REVOLVER. Caliber .28. Three-and-one-half-inch, octagonal, rifled barrel. There are five chambers, in a long, fluted cylinder, for loose ammunition. Stamped upon the top of the barrel are the words "Colt's Pt. 1855, Address Col. Colt, Hartford, Ct., U.S.A." Various parts of the revolver are marked with the number "26767." In other respects this piece is similar to No. 1.

U. S. A.

No. 8. (N4200). PERCUSSION REVOLVER. Caliber .31.

Four-and-one-half-inch, round, rifled barrel. Has five chambers in a long, fluted cylinder. All metal parts are made of steel, and were formerly silver-plated. The grip-plates are of smooth ivory. Stamped upon the top of the barrel are the words "Address Col. Colt New-York U.S.A.", while the cylinder is marked "Patented Sept. 10th 1850." Various parts of the revolver bear the number "6499." See No. 1 for further details regarding this model.

U. S. A.

No. 9. (N4199). PERCUSSION REVOLVER. Caliber .31. Four-and-one-half-inch, round, rifled barrel. There are five chambers in a long, fluted cylinder which is marked "Patented Sept. 10th, 1850." The top of the barrel is stamped "Address Col. Colt New-York U.S.A." Various parts bear the number "4628." In other particulars this revolver is similar to No. 1.

U. S. A.

No. 10. (N4203). PERCUSSION REVOLVER. Caliber .28. Three-and-one-half-inch, octagonal, rifled barrel. Has the usual five chambers which in this piece are in a smooth cylinder. All metal parts are made of steel and are finely engraved. The grip is of ivory with a helmeted, Greek head in high relief. Stamped upon the top of the barrel are the words "Colt's Pt. 1855, Address Col. Colt Hartford, Ct., U.S.A.", while the cylinder is marked "Patented Sept. 10th, 1850." Various parts of the piece bear the number "26522." See No. 1 for a more detailed description of this model.

U. S. A.

No. 11. (N4157-59a-e). PAIR OF PERCUSSION REVOLVERS. Caliber .28. Three-and-one-half-inch, octagonal, rifled barrels. Each of the fluted cylinders contains five chambers. All metal parts are made of steel, and are heavily silver-plated. The grip-plates are made of checkered ivory. The top of each barrel is marked "Colt's Pt. 1855, Address Col. Colt, Hartford, Ct., U.S.A.", while each cylinder is marked "Patented Sept. 10th 1850." One revolver bears the number "25057" and the other "25067." The accessories in the original case consist of a bullet-mold, powder-flask, cleaning-rod, and a combined screw-driver and nipple-wrench. This is a set of the model of 1855 revolvers. See No. 1.

U. S. A.

No. 12. (1353). PERCUSSION REVOLVER IN CASE. Caliber .28. Three-and-one-half-inch, octagonal, rifled barrel. There are five chambers in the engraved cylinder. This model 1855 revolver was carried by Lieutenant Thomas of the 82nd Illinois Volunteers during the four years of the Civil War. The accessories in the case consist of a screw-driver, powder-flask, bullet-mold and a box of percussion-caps. Various parts of the revolver bear the number "8214." See No. 1.

U. S. A.

No. 13. RIM-FIRE REVOLVER. Caliber .38. Two-and-one-quarter-inch, round, rifled barrel. Has five chambers in the cylinder for rim-fire, metallic cartridges. Pressing upon the catch in front of the cylinder allows the cylinder-pin to be withdrawn forward for the removal of the cylinder. There is a groove in the rear and on the right side of the frame for the insertion of the cartridges into the chambers. Specimens of this kind are sometimes known as "Colt's New Pocket Revolver." The top of the barrel is marked "Colt's Pt. F. A. Mfg. Co. Hartford, Ct., U.S.A.", while on the side are the words "Colt New 38." Upon the inside of the grip is the number "7457."

U. S. A.

No. 14. (N4214). RIM-FIRE REVOLVER. Caliber .41. Two-and-one-quarter-inch, round, rifled barrel. There are five chambers in the fluted cylinder. Provided with a front sight, while a groove in the rear of the frame serves as a back sight. The cylinder-pin withdraws forward and may be used as an ejector. All metal parts were formerly nickel-plated. The grip-plates are of polished ivory. Stamped upon the top of the barrel are the words "Colt's Pt. F.A. Mf'g. Co., Hartford, Ct., U.S.A.", while upon the bottom is the number "5058."

U. S. A.

No. 15. (N6743). RIM-FIRE REVOLVER. Caliber .38. Two-and-one-quarter-inch, round, rifled barrel. Has five chambers in a fluted cylinder for rim-fire, metallic cartridges. Pressing on the catch in front of the cylinder allows the cylinder-pin to be withdrawn forward for the removal of the cylinder. All metal parts are blued, and the grip-plates are of ivory. The top of the barrel is stamped "Colt's Pt. F.A. Mfg. Co., Hartford, Ct., U.S.A." The words "Pet Colt" are engraved upon the barrel. On the frame beneath the grip is the number "384." This is another example of "Colt's New Pocket Revolver."

U. S. A.

No. 16. (N6744). RIM-FIRE REVOLVER. Caliber .41. Two-and-one-quarter-inch, round, rifled barrel. There are five chambers in the fluted cylinder. All metal parts are nickel-plated and engraved with scroll designs. The grip-plates are made of smooth ivory. Upon the top of the barrel are the words "Colt's Pt. F.A. Mfg. Co. Hartford, Ct., U.S.A.", and on the side "Colt New 41." The under-side of the barrel is stamped with the number "3588."

U. S. A.

No. 17. (N4216). RIM-FIRE PISTOL. Caliber .41. Two-and-one-half-inch, round, rifled barrel. This pistol fires a single, rim-fire, metallic cartridge. There is a sight in front, while the hammer serves as a rear sight. Bringing the pistol to half-cock, allows the barrel to be swung to the right in order to load. The piece is provided with a self-ejector. The frame is of nickel-plated brass and the barrel is plated with the same metal. Has smooth, ivory grip-plates. The left side of the frame is stamped ".41 Cal.", and the top of the barrel "—COLT—". Upon the inside of the grip is the number "35582." No patent date is given.

U. S. A.

No. 18. (N4217). RIM-FIRE PISTOL. Caliber .41. Two-and-one-half-inch, round, rifled barrel. The grip-plates are made of polished wood. In markings and in all other respects this pistol is similar to that of No. 17. The serial number is "13928."

U. S. A.

No. 19. (N4212). RIM-FIRE REVOLVER. Caliber .22. Two-and-three-eighths-inch, round, rifled barrel. Has seven chambers in a smooth cylinder for rim-fire cartridges. A notch in the top of the hammer serves as a rear sight. This single-action revolver has neither ejector nor trigger-guard. The chambers are loaded from the rear and on the right side of the frame. Rotating a catch beneath the barrel to the right allows the latter to be drawn forward from the cylinder-pin. The grip is made of wood and the frame of brass. The top of the barrel is stamped "Colt's Pt. F.A. Mf'g. Co., Hartford, Ct., U.S.A." Upon the butt is the number "10852."

U. S. A.

No. 20. (N4211). RIM-FIRE REVOLVER. Caliber .22. Two-and-one-quarter-inch, round, rifled barrel. There are seven chambers in the fluted cylinder for rim-fire cartridges. Has a front sight, while a groove in the rear of the frame serves as a back sight. There is no ejector or trigger-guard. The frame is made of brass. Upon the right side of the barrel the words "Little Colt" are engraved, and stamped upon the top, "Colt's Pt. F.A. Mf'g Co., Hartford, Ct., U.S.A." The number "743" appears upon the butt.

U. S. A.

No. 21. (N6690). RIM-FIRE REVOLVER. Caliber .22. Two-and-three-eighths-inch, round, rifled barrel. Has seven chambers in a smooth cylinder. Rotating a catch in front of the trigger to the right allows the barrel to be removed from the cylinder-pin. There is no ejector. A notch in the top of the hammer serves as a rear sight. Upon the top of the barrel are the words "Colt's Pt. F.A. Mfg. Co., Hartford, Ct., U.S.A." Barrel and butt bear the number "83858."

U. S. A.

No. 22. (N6689). RIM-FIRE REVOLVER. Caliber .22. Two-and-seven-eighths-inch, round, rifled barrel. There are seven chambers in the cylinder. Rotating a catch in front of the trigger to the right allows the barrel to be removed from the cylinder-pin. Provided with a spring-ejector. A notch in the top of the hammer serves as a rear sight. Upon the top of the barrel are the words "Colt's Pt. F.A. Mfg. Co., Hartford, Ct., U.S.A." Barrel and butt bear the number "565."

U. S. A.

No. 23. (N6691). RIM-FIRE REVOLVER. Caliber .22. Two-and-one-quarter-inch, flat, rifled barrel. Has seven chambers in the cylinder. Pressing a catch on the left side of the frame in front of the cylinder enables the cylinder-pin to be withdrawn forward for the removal of the cylinder. The cylinder-pin is used as an ejector. Stamped upon the top of the barrel are the words "Colt's Pt. F.A. Mfg. Co., Hartford, Ct., U.S.A.", and on the left side, "Colt New 22." The butt bears the number "15277."

U. S. A.

No. 24. (N4210). RIM-FIRE REVOLVER. Caliber .22. Two-and-one-half-inch, round, rifled barrel. There are seven chambers in the smooth cylinder. Has no ejector. Rotating a catch beneath the barrel to the right, allows the barrel to be drawn forward from the cylinder-pin. The grip is made of wood. Stamped upon the top of the barrel are the words "Colt's Pt. F.A. Mf'g. Co., Hartford, Ct., U.S.A." The butt bears the number "97468."

U. S. A.

No. 25. (N6721). PERCUSSION REVOLVER. Caliber .36. Four-and-one-half-inch, round, rifled barrel. Has five chambers in a fluted cylinder. This is a finely made presentation piece in the original case, which has the form of a red leather book trimmed with gold. The title on the back of the "volume" reads "Colt on the Constitution Higher Law & Irrepressible Conflict. Jan. 1st, 1861." The trigger-guard and back-strap are silver-plated and the tip of the lever ramrod is knurled. Stamped upon the top of the barrel are the words "Saml. Colt New-York U.S. America," and on the left side of the frame, "Colts Patent," and "36 Cal." The back-strap is engraved with the name "Horace Ripley," the man to whom the set was originally presented. Various parts of the revolver bear the number "26780." The case contains all of the original accessories: percussion-caps, small cases of cartridges in combustible envelopes, bullet-mold, nipple wrench and screw driver, and a powder flask.

U. S. A.

No. 26. (N4208). RIM-FIRE REVOLVER. Caliber .36. Five-and-one-half-inch, round, rifled barrel. Has five chambers for rim-fire, metallic cartridges. This belt pistol was altered from loose ammunition, and then provided with a spring ejector. Trigger-guard and back-strap are made of silver-plated brass. All steel parts were formerly blued. The top of the barrel is stamped "Colt's Pt. F.A. Mf'g. Co., Hartford, Ct., U.S.A.", and the frame is marked "Pat. July 25, 1871, Pat. July 2, 1872." The cylinder and the rest of the parts are stamped with the number "13734."

U. S. A.

No. 27. (N4215). RIM-FIRE REVOLVER. Caliber .41. Three-inch, round, rifled, steel barrel. Has four chambers in a "clover leaf" cylinder for rim-fire cartridges. There is a sight in front, while a groove in the rear of the frame serves as a back sight. An ejector is carried beneath the barrel. The trigger is of the sheath variety. All metal parts are nickel-plated, and the grip is of polished rosewood. Stamped upon the top of the barrel are the words "Colt's House Pistol, Hartford, Ct., U.S.A.", and on top of the frame, "Pat. Sep. 19, 1871." Barrel and butt bear the number "5825." It was with a "clover leaf" revolver of this kind that E. S. Stokes killed James Fisk in 1872. Fisk was a financier who at one time was associated with Jay Gould in the directorate of the Erie Railway. These men carried financial buccaneering to extremes, their activities including open alliance with the Tweed ring, the wholesale bribing of legislatures and the buying of judges. Their attempt to corner the gold market resulted in the fateful Black Friday of September 24th, 1869. Fisk was shot and killed in New York City by Stokes, who was a former business associate of his, on January 6th, 1872.

U. S. A.

No. 28. (N4213). RIM-FIRE REVOLVER. Caliber .41. Two-and-one-half-inch, round, rifled barrel. Has five chambers in a smooth cylinder for rim-fire cartridges. There is no ejector or trigger-guard. The frame is made of engraved brass, and all steel parts were formerly blued. Stamped upon the top of the barrel are the words "Colt's House Pistol, Hartford, Ct., U.S.A.", and on top of the frame, "Pat. Sept. 19, 1871." The butt bears the number "8941."

U. S. A.

No. 29. (N6244). RIM-FIRE REVOLVER. Caliber .38. One-and-three-eighths-inch, octagonal barrel, rifled with seven grooves. There are only four chambers in the cylinder. The frame is made of brass, which formerly was plated with a white metal. The other metal parts are made of blued steel. A ramrod, which also serves as a cylinder-pin, is carried beneath the barrel. Stamped upon the left side of the barrel is the word "Colt." The number "1837" appears upon the butt. No other data are given.

U. S. A.

No. 30. (N4219). RIM-FIRE PISTOL. Caliber .41. Two-and-one-quarter-inch, round, rifled steel barrel having a ridge on top. Fires a rim-fire, metallic cartridge. Has a front sight, while a groove in the ear of the hammer serves as a rear sight. Bringing the pistol to half-cock and drawing the small, rough button on the right side of the frame to the rear, allows the barrel to be turned to the left in order to load. This form of Colt was sometimes known as a "Deringer" and differs from a similar one made by the National Arms Company in having a self-ejector. The top of the barrel is stamped "Colt's Pt. F.A. Mf'g. Co., Hartford, Ct., U.S.A. No. 1." The grip is of iron and slightly checkered. Upon the bottom of the barrel is the number "3328."

U. S. A.

No. 31. (N4218). RIM-FIRE PISTOL. Caliber .41. Two-and-one-quarter-inch, round, rifled barrel having a ridge on top. The top of the barrel is stamped "Colt's Pt. F.A. Mf'g. Co., Hartford, Ct., U.S.A. No. 2." The grip is made of wood and is checkered, while the under-side of the barrel bears two English proof-marks and the number "7271." The action is shown open. In all other respects this pistol is similar to No. 30.

U. S. A.

No. 32. (N4183). PERCUSSION REVOLVER. Caliber .36. Six-and-one-half-inch, octagonal, rifled barrel. Has five chambers in the cylinder which is turned down at the back. There is a stage-coach holdup engraved upon the cylinder which also bears the words "Colts Patent No. 6323." Various other parts of the revolver are stamped with the number "16323." The sides of the barrel near the cylinder are engraved, and the top of the barrel is stamped "Address Col. Saml. Colt New-York, U.S. America." This form of revolver was also known as the "belt pistol."

U. S. A.

No. 33. (N6679). PERCUSSION REVOLVER. Caliber .36. Five-and-one-half-inch, rifled steel barrel. There are five chambers in the cylinder. Various parts of this revolver bear the number "3574." In all other respects this specimen is similar to No. 32.

U. S. A.

No. 34. (N3303). PERCUSSION REVOLVER. Caliber .36. Five-and-one-half-inch, round, rifled, steel barrel. There are five chambers in the semi-fluted cylinder which is marked with the words "Pat. Sept. 10th 1850." Trigger-guard and back-strap are made of silver-plated brass. All steel parts are blued, and the grip-plates are of smooth, polished walnut. This model was known as the "police" or "belt" pistol. Stamped upon the top of the barrel are the words "Address Col. Saml. Colt New-York U.S. America." Various parts of the revolver bear the number "13182."

U. S. A.

No. 35. (N4172). PERCUSSION REVOLVER. Caliber .36. Five-and-one-half-inch, round, rifled barrel. Various parts of this revolver are stamped with the number "18553." In other respects it is the same as No. 34.

U. S. A.

No. 36. (N4171). PERCUSSION REVOLVER. Caliber .36. Six-and-one-half-inch, round, rifled barrel. Trigger-guard and back-strap are made of brass, which was formerly silver-plated. All steel parts were formerly blued. Barrel and cylinder are marked as those of No. 34. Various parts of the revolver are stamped with the number "24367."

U. S. A.

No. 37. (N4186). PERCUSSION REVOLVER. Caliber .31. Six-inch, octagonal, rifled, steel barrel. The front sight is missing. The cylinder, which contains five chambers, is partly engraved with a scene representing a stage-coach holdup and is marked "Colts Patent No. 174464." All other parts of the revolver bear the same number. The top of the barrel is stamped "Address Saml. Colt New York City." This is another example of the old model belt pistol.

U. S. A.

No. 38. (N4185). PERCUSSION REVOLVER. Caliber .31. Six-inch, octagonal, rifled, steel barrel. There are six chambers in the cylinder for loose ammunition. Trigger-guard and back-strap are made of brass, and were formerly silver-plated. The frame and hammer are case-hardened in variegated colors. Cylinder and barrel were formerly blued. The cylinder is engraved with a scene representing a stage-coach holdup, and is stamped "Colts Patent No. 202272." All other parts of the revolver bear the same number. Stamped upon the top of the barrel are the words "Address Saml. Colt Hartford, Ct."

U. S. A.

No. 39. (N6687). PERCUSSION REVOLVER. Caliber .31. Six-inch, octagonal, rifled, steel barrel. Has five chambers in the cylinder. The trigger-guard is of slightly different shape than that of No. 38. Cylinder and other parts of the revolver bear the number "181932." Markings and other characteristics are the same as those of No. 38.

U. S. A.

No. 40. (N4188). PERCUSSION REVOLVER. Caliber .31. Six-inch, octagonal, rifled, steel barrel. There are five chambers in the cylinder, which is engraved with a scene representing a stage-coach holdup. The cylinder is also stamped "Colts Patent 5538." All other parts of the revolver bear the same serial number. Upon the top of the barrel are the words "ADDRESS COL. COLT LONDON." Each chamber and also the barrel are stamped with English proof-marks. All metal parts of this revolver are made of blued steel, wherein it differs from those made in the United States which had brass trigger-guards and back-straps.

England.

No. 41. (N4184). PERCUSSION REVOLVER. Caliber .31. Six-inch, octagonal, rifled, steel barrel. The cylinder is marked "Colts Patent 1776." Various other parts of the revolver also bear the same number. In other respects it is identical with No. 40.

England.

PLATE 102.

COLT METALLIC CARTRIDGE REVOLVERS

No. 1. (N6737). CENTER-FIRE REVOLVER. Caliber .32. Seven-and-one-half-inch, round, rifled, steel barrel. There are six chambers in a fluted cylinder. This model 1875 Colt revolver has a .32 caliber barrel on a frame which was intended to carry a .45 caliber one. The grip-plates are made of smooth walnut. The revolver is single-action and has a spring ejector and a gate located on the right side. The barrel is stamped "Colt's Pt. F.A. Mfg. Co., Hartford, Ct., U.S.A.", and "32 W.C.F., and on the frame are the words "Pat. Sept. 19, 1871, July 2, '72, Jan. 19, '75." Various parts of the piece bear the serial number "329762."

U. S. A.

No. 2. (N6730). CENTER-FIRE REVOLVER. Caliber .45. Seven-and-one-half-inch, round, rifled barrel. The top of the barrel is stamped "Colt's Pt. F.A. Mfg. Co., Hartford, Ct., U.S.A.", while the frame is marked "Pat. Sept. 19, 1871, Pat. July 2, -72, Pat. Jan. 19, -75." The gate is stamped with the number "895," while the rest of the parts bear the number "66708." In all other respects this revolver is similar to No. 1.

U. S. A.

No. 3. (N4223). CENTER-FIRE REVOLVER. Caliber .44. Seven-and-one-half-inch, round, rifled barrel. Has six chambers in a fluted cylinder for center-fire cartridges. This is a single-action United States army revolver, model of 1872, which has been changed from the rim-fire system to the center-fire. All metal parts are made of blued steel. Stamped upon the top of the barrel are the words "Colt's Pt. F.A. Mf'g. Co., Hartford, Ct., U.S.A." The butt bears the number "31158," while the rest of the parts are stamped with the number "3762."

U. S. A.

No. 4. (N6731). CENTER-FIRE REVOLVER. Caliber .44. Five-and-one-half-inch, round, rifled barrel. Has six chambers in a fluted cylinder. This is said to be a Colt English artillery model of 1875. Each chamber bears an English proof-mark. A spring ejector and a gate are carried on the right side. The grip-plates are made of polished walnut. The top of the barrel is marked "Colt's Pt. F.A. Mfg. Co., Hartford, Ct., U.S.A.", and the frame is stamped "Pat. Sept. 19, 1871, Pat. July 2, 1872." The gate bears the number "661," while the remaining parts are marked "17866."

U. S. A.

No. 5. (N6733). CENTER-FIRE REVOLVER. Caliber .45. Five-and-one-half-inch, round, rifled barrel. This is a United States artillery model of 1879. It has walnut grip-plates. A spring ejector and a loading gate are located on the right side. The barrel is stamped "Colt's Pt. F.A. Mfg. Co., Hartford, Ct., U.S.A.", while the frame is marked "Pat. Sept. 19, 1871, July 2, '72, Jan. 19, '75," and also with the letters "U.S." The gate bears the number "593," the frame "137672," and the trigger-guard "10219."

U. S. A.

No. 6. (N6734). CENTER-FIRE REVOLVER. Caliber .45. Four-and-three-quarters-inch, round, rifled barrel. This is another example of the United States army artillery model of 1873. Grip, ejector and gate are the same as those of No. 5. The barrel is stamped "Colt's Pt. F.A. Mfg. Co., Hartford, Ct., U.S.A." The frame is marked with the usual patent dates peculiar to this model. The gate bears the number "6235," while the rest of the parts are stamped with the number "108409."

U. S. A.

No. 7. (N6732). CENTER-FIRE REVOLVER. Caliber .45. Seven-and-one-half-inch, round, rifled, steel barrel. Has six chambers in the fluted cylinder. This is a single-action, United States cavalry revolver, model of 1875. The grip-plates are of checkered ebonite.

Gate and spring ejector are the same as those of No. 5. The barrel is stamped "Colt's Pt. F.A. Mfg. Co., Hartford, Ct., U.S.A. 45 Colt," and the frame "Pat. Sept. 19, 1871, July 2, '72, Jan. 19, '75." Various parts of the revolver bear the number "188194."

U. S. A.

No. 8. (N6736). CENTER-FIRE REVOLVER. Caliber .38. Seven-and-one-half-inch, round, rifled barrel. There are six chambers in the fluted cylinder. This is a target model Colt, having a very sensitive trigger. Gate and ejector are the same as those of No. 5. The piece is marked "Colt's Pt. F.A. Mfg. Co., Hartford, Ct., U.S.A.", and "Pat. Sept. 19, 1871, July 2, '72, Jan. 19, '75." Various parts of the revolver bear the number "331735."

U. S. A.

No. 9. (N4231). CENTER-FIRE REVOLVER. Caliber .44. Five-and-one-half-inch, rifled barrel. Has six chambers in a semi-fluted cylinder. The grip is of checkered ebonite. This is a model of 1875 United States army revolver. The barrel is not stamped with the place of manufacture. The frame bears the words "Pat. Sept. 19, 1871, July 2, '72, Jan. 19, '75." The butt is marked with the army number "693" and "186938," the guard "187205," and the barrel "139129."

U. S. A.

No. 10. (N4228). CENTER-FIRE REVOLVER. Caliber .45. Four-and-three-quarters-inch, round, rifled barrel. There are six chambers in the semi-fluted cylinder. The grip-plates are of checkered ebonite. Barrel and cylinder are blued, while the rest of the parts are case-hardened in mottled colors. The cylinder-pin withdraws from in front. This model of 1875 revolver is stamped upon the barrel "45 Colt," and on the top, "Colt's Pt. F.A. Mf'g. Co., Hartford, Ct., U.S.A." All parts bear the serial number "296703."

U. S. A.

No. 11. (N6717). CENTER-FIRE REVOLVER. Caliber .41. Five-and-one-half-inch, round, rifled barrel. Has six chambers. The grip-plates are of checkered ebonite. Stamped upon the barrel are the words "Colt's Pt. F.A. Mfg. Co., Hartford, Ct., U.S.A.", and "41 COLT." The frame is marked "Pat. Sept. 19, 1871, July 2, '72, Jan. 19, '75." Various parts of the revolver bear the number "262055."

U. S. A.

No. 12. (N6735). CENTER-FIRE REVOLVER. Caliber .38. Four-and-three-quarters-inch, round, rifled barrel. This United States artillery model of 1876 has six chambers in a fluted cylinder. The grip-plates are of checkered ebonite. The barrel is stamped "Colt's Pt. F.A. Mfg. Co., Hartford, Ct., U.S.A.", "38 W.C.F.", and the frame, "Pat. Sept. 19, 1871, July 2, '72, Jan. 19, '75." The gate bears the number "760," while the rest of the parts are stamped with the number "217637."

U. S. A.

No. 13. (N4227). CENTER-FIRE REVOLVER. Caliber .32. Five-and-one-half-inch, round, rifled barrel. The frame is case-hardened in mottled colors, the barrel and guard are blued, while the grip is of mother-of-pearl, bearing a long-horned steer's head in high relief. This is probably a presentation piece. The top of the barrel is stamped "Colt's Pt. F.A. Mf'g. Co., Hartford, Ct., U.S.A." All parts bear the serial number "281389."

U. S. A.

No. 14. (N6718). CENTER-FIRE REVOLVER. Caliber .45. Four-inch, round, rifled barrel. There are six chambers in the cylinder. The grip-plates are of polished ivory. This revolver carries no ejector but is provided with the usual gate upon the right side. The top of the barrel is stamped "Colt's Pt. F.A. Mfg. Co., Hartford, Ct., U.S.A.", and the side "45 COLT." Upon the frame are the usual patent dates for this model. The gate is stamped with the number "5326," while the rest of the parts bear the number "80553."

U. S. A.

No. 15. (N3316). CENTER-FIRE REVOLVER. Caliber .44. Seven-and-one-half-inch, round, rifled barrel. Has the usual six chambers. All steel parts are blued. The grip-plates are of checkered ebonite, and the butt is provided with an iron ring. This is said to be an 1880 model United States army revolver. The top of the barrel is stamped "Colt's Pt. F.A. Co., Hartford, Ct., U.S.A." The number "46106" appears upon the butt.

U. S. A.

No. 16. (N4229). CENTER-FIRE REVOLVER. Caliber .44. Seven-and-one-half-inch, round, rifled barrel. All metal parts are nickel-plated. Has a smooth, ivory grip. The barrel bears the words "Colt

Frontier Six Shooter," and the side, "Colt's Pt. F.A. Mf'g. Co., Hartford, Ct., U.S.A." The butt is stamped with the number "7161." All other characteristics are the same as those of No. 15.

U. S. A.

No. 17. (N4232). CENTER-FIRE REVOLVER. Caliber .45. Six-inch, round, rifled barrel. Has six chambers. All steel parts are blued. The grip is of checkered ebonite, and the butt is provided with an iron ring. This model 1892 United States army revolver was made for service in the Philippine Islands. The trigger and trigger-guard were made large so as to allow these parts to be used with a gauntlet-covered hand when fighting in the jungle. Stamped upon the top of the barrel are the words "Colt's Pt. F.A. Mf'g. Co., Hartford, Ct., U.S.A.", and on the side, "45 Colt." The number "45807" appears upon the butt.

U. S. A.

No. 18. (N6714). CENTER-FIRE REVOLVER. Caliber .45. Seven-and-one-half-inch, round, rifled barrel. Has the usual six chambers in the cylinder. This is a Colt Special target revolver, United States officer's model, with a shoulder stock. This revolver has a strap over the top of the cylinder, a spring ejector and a fixed rear sight. It is single-action and has a very sensitive trigger. The top of the barrel is stamped "Colt's Pt. F.A. Mfg. Co., Hartford, Ct., U.S.A." Various other parts bear the number "317012."

U: S. A.

No. 19. (N6738). CENTER-FIRE REVOLVER. Caliber .38. Five-and-one-half-inch, round, rifled, steel barrel. Has six chambers in the fluted cylinder. This is a Colt "Frontier Model." It has a strap swivel in the butt. All steel parts, with the exception of the trigger, are nickel-plated. The barrel is stamped "Colt's Pt. F.A. Mfg. Co., Hartford, Ct., U.S.A.", and "38 W.C.F." No patent date is given. The gate is stamped with the number "122," and the butt with "40704."

U. S. A.

No. 20. (N4234). CENTER-FIRE REVOLVER. Caliber .41. Six-inch, round, rifled barrel. There are six chambers in the cylinder. Barrel and cylinder are blued, while the rest of the parts are case-hardened in variegated colors. The side of the barrel is stamped "Colt D.A. .41," and the top "Colt's Pt. F.A. Mf'g. Co., Hartford, Ct., U.S.A."

This is a model 1875 belt pistol, all parts of which are marked with the number "150255."

U. S. A.

No. 21. (N6740). CENTER-FIRE REVOLVER. Caliber .38. Four-and-one-half-inch, round, rifled barrel. This six-chambered Colt revolver was made for the United States Express Company. All steel parts are nickel-plated. A spring ejector and a gate are located on the right side. The grip is of checkered ebonite. The top of the barrel is stamped "Colt's Pt. F.A. Mfg. Co., Hartford, Ct., U.S.A.", and also with the express company's number, "93748," and with the letters "U.S.X." The regular serial number of this revolver has been partly erased.

U. S. A.

No. 22. (N6739). CENTER-FIRE REVOLVER. Caliber .41. Four-and-one-half-inch, round, rifled barrel. This is a type of Colt pocket revolver which has six chambers in a fluted cylinder. All metal parts are silver-plated and engraved with scrolls. The grip is of mother-of-pearl. The top of the barrel is marked "Colt's Pt. F.A. Mfg. Co., Hartford, Ct., U.S.A." Trigger-guard and frame bear the number "57427."

U. S. A.

No. 23. (N6746). CENTER-FIRE REVOLVER. Caliber .41. Five-inch, round, rifled barrel. This is a type of Colt police revolver and has six chambers in a fluted cylinder. All steel parts are blued. There is no ejector, although the cylinder-pin may be used as one. The barrel is stamped "Colt's Pt. F.A. Mfg. Co., Hartford, Ct., U.S.A.", and "COLT D.A. 41." The gate bears the number "310," and the rest of the parts are stamped with the number "2248."

U. S. A.

No. 24. (N6741). CENTER-FIRE REVOLVER. Caliber .38. Four-and-one-half-inch, round, rifled barrel. Has six chambers in a fluted cylinder. This is another Colt revolver which was made for the United States Express Company. All steel parts are nickel-plated. The barrel is stamped "Colt's Pt. F.A. Mfg. Co., Hartford, Ct., U.S.A.", and also with the express company's number, "8738," and "U.S.X." The serial number is "8738."

U. S. A.

No. 25. (N4233). CENTER-FIRE REVOLVER. Caliber .38. Three-and-one-half-inch, round, rifled, steel barrel. Has six chambers. The cylinder-pin is withdrawn from in front and serves as an ejector. All metal parts, except the trigger and hammer, are nickel-plated. The side of the barrel is stamped "Colt D.A. 38," and the top "Colt's Pt. F.A. Mf'g. Co., Hartford, Ct., U.S.A." Upon the frame are the patent dates, "Sept. 18, 1871, Sept. 15, 1874, and Jan. 19, 1875." All parts bear the serial number "81512."

U. S. A.

No. 26. (N6742). CENTER-FIRE REVOLVER. Caliber .38. Two-and-one-half-inch, round, rifled barrel. There are six chambers in the cylinder. All steel parts were formerly nickel-plated. The cylinder-pin, which may be withdrawn from in front, serves as an ejector. The barrel is stamped "Colt's Pt. F.A. Mfg. Co., Hartford, Ct., U.S.A.", and "COLT D.A. 38." Various parts of the revolver bear the number "18785."

U. S. A.

No. 27. (N6715). CENTER-FIRE REVOLVER. Caliber .44. Seven-and-one-half-inch, round, rifled barrel. Has six chambers in a fluted cylinder. This is an example of the Bisley model of the Colt revolver. The grip-plates are of checkered ebonite and a spring ejector is located on the right side. The piece is single-action and is stamped upon the barrel, "Colt's Pt. F.A. Mfg. Co., Hartford, Ct., U.S.A. (Bisley Model) COLT FRONTIER SIX SHOOTER." Upon the frame are the words "Pat. Sept. 19, 1871, July 2, '72, Jan. 19, '75." Various parts of the revolver are marked with the number "276415."

U. S. A.

No. 28. (N6748). CENTER-FIRE REVOLVER. Caliber .44. Seven-and-one-half-inch, round, rifled barrel. This is also a Bisley model. The barrel is stamped "Colt's Pt. F.A. Mfg. Co., Hartford, Ct., U.S.A. (BISLEY MODEL), 44 Russian Ctg." The frame bears the words "Pat. Sept. 19, 1871, July 2, '72, Jan. '75." Various parts of the piece are marked with the number "296761."

U. S. A.

No. 29. (N6716). CENTER-FIRE REVOLVER. Caliber .45. Four-and-three-quarters-inch, round, rifled barrel. Has the usual six chambers in the fluted cylinder. This is another form of the Bisley model of the Colt revolver. The barrel is stamped "Colt's Pt. F.A. Mfg. Co., Hartford, Ct., U.S.A. (BISLEY MODEL), 45 COLT." The frame is marked "Pat. Sept. 19, 1871, July 2, '72, Jan. 19, '75." The gate bears the number "227," while the rest of the parts are stamped with the number "214264."

U. S. A.

No. 30. (N4226). CENTER-FIRE REVOLVER. Caliber .38. Five-and-one-half-inch, round, rifled barrel. Has six chambers. This is still another form of the Bisley model, the side of the barrel of which is marked "(Bisley Model) 38 W.C.F.", and the top, "Colt's Pt. F.A. Mf'g. Co., Hartford, Ct., U.S.A." This is a model of 1875, various parts of which bear the number "219159."

U. S. A.

No. 31. (N6747). CENTER-FIRE REVOLVER. Caliber .32. Four-and-three-quarters-inch, round, rifled barrel. This six-chambered Bisley model has a .32 caliber barrel in a frame which was intended to carry a .44 caliber one. All steel parts are blued, except the frame, which is case-hardened in mottled colors. The barrel is stamped "Colt's Pt. F.A. Mfg. Co., Hartford, Ct., U.S.A., (BISLEY MODEL) 32 W.C.F." The gate bears the number "4401," while the rest of the parts are stamped with the number "295675."

U. S. A.

No. 32. (N4224). CENTER-FIRE REVOLVER. Caliber .38. Six-inch, round, rifled barrel. There are six chambers in the fluted swing-out cylinder. The empty shells are ejected by pushing the cylinder-pin back. The revolver, which is double-acting, is provided with a rebounding hammer. The butt is stamped "U.S.N. 38 D.A. No. 3812, P.W.W.K. 1889," and the top of the barrel "Colt's Pt. F.A. Mfg. Co., Hartford, Ct., U.S.A. Patented Aug. 5th, 1884, November 6th, 1888." The serial number is "3769."

U. S. A.

No. 33. (N4225). CENTER-FIRE REVOLVER. Caliber .38. Six-inch, round, rifled barrel. Has six chambers. Pulling back on the lug in the rear of the cylinder, allows the latter to be swung out to the left. The empty shells are ejected by pushing back on the cylinder-pin. The butt is stamped "U.S. Army Model 1894," and the top of the barrel "Colt's Pt. F.A. Mf'g. Co., Hartford, Ct., U.S.A., Patented Aug. 5, 1884, Nov. 6, '88, Mar. 5, '95." The serial number is "5817."

U. S. A.

No. 34. (N4230).   CENTER-FIRE REVOLVER.   Caliber .38. Six-inch, round, rifled barrel.   The butt is stamped "U.S.M.C." (United States Marine Corps), and is provided with a ring.   The barrel is marked "Colt's Pt. F.A. Mf'g. Co., Hartford, Ct., U.S.A., Patented Aug. 5, 1884, Nov. 6, '88, Mar. 5, '95."   In other respects this model 1894 navy revolver is the same as No. 33.   The serial number is "1812."

U. S. A.

No. 35. (N6720).   CENTER-FIRE REVOLVER.   Caliber .38. Six-inch, round, rifled barrel.   Has the usual six chambers.   All steel parts are blued.   The top of the barrel is stamped "Colt's Pt. F.A. Mfg. Co., Hartford, Ct., U.S.A. Pat'd. Aug. 5, 1884, June 5, 1900, July 4, 1905," and the left side, "COLT ARMY SPECIAL 38."   The serial number is "316059F."

U. S. A.

No. 36. (N6719).   CENTER-FIRE REVOLVER.   Caliber .45. Seven-and-one-half-inch, round, rifled barrel.   There are six chambers in the fluted cylinder.   All steel parts are blued.   Stamped upon the top of the barrel are the words "Colt's Patent Fire Arms Manufacturing Co., Hartford, Ct., U.S.A. Pat. Aug. 5, 1884, June 5, 1900," and the side of the barrel, "NEW SERVICE 45 COLT."   There is a ring in the butt.   This revolver has the usual swing-out cylinder and the solid frame.   The serial number is in the form of "6204" stamped over "4."

U. S. A.

No. 37. (N6725).   CENTER-FIRE REVOLVER.   Caliber .41. Six-inch, round, rifled barrel.   Has six chambers.   All steel parts were formerly blued.   Stamped upon the top of the barrel of this double-action revolver are the words "Colt's Pt. F.A. Mfg. Co., Hartford, Ct., U.S.A. Patented Aug. 5, 1884, Nov. 6, '88, Mar. 5, '95," and on the left side, "COLT D.A. 41."   The frame bears the number "411" and the butt "70374."

U. S. A.

No. 38. (N6722).   CENTER-FIRE REVOLVER.   Caliber .41. Four-and-one-half-inch, round, rifled barrel.   Has five chambers in a fluted cylinder.   All steel parts were formerly blued.   Pulling back on a catch on the left side of the frame releases the cylinder and allows the latter to swing out to the side for ejection or for loading.   The barrel is stamped "Colt's Pt. F.A. Mfg. Co., Hartford, Ct., U.S.A. Pat. Aug.

5, '84, Nov. 6, '88, Mar. 5, '95," and the side, "COLT D.A. 41." The frame bears the number "130" and the butt "220518."

U. S. A.

No. 39. (N6723). CENTER-FIRE REVOLVER. Caliber .38. Three-inch, round, rifled barrel. Has six chambers in a fluted cylinder which swings out to the left side for ejection and for loading. All metal parts are of blued steel. The grip-plates are made of mother-of-pearl. Upon the top of the barrel are the words "Colt's Pt. F.A. Mfg. Co., Hartford, Ct., U.S.A. Pat. Aug. 5, '84, Nov. 6, '88, Mar. 5, '95," and on the side, "Colt D.A. 38." The frame bears the number "1164 N."

U. S. A.

No. 40. (N6724). CENTER-FIRE REVOLVER. Caliber .41. Three-inch, round, rifled barrel. Has six chambers. All steel parts are blued. This revolver has a .41 caliber barrel in a frame which was intended to carry one of .44 caliber. The barrel bears the same patent dates as that of No. 39, and in addition the words "COLT. D.A. 41." The cylinder-catch bears the number "887" and the butt "185590."

U. S. A.

No. 41. (N4222). CENTER-FIRE REVOLVER. Caliber .38. Four-and-three-eighths-inch, round, rifled barrel. Has five chambers, in a fluted cylinder, which are loaded from the rear and on the right side of the frame. The revolver is single-acting and has no trigger-guard. An ejector is located upon the right side. All metal parts are nickel-plated. Upon the side of the barrel are the words "New Police .38," and upon the top, "Colt's Pt. F.A. Mf'g. Co., Hartford, Ct., U.S.A." No patent date is given. The butt is marked with the number "14481."

U. S. A.

No. 42. (N4220). CENTER-FIRE REVOLVER. Caliber .36. Two-and-one-quarter-inch, round, rifled barrel. Has six chambers. The cylinder-pin is withdrawn from in front and serves as an ejector. All metal parts were formerly blued. The side of the barrel bears the words "Colt's House Pistol," and the top, "Colt's Pt. F.A. Mf'g. Co., Hartford, Ct., U.S.A." No patent date is given. The serial number "18999" appears upon the butt.

U. S. A.

No. 43. (N6727). CENTER-FIRE REVOLVER. Caliber .32.

Six-inch, round, rifled barrel. All steel parts are blued. Has six chambers. Pulling back on a catch on the left side of the frame releases the
cylinder and allows the latter to swing out to the side for ejection or
for loading. This is a double-acting police revolver. The top of the
barrel is stamped "Colt's Pat. Fire Arms Mfg. Co., Hartford, Conn.,
U.S.A. Pat'd. Aug. 5, 1884, June 5, 1900," and the left side "COLT
NEW POLICE 32." The serial number is "27347."

U. S. A.

No. 44. (N6728). CENTER-FIRE REVOLVER. Caliber .32.
Four-inch, round, rifled barrel. Has six chambers. This is another
example of the Colt police revolver. The cylinder swings out to the
left side for ejection and for loading. Upon the top of the barrel are
the words "Colt's Pat. Fire Arms Mfg. Co., Hartford, Ct., U.S.O.
Pat'd. Aug. 5, 1884, June 5, 1900," and on the side, "Colt New Police
32." The frame bears the number "35744."

U. S. A.

No. 45. (N6729). CENTER-FIRE REVOLVER. Caliber .38.
Four-and-three-quarters-inch, round, rifled barrel. The revolver is
double-acting and has five chambers. This Colt is of Spanish make
and is stamped upon the barrel "ULTIMO MODELO COLT CON
ESTRELLA FABRICADO CON ACERO BESSEMER QUIN-
TANA HARMANOS Y Ca BILBAO," the translation of which is:
"Latest Model Colt with Star Made from Bessemer Steel by Quintana
Brothers and Company at Bilbao." The grip-plates bear a coat-of-arms
and the name "Vizcaya." This revolver was made for the Spanish
armored cruiser of that name. This vessel, together with some others
under Admiral Cervera, was destroyed by the American fleet in the
Battle of Santiago, July 3,1898. The revolver was made in Bilbao,
capital of the province of Biscay, and famous from the earliest times for
the excellent quality of its steel blades.

Spain.

No. 46. (N6726). CENTER-FIRE REVOLVER. Caliber .32.
Four-inch, round, rifled barrel. Has six chambers. All steel parts are
blued. This is a Colt police revolver. The barrel is marked "Colt's
Pt. F.A. Mfg. Co., Hartford, Ct., U.S.A. Pat'd. Aug. 5, 1884, June 5,
1900, July 4, 1905," and "POLICE POSITIVE SPECIAL 32-20
W.C.F." The serial number is "93444."

U. S. A.

PLATE 103.

COLT AUTOMATIC PISTOLS

No. 1. (N4236). AUTOMATIC PISTOL. Caliber .38. Six-inch, rifled barrel. The magazine carries seven rimless, smokeless cartridges. This pistol fires a 105-grain bullet with a muzzle velocity of 1,300 feet per second. This is an example of the first model United States army automatic, patented April 20, 1897. These pistols were sold to the government at $30.00 each. This one has the grooves in the slide at the rear. The rear sight is also a safety-device, when it is up, as shown, the pistol can be fired, but not when the sight is down. The grip-plates are smooth. The left side of the slide is stamped "Browning's Patent, Pat'd. April 20, 1897, Colt's Patent Fire Arms Mfg., Hartford, Conn., U.S.A.", and the right side, "Automatic Colt, Calibre 38 Rimless Smokeless." The pistol is also marked with the serial number "98" and with the letters "U.S."

U. S. A.

No. 2. (N4235). AUTOMATIC PISTOL. Caliber .38. Six-inch, rifled barrel. This is an example of the second model United States army automatic. This one has the grooves in the slide at the front, and the grip-plates are checkered. The left side of the slide is stamped "Browning's Patent, Pat'd. April 20, 1897, Colt's Patent Fire Arms Mfg. Co., Hartford, Conn., U.S.A.", and the right side, "Automatic Colt, Calibre 38 Rimless Smokeless." The pistol bears the number "1700" and the letters "U.S." Rear sight, magazine and other features are the same as those of No. 1.

U. S. A.

No. 3. (N7029). AUTOMATIC PISTOL. Caliber .38. Six-inch, rifled barrel. This is an example of the first United States navy model of Colt's automatic pistol. It has a smooth, wooden grip, but no strap loop. The rear sight acts as a safety-device. See No. 1. Only 200 pistols of this model were made for the United States government for trial. The left side of the slide is marked "Browning's Patent, Pat'd. April 20, 1897," "Colt's Patent Fire Arms Mfg. Co., Hartford, Conn., U.S.A.", and the right side, "Automatic Colt Calibre 38 Rimless Smokeless," and the number "1002." The left side of the frame bears the letters "U.S.N." and the number "2."

U. S. A.

No. 4. (N7031). AUTOMATIC PISTOL. Caliber .38. Six-inch, rifled barrel. The magazine carries seven rimless and smokeless cartridges. This is an example of the first commercial model of the Colt automatic pistol, sometimes known as the "sporting model." This model is now obsolete. The grooves in the slide are in front. This pistol has vulcanite grip-plates, a low hammer, but no safety-device. The markings and patent dates are identical with those on No. 3. The left side of the frame bears the number "8073."

U. S. A.

No. 5. (N7027). AUTOMATIC PISTOL. Caliber .38. Four-and-one-quarter-inch, rifled barrel. The magazine carries seven rimless and smokeless cartridges. This is an example of the first commercial model of the Colt automatic pistol having the short barrel. It has no belt loop and no safety-device. The slide over the barrel has the grooves in the rear. The grip-plates are of checkered ebonite. Upon the left side of the slide are the words "Browning's Patent, Apr. 20, 1897, Sept. 9, 1902. Colt's Patent Fire Arms Mfg. Co., Hartford, Conn., U.S.A.", and the right side "Automatic Colt, Calibre 38 Rimless Smokeless." The serial number is "19739."

U. S. A.

No. 6. (N7028). AUTOMATIC PISTOL. Caliber .38. Six-inch, round, rifled barrel. This is an example of the commercial military model of Colt's automatic pistol. The slide has a notch on the left side which engages a catch so that when the slide is drawn back the catch retains it in that position until released. The drawing back of the slide in this manner permits of easy inspection of barrel and magazine. This model has a lanyard loop. The left side of the slide is marked "Patented Apr. 20, 1897, Sept. 9, 1902. Colt's Pt. F.A. Mfg. Co., Hartford, Ct., U.S.A.", and the right side, "Automatic Colt, Calibre 38 Rimless Smokeless." The serial number is "37125."

U. S. A.

No. 7. (N7030). AUTOMATIC PISTOL. Caliber .38. Five-inch, rifled barrel. This is an example of the model 1902 United States military pistol. Only 300 of these were made for the government. The slide over the barrel is checkered in front. This pistol is equipped with safety stop, strap swivel and a low hammer. The grip-plates are of checkered ebonite. Upon the left side of the slide are the words "Browning's Patent, Pat'd. April 20, 1897. Colt's Patent Fire Arms Mfg. Co.,

Hartford, Conn., U.S.A.", and on the right, "Automatic Colt, Calibre 38 Rimless Smokeless Model 1902." The trigger-guard bears the letters "U.S.", and the serial number is "15064."

U. S. A.

No. 8. (N7033). AUTOMATIC PISTOL. Caliber .45. Five-inch, rifled barrel. This is an example of one of the commercial models of Colt's automatic pistol. It is provided with a lanyard loop, and the hammer is of special form. A catch on the left side of the frame under the slide keeps the latter back to permit inspection of the inside of the barrel and magazine. The grip-plates are made of checkered walnut. The left side of the pistol bears the words "Patented Apr. 20, 1897, Sept. 9, 1902, Colt's Pt. F.A. Mfg. Co., Hartford, Ct., U.S.A.", and the right side, "Automatic Colt, Calibre 45 Rimless Smokeless." The serial number is "598."

U. S. A.

No. 9. (N7032). AUTOMATIC PISTOL. Caliber .45. Five-inch, rifled barrel. This is an example of one of the second commercial models of Colt's automatic pistol. It has no loop for the lanyard, but has the regulation hammer. A catch on the left side of the frame under the slide keeps the latter back to permit inspection of the inside of the barrel and magazine. The grip-plates are of checkered walnut. The left side of the slide is marked "Patented Apr. 20, 1897, Sept. 9, 1902, Dec. 19, 1905. Colt's Pt. F.A. Mfg. Co., Hartford, Ct., U.S.A.", and the right side, "Automatic Colt Calibre 45 Rimless Smokeless." The serial number is "5063."

U. S. A.

No. 10. (N7034). AUTOMATIC PISTOL. Caliber .45. Five-inch, rifled barrel. This is an example of the second United States government .45 model automatics. Only 200 of these pistols were made for the government for trial. A safety-device is located in the back-strap below the hammer. In order to fire the pistol, the trigger must be pulled and the safety-device pressed inward by the ball of the thumb simultaneously. This pistol also has a safety stop for the slide. The left side is stamped "Patented, Apr. 20, 1897, Sept. 9, 1902, Dec. 19, 1905. Colt's Patent Fire Arms Mfg. Co., Hartford, Conn., U.S.A.", and the right, "Automatic Colt Calibre 45 Rimless Smokeless." The serial number is "3."

U. S. A.

No. 11.. (N7035). AUTOMATIC PISTOL. Caliber .45. Five-inch, rifled barrel. This is an example of the first caliber .45 automatic pistols made for the United States government. Only twenty-four were made for the government for trial and only twelve were ever sold, which makes pistols of this model very rare. The pistol here illustrated has a luminous fore-sight, a button release for the magazine, and a safety-device in the back-strap. There is also a safety stop for the slide. The opening in the slide for the ejection of the empty shells is located on top. The left side of the slide is stamped "Patented Apr. 20, 1897, Sept. 9, 1902, Dec. 19, 1905. Colt's Pt. F.A. Mfg. Co., Hartford, Ct., U.S.A.", and the right side, "Automatic Colt Calibre 45 Rimless Smokeless." The serial number is "12."

U. S. A.

No. 12. (N7025). AUTOMATIC PISTOL. Caliber .45. Five-inch, rifled barrel. This is an example of the latest type, model 1911, United States army automatic. Over 375,000 of these were made for service during the World War. Owing to the great demand, the time was not taken to blue-polish the steel of these pistols. This pistol has a button release for the magazine, a safety-device in the back-strap, a safety-stop in the slide, and the pistol can also be made safe by bringing the hammer to half-cock. The opening in the slide for the ejection of the empty shells is located on top. The grip-plates are of checkered walnut. The left side of the slide is stamped "Patented Apr. 20, 1897, Sept. 9, 1902, Dec. 19, 1905, Feb. 14, 1911, Aug. 19, 1913. Colt's Pt. F.A. Mfg. Co., Hartford, Ct., U.S.A.", and also "United States Property," while the right side bears the words "MODEL OF 1911 U.S. ARMY." The serial number is "456934."

U. S. A.

No. 13. (N7154). AUTOMATIC PISTOL. Caliber .45. Five-inch, rifled barrel. This is an officer's pistol of the 1911 model United States army automatic. It differs from the regulation arm in having all steel parts polished and blued. This, and other Colt automatic pistols, may be taken apart without the aid of tools, as shown in the illustration. The most expensive part of the pistol is the receiver, the part to which grip-plates, barrel and slide are attached, this part amounting to almost one-half of the entire cost of the weapon. The pistol here shown was carried by a United States Ordnance officer during the World War. The left side is marked "Patented Apr. 20, 1897, Sept. 9, 1902, Dec. 19, 1905, Feb. 14, 1911. Colt's Pt. F.A. Mfg. Co., Hartford, Ct., U.S.A.", and the right side, "Colt Automatic Calibre 45. Government Model." The serial number is "C3247."

U. S. A.

PLATE 104.

REMINGTON REVOLVERS AND PISTOLS

No. 1. (N3304). PERCUSSION REVOLVER. Caliber .44. Eight-inch, octagonal, rifled barrel. There are six chambers for loose ammunition. A lever ramrod is carried beneath the barrel and the cylinder-pin is withdrawable from in front. The barrel is heavily blued, and the trigger-guard is of brass. This type of weapon was known as the "Remington-Beals Army Revolver." The top of the barrel is stamped "Beals' Patent, Sept. 14, 1858, Manufactured by Remingtons' Ilion, N.Y." The bottom of the barrel is marked with the serial number "1447."

U. S. A.

No. 2. (N4115). PERCUSSION REVOLVER. This revolver is similar in all respects to No. 1, except for the serial number, which in the present case is "975."

U. S. A.

No. 3. (N4116). PERCUSSION REVOLVER. Caliber .36. Seven-and-one-half-inch, octagonal, rifled barrel. The serial number "3827" is marked upon the bottom of the barrel. In all other respects the revolver is similar to No. 1.

U. S. A.

No. 4. (N4110). PERCUSSION REVOLVER. Caliber .44. Eight-inch, octagonal, rifled barrel. Has six chambers in the cylinder. All steel parts, with the exception of the hammer, are heavily blued. This weapon is known as the new model "Remington-Beal's Army Revolver." The top of the barrel is stamped "Patented Sept. 14, 1858, E. Remington & Sons, Ilion, New York, U.S.A., New Model," and the bottom, "117068."

U. S. A.

No. 5. (N2869). PERCUSSION REVOLVER. Caliber .44. Eight-inch, octagonal, rifled barrel. This is another example of the new model Remington-Beal's revolver. Its serial number is "3426." In all other respects it is similar to No. 4.

U. S. A.

No. 6. (N4113). PERCUSSION REVOLVER. Caliber .44. Eight-inch, octagonal, rifled barrel. Has six chambers in the cylinder. The top of the barrel is marked as No. 4. A screw in the frame behind the hammer acts as a safety-device, since it prevents the revolver from being cocked. This model was issued to state troops. The serial number "33741" appears upon the bottom of the barrel.

U. S. A.

No. 7. (N4109). PERCUSSION REVOLVER. Caliber .36. Seven-and-three-eighths-inch, octagonal, rifled barrel. Has six chambers for loose ammunition. The top of the barrel is stamped as No. 4, while the bottom bears the serial number "36702."

U. S. A.

No. 8. (N4114). PERCUSSION REVOLVER. Caliber .36. Six-and-one-half-inch, octagonal, rifled barrel. There are six chambers in the cylinder, which is fluted. The top of the barrel is stamped "Manufactured by Remingtons, Ilion, N.Y. Riders' Pt., Aug. 17, 1858, May 31, 1859." The patent of August 17, 1858, was issued to Joseph Rider of Newark, Ohio, for "so combining the parts of the lock of a revolver that a single spring serves for all purposes." The second patent was also issued to Rider "for internal improvements in the lock of a double-action arm." The bottom of the barrel is stamped with the serial number "39."

U. S. A.

No. 9. (N4111). PERCUSSION REVOLVER. Caliber .45. Eight-inch, octagonal, rifled barrel. Has the usual six chambers. The top of the barrel is stamped "Patented Dec. 17, 1861, Manufactured by Remington's Ilion, N.Y.", while the bottom bears the number "14620."

U. S. A.

No. 10. (N4112). PERCUSSION REVOLVER. Caliber .44. Eight-inch, octagonal, rifled barrel. This revolver is fitted with an arrangement whereby the cylinder-pin can be withdrawn forward without lowering the loading lever. The trigger-guard is of brass, as is usual in this type of Remington. The top of the barrel is marked "Patented Dec. 17, 1861, Manufactured by Remingtons' Ilion, N.Y.", and the bottom bears the number "5545."

U. S. A.

No. 11. (N4640). RIM-FIRE REVOLVER. Caliber .38. Six-

and-one-half-inch, octagonal, rifled barrel. Has six chambers for rim-fire cartridges. The now obsolete lever ramrod is still carried beneath the barrel. This is an 1881 model Remington belt pistol. The rear of the cylinder is covered by a steel disc which must be removed before the cartridges can be inserted. The piece is double-acting and has marked upon the top of the barrel "Patented Sept. 14, 1858, E. Remington & Sons, Ilion, New York, U.S.A., New Model." Upon the bottom of the barrel is the number "5470."

U. S. A.

No. 12. (N5905). PERCUSSION REVOLVER IN CASE. Caliber .44. Eight-inch, octagonal, rifled barrel. The top of the barrel is marked as No. 11. This revolver has never been fired, and all accessories are as they were originally supplied by the factory. The accessories consist of a box of percussion-caps, a package of cartridges, a screw driver and a bullet mold. Upon the bottom of the barrel is the serial number "1075."

Donor: Mrs. W. W. Coleman.

U. S. A.

No. 13. (N4136). RIM-FIRE REVOLVER. Caliber .45. Eight-inch, octagonal, rifled barrel. Has five chambers for rim-fire cartridges. The trigger-guard is of silver-plated brass, and all steel parts were formerly blued. This revolver is equipped with a recoil shield at the rear of the cylinder. When metallic cartridges came into use, it was believed that if the rear of the cartridge was fully exposed, the explosion would act rearward instead of forward, hence the recoil shield was relied upon to back the cartridges. The barrel is stamped "Patented Sept. 14, 1858, E. Remington & Sons, Ilion, New York, U.S.A. New Model." Upon the bottom of the barrel are the numbers "4037" and "142287."

U. S. A.

No. 14. (N4131). CENTER-FIRE REVOLVER. Caliber .45. Eight-inch, octagonal, rifled, steel barrel. Has six chambers for center-fire, metallic cartridges. This revolver was remodeled from the loose-ammunition system, and carries the old type of lever ramrod beneath the barrel which is useless for center-fire cartridges. The top of the barrel is stamped "Patented, Sept. 14, 1858, E. Remington & Sons, Ilion, New York, U.S.A., New Model." The barrel bears the number "93471," while the inside of the grip is marked "94156."

U. S. A.

No. 15. (N4129). CENTER-FIRE REVOLVER. Caliber .36. Seven-and-three-eighths-inch, octagonal, rifled barrel. Has six chambers in the cylinder. This revolver was remodeled from the loose-ammunition system by the government. A side ejector and a gate at the rear of the cylinder were added. The top of the barrel is stamped "Beals' Patent Sept. 14, 1858, Manufactured by Remingtons' Ilion, N.Y." The barrel bears the old serial number, "14707," and the new one "911."

U. S. A.

No. 16. (N4130). CENTER-FIRE REVOLVER. Caliber .36. Seven-and-three-eighths-inch, octagonal, rifled barrel. Has six chambers in the cylinder. This revolver was remodeled in the same way as No. 15. The top of the barrel bears the figure of an anchor, denoting naval property, and is also stamped with the words "Patented, Sept. 14, 1858, E. Remington & Sons, Ilion, New York, U.S.A., New Model." The barrel bears the old serial number "32260" and the new one "72."

U. S. A.

No. 17. (N4139). CENTER-FIRE REVOLVER. Caliber .44. Seven-and-one-half-inch, round, rifled barrel. There are six chambers in a fluted cylinder. A spring ejector is located upon the left side and a gate in the rear of the frame allows for the insertion of cartridges in the rear of the cylinder. The top of the barrel is stamped "E. Remington & Sons, Ilion, N.Y., U.S.A.", and the inside of the grip bears the number "3610."

U. S. A.

No. 18. (N4105). RIM-FIRE PISTOL. Caliber .50. Eight-and-one-half-inch, round, rifled barrel. This is a model 1866 Remington navy pistol. This model differs from the model of 1867 in having the trigger in a sheath and in having no trigger-guard. The top of the barrel is stamped with the figure of an anchor. Barrel and lock-plate are covered with a coating of tin to prevent corrosion by sea water. The lock-plate is stamped "Remingtons Ilion, N.Y., U.S.A., Pat. May 3d, Nov. 15th, 1864, April 17th, 1866." The serial number "3239" appears upon the bottom of the barrel. Only 500 pistols of this model are reputed to have been made.

U. S. A.

No. 19. (N6796). RIM-FIRE PISTOL. Caliber .50. Eight-and-one-half-inch, round, rifled barrel. This is another example of the model 1866 Remington navy pistol. See No. 18. No serial number is given upon the pistol here shown. Markings and other characteristics are the same as those of No. 18.

U. S. A.

No. 20. (N4107). RIM-FIRE PISTOL. Caliber .50. Seven-inch, round, rifled barrel. The rear sight is a V-shaped notch on top of the breech-block. This, the model of 1867, was the second of the so-called Remington navy pistols and fired the same kind of cartridge as the model of 1866. The model 1867 differed from the previous one in having a trigger-guard and a shorter barrel. The government bought 500 of these for use in the navy in 1867. Later many were also issued to the army. The one here illustrated bears the mark of an anchor upon the barrel, while the lock-plate is stamped "Remingtons Ilion, N.Y., U.S.A., Pat. May 3d, Nov. 15th, 1864, April 17th, 1866." The serial number "6293" is marked upon the bottom of the barrel.

U. S. A.

No. 21. (N4108). CENTER-FIRE PISTOL. Caliber .50. Seven-inch, round, rifled barrel. This is another example of the model 1867 Remington navy pistol. See No. 20. The top of the barrel is stamped with the figure of an anchor and the letters "I.E.B." The lock-plate is marked in the same manner as No. 20 and the bottom of the barrel bears the number "4139."

U. S. A.

No. 22. (N4585). RIM-FIRE PISTOL. Caliber .22. Ten-inch, half-octagonal, rifled barrel. Has a silver fore-sight and an adjustable rear sight. This is a target pistol having the "Remington-Rider" action and is modeled after the model 1866 navy pistol. The lock-plate is marked "Remingtons Ilion, N.Y., U.S.A., Pat. May 3d, Nov. 15th, 1864, April 17th, 1866." The bottom of the barrel bears the serial number "22."

U. S. A.

No. 23. (N4584). RIM-FIRE PISTOL. Caliber .22. Ten-inch, half-octagonal, rifled barrel. Has front and rear sights. This is another example of the Remington target pistol. The side of the barrel is stamped with the letters "H.F.B.", probably the initials of a former owner. The grip is checkered. The lock-plate is marked in the same manner as that of No. 22. The numbers "598" and "375" appear upon the inside of the grip.

U. S. A.

No. 24. (N4106). CENTER-FIRE PISTOL. Caliber .50. Eight-inch, round, rifled barrel. This is a model 1871 Remington pistol. It was intended for army use only and used center-fire copper cartridges. This is really an alteration of the 1867 model Remington pistol. Upon recommendation by a board of officers of the Ordnance Department, the Remington company changed the design of the 1867 model so as to improve the balance and grip. Even at this day the frame and handle of this pistol are much sought for by target shooters because of the exquisite balance and the fine grip that these afford when provided with a modern barrel. The inside of the grip of the specimen here shown is stamped with the numbers "3034" and "3323." The maker's name is not given upon this pistol, although it was made by the Remington company.

U. S. A.

No. 25. (N1805). CENTER-FIRE PISTOL. Caliber .50. Eight-inch, round, rifled barrel. This is another example of the model 1871 Remington army pistol. See No. 24. The lock-plate is stamped "Remington's, Ilion, N.Y. U.S.A. Pat. May 3d, Nov. 15th, 1864, April 17th, 1866." Upon the inside of the grip are the numbers "2628" and "2659." The left side of the stock is stamped with a government inspector's stamp.

U. S. A.

No. 26. (N778). CENTER-FIRE PISTOL. Caliber .55. Eight and-one-quarter-inch, octagonal, rifled, steel barrel. This pistol is provided with brass lock-plates and an ivory grip. It has a sheath trigger like the first model of this series, the model 1866. See No. 18. The pistol here shown was probably not made by the Remington Company. The left side of the lock-plate is engraved with the inscription "M.B. En Vilia Garcia. Mayo. 20 De 1873." The right side bears the words "Numero 4 Por. Pascual Valdez." No maker's name or serial number are given. This pistol saw service in the wars and insurrections in Cuba about 1873.

?

No. 27. (N4587). CENTER-FIRE PISTOL. Caliber .38. Five-and-one-half-inch, round, rifled, double barrels. This pistol is provided with a large checkered grip which is fitted with an iron ring in the butt. One trigger serves to fire both barrels. When the pistol is cocked and the trigger is pulled, the right barrel is discharged first. Pulling the trigger a second time fires the left barrel. Stamped upon the left side of the lock-plate are the words "Brevet Remington" and the number "615," while the right side is marked with the Liége proof-mark and "E.M. and L. Nagant, Liége, Brevet Nagant." Various parts of the pistol bear the serial number "615."

Belgium.

No. 28. (N4586). CENTER-FIRE PISTOL. Caliber .45. Five-and-one-quarter-inch, round, rifled, steel barrel which is flared at the muzzle. There is no rear sight. The trigger-guard is fitted with an auxiliary grip for the third finger. Barrel and action are stamped with the Liége proof-mark. Pistols of this model are said to have been used by the French gendarmes. The under-side of the barrel is stamped with the number "11.0." No maker's name is given upon this imitation Remington.

Belgium.

PLATE 105.

REMINGTON, AND SMITH AND WESSON
REVOLVERS AND PISTOLS

No. 1. (N807). PERCUSSION REVOLVER. Caliber .32. Three-inch, octagonal, rifled barrel. There are five chambers in the cylinder for loose ammunition. The trigger-guard is made of German silver and the grip is smooth and made of ebonite. Stamped upon the top of the barrel are the words "F. Beals' Patent, June 24, '56, & May 26, '57, Manufactured by Remingtons' Ilion, N.Y." For details regarding these patents, see No. 4. The inside of the grip bears the serial number "23."

U. S. A.

No. 2. (N4510.) PERCUSSION REVOLVER. Caliber .32. Three-inch, octagonal, rifled barrel. Has five chambers in the cylinder. Trigger-guard and grip are the same as those of No. 1. The markings upon the top of the barrel are the same as those on No. 1. This revolver, and similar ones of this model, has an outside pawl turning the cylinder. Upon the bottom of the grip is the number "361."

U. S. A.

No. 3. (N4515). PERCUSSION REVOLVER. Caliber .32. Three-inch, octagonal, rifled barrel. Has five chambers. All steel parts of this double-acting revolver were formerly blued. The trigger-guard is of German silver and the grip of checkered ebonite. Stamped upon the top of the barrel are the words "Manufactured by Remingtons, Ilion, N.Y. Riders' Pt., Aug. 17, 1858, May 3, 1859." The first patent of Joseph Rider was concerned in "so combining the parts of the lock of a revolver that a single spring serves for all purposes," while the second patent had "internal improvements in the lock of a double-action arm." These peculiarities are exhibited in the specimen here shown. The barrel and frame are stamped with the serial number "1513."

U. S. A.

No. 4. (N4528). PERCUSSION REVOLVER. Caliber .32. Four-inch, octagonal, rifled barrel. Has five chambers in an etched cylinder. This revolver is single-acting. The grip is made of checkered ebonite. Stamped upon the top of the barrel are the words "Beal's Patent 1856 & 57 & 58, Manufactured by Remingtons' Ilion, N.Y." The first patent had a cylinder-rotating mechanism showing on the left side of the frame; the second had a spring catch for locking the cylinder, of form to conform to the periphery of the cylinder and projecting in front of the recoil shield and to one side of the cylinder; the third patent covered a cylinder-pin locked in place when the loading lever is in place and withdrawable forward when the lever is down. All three of these patent features are shown in this specimen, the serial number of which is "862."

U. S. A.

No. 5. (N4531). PERCUSSION REVOLVER. Caliber .32. Four-and-one-half-inch, octagonal, rifled barrel. There are five chambers in the cylinder. Has a brass trigger-guard. A lever ramrod is carried beneath the barrel, and the cylinder-pin is withdrawable from in front. All steel parts were formerly blued. The top of the barrel is stamped "Patented Sept. 14, 1858, Remington & Sons, Ilion, New York, U.S.A., New Model," and the side of the frame is engraved "American Express Co. No. 30." Express companies frequently furnished arms to their messengers, and this is an example of the Remington revolver issued. The serial number "10567" appears upon the bottom of the barrel.

U. S. A.

No. 6. (N4554). PERCUSSION REVOLVER. Caliber .32. Three-and-three-eighths-inch, octagonal, rifled barrel. Has five chambers in the cylinder. There is no trigger-guard, and the cylinder-pin is withdrawable from in front. All steel parts of the revolver are heavily blued. The top of the barrel is stamped "Patented Sept. 14, 1858, Remington & Sons, Ilion, New York, U.S.A., New Model," and the bottom bears the serial number "12851."

U. S. A.

No. 7. (N4641). RIM-FIRE REVOLVER. Caliber .38. Four-and-one-half-inch, octagonal, rifled barrel. Has five chambers. All steel parts were formerly nickel-plated. This model is known as the Remington police revolver. It was converted from the loose-ammunition system to that of the metallic cartridge by cutting off the rear of the old cylinder and supplying a steel disc to cover it. This disc not only filled up the shortage but also gave the appearance of a safe and solid breech. The cartridges could be inserted at either breech or muzzle of the cylinder. The top of the barrel is stamped "Patented Sept. 14, 1858, March 17, 1863, E. Remington & Sons, Ilion, New York, U.S.A. New Model." The serial number "2799" appears upon the bottom of the barrel.

U. S. A.

No. 8. (N4642). RIM-FIRE REVOLVER. Caliber .32. Three-and-one-half-inch, octagonal, rifled barrel. Has five chambers for rim-fire cartridges. All steel parts of the weapon are ornamented with scroll designs. This model is known as the Remington-Beals pocket revolver. It was converted from the loose-ammunition system to that of the metallic cartridge in the same manner as shown described under No. 7. Stamped upon the top of the barrel are the words "Patented Sept. 4, 1858, E. Remington & Sons, Ilion, New York, U.S.A. New Model," while the bottom bears the number "19538."

U. S. A.

No. 9. (N4643). RIM-FIRE REVOLVER. Caliber .32. Two-and-one-quarter-inch, octagonal, rifled barrel. Has five chambers in the cylinder. All steel parts were formerly nickel-plated and the grip is of checkered ebonite. The revolver, which is double-acting, was converted from the loose-ammunition system to that of the metallic cartridge in the same manner as shown and described under No. 7. No maker's name is given upon this Remington revolver. Cylinder and disc bear the number "1408," while the grip is stamped with the number "3."

U. S. A.

No. 10. (N5639). CENTER-FIRE REVOLVER. Caliber .38. Three-and-three-quarters-inch, octagonal, rifled barrel. Has five chambers in a fluted cylinder. There is no trigger-guard. The left side of the barrel is fitted with a spring ejector and the grip-plates are of checkered ebonite. The top of the barrel is stamped "E. Remington & Sons, Ilion, N.Y., Pat. W.S. Smoot, Oct. 21, 1873." No serial number is given.

U. S. A.

No. 11. (N4676). RIM-FIRE REVOLVER. Caliber .36. Three-and-three-quarters-inch, octagonal, barrel. There are five chambers in a fluted cylinder. This revolver is equipped with a spring ejector, but it has no trigger-guard. The grip-plates are of checkered ebonite. The top of the barrel is stamped "E. Remington & Sons, Ilion, N.Y., Pat. W.S. Smoot, Oct. 21, 1873." No serial number is given.

U. S. A.

No. 12. (N4679). RIM-FIRE REVOLVER. Caliber .38. Two-and-three-quarters-inch, octagonal, smooth-bore barrel. There are five chambers in a fluted cylinder. Has a spring ejector, but no trigger-guard. The grip is of mother-of-pearl, and the body and barrel are engraved with scroll designs. All metal parts were formerly silver-plated. The top of the barrel is stamped "E. Remington & Sons, Ilion, N.Y., Pat. W.S. Smoot, Oct. 21, 1873." The inside of the grip is marked with the number "2530."

U. S. A.

No. 13. (N4583). RIM-FIRE PISTOL. Caliber .38. Four-inch, half-octagonal, rifled barrel. Has front and rear sights. This pistol has the Remington-Rider breech-block, which, when drawn back for the insertion of the cartridge, acts at the same time as a safety-device. The top of the barrel is stamped "Remington's, Ilion, N.Y., Patd. Oct. 1, 1861, Nov. 15, 1861," and the bottom bears the serial number "1330."

U. S. A.

No. 14. (N774). RIM-FIRE PISTOL. Caliber .38. Four-inch, half-octagonal, rifled barrel, equipped with front and rear sights. Barrel and body are engraved with scroll designs. The barrel is silver-plated,

while the body is plated with gold. The grip is of mother-of-pearl.
Action and markings of this pistol are the same as those of No. 13. The
serial number is "1787."

U. S. A.

No. 15. (N3833). RIM-FIRE PISTOL. Caliber .41. Two, three-
inch, superimposed, rifled, steel barrels. The hammer fires each barrel
in succession. Rotating a lever upon the right side of the frame through
180 degrees releases a catch which allows the pistol to be opened or
"broken" by swinging the barrels upward. A hand ejector is located
upon the left side between the barrels. The upper barrel is stamped
"E. Remington & Sons, Ilion, N.Y. Elliott's Patent, Dec. 12, 1865."
The lower barrel bears the serial number "564." Pistols of this model,
as Nos. 15, 16, 17, and 18, are often known as Remington "Derringers"
and were so called by the makers.

U. S. A.

No. 16. (N4600). RIM-FIRE PISTOL. Caliber .41. Two, three-
inch, superimposed, rifled, steel barrels. The action of this Remington-
Elliott pistol is shown open. Markings and other characteristics are the
same as those of No. 15. The serial number is "838."

U. S. A.

No. 17. (N4601). RIM-FIRE PISTOL. Caliber .41. Two, three-
inch, superimposed, rifled barrels. Barrels and body are engraved with
scroll designs and silver-plated. The grip-plates are of mother-of-pearl.
In other respects this pistol is identical with No. 15. The serial number is "4690."

U. S. A.

No. 18. (N2170). RIM-FIRE PISTOL. Caliber .41. Two, three-
inch, rifled, steel barrels. The upper barrel is stamped "Remington
Arms Co., Ilion, N.Y.", and the lower bears the number "105." In
other respects this piece is similar to No. 15.

U. S. A.

No. 19. (N4906). RIM-FIRE PISTOL. Caliber .22. Four, three-
inch, rifled, steel barrels bored out of one block. This pistol resembles
the early pepper-box type, and is opened or "broken" by pulling forward
upon the small catch in front of the ring-trigger. The pistol has a concealed revolving hammer which fires the four barrels in succession.
The body is engraved with scroll designs and is gold-plated. The frame

is similarly engraved but is silver-plated. The grip is of ivory. Stamped upon the pistol are the words "Manufactured by E. Remington & Sons, Ilion, N.Y., Elliott's Patents May 29, 1860, Oct. 1, 1861." The inside of the frame bears the serial number "373."

U. S. A.

No. 20. (N4905). RIM-FIRE PISTOL. Caliber .38. Four, three-and-three-eighths-inch, rifled barrels bored out of one solid steel block. Equipped with front and rear sights. This pistol is loaded, opened or "broken" and fired in the same manner as that of No. 19. It is also marked in the same way. The serial number is "5693."

U. S. A.

No. 21. (N4580). RIM-FIRE PISTOL. Caliber .41. Two-and-one-half-inch, half-octagonal, rifled barrel. The striker upon the hammer serves as a rear sight. In firing, the hammer receives the full effect of the recoil. All metal parts are made of steel, which were formerly blued. The top of the barrel is stamped "Remingtons, Ilion, N.Y., Pat. Aug. 27, 1867," while the lower part of barrel and body bear the number "3877."

U. S. A.

No. 22. (N3834). RIM-FIRE PISTOL. Caliber .32. Three-inch, octagonal, rifled barrel. This type is known as the magazine pistol. Cartridges were inserted into the magazine beneath the barrel. A coiled helical spring in the magazine forced the cartridges into the space in the rear of the pistol between the breech and breech-block. After the pistol was fired, the breech-block was pulled back by the thumb, where-upon the empty shell was ejected and the pistol became cocked. Releas-ing the breech-block caused it to move forward and to insert a cartridge into the breech. The pistol was then again ready for firing. The top of the barrel is stamped "E. Remington & Sons, Ilion, N.Y. Rider's Pat. Aug. 16, 1871." No serial number is given.

U. S. A.

No. 23. (N4914). RIM-FIRE PISTOL. Caliber .32. Three-inch, octagonal, rifled barrel. Barrel and body are engraved with scroll designs, and are silver-plated. The grip is of mother-of-pearl. In other respects this magazine pistol is identical with No. 22 shown above.

U. S. A.

No. 24. (N4582). RIM-FIRE PISTOL. Caliber .22. Three-and-

one-quarter-inch, round, rifled barrel. All parts are made of steel.
This pistol is provided with front and rear sights and an ivory grip.
The pistol is single-shot and the cartridge is inserted into the breech.
This specimen bears neither manufacturer's name nor serial number;
it was, however, made by Remington. See No. 25.

U. S. A.

No. 25. (N4881). RIM-FIRE PISTOL. Caliber .15. Three-inch,
half-octagonal, smooth-bore, brass barrel. The body is also of brass and
is slightly engraved. Hammer and trigger are of steel. Equipped with
a front and rear sight. The left side of the barrel is stamped "Rider's
Pt., Sept. 13, 1859." No serial number is given. Made by Remington.

U. S. A.

No. 26. (N4652). RIM-FIRE REVOLVER. Caliber .22. Three-
and-one-eighth-inch, rifled barrel. Has seven chambers in a smooth
cylinder. The rear sight also serves as a cylinder stop. This feature
was patented by Smith, Horace and Wesson, July 5, 1859. It is described as a "wedge on top of the nose of the hammer, a spring in the
top strap, and a stop bolt, all for the purpose of furnishing a simple
cylinder stop." The revolver is opened or "broken" by tipping the
barrel upward, as is illustrated in No. 28. The top of the barrel is
stamped "Smith & Wesson, Springfield, Mass." The serial number
"86372" appears upon the butt.

U. S. A.

No. 27. (N4657). RIM-FIRE REVOLVER. Caliber .22. Three-
and-one-eighth-inch, rifled, steel barrel. The cylinder has seven chambers. Pressing upward on the catch in front of the trigger allows the
revolver to be opened or "broken" by tipping the barrel upward. See
No. 28. This revolver was made by Smith and Wesson but is not so
marked. The inside of the grip bears the number "137."

U. S. A.

No. 28. (N4653). RIM-FIRE REVOLVER. Caliber .22. Three-
and-one-eighth-inch, rifled barrel. Has seven chambers in the cylinder.
The revolver is opened or "broken" as shown in the illustration. All
metal parts of the revolver are silver-plated. The top of the barrel is
stamped "Smith & Wesson, Springfield, Mass. Pat. Apr. 3, '55, July 5,
'59, Dec. 18, '60." The butt is stamped with the serial number "90380."

U. S. A.

No. *29*. (N4636). RIM-FIRE REVOLVER. Caliber .22. Three-and-one-eighth-inch, rifled barrel. Has seven chambers in the cylinder. Pressing upon a catch in front of the trigger enables the cylinder-pin to be withdrawn for the removal of the cylinder. The body is of brass. The top of the barrel is stamped "Made for Smith & Wesson By Rollin White Arms Co., Lowell, Mass." Rollin White held the patent for boring the chambers of the cylinder clear through for the purpose of loading at the rear. This part of the patent was purchased later by Smith and Wesson for use in their revolvers. The butt bears the number "5637."

U. S. A.

No. *30*. (N4648). RIM-FIRE REVOLVER. Caliber .31. Six-inch, octagonal, rifled barrel. Has six chambers for rim-fire cartridges. The rear sight also serves as a cylinder stop, a feature patented by Smith Horace and Wesson. See No. 26. This revolver is also opened or "broken" by tipping the barrel upward. Stamped upon the top of the barrel are the words "Smith & Wesson, Springfield, Mass." The serial number "8849" appears upon the butt.

U. S. A.

No. *31*. (N3356). RIM-FIRE REVOLVER. Caliber .31. Six-inch, octagonal, rifled barrel. Has six chambers in the cylinder. The top of the barrel is stamped "Smith & Wesson, Springfield, Mass.", and the butt bears the serial number "12727." Other features are the same as those of No. 26.

U. S. A.

No. *32*. (N4651). RIM-FIRE REVOLVER. Caliber .31. Three-and-one-half-inch, octagonal, rifled barrel. Has five chambers in the cylinder. The top of the barrel is stamped "Smith & Wesson, Springfield, Mass. Patd. Apr. 3, 1855, July 5, 1859 & Nov. 21, 1865." The last patent was issued for a "cylinder having no center pin which bears at front and rear upon two adjustable center screws." The butt bears the serial number "37402." Other characteristics are the same as those of No. 26.

U. S. A.

No. *33*. (N4650). RIM-FIRE REVOLVER. Caliber .31. Three-and-one-half-inch, rifled barrel. There are five chambers in a fluted cylinder. The top of the barrel is stamped "Smith & Wesson, Springfield, Mass., Pat. Apr. 3, '55, July 5, '59 & Nov. 21, 1865." See Nos. 26 and 33, which are similar to the specimen here shown. The serial number is "43231."

U. S. A.

No. 34. (N4649). RIM-FIRE REVOLVER. Caliber .31. Three-and-one-half-inch, rifled barrel. There are five chambers in the cylinder. In this piece the rear sight is absent and does not serve as a cylinder stop. Pressing upward on the catch in front of the trigger allows the revolver to be opened or "broken" by tipping the barrel upward. The cylinder may then be removed. All steel parts are blued. The top of the barrel is stamped "Smith & Wesson, Springfield, Mass. Patd. Apr. 3, July 5, 1859 & Nov. 21, 1865." The butt bears the number "22578."

U. S. A.

No. 35. (N4141). CENTER-FIRE REVOLVER. Caliber .44. Six-and-one-half-inch, rifled barrel. Has six chambers in a fluted cylinder. Pulling up on a catch above the trigger, allows the revolver to be opened or "broken" by tipping the barrel downward. "Breaking" the revolver causes the ejector to rise in the rear of the cylinder which throws out all empty shells in one operation. This Smith and Wesson revolver is known as the 1871 model. The top of the barrel is stamped "Smith & Wesson, Springfield, Mass., U.S.A. Pat'd. Jan. 17 & 24, '65. July 11, '65. Aug. 24, '69. Apr. 20, '75. Feb. 20 & Dec. 18, 1877. Reissue July 25, 1871." The rear of the cylinder bears the serial number "10165."

U. S. A.

No. 36. (N4143). CENTER-FIRE REVOLVER. Caliber .45. Seven-inch, rifled, steel barrel. Has six chambers in the cylinder, which is fluted. Pulling back on a catch in front of the trigger allows the revolver to be opened or "broken" by tipping the barrel downward. Ejection of the empty shells is the same as for No. 35. This Smith & Wesson revolver is known as the "Schofield" model. The left side of the barrel is stamped "Smith & Wesson, Springfield, Mass., U.S.A. Pat. Jan. 17th & 24th '65, July 11th, '65, Aug. 24th, '69, July 25th, '71," and the right side "Schofield's Pat. Apr. 22d, 1873." The cylinder bears the number "1310" and the butt "175."

U. S. A.

PLATE 106.

VOLCANIC PISTOLS AND SMITH AND WESSON REVOLVERS

No. 1. (N4869). MAGAZINE PISTOL. Caliber .36. Eight-inch, octagonal, rifled, steel barrel. This specimen is fitted with a pistol-grip carbine stock made out of one piece of black walnut. The rear sight is of unusual form. The body is of unornamented brass. This is a Volcanic pistol. For further details, see No. 5. The top of the barrel is stamped "The Volcanic Repeating Arms Co., Patent New Haven, Conn., Feb. 14, 1854." The trigger-guard bears the serial number "263."

U. S. A.

No. 2. (N4870). MAGAZINE PISTOL. Caliber .36. Eight-inch, octagonal, rifled, steel barrel. This pistol is provided with a detachable carbine stock. The rear sight may be adjusted by means of a thumb-screw. Barrel and trigger-guard were formerly blued. See No. 5 for detailed information regarding the Volcanic pistol. The top of the barrel is stamped "The Volcanic Repeating Arms Co., Patent New Haven, Conn., Feb. 14, 1854." The trigger-guard bears the serial number "469."

U. S. A.

No. 3. (N4873). MAGAZINE PISTOL. Caliber .36. Six-inch, octagonal, rifled barrel. Has an unornamented brass body. Barrel and magazine are blued. In other respects this Volcanic pistol is similar to No. 5. The top of the barrel is stamped "Patent Feb. 14, 1854, New Haven, Conn." No maker's name is given. This pistol was made by the Volcanic Repeating Arms Co. for Smith & Wesson. The inside of the grip bears the serial number "2066."

U. S. A.

No. 4. (N4876). MAGAZINE PISTOL. Caliber .31. Three-and-one-half-inch, octagonal, rifled barrel. The brass body is engraved with scrolls. All steel parts, except the hammer, are blued. This is another Volcanic pistol, detailed information regarding which is given under No. 5. The top of the barrel is stamped "New Haven, Conn., Patent Feb. 14, 1854." No maker's name is given. These pistols were made by the Volcanic Repeating Arms Company of New Haven, for Smith and Wesson. The serial number "1236" appears inside of the grip.

U. S. A.

No. 5. (N6789). MAGAZINE PISTOL. Caliber .31. Four-inch, half-octagonal, rifled, steel barrel. The cartridges of these pistols had no shell; the hollow in the rear end of the bullet itself was filled with mercury-fulminate covered by a cork or copper disc, which was pierced by the firing-pin when the pistol was discharged. In consequence of this, no shell had to be extracted. The fulminating cartridges were carried in the magazine located beneath the barrel and carried into the breech by means of the lever which served at the same time as a trigger-guard. This is one of the first type of pistols to bear the name of Smith and Wesson. They were not made by Smith and Wesson or patented by them, as is sometimes supposed, but were made for them by the Volcanic Repeating Arms Company of New Haven, Connecticut. The top of the barrel is stamped "Cast-Steel Smith & Wesson, Norwich, Ct., Patent Dec. 25, 1849." Upon the inside of the grip is the serial number "8."

U. S. A.

No. 6. (N4871). MAGAZINE PISTOL. Caliber .36. Eight-inch, octagonal, rifled barrel. The body is of unornamented brass. Barrel, magazine and trigger-guard are blued. This pistol is in excellent condition. In other respects it is similar to No. 5, which see for detailed information concerning the Volcanic pistol. Stamped upon the top of the barrel are the words "The Volcanic Repeating Arms Co., Patent New Haven, Conn., Feb. 14, 1854," which company made the pistols for Smith and Wesson. Trigger and inside of the grip bear the serial number "1696."

U. S. A.

No. 7. (N4875). MAGAZINE PISTOL. Caliber .31. Four-inch, half-octagonal, rifled, steel barrel. The body is engraved with scroll designs. The sides of the body and the grip-plates are made of ebonite. In other respects this pistol is similar to No. 5. The barrel is marked "Cast-Steel Smith & Wesson Norwich, Ct., Patent Feb. 14, 1854." Upon the inside of the grip is the serial number "B 10."

U. S. A.

No. 8. (N808). CENTER-FIRE REVOLVER. Caliber .32. Three-and-one-half-inch, rifled barrel. There are five chambers in a fluted cylinder. The revolver is fitted with a concealed hammer. The weapon, which is double-acting, is opened or "broken" by pressing down upon a small catch behind the rear sight. The revolver cannot be fired unless the large safety-device in the rear of the grip is pressed inward at the same time the trigger is pulled. The engraving and silver mounting were executed by Tiffany of New York. The top of the barrel is stamped "Smith & Wesson, Springfield, Mass., U.S.A. Pat'd. Feb. 20, '77, Dec. 18, '77, May 11, '80, Sept. 11, '83, Oct. 2, '83, Aug. 4, '85." The cylinder is stamped with the serial number "41783." This revolver was carried by Mr. Rudolph J. Nunnemacher on his travels. It is shown in its original case, with cleaning rod and cartridge box.

U. S. A.

No. 9. (N4874). MAGAZINE PISTOL. Caliber .31. Six-inch, octagonal, rifled barrel. This pistol is provided with a round, checkered grip. The body is of brass, which is silver-plated and engraved with scrolls. Operating the lever in this pistol, cocks it at the same time. The firing-pin is contained in the bolt which shoves the cartridge into the breech. This pistol fired a metallic cartridge, and its action is similar to the pistol made by the Volcanic Repeating Arms Company. The top of the barrel is stamped "Venditti E Ci. Langusi." No serial number is given.

Italy?

No. 10. (N4872). MAGAZINE PISTOL. Caliber .36. Seven-and-three-quarters-inch, rifled barrel. The body is engraved with scroll designs and the grip-plates are of ebony. Operating the lever in this pistol cocks it at the same time, which is not the case with some of the other Volcanic pistols shown here. This pistol was made under W. A. Steven's patent. In other respects it is similar to No. 5. Stamped upon the top of the barrel are the words "Smith & Wesson, Patent Norwich, Ct." The inside of the grip is marked with the serial number "218."

U. S. A.

No. 11. (N2952). CENTER-FIRE REVOLVER. Caliber .44. Seven-inch, rifled, steel barrel. Has six chambers in a fluted cylinder. Pulling up on a catch above the trigger, allows the revolver to be opened or "broken" by tipping the barrel downward. "Breaking" the revolver causes an ejector to rise in the rear of the cylinder which throws out all empty shells in one operation. The weapon is covered with a coating of black enamel and is known as the Russian model. The top of the barrel is stamped "Smith & Wesson, Springfield, Mass., U.S.A. Pat.

July 10, '60, Jan. 17, Feb. 17, July 11, '65, & Aug. 24, '69. Russian Model." The cylinder bears the serial number "1224."

U. S. A.

No. 12. (N4845). CENTER-FIRE REVOLVER. Caliber .32. Three-and-one-half-inch, round, rifled, steel barrel. Has five chambers in a fluted cylinder. The revolver is opened or "broken" in the same manner as described for No. 11. This is a double-acting, model 1882, Smith and Wesson revolver. The top of the barrel is marked "Smith & Wesson, Springfield, Mass., U.S.A. Pat'd. Jan. 24, '65, July 11, '65, Aug. 24, '69, Reissue July 25, '71, May 11, '80, Jan. 3, '82." The butt bears the serial number "99497."

U. S. A.

No. 13. (N4142). CENTER-FIRE REVOLVER. Caliber .44. Five-and-one-quarter-inch, rifled barrel. There are six chambers in a fluted cylinder. The trigger-guard is fitted with an auxiliary grip for the third finger. The revolver is single-acting and is opened or "broken" as No. 11. All metal parts are nickel-plated. This is a Russian model Smith and Wesson revolver and the top of the barrel bears a Russian inscription. The rear of the cylinder bears the serial number "573."

U. S. A.

No. 14. (N4843). CENTER-FIRE REVOLVER. Caliber .38. Five-inch, round, rifled barrel. Has five chambers in a fluted cylinder. The revolver is opened or "broken" as described under No. 11. The piece, which is double-acting, has a mother-of-pearl grip and nickel-plated metal parts. This specimen is known as the "Euskaro" or the Spanish model. The term "Euskaro" is probably a variation of "Éuscaro," which means anything pertaining to the Basques or to their language. The top of the barrel is marked "Smith & Wesson's Cartridges Are Those That Fit Best The 'Euskaro' Revolver." The butt-plate bears the serial number "119965." The revolver here shown was bought in a new condition in 1904 in Mexico City.

U. S. A.

No. 15. (N4144). CENTER-FIRE REVOLVER. Caliber .38. Six-inch, round, rifled, steel barrel. There are six chambers in a fluted cylinder. The revolver is double-acting. Pressing upon the roughened catch in back of the cylinder enables the latter to be thrown out to the left for ejection of the empty shells or for loading. This is a model 1901 navy revolver. The top of the barrel is stamped "Smith & Wesson,

Springfield, Mass., U.S.A. Pat'd. Apr. 9, '89, Mar. 27, '94, May 21, '95, July 16, '95, Aug. 4, '96, Dec. 22, '96, Oct. 14, '98, Oct. 8, '01, Dec. 17, '01. S. & W. .38 Mil." The butt-plate is stamped with an anchor and "U.S.N. 38 D.A. No. 1122 J. A. D." The grip bears the serial number "25102." The revolver has a rebounding hammer.

U. S. A.

PLATE 107.

MISCELLANEOUS PISTOLS AND REVOLVERS

No. 1. (N6800). PERCUSSION PISTOL. Caliber .64. Eleven-and-one-half-inch, octagonal, rifled, steel barrel. The carbine stock is hollow, and contains a bullet mold. When the lever upon the right of the frame is released, it may be swung upward and in so doing pushes the barrel slightly forward. After this operation the barrel can be tipped upward and access may be had to the breech so that it may be charged with a paper cartridge. The hammer is large and ring-shaped. The bottom of the barrel bears an undecipherable name and the number "17." Frame and carbine stock are engraved "H.G. 17."

Germany.

No. 2. (N4590). PERCUSSION PISTOL. Caliber .56. Sixteen-inch, round, tapering, smooth-bore, steel barrel. This pistol was made under Alonzo D. Perry's patent of January 16, 1855. It has a breech-loading mechanism adapted both to paper cartridges and to loose ammunition. The trigger-guard served also as a lever. One of the merits of the Perry action was the gas-tight joint, in which respect it was better than the Hall and Sharp's mechanisms. Furthermore, the Perry action was provided with an automatic device for removing exploded caps. A number of Perry pistols in both army and navy sizes were offered to the government but were not accepted, although they were highly commended by some contemporaneous arms experts. Poor financing caused the Perry Patent Arms Company of Newark, New Jersey, to fail and today their arms have become a rarity. The specimen here shown is stamped with the maker's name and with the number "4."

U. S. A.

No. 3. (N4397). PERCUSSION PISTOL. Caliber .38. Twelve-inch, octagonal, rifled, steel barrel. This muzzle-loading target pistol is fitted with a shoulder stock and a telescopic sight. It also has an under-hammer and a saw-handle grip. The pistol is very heavy, weighing seven-and-one-half-pounds with the shoulder stock. Stamped upon the top of the barrel are the words "W. Billinghurst, Rochester, N.Y." No serial number is given.

U. S. A.

No. 4. (N6330). PERCUSSION PISTOL. Caliber .52. Fourteen-and-three-eighths-inch, smooth-bore, octagonal barrel. The main spring consists of a thick, spiral, brass spring wound about the axis of the hammer. This pistol has front and rear sights, but the latter is useless, since both hammer and nipple interfere when aiming. The stock is of polished hard wood. The ramrod is missing. This pistol was probably a match-lock which was remodeled to the percussion system. No marks of identification are given.

Japan.

No. 5. (N4104). CENTER-FIRE REVOLVER. Caliber .50. Nine-and-three-quarters-inch, round, rifled barrel. This is a United States Springfield experimental breech-loading pistol. It fires a center-fire, metallic cartridge of the huge caliber given above and is equipped with the regular Springfield lock of the model 1873 United States army rifle. Upon the lock-plate are the United States eagle and the words "U.S. Springfield 1863," indicating that the plate came from a muzzle-loading Springfield of Civil War days. The pistol, which is very heavy and unwieldy, bears the number "48392" upon the left side of the barrel. It is said that after several of these pistols were completed, one was given to a sentry to shoot as a trial. "After firing the piece the soldier began to run and has not been seen or heard from since."

U. S. A.

No. 6. (N4591). PERCUSSION PISTOL. Caliber .50. Ten-inch, half-octagonal, rifled barrel. The rear sight consists of three leaves numbered 1, 3 and 5. When the lever, which acts at the same time as a trigger-guard, is lowered, the barrel tips downward for loading, as shown in the illustration. This pistol used a paper cartridge. The left side of the frame is stamped "Edward Maynard, Patentee, May 27, 1851, Dec. 6, 1859," while the right side is marked "Manufactured by Mass. Arms Co. Chicopee Falls."

U. S. A.

No. 7. (N762). NEEDLE-GUN. Caliber .65. Nine-inch, round, smooth-bore, steel barrel. This is undoubtedly a remodeled English "Tower" piece of the 18th century, from which all marks of identification have been removed. When the steel back-strap is pulled upward, as shown, the breech-block is drawn back for the admission of the cartridge, and the hammer is cocked in the same operation. Bullet and charge are forced into the breech when the back-strap is lowered. By pulling the trigger the needle or firing-pin is released and shoots forward into a patch of detonating powder in the center of the cartridge. This pistol formerly carried a swivel ramrod. No maker's name or proof-marks are given.

England.

No. 8. (N6329). PERCUSSION-CAP PISTOL. Caliber .63. Three-and-one-half-inch, round, tapering, smooth-bore, brass barrel. Has an under-hammer. This type is sometimes known as a "bludgeon pistol." All parts are made of brass. The barrel may be unscrewed from the rest of the pistol. The grip is in the form of an eagle's head and is hollow, so that it may serve as a container for shot, powder and caps. An English proof-mark appears upon the barrel.

England.

No. 9. (N5637). TARGET PISTOL. Caliber .31. Fifteen-inch, round, rifled, steel barrel. This pistol, which uses center-fire cartridges, is equipped with a shoulder stock and a sheath trigger. A safety-device is located on the left side of the frame. Pressing upon a catch in front of the trigger enables the barrel to be turned to the right for loading or for the ejection of the empty shell. A hand ejector is located upon the right side of the barrel. Equipped with peep sights. The left side of the barrel is stamped "Frank Wesson, Worcester, Mass., Patented May 31, 1870." The stock bears the number "2320" and the pistol "1300."

U. S. A.

No. 10. (N4868). CENTER-FIRE REVOLVER. Five chambers for caliber .38 center-fire cartridges, the caliber of the twenty-four-inch, round, smooth-bore, brass barrel is .535. The top of the frame is stamped "Pat. Mar. 28, '71, Jan. 5, '86. .38 Cal. Center Fire," and the barrel bears the number "1301." This revolver was made under the Hinsdale patent. The butt-plate is bored for a shoulder stock, indicating that the revolver was used for target practice. The piece is single-acting. Designed for the use of shot-cartridges, for birds, etc.

U. S. A.

No. 11. (N4687). RIM-FIRE REVOLVER. Six chambers for caliber .38 rim-fire cartridges, the caliber of the twenty-four-inch, round, smooth-bore, brass barrel is .535. The top of the frame is stamped "Hinsdale, .38 Cal. Rim Fire. Pat. Mch. 28, '71, May 27, '79." The barrel bears the number "8003." The butt-plate is bored for a shoulder stock. Like No. 10, this revolver is also single-acting. Designed for the use of shot-cartridges, for birds, etc.

U. S. A.

No. 12. (N7070). PERCUSSION PISTOL. Caliber .35. Six-and-one-quarter-inch, smooth-bore, octagonal, steel barrel. This target pistol formerly was fitted with a removable breech-block which is now missing. Frame and trigger-guard are engraved, and the grip is fluted. Barrel and frame bear the letters "H D" and the numeral "90"; they also bear an undecipherable proof-mark which is probably Austrian. No maker's name is given.

Austria?

No. 13. (N4574). RIM-FIRE PISTOL. Caliber .22. Eight-and-three-quarters-inch, half-octagonal, smooth-bore, steel barrel. This is a breech-loading target pistol which is fitted with an auxiliary grip for the third finger. The barrel is inlaid with gold, while the stock is of ebony and elaborately carved with excised figures of leaves. The steel trigger-guard and butt-plate bear similar figures. Upon the left side of the barrel is the Liége proof-mark. No maker's name is given. The original cleaning rod is shown below the pistol.

France.

No. 14. (N2788). RIM-FIRE PISTOL. Caliber .22. Nine-and-three-quarters-inch, half-octagonal, smooth-bore, steel barrel. This breech-loading target pistol is fitted with an auxiliary grip for the third finger. It has an adjustable rear sight, but the fore-sight is missing. Butt-plate, trigger-guard and hammer are engraved. Upon the left side of the barrel is the Liége proof-mark. No maker's name is given.

Belgium.

No. 15. (N2787). RIM-FIRE PISTOL. Caliber .35. Eight-inch, octagonal, smooth-bore, steel barrel. This is a form of a "French gallery pistol." It is fitted with an auxiliary grip for the third finger and fires rim-fire cartridges which are loaded through the breech. The fore-stock is carved to resemble the head of a dog. Trigger-guard and hammer are engraved with scrolls, and the grip is fluted. The bottom of the barrel bears the serial number "18." No maker's name is given.

France.

No. 16. (N4568). RIM-FIRE PISTOL. Caliber .22. Seven-and-one-half-inch, octagonal, smooth-bore, steel barrel. This is another French gallery pistol. It fires rim-fire cartridges which are loaded at the breech. The grip is of ebony and bears carved, excised figures of leaves. The trigger is of the concealed folding variety. The frame is stamped with the number "21" and the barrel with "3903." The Liége proof-mark also appears upon the left side of the barrel.

France.

No. 17. (N4385). PERCUSSION PISTOL. Caliber .22. Eight-inch, octagonal, smooth-bore, steel barrel. This is a muzzle-loading target pistol which has the nipple for the percussion-cap directly in the rear of the barrel. The trigger-guard is fitted with an auxiliary grip. The frame is engraved and the grip is fluted. No maker's name or serial number are given. This is said to be an Austrian piece.

Austria?

No. 18. (N4386). PERCUSSION PISTOL. Caliber .22. Five-and-three-quarters-inch, half-octagonal, cannon-shaped barrel. This smooth-bore, muzzle-loading target pistol has the nipple for the percussion-cap placed directly in the rear of the barrel. The back-strap is in the form of a spring which throws the hammer forward when the pistol is discharged. Frame and trigger-guard are engraved. No maker's name or serial number are given.

Austria.

No. 19. (N4147). PERCUSSION REVOLVER. Caliber .36. Seven-and-one-half-inch, round, rifled barrel. This Confederate revolver has already been described under PLATE 100, No. 9.

U. S. A.

No. 20. (N4148). PERCUSSION REVOLVER. Caliber .36. Seven-inch, octagonal, rifled barrel. There are six chambers in the cylinder. The frame is of brass and is cast in one piece. This is a Confederate revolver of the Whitney pattern, made by Speller and Brothers of Augusta, Georgia, in 1863. The maker's name is not given upon this specimen. The serial number "703" appears upon the bottom of the barrel and upon the butt.

U. S. A.

No. 21. (N4146). PERCUSSION REVOLVER. Caliber .36. Seven-and-one-half-inch, half-octagonal, rifled barrel. Has six chambers in the cylinder. This is a Confederate revolver which was made in the Augusta Pistol Factory in Augusta, Georgia. In 1862 this factory, which was owned by Smith and Regdon, was called the Ansley Factory. The top of the barrel bears the letters "C.S.A." and various parts bear the serial number "2330." The revolver has double the number of cylinder-stops than are usually found upon revolvers of this period.

U. S. A.

No. 22. (N4607). PERCUSSION PISTOL. Caliber .65. Seven-and-one-quarter-inch, octagonal, smooth-bore, steel barrel. Fires a cartridge with a combustible paper envelope. Pressing a button in front of the trigger allows the barrel to be swung to the left for loading. Lock-plate, hammer and frame are engraved with scroll designs. The barrel is damaskeened. A cap-box is carried in the butt. Engraved upon the top of the barrel are the words "Fini Par Le Page Moutier Arqr. Du Roi," which means "Made by Le Page Moutier Armorer to the King." The frame bears the serial number "14."

France.

No. 23. (N771). PERCUSSION PISTOL. Caliber .20. Eight-inch, round, smooth-bore barrel. This is a "parlor-pistol." The hammer actuates a piston within the barrel. When the trigger is pulled, the piston flies forward and strikes a percussion-cap placed in the breech of a two-and-one-half-inch barrel which is contained in the larger barrel. A hole is cut in the under-side of the larger barrel so that the cap may be placed in position upon the smaller barrel. The lock-plate is engraved and bears the name "Sharpe." This is probably a remodeled dueling pistol.

England.

No. 24. (N6791). PERCUSSION PISTOL. Caliber .18. Four-and-one-half-inch, octagonal, smooth-bore barrel. This muzzle-loading target pistol has the nipple for the percussion-cap placed directly in the rear of the barrel. The frame is engraved and the grip is fluted. No maker's name or serial number are given. This is probably a German or an Austrian piece.

Germany?

No. 25. (N4588). PERCUSSION PISTOL. Caliber .36. Six-inch, half-octagonal, rifled barrel. This pistol fires a paper cartridge which is forced into the breech by means of a loading lever located behind the trigger-guard. The cartridge is placed into the pistol from the right side. The barrel may be removed by releasing the pin in front of the nipple. Upon the top of the barrel are the words "W. W. Marston, New York, Patented 1850." Frame and barrel bear the number "248."

U. S. A.

No. 26. (N4581). PERCUSSION PISTOL. Caliber .31. Five-inch, round, tapering, rifled barrel. This Sharp's pistol fires a paper cartridge. The falling breech-block for loading at the rear of the barrel with either paper cartridges or loose powder and ball, as in the Sharp's rifle, was applied to a limited extent to pistols also. Such pistols for army and navy use were offered to the government, but were declined. Later, the same system was used for metallic cartridges; this system being also declined by the government. Sharp's single-shot breech-loading pistols are now rarely found. They were apparently hand-made and occur with or without a fore-end of either wood or metal, and with wood or metal handles of varying shapes. The left side of this specimen is stamped "Sharp's Patent Arms Mfed. Fair Mount, Phila. Pa." The serial number is "102."

U. S. A.

No. 27. (N4572). RIM-FIRE TARGET PISTOL. Caliber .33. Nine-and-one-quarter-inch, smooth-bore, octagonal, steel barrel. There is an auxiliary grip on the trigger-guard. The breech-block and action are similar to those of the Remington single-shot navy pistol, model of 1866. The top of the barrel bears a French proof-mark in the form of a crown over the letters "AF." The bottom of the barrel bears the serial number "3233." No maker's name is given.

France.

No. 28. (N2789). CENTER-FIRE TARGET PISTOL. Caliber .34. Eight-and-three-quarters-inch, smooth-bore, octagonal, steel barrel. The trigger-guard has an auxiliary grip for the third finger. Releasing a lever on the left side of the frame allows the breech-block to be swung to the right for loading. Frame and trigger-guard are engraved with scrolls and the grip is fluted. Barrel and body bear the name "Loron & G." The serial number "405" appears upon the barrel.

France?

No. 29. (N4593). RIM-FIRE TARGET PISTOL. Caliber .22. Nine-inch, round, rifled, steel barrel. The holes in the back-strap serve for the attachment of a shoulder stock. There is an auxiliary grip or spur upon the trigger-guard. The rear sight is adjustable for deflection and elevation. Pressing a catch upon the left side of the frame allows the barrel to drop downward for loading or for removal of the empty shell. The barrel is stamped "J. Stevens & Co., Chicopee Falls, Mass., Pat. Sept. 6, 1864." Trigger-guard and breech bear the number "11745."

U. S. A.

No. 30. (N7148). CENTER-FIRE REVOLVER. Caliber .45. Seven-and-one-half-inch, round, rifled barrel. This is a single-action, Colt "Army" revolver which has been equipped with an automatic cartridge ejector by some inventor. A device consisting of a cylinder and piston is attached to the lower part of the barrel as shown. A fine hole is bored through the barrel near the piston head. When the revolver is fired, a portion of the gases from the exploding cartridge escapes through the fine hole and causes the rod which is attached to the piston to fly back and eject a shell. However, the invention has the disadvantage of throwing out a loaded cartridge if the revolver is fired when all chambers are full. The top of the barrel is stamped "Colt's Pt. F.A. Mfg. Co., Hartford, Ct., U.S.A.", and the side "45 Colt." Upon the frame are the words "Pat. Sept. 19, 1871, July 2, '72, Jan. 19, '75." Various parts of the revolver bear the serial number "196516."

U. S. A.

No. 31. (N4594). RIM-FIRE TARGET PISTOL. Caliber .22. Ten-inch, round, rifled, steel barrel. A groove in the butt is for the attachment of a shoulder stock. Pressing a catch upon the left side of the frame allows the barrel to drop downward for loading or for removal of the empty shell. The barrel is stamped "J. Stevens & Co., Chicopee Falls, Mass. Pat. Sept. 6, 1864." Upon the butt is the serial number "1583."

U. S. A.

No. 32. (N4593). RIM-FIRE TARGET PISTOL. Caliber .22. Ten-and-three-quarters-inch, half-octagonal, rifled barrel. The trigger-guard also serves as a loading lever; when it is pulled down the barrel drops downward for loading. The breech is fitted with a self-extractor. Stamped upon the left side of the barrel are the figures "0.2 gr. N.G.

P.M/71." and "1.8 gr Bl."   These letters and figures refer to the weight of the charge of powder and to the weight of the bullet respectively.   The left side of the frame is marked "D.R.Pat. 73355," meaning "Deutsches Reich Patent No. 73355."   The barrel bears the double proof-marks of Germany.   The serial number of the pistol is "6134." No maker's name is given.

Germany.

No. 33. (N7093). CENTER-FIRE SHOTGUN PISTOL. Caliber .607.   Twenty gauge.   Ten-inch, smooth-bore, double barrels.   Has a roughened rib on top between the barrels to minimize reflection in aiming.   Pushing the lever on top of the grip to the right releases the barrels so that they may be tipped downward for loading.   Has a self-ejector.   Breaking the piece also cocks it.   When the piece is cocked it is also made safe, the safety being a small slide on the back-strap. Pushing the slide upward makes the gun ready for firing.   The piece is rather heavy and not well balanced, weighing four-and-one-half pounds.   Weapons of this kind are intended to be carried in automobiles for protection against holdups, or to be used for offices or banks.   The gun shoots a wide pattern and will fire any twenty-gauge shell loaded with shot, buckshot, or a single ball.   The barrels are stamped "Smokeless Powder Steel Made in U.S.A.", and the frame, "Auto Burglar Gun Ithaca Gun Co., Ithaca, N.Y."   The serial number is "361429."

U. S. A.

No. 34. (N6332). CENTER-FIRE PISTOL. Caliber .45. Four, six-and-one-quarter-inch, round, rifled barrels.   Has a concealed hammer which fires the four barrels in succession.   Compressing the catch above the grip, allows the barrels to drop downward and also ejects the empty shells at the same time.   Each of the barrels and the breech-blocks are stamped with English proof-marks.   No maker's name is given.   This pistol is described by Greener in his noted book, "The Gun," and called by him the "mitralleuse."   Similar pistols were made by Charles Lancaster of London, and by Braendlin of Birmingham. The one here shown bears the serial numbers "19" and "5422."

England.

PLATE 108.

POWDER TESTERS, FIRE LIGHTERS, SIGNAL PISTOLS,
CAP TESTERS AND TRAP GUNS

No. 1. (N5635). POWDER TESTER. Sometimes also known

as an "éprouvette," or "éprouvette à poudre." A known amount of powder was placed into the barrel and the charge ignited by means of a slow-burning match held to the touch-hole. The force of the explosion turned the graduated wheel, which is held to the muzzle by means of a spring, through a certain number of divisions. The amount of rotation of the wheel was held to be an indication of the strength of the powder. No maker's name or marks of identification are given.

?

No. 2. (N3282). POWDER TESTER. This tester resembles the flint-lock pistol of the same period. The brass wheel is graduated from 1 to 8. The strength of the powder was determined in the same manner as described under No. 1. The éprouvette is engraved with the words "Ketland & Co., London."

England.

No. 3. (N684). POWDER TESTER. In this éprouvette, resistance to rotation of the graduated disc is offered by means of a spring placed under the muzzle. The strength of the powder was determined in the same manner as described under No. 1. This tester was obtaned in Germany. No maker's name is given.

Germany.

No. 4. (N1161). FIRE LIGHTER. Enclosed in an iron box an inch-and-one-half long. When closed, the lighter could be carried in the pocket. The cock was released by pressing a small button on the side of the box. Tinder was carried in the circular hole in front of the steel. The bottom of the box is fitted with a brass ring and ornamented with a lotus flower of the same metal.

Japan.

No. 5. (N1162). FIRE LIGHTER. Of the rectangular type. The tinder was carried in the square box covered by the iron lid. The cock is released by pressing the small roughened trigger upon the left side of the lighter.

Japan.

No. 6. (N3372). POWDER TESTER. A known amount of powder was placed in the upright barrel and the bottom, which is attached to the pointer, was closed. The flint-lock above the barrel was next cocked and then discharged by pulling the trigger. The force of the explosion caused the pointer to move over a certain number of divisions on the scale which is numbered from 1 to 13. The amount of the rotation of the indicator was assumed to be a measure of the strength of the powder. By releasing the spring above the barrel and swinging it to the left, the pointer could be reset to number 1 on the scale. This powder tester was obtained in Mexico.

Donor: F. Pearson.

Mexico.

No. 7. (N6345). POWDER TESTER. This éprouvette has a vertical barrel. The notched wheel is numbered from 1 to 12. A thumb screw serves to compress the spring so that the graduated wheel may be released. The lock is of the Miguelet variety and the flash-pan of the tester is ribbed as in the case of the snaphaunce firearms. The strength of the powder was determined in the same manner as described under specimens Nos. 1 and 6.

Spain.

No. 8. (N4938). POWDER TESTER. There is no graduated scale upon this éprouvette. The number of notches over which the serrated wheel travels is taken as an indication of the strength of the powder. See Nos. 1 and 6 for more detailed information concerning powder testers. The specimen here shown was made in the United States and belongs to the period of about 1808.

U. S. A.

No. 9. (N7072). FIRE LIGHTER. Mounted on wheels. The compartment for spare tinder formerly had a door upon the left side. This lighter has two barrels fitted with touch-holes so that it could be used as a hand cannon. The upper jaw of the cock, the trigger and other essential parts are missing. The left side of the body is engraved "G.F.G." No maker's name is given.

?

No. 10. (N7066). FIRE LIGHTER. This lighter carries an oil cup. See No. 13. Spare tinder was carried in an iron compartment below the grip. The lighter was obtained in Nuremberg.

Germany.

No. 11. (N4939). FIRE LIGHTER. Of the early English type. Extra tinder was carried in the square hollow of the lighter which is closed by an iron door. The upper jaw of the cock and the trigger of the lighter are missing. No marks of identification are given.

England.

No. 12. (N7067). FIRE LIGHTER. All parts are made of iron. This lighter has no extra compartment for spare tinder. This specimen was obtained in Nuremberg, and bears no maker's marks.

Germany.

No. 13. (N831). FIRE LIGHTER. This lighter carries an oil cup on the right side. Oil was added to the smoldering fire in the tinder box in order to get a flame more rapidly and also to save exertion in blowing the tinder. Extra tinder was carried in the hollow of the lighter which is closed by a square sliding door.

England.

No. 14. (N4937). FIRE LIGHTER. This large lighter has a compartment with a door on the left side. Spare tinder, in the form of charred linen, was kept in this compartment. This lighter is not equipped with an oil cup. It is of English manufacture, but no maker's name is given.

England.

No. 15. (N4936). SIGNAL PISTOL. Essential parts belonging to the front and rear portions of this pistol are missing, so that its method of operation is difficult to ascertain. No maker's name or serial number are given.

?

No. 16. (N755). PERCUSSION-CAP TESTER. The percussion-cap was placed upon the nipple of the iron tube, while some powder was placed in the brass barrel which is attached at right angles to the iron tube. Upon firing the tester, the spark from the exploding cap traveled down the four-inch iron tube and ignited the powder in the barrel, assuming, of course, that the percussion-cap was of good quality. The lock-plate is engraved "E.I.C. Percussion Cap Provett." No maker's name is given.

France?

No. 17. (N4932). SIGNAL PISTOL. This pistol has a total length of nine inches. All moving parts are of steel, the rest being of bronze. This is an example of the Very signal pistol. See Nos. 18 and 19. The left side of the barrel is stamped "U.S. N.Y.W. W.N.J. 1864." The serial number is "35."

U. S. A.

No. 18. (N7074). SIGNAL PISTOL. The dimensions and material of which this Very pistol is composed are the same as those of No. 17. The specimen here illustrated shows the tube containing the colored fire held in place in the barrel by means of the steel clamp in front of the trigger. This is another form of signal pistol which was invented by Lieutenant Very of the United States navy. The bottom of the pistol is stamped "U.S. Army Signal Pistol A.J.M. 1861." The specimen here shown was found by Joseph Broughton of the Union Army at Raccoon Ford, Virginia, during Hooker's campaign before the battle of Chancellorsville, May 2-3, 1863.

Donor: H. L. Broughton.

U. S. A.

No. 19. (N4933). SIGNAL PISTOL. One-and-five-eighths-inch, tapering brass barrel. The paper cartridge containing the colored fire was clamped into the barrel by means of the steel lever situated in front of the trigger. Cartridges containing various colored lights could be shot into the air by the use of this pistol and signals given by this means. The pistol here shown is also the invention of Lieutenant Very of the United States navy. Upon the butt-plate are the words "U.S. Army Signal Pistol 1862 A.J.M." The inside of the grip bears the serial number "55."

U. S. A.

No. 20. (N4931). SIGNAL PISTOL. Four-inch, round, brass barrel having a diameter of one-and-one-half-inches. The barrel may be swung either to the left or to the right for loading. This pistol uses center-fire cartridges. Stamped upon the right side of the pistol are the words "Ord. Dept. U.S. N.Y.W. 1883 J.A.H." The pistol also bears the figure of an anchor, indicating ownership by the navy.

U. S. A.

No. 21. (N4929). SIGNAL PISTOL. Twelve-inch, round, steel barrel having a diameter of three-quarters of an inch. The barrel swings downward for loading when the catch, located in front of the trigger, is pulled. This pistol is made for the center-fire system. The specimen here shown is said to have been used by the French navy. It bears the serial number "37."

France?

No. 22. (N7023). SIGNAL PISTOL. Five-and-three-quarters-inch, round barrel. All metal parts are made of brass, with the exception of the mainspring, ejector, trigger, and breech-strap, which are of steel. The bore is one-and-one-eighth-inches in diameter. Upon the right side of the frame are numerous British proof-marks and the words "Webley & Scott, Ltd., London & Birmingham III." The serial number is "58390." This pistol is said to have been used on the steamship "Manley."

England.

No. 23. (N4930). SIGNAL PISTOL. The charge is contained in a two-and-one-half-inch, center-fire, metallic shell resembling that of a shotgun. The shell is held in place in the frame by means of a spring. An ejector in the form of a brass button is located on the right side of the pistol. The back-strap bears the words "Mario Cresta Hamburg." No serial number is given.

Germany.

No. 24. (N4940). TRAP PISTOL. Caliber .32. Two, three-and-one-half-inch, octagonal, smooth-bore, steel barrels. Uses loose ammunition. The nipples for the percussion-caps are placed upon the under-side of the barrels. This pistol is sometimes known as the "Kangaroo Catcher." The hooks in front of the barrels were baited and the trap pistol was cocked. When the animal pulled at the bait, it received both charges of the barrels. The cloth cover serves to protect the barrels from rain and also to hide them from the animal. The top of the pistol is marked "F. Reuthe's Patent," and the rib in front of the muzzle is dated 1857.

U. S. A.?

No. 25. (N764). TRAP PISTOL. This is another Reuthe's Kangaroo Catcher described under No. 24. It shows the top view of the pistol with the cover removed. Markings, date and other features are the same as those of No. 24.

U. S. A.

No. 26. (N4941). TRAP PISTOL. Caliber .38. Two, four-inch, smooth-bore, steel barrels. Uses loose ammunition. Percussion-caps are placed over the nipples at the breech of the barrels to fire the charges. A string or wire is fastened to the trigger which is located on top of the barrels. When the string is pulled the barrels point in the direction of the pull, being mounted on a swivel shown below the gun, and the charges are fired. The piece is stamped with the patent date of December 18, 1883, but no maker's name is given.

U. S. A.

PLATE 109.

MISCELLANEOUS GUNS AND FREAKS

No. 1. (N6234). FLINT-LOCK PISTOL. Caliber .44. Seven-and-three-quarters-inch, round, smooth-bore barrel. All parts are made of iron. The hammer is of the early or goose-neck variety and the shape of the trigger is just opposite that of the modern ones. A ram-rod was formerly carried in the thimble beneath the barrel. The grip is slightly ornamented with incised designs and has a hemispherical cap brazed to it on the outer end. No maker's name or other marks of identification are given upon this pistol, which is said to be of Scotch origin.

Scotland?

No. 2. (N4569). CENTER-FIRE PISTOL. Caliber .22. Ten-and-one-half-inch, half-octagonal, rifled, steel barrel. The breech-block of this target pistol can be swung to the left for loading. An ejector is situated in front of the breech-block. The left side of the barrel is stamped "0.2 gr 11. G.P.M/71. 1.8 gr Bl.", which indicates the weight of the charge of powder and the weight of the leaden bullet respectively. This pistol bears the double German proof-marks.

Germany.

No. 3. (7068). CENTER-FIRE PISTOL. Caliber .22. Seven-and-one-quarter-inch, half-octagonal, rifled barrel. The breech-block of this target pistol can be swung to the left for loading. An ejector is located in front of the breech-block. Has a sheath trigger. The left side of the barrel is stamped "0.05 gr. N.G.P.M/71. 1 gr. Bl.", which indicates the weight of the charge of powder and the weight of the leaden bullet. The breech-block is stamped with the number "2" and the bottom of the barrel bears the double German proof-mark.

Donor: Howard Hildebrand.

Germany.

No. 4. (N4570). RIM-FIRE PISTOL. Caliber .22. Four-and-one-half-inch, half-octagonal, smooth-bore barrel. The breech-block in this pistol is missing. The pistol was used for target practice and is equipped with a hand ejector. The grip is of checkered ebonite and the trigger is of the sheath variety. Upon the under-side of the barrel is a German proof-mark and on the inside of the grip is the number "43."

Germany.

No. 5. (N5702). RIM-FIRE PISTOL. Caliber .22. Three-and-three-quarters-inch, half-octagonal, smooth-bore barrel. The breech-block of this pistol can be swung to the left for loading or for removal of the empty cartridge. An ejector is located in front of the breech-block. Barrel and frame are nickel-plated. No serial number or maker's name are given, but the under-side of the barrel bears the German proof-mark.

Donor: Charles E. Crawford.

Germany.

No. 6. (N4571). RIM-FIRE PISTOL. Caliber .22. Three-inch, half-octagonal, smooth-bore barrel. The breech-block in this pistol can be swung to the left for loading or for removal of the empty cartridge. An ejector is located in front of the breech-block. The under-side of the barrel bears the German proof-mark, while grip and breech-block bear the number "2."

Germany.

No. 7. (N4882). CENTER-FIRE PISTOL. Caliber .25. Two, two-and-one-half-inch, superimposed, rifled, steel barrels. Has a square cartridge holder containing four chambers. The concealed folding trigger fires first the upper chamber then the second one. The cartridge holder is then revolved through 180 degrees by hand and the process is repeated upon the remaining two chambers. A hand ejector is carried in the grip. Barrel and holder bear German proof-marks. The cartridge holder is stamped "Patent, Deutschland 98382, Belgien 134215, England 11998, Russland, Nord America." The barrels bear the serial number "278" and the holder "1822."

Germany.

No. 8. (N6339). CENTER-FIRE PISTOL. Caliber .25. Four, two-and-five-eighths-inch, superimposed, round, rifled barrels. Pulling the trigger automatically raises the whole block of barrels successively, ejects the fired shell, and brings the loaded barrel under the hammer until all four barrels have been fired. A safety-device is situated upon the left side of the frame. The right grip-plate is stamped "Reform

Pistol D.R.P. 177023." D.R.P. stands for "Deutsches Reich Patent." Breech-block and barrels bear German proof-marks.

Germany.

No. 9. (N4927). PIN-FIRE PISTOL. Caliber .22. Six, smooth-bore, steel barrels in a one-and-one-half-inch, steel cylinder. This revolver is of the pepper-box type, and is a combined brass knuckle, dagger and revolver. The dagger is three inches long and is wavy like a Javanese kris. The folding-trigger cocks, revolves and fires at a single action. Knuckle and dagger may be folded about the cylinder so that the whole outfit may be carried in the pocket. The cylinder bears the Liége proof-mark and the frame is stamped "Bolne-Bar Brevetté." The serial number is "161."

Belgium.

No. 10. (N4924). PIN-FIRE PISTOL. Caliber .30. Ten, one-and-three-eighths-inch, rifled, steel barrels bored out of one block. Because of its shape, this piece is sometimes called a "harmonica" pistol. The pin-fire cartridges are inserted in the breech and held there by a perforated frame extending across the barrels. In firing, the barrels travel from right to left. This motion is automatic and is accomplished by pulling the trigger. A hand ejector is carried in the grip. The barrels and frame bear the name "A. Jarre, Bvt., S.G.D.G." The serial number is "485." The Liége proof-mark is stamped upon the block of the barrels.

Belgium.

No. 11. (N4923). PILL-LOCK PISTOL. Caliber .30. Three-and-three-quarters-inch, octagonal, rifled barrel. Ten chambers are bored in the periphery of a steel disc. These chambers are loaded with loose ammunition. Upon the right side of the disc are ten small holes bored at right angles to and communicating with each chamber. Each of these small holes receives a pellet of fulminate which is exploded one at a time by the side-hammer when the trigger is pulled. The pistol is double-acting and carries a safety-device under the side-hammer. It is stamped "Système à Noel, J.P. Gouery, Paris," and bears the serial number "8."

France.

No. 12. (N4388). RIM-FIRE PISTOL. Caliber .22. Four-inch, octagonal, smooth-bore barrel. This is a form of French gallery pistol.

It is fitted with a trigger-guard and fires rim-fire cartridges which are loaded into the breech of the barrel. The frame is engraved with scrolls. The bottom of the barrel is stamped "391 Mariette" and with an undecipherable word.

France.

No. 13. (N6336). CENTER-FIRE PISTOL. Caliber .32. Has eight barrels bored in a three-inch block, each barrel being rifled with four grooves. The barrels diverge from the breech in order to give a scattering effect to the shots. The gun is loaded by means of chargers containing eight cartridges each. Ejection of all eight shells occurs when the piece is opened or "broken" by first releasing a catch and then pushing the barrels downward. One trigger serves all barrels. There are two hammers, each one of which has four noses for firing four cartridges. If but one hammer is cocked four charges are fired and if two are cocked all eight cartridges are set off at one time. The trigger-guard is missing. There are no marks of identification upon this piece, which is apparently an inventor's model.

U. S. A.?

No. 14. (N5640). CENTER-FIRE REVOLVER. Caliber .30. One-and-three-quarters-inch, round, rifled barrel. Seven cartridges are inserted in the periphery of a ring-shaped cylinder in a manner similar to the spokes of a wheel, the base of each cartridge resting near the hub or center of the ring. The concealed trigger cocks, revolves and fires at a single action. The revolver is intended to be held concealed in the palm of the hand and is fired by applying pressure by means of the ball of the thumb against the curved lever in the back of the weapon, the fingers of the hand resting against the two lugs in front of the revolver. The revolver bears the words "The Protector. Pat. Mch. 6, '83, Aug. 29, '93. Chicago Fire Arms Co., Chicago, Ill." The serial number is "8154."

U. S. A.

No. 15. (N4918). RIM-FIRE REVOLVER. Caliber .22. One-and-three-eighths-inch, half-octagonal, smooth-bore barrel. The ring-shaped cylinder has a capacity of ten cartridges. The method of loading and the action of this pistol are the same as those described under No. 14. Cylinder and cover, which have been removed, are shown below the revolver. Stamped upon the weapon are the words "Le Protector. Système E. Turbiaux. Bte. S.G.D.G. En France Et A L'Etranger, Paris." The serial number is "8924."

France.

No. 16. (N4921). RIM-FIRE REVOLVER. Caliber .22. One-and-three-quarters-inch, rifled barrel. The trigger, which is on top of the barrel, cocks, revolves and fires at a single action. There are five chambers in the cylinder. This revolver is intended to be held concealed in the palm of the hand and the trigger operated by the forefinger. The weapon goes under the name of "Little All Right." It was made in Lawrence, Mass., and patented January 18, 1876. The front of the frame bears the serial number "813" but no maker's name is given.

U. S. A.

No. 17. (N4925). KNIFE-PISTOL. Caliber .30. Three-inch, smooth-bore, double, steel barrels, which are fired by means of percussion-caps. The knife-blade is four inches in length. Triggers and knife may be folded up so that the pistol may be easily carried in the pocket. The plates covering the frame are of engraved German silver. The barrels bear no proof-marks, neither are maker's name nor shop number given.

Belgium.

No. 18. (N4926). KNIFE-PISTOL. Caliber .35. Four-inch, smooth-bore, damaskeened barrel. Has a knife-blade four inches in length. The pistol is also equipped with a button-hook and a corkscrew, the latter serving also as a trigger. The handle of the knife is inlaid with mother-of-pearl. Upon the barrel is the Liége proof-mark, while the blade is stamped with an undecipherable name.

Belgium.

No. 19. (N765). PIN-FIRE PISTOL. Caliber .30. Three-and-one-quarter-inch, round, rifled barrel having a breech-block which tilts back for the admission of a pin-fire cartridge. The blade of this knife-pistol is two-and-one-half-inches in length. Pulling the trigger releases a stiff spring on top of the barrel which fires the pistol. When folded up the weapon may easily be carried in the pocket. The left side is stamped "C.F. Brevetté 19." No maker's name is given.

France.

No. 20. (N4920). RIM-FIRE PISTOL. Caliber .22. Four, one-and-three-eighths-inch, smooth-bore, brass barrels, which are fired in succession by a concealed, revolving trigger. This pistol is intended to be held concealed in the palm of the hand and is fired by drawing the square slide situated in the rear and below the barrels upward by one of the fingers. Pressing back the catch above the barrels allows the latter to be tipped downward for loading. See No. 21. The pistol carries a flat, steel ejector. No maker's name is given, but this pistol is known as the "Unique" and was made by the C. S. Shatuck Arms Co., of Hatfield, Massachusetts. It bears the serial number "79."

U. S. A.

No. 21. (N4919). CENTER-FIRE PISTOL. Caliber .32. Four, one-and-one-half-inch, smooth-bore, brass barrels, which are fired in succession by a concealed, revolving trigger. This pistol is manipulated in the same manner as described for No. 20. The side plate is removed and the barrels are tipped downward. The left side of the pistol is stamped "Unique, C.S. Shatuck Arms Co., Hatfield, Mass." The serial number is "211."

U. S. A.

No. 22. (N4922). RIM-FIRE PISTOL. Caliber .22. One-and-three-quarters-inch, rifled barrel. This "lady's pistol" fires one rim-fire cartridge at a time. The pistol is so small that it may be readily carried in a woman's handbag. All metal parts are nickel-plated and the grip is of mother-of-pearl. Has a folding-trigger. The barrel tips downward for loading when a catch upon the left side is pushed forward. The words "Hopkins & Allen Arms Co., Norwich, Conn., U.S.A." appear upon the barrel. The butt bears the serial number "154." This is a 1912 model.

U. S. A.

No. 23. (N4908). CENTER-FIRE PISTOL. Caliber .32. Two-and-three-quarters-inch, smooth-bore, steel barrel, which unscrews to load. This pistol fires but a single shot at one time. The weapon is intended to be held concealed in the palm of the hand and is fired by holding the roughened ebonite knob against the ball of the thumb and drawing down the ring around the barrel by means of the two fingers of the hand. The barrel is stamped "Pat. 5-2-'05." No maker's name is given.

U. S. A.

No. 24. (N5701). CENTER-FIRE PISTOL. Caliber .41. Two-and-one-half-inch, round, rifled, steel barrel. This pistol is of the Der-

inger type and fires either rim or center-fire cartridges from the breech. All metal parts, with the exception of the hammer, are silver-plated. The top of the barrel is stamped "Williamson's Pat. Oct. 2, 1866, New York." Barrel and frame bear the serial number "20173."
Donor: C. L. Fortier.

U. S. A.

No. 25. (N4907). CENTER-FIRE PISTOL. Caliber .41. Two-and-one-half-inch, round, rifled barrel. This pistol is of the Deringer type and fires either rim or center-fire cartridges from the breech. A metallic cartridge with a three-sixteenths-inch nipple, upon which a percussion-cap may be placed, is shown above the pistol. All metal parts, with the exception of the trigger, barrel and hammer, are heavily gold-plated. The barrel is stamped "Williamson's Pat. Oct. 2, 1866, New York," and also with the serial number "3068."

U. S. A.

No. 26. (N7147). PERCUSSION PISTOL. Caliber .47. Five-and-five-eighths-inch, round, smooth-bore barrel, which may be unscrewed for loading. The percussion-cap is placed upon the under-side of the chamber behind the barrel. A long, flat, wedge-shaped spring below the barrel fires the cap by snapping against it when the trigger is pulled. The grip is made of black walnut and bears the letters "D S" which are cut into the butt. No other marks of identification are given.

U. S. A.

No. 27. (N7071). TOY PISTOL. All parts are made of cast-iron. This pistol is sometimes known as the "penny pistol" because when a one-cent piece is placed in the slot behind the hammer, and the trigger is pulled, the coin is sent whizzing through the air by the force of the impact. No maker's name is given.

U. S. A.

No. 28. (N4460). PEPPER-BOX PISTOL. Caliber .28. Six, two-inch, smooth-bore barrels bored in a fluted cylinder. The cylinder cannot revolve. This pepper-box was used as a trap pistol to shoot kangaroo. The butt was fastened to a tree and the trigger attached by a string to a tuft of grass. Pulling the string released the hammer and fired all six charges at once. Made by North and Cauch of Middletown, Connecticut. The butt-plate bears the serial number "29."

U. S. A.

No. 29. (N6248). RIM-FIRE PISTOL. Caliber .22. Three-and-one-quarter-inch, smooth-bore, round barrel. This model is sometimes known as the "vest pocket" pistol. The top of the barrel is stamped "Remingtons, Ilion, N.Y., Patent Oct. 1, 1861," and the bottom bears the number "4809."
Donor: Gustav Marx.

U. S. A.

No. 30. (N7069). RIM-FIRE PISTOL. Caliber .22. Two-and-one-quarter-inch, round, smooth-bore barrel. The empty shell is ejected upon pulling the hammer back. The top of the barrel is stamped "Rodier Pat. Sep. 15, '74," and the grip bears the serial number "613." The inventor, Lewis C. Rodier, lived in Springfield, Massachusetts, and designed various mechanisms in firearms.

U. S. A.

No. 31. (N4943). RIM-FIRE PISTOL. Caliber .22. Two-and-one-half-inch, octagonal, smooth-bore barrel. This is a toy pistol and is intended to fire blank cartridges only. The frame is of cast-iron and bears the name "Climax, Pat. Apr. 6, 1880." No maker's name is given. This pistol is very dangerous to the user.

U. S. A.

No. 32. (N4577). TOY PISTOL. Two-and-three-eighths-inch, round, iron barrel. Trigger-guard, barrel and frame are made from one piece of cast-iron. This pistol fires percussion paper discs of the Fourth of July variety. The top of the barrel is stamped "Boss, Pat. Apr. 22, '73." No maker's name or serial number are given.

U. S. A.

No. 33. (N4576). RIM-FIRE PISTOL. Caliber .25. Two-and-three-eighths-inch, cast-iron, smooth-bore barrel. Barrel, trigger-guard and body are cast in one piece. The hammer is also cast-iron. Upon the bottom of the barrel is the number "38" but no maker's name is given. This is a dangerous toy because, when fired, its fragile construction might result in serious injury to the user.

U. S. A.

No. 34. (N767). RIM-FIRE PISTOL. Caliber .22. Two-and-three-eighths-inch, cast-iron, smooth-bore barrel. This is an all-cast-iron affair like No. 33 and equally as dangerous. The bottom of the barrel bears the number "341" but no maker's name is given.

U. S. A.

No. 35. (N4579). RIM-FIRE PISTOL. Caliber .22. Two-inch, cast-iron, smooth-bore barrel. Trigger and hammer are also of cast-iron. No maker's name is given. The inside of the grip is stamped with the serial number "2638." Pistols of this kind were intended to fire blank cartridges only, and even for this purpose they were extremely dangerous because of their fragile construction.

U. S. A.

No. 36. (N4578). RIM-FIRE PISTOL. Caliber .22. Two-and-one-quarter-inch, round, smooth-bore steel barrel. The empty shell is ejected upon pulling the hammer back. The top of the barrel is stamped "Rodier Pat. Sep. 15, '74," and the grip bears the number "131." See No. 30.

U. S. A.

No. 37. (N4575). TOY PISTOL. Two-and-one-quarter-inch, iron barrel. The grip and body is of brass. Hammer and trigger are made from one piece of iron. This piece fires percussion paper discs. The top of the barrel is stamped "1776." No maker's name or serial number are given.

U. S. A.

PLATE 110.

*WORLD WAR PISTOLS AND REVOLVERS
AND EARLY AUTOMATICS*

No. 1. (L1821). CENTER-FIRE REVOLVER. Caliber .455. Seven-and-one-half-inch barrel, rifled with six grooves. Trigger and back-strap are knurled. This is a Colt "New Service" revolver, using the .455 British Eley cartridge. The grip-plates are of checkered walnut. This revolver was used by Captain Shaw of the Canadian artillery during the World War and bears two notches on the right grip-plate indicating the number of victims taken. Stamped upon the top of the barrel are the words "Colt's Pt. F.A. Mfg. Co., Hartford, Ct. U.S.A. Pat'd Aug. 5, 1884, June 5, 1900, July 4, 1905," and on the side, "New Service 455 Eley." The top of the frame over the cylinder is engraved "Capt. W.E.V. Shaw, Canadian Field Artillery." The serial number is "75239."

Loaned by Capt. W. E. V. Shaw.

U. S. A.

No. 2. (L1487). CENTER-FIRE REVOLVER. Caliber .45. Five-and-one-half-inch, round barrel, rifled with six grooves, making one turn in 14.66 inches. This is a United States Army, Smith and Wesson double-action revolver, model of 1917. Like the Colt described above, it also has six chambers. This revolver, together with the similar 1917 model Colt, is chambered to fire .45 caliber pistol cartridges model of 1911, such as are used in the automatic pistol also of the 1911 model. These cartridges are of the rimless type and, therefore, it is necessary to provide some means to serve as an abutment against which the ejector of the revolver can act. Ammunition for the revolver is, therefore, furnished in the form of clips, each clip holding three cartridges. The clip is made of tempered spring steel and the cartridges are not rigidly held in it, but can be removed therefrom by springing them through a narrowed slot so that they are retained against accidental release from the clip. After firing, the clips may be saved and be reloaded by hand with automatic pistol cartridges. The revolver is not suited to fire .45 rim cartridges such as were used in the earlier army revolvers of the 1909 model. It will, however, fire automatic pistol cartridges of the 1911 model without clips. In this case the ejector will not remove the empty shells. They may be removed singly with the finger nail, the rim of another cartridge, or may be pushed out by means of a small rod. Revolvers of this kind have been fired eighteen times in thirty-five seconds, using clipped ammunition and beginning and ending with the cylinder closed and chambers empty. At twenty-five feet the velocity is 806 feet per second. The drift of the bullet is to the left and is more than neutralized by the pull of the trigger when firing from the right hand. At the short range at which this weapon is ordinarily used the drift is practically negligible.

This revolver saw service in the American Expeditionary Forces in France, where it was carried by Sergeant Harry Korbel. Stamped upon the top of the barrel are the words "Smith & Wesson, Springfield, Mass., U.S.A., Patented Dec. 17, 1901, Feb. 6, 1906, Sep. 14, 1909," on the side, "S. & W. D.A. 45," and on the bottom, "United States Property." The butt-plate is marked "U.S. Army Model 1917 No. 5905." Various parts bear the serial number "9616."

Loaned by Sergeant Harry Korbel.

U. S. A.

No. 3. (L1351). CENTER-FIRE REVOLVER. Four-and-five-eighths-inch, round barrel. This is an 1873 model gendarme revolver which was obtained by Dr. Philip W. Place, while a First Lieutenant in the United States Medical Corps with the A. E. F. Various parts of the piece bear the serial number "H71843." Before any measurements could be made upon the revolver it was returned to Dr. Place at his request.

Formerly Loaned by Dr. P. W. Place.

France.

No. 4. (N4934). AUTOMATIC PISTOL. Caliber 8 mm. (.315). Five-and-one-half-inch, octagonal, rifled barrel. The magazine in front of the ring-trigger holds five cartridges. This pistol may be said to be the transition type to the modern automatic. In this semi-automatic pistol the recoil opens the breech and ejects the empty shell. The closing of the breech and the introduction of a new cartridge into the chamber must, however, be accomplished by means of the ring-trigger. This trigger is in the form of a lever, which, when pulled back, pushes the uppermost cartridge from the magazine and the cylindrical lock into the breech of the barrel. At the same time the spiral spring of the striker is compressed and, when the small trigger within the ringed one is pulled, the striker fires the cartridge. The resulting recoil throws the cylinder back and the spring within the magazine forces up a new cartridge which is then pushed into the breech, as has been described. This pistol is equipped with a front sight and an adjustable sight in the rear. When the small safety, located just beneath the cylinder, is pushed to the left, the striking-bolt cannot be released by means of the small trigger, and consequently the pistol cannot be discharged. The right side of the frame bears the words "Patent Bittner." The upper part of the barrel bears the Austrian proof-mark, while the lower part is stamped with the numbers "1702.98." This pistol was patented about 1893 by Gustav Bittner of Weipert, Bohemia.

Austria.

No. 5. (N4935). AUTOMATIC PISTOL. Caliber .32. Six-and-one-half-inch, round, rifled barrel. The magazine in the grip holds ten center-fire cartridges. Equipped with a front sight and an adjustable rear sight which may be elevated from 100 to 500 yards. This is a true automatic, since the force of the recoil opens the breech, extracts the empty case, cocks the pistol, reloads the chamber with the top cartridge from the magazine and closes the breech, leaving the pistol ready to be fired on again pressing the trigger. Pressing upward on the first lever on the left side of the frame makes the pistol safe. Various parts of the weapon are marked with the serial number "421." Barrel and frame are stamped with proof-marks in the form of double crowns over the letter "U." No maker's name is given.

Germany.

No. 6. (L1709). (18981). AUTOMATIC PISTOL. Caliber 9 mm. (.354). Four-inch, round, tapering barrel rifled with six grooves. The magazine is carried in the grip and has a capacity of eight cartridges. With the magazine fully loaded the pistol weighs 990 grams (2.18 pounds). The muzzle velocity is 310 meters per second (1017.1 feet per second), and the extreme range 1,500 meters (4,921.2 feet). The official designation of this weapon in the German army is "Die Pistole 08," the numerals indicating the year in which the pistol was adopted. The arm is, however, more frequently called the "Luger" or "Luger Parabellum." This pistol has probably the best grip of any automatic made so far. The rear portion of the front sight is milled and the tip is polished to facilitate aiming. The specimen is shown with the carbine stock attached and with a large "snail" magazine in place in the grip. This magazine has a capacity of thirty-two cartridges and when used with the Luger, as illustrated, served as a supplementary automatic rifle. Originally magazines of this kind were intended to be used with the long-barreled Luger but later the device was also used with the short-barreled pistol. The snail magazine can be loaded by hand, but as the number of cartridges inserted increases, it becomes correspondingly difficult to put the last few into the magazine. To overcome this difficulty a loading lever was sometimes used. The usual magazine with two cartridges are displayed below the pistol. Stamped upon the top of the pistol is the word "Erfurt" over a crown, the year of manufacture, "1916," and the number "76." The side of the barrel bears two German proof-marks and the serial number "2876." The weapon here shown was taken by Captain Shaw at Vimy Ridge, in April, 1917.

Pistol and Stock Loaned by Captain W. E. V. Shaw.

Germany.

No. 7. (L1627). AUTOMATIC PISTOL. Caliber 9 mm. (.354). Four-inch, round, tapering barrel. Action and other characteristics are the same as described for No. 6. The Luger here shown was taken from a German officer at the battle of Château Thiery, in 1918. Upon the top of the barrel is a monogram consisting of the letters "L.W.M."

(probably Luger Waffen Manufactur), and the year of manufacture, "1915." The serial number is "5838."

Loaned by Christo Ganchoff.

Germany.

No. 8. (L1710). "SNAIL" MAGAZINE. This magazine has a capacity of thirty-two 9 mm. cartridges and is intended to be used with the Luger automatic pistol. See No. 7. Upon the bottom of the magazine is a folding lever which operates a spring that forces the cartridges into the chamber of the pistol when firing is going on. The lever also serves as an indicator to the number of cartridges in the magazine; the circumference of the base of the magazine bearing the numbers 12, 17, 22, 27 and 32. As the cartridges are discharged the lever swings over these figures and indicates the number of cartridges that are left. This magazine was found after the battle of Amiens, in 1918, and bears the serial number "392683."

Loaned by Captain W. E. V. Shaw.

Germany.

No. 9. (11480). AUTOMATIC PISTOL. Caliber 7.65 mm. (.301). Three-and-three-eighths-inch, rifled barrel. This is a relatively cheap form of pistol, all parts of which are rather crudely made. The left side of the slide bears a partly undecipherable inscription, part of which reads "Cal. 7.65 Eibar 1912 'Vesta' Patent." This pistol, which was made in Eibar, Spain, bears the serial number "75412." The magazine is missing.

Spain.

No. 10. (N7026). AUTOMATIC PISTOL. Caliber .25. Two-and-one-eighth-inch, rifled barrel. The magazine has a capacity of six cartridges. The pistol, which has a concealed hammer, may be taken apart without the aid of tools. A safety-device is located on the left side. There are no sights of any kind. The grip-plates are of checkered ebonite and bear the name "Orbea." Upon the left side of the pistol is stamped "Orbea Y Cia. S. En C. Eibar (España) Pistol Automatica Precision." The serial number is "20." Made in Eibar.

Spain.

No. 11. (N2069). AUTOMATIC PISTOL. Caliber 7.63 mm. (.30). Four-inch, rifled barrel. Carries five cartridges either loose or in the clip in the magazine. The bullets weigh 85 grains each and can be fired at the rate of 120 per minute, allowing time for the refilling of the magazine. The muzzle velocity is 1,400 feet per second, and the weapon is sighted to 1,000 yards. The military Mauser carries ten cartridges in the magazine and has a longer barrel. See No. 12. The right side of the weapon here shown is marked "Waffenfabrik Mauser Oberndorf a Neckar," and the left side, "Von Lengerke & Detmold New York." The pistol is also stamped with German proof-marks and the serial number "29575."

Germany.

No. 12. (L1628). AUTOMATIC PISTOL. Caliber 9 mm. (.354). Five-and-one-half-inch, rifled barrel. This is the Mauser military model. The magazine has a capacity of ten cartridges. The muzzle velocity is 1,070 feet per second. This pistol has a long range rear sight, graduated from 50 to 1,000 meters. The barrel is stamped with German proof-marks and the words "Waffenfabrik Mauser Oberndorf A/N." The serial number is "215113." Upon the right lock-plate are the words "Waffenfabrik Mauser Oberndorf A. Neckar." This pistol was obtained in Germany, in 1918.

Loaned by Christo Ganchoff.

Germany.

No. 13. (L1712). AUTOMATIC PISTOL. Caliber 7.63 mm. (.300). Three-and-three-eighths-inch, rifled barrel. The safety is on the left side in the form of a roughened catch, which, when pulled down, makes the pistol safe. To fire, a small button below the catch is pressed that causes the safety device to spring upward and makes the pistol ready for firing. The left side of the slide is marked "Waffenfabrik Mauser A.G. Oberndorf A.N. Mauser's Patent," together with the serial number "130024." The pistol here shown was taken from a German non-commissioned officer at Amiens, in 1918.

Loaned by Captain W. E. V. Shaw.

Germany.

No. 14. (18976). AUTOMATIC PISTOL. Caliber 7.63 mm. (.300). Three-and-five-eighths-inch, rifled barrel. This is a "Dreyse" automatic pistol. The left side of the frame is marked "Dreyse Rheinische Metallwaaren & Maschinonfabrik Abt. Sömmerda." (Dreyse Rhine Metal and Machine Works, Sömmerda Branch). Various parts of the pistol are stamped with German proof-marks in the form of a crown over the letter "N." The serial number is "173120."

Germany.

No. 15. (18977, 28094, L1713). SIGNAL PISTOL AND SIGNAL CARTRIDGES. Nine-inch barrel having a muzzle-diameter of one-and-one-sixteenth-inches. The two short shells, resembling those of a shotgun, contain signal fire-works known as "golden showers." The longer shells are "parachute-flare" cartridges. The bases of these cartridges are filled with an explosive which is set off by means of a slow-burning fuse ignited from the original charge of the pistol. When the cartridge reaches an altitude of about 300 feet the charge in the base explodes and simultaneously ejects and ignites a stick of magnesium to the upper end of which is attached a small silk parachute that is kept floating partly by the heat from the burning magnesium below. The magnesium burns with an intensely bright, white light similar to the flashlight used by the photographer. A signal light of this kind will burn about two minutes and light up a considerable area, making any objects therein stand out very conspicuously. The pistol bears various German proof-marks, together with the serial number "27235."

Cartridges Loaned by Captain W. E. V. Shaw.

Germany.

FORSYTH PERCUSSION OR "DETONATING" SHOTGUN

The inventor of the percussion method of ignition of the powder-charge of a firearm was the Rev. Alexander John Forsyth, LL.D., a Scotch Presbyterian clergyman, who was minister of the parish of Belhelvie, a seacoast village twelve miles north of the city of Aberdeen, Scotland, from 1790 to 1843, the latter year that of his death.

His father having also been a clergyman and the minister of the same charge from 1743 to 1790, the son succeeded the father on the latter's death, spent his life in the parish, and died in the same house, the parish manse, in which he was born. Father and son together thus served the same church for an even century.

Forsyth graduated from the University of Aberdeen in 1786. The manse in which his entire life was spent—save for absences in London when developing his invention—being near the shore, and both the waters and the inland fields and forests then abounding in water-fowl, upland birds and small game, Forsyth hunted assiduously from his youth and developed a lifelong devotion to firearms and their use. Never marrying, his profession allowed him considerable leisure for studies in the chemistry of the time, his favorite pursuit, through which, together with his recreation of shooting, his great invention came about.

While the flint-lock arms of the time had reached their highest degree of mechanical perfection, embodied in the products of the leading English gunmakers—the Mantons, Eggs, Turner, Nock, Wilson, Purdey, and others—there were inevitable limitations inherent in the flint-lock system of ignition. These were the uncertainty of the flint striking fire every time, the exposure of the priming to the effects of dampness and moisture, despite the best construction of the priming-pan, and the flash and smoke from the priming immediately preceding the firing of the charge itself; from which latter arose alike the fact and the saying, still extant, as to the game escaping "at the flash of the gun." Thus, Napoleon had offered a large reward for some improved method of ignition which should be applicable to the arms of his troops. It remained for the humble parish minister of an obscure Scotch village to make the epochal discovery.

As early as 1793, when but twenty-five years of age, Forsyth was considering some such improvement.[166] Various experiments with chlorate of potash and fulminate of mercury, known at the time as "detonating powders," demonstrated the possibility of their being ignited by a blow. By December, 1805, Forsyth had made—for he was an amateur blacksmith as well as a chemist, and had a small building equipped with forge and anvil attached to the manse—a workable gun-lock in which a minute quantity of fulminate of mercury deposited in a touch-hole was struck and ignited by a firing-pin the size of a needle, driven by a blow from the "hammer" of the gun. With a fowling-piece operating on this system he shot through the hunting season of 1805. The arm was literally the marvel of all who beheld it, the sole representative of its kind in all the world of arms of the time and embodying the greatest single improvement in firearms ever conceived. Making a trip to London in the spring of 1806 to visit some friends, he took his gun along to show them. The piece coming to the notice of Lord Moira,[167] a member of the Privy Council of George III, and Master of Ordnance and Constable of the Tower, he persuaded Forsyth to remain in London for a time and with the facilities and workmen of the Tower at his service, to plan the application of his invention to the

---

[166]Forsyth's Statement to the House of Commons, 1840. Quoted in Memoir by Maj. Gen. Sir Alexander John Forsyth Reid, K.C.B., M.A., LL.B., entitled "The Rev. Alexander John Forsyth and His Invention of the Percussion Lock," Aberdeen University Press, 1909. For convenience, further references to this work will mention it as the Memoir.

[167]Francis Rawdon, 1754-1826, who in 1790 took the additional surname of Hastings, was educated at Harrow, entered the army, was a lieutenant in the Fifth British Foot regiment at Bunker Hill, rose to lieutenant-colonel, became adjutant-general of the British forces in the colonies, in 1783 was created Baron Rawdon, in 1793 succeeded to the title of Earl of Moira, in 1806 became Master-General of Ordnance, and in 1813 Governor-General of India.

firearms of the British army. Forsyth accordingly remained in London, engaged upon these tasks, until the summer of 1807.

Moira being superseded in office in 1807 by the Earl of Chatham,[168] the latter proved hostile alike to Forsyth's ideas and his labors, and ordered him to discontinue his experiments and to leave the Tower, and Forsyth accordingly returned to his Scottish parish. Before leaving London, however, he organized a company for the manufacture of fire-arms equipped with his invention. A patent being awarded him on July 4th,[169] the firm of Forsyth & Co., began the manufacture of arms at No. 10 Piccadilly, whence it removed in 1818 to No. 8, Leicester Square, where it remained in business until 1852, nine years after Forsyth's death.

Forsyth's occupancy of his pastorate at Belhelvie remained uninterrupted until his death, June 11th, 1843, in the seventy-fifth year of his age. His last years were marked by several recognitions of his services in inventing the improvement which rapidly superseded the imperfections of the flint-lock system. In 1840 (July 14th) the member of Parliament for Aberdeen proposed in the House of Commons that Forsyth be appropriately reimbursed for his invention, which had in the meantime been adopted for the arms of the British army. Lord Brougham, a relative of Forsyth, spoke on behalf of the petition, and after a long delay, on April 21st, 1842, the Lords of the Treasury ordered a grant of £200 as a "remuneration as the original Inventor of percussion for small arms." Public opinion appears to have been to the effect that such reward was pitifully inadequate, and eighteen months later (October, 1843), four months after Forsyth's death, the Treasury voted an addition £1,000 to Forsyth's surviving relatives.[170] In the year 1834 he was presented by Aberdeen University with an honorary degree of Doctor of Laws, the wording of the award stating that it was "as a mark of esteem for his private character and for his attainments in Science." In June, 1841, he was given a public dinner by the people of his parish and was presented with a service of silver inscribed "As a mark of regard and esteem for his unwearied services as their Pastor for upward of fifty years." Among his contributions to the welfare

---

[168]John Pitt, 1756-1835, second Earl of Chatham, son of the great premier and brother of the great Premier, First Lord of the Treasury and Chancellor of the Exchequer. Commanding the British force of the Walcheren expedition which was cut to pieces by Napoleon on the River Scheldt in 1809, he was blamed throughout England for the disaster. In 1820 he was appointed Governor of Gibraltar, where he died.

[169]Abridgements of the Specifications relating to Fire-Arms and other Weapons, Ammunition and Accoutrements; Geo. E. Eyre and Wm. Spottiswoode, Printers; the Great Seal Patent Office, London, 1859.

[170]The Memoir.

of his parish were the establishment of a Parish Savings Bank and his personal practice among his people of vaccination against smallpox, upon the discovery of the treatment by his personal friend, the famous English physician, Edward Jenner.[171]

The gun here shown was manufactured at the Leicester Square establishment, probably soon after removal thither.[172]

The dominant idea in all of Forsyth's applications of his discovery being that of providing upon the arm itself a supply of priming-explosive from which an adequate small amount should be available for each shot, he produced at least three forms of priming-magazines, each adapted to deposit an ignition-charge of loose fulminate in the tube or "touch-hole," where it was exploded by the fall of the hammer. One early form of magazine took the place, on the gun-lock-plate, of the steel against which the flint struck on the common flint-lock, depositing the fulminate in a touch-hole substituted for the pan. A double-barreled shotgun was thus equipped, made by Forsyth & Co., is among the collection of arms in the United States National Museum at Washington, loaned by Major Jerome Clark of the U. S. Army. It undoubtedly represents one of Forsyth's earliest productions for application to the then regulation flint-lock arms of the British army. A second form—applied only to pistols as far as is known—had a small, box-like magazine above the front end of the lock-plate, which was linked to the hammer and slid to and fro with the movement of the latter. Cocking the hammer drew the magazine over the touch-hole, in which it deposited the fulminate and, on the fall of the hammer, it slid forward.[173]

The third and most highly-developed form of the Forsyth priming-magazine is on the gun here shown. After loading the piece and cocking it, the revolvable magazine is given a half-turn by the shooter, when it deposits the fulminate in a touch-hole in its axis, communicating with the charge in the barrel. Returning the magazine to its former position presents under the hammer the head of a spring-upheld firing-pin. The blow of the hammer upon this strikes the lower end of the pin into the deposited charge in the internal touch-hole, firing the piece.

The serious defect in Forsyth's method of percussion-ignition lay in the quantity of highly-sensitive and inflammable fulminate contained in the magazine in immediate proximity to the touch-hole and the explosion occurring there, there being always a decided possibility of the

---

[171]For these and many other details of his life, see The Memoir.
[172]See text of inscription in original gun-case, Plate 113.
[173]Illustrated in the Memoir. Also in "A Century of Guns," H. J. Blanch, London, 1909, Plate II.

explosion of the contents of the magazine as well. That such an occurrence was recognized as not unlikely, is evident from the fact that in all forms of Forsyth's magazines a safety-vent was provided. This was covered by a cap of leather or horn designed to be blown away on such accidental explosion of the contents of the magazine without injury to either the gun or the shooter. In the view of the left-hand lock-mechanism, Plate 112, there may be seen on the lowest face of the magazine a small pivot-button and screw-head, which held this safety-vent cover in place.

PLATES 111, 112 and 113.

FORSYTH PERCUSSION OR "DETONATING" SHOTGUN

Bore .691 of an inch, 15 gauge. Barrels thirty-one-and-one-quarter-inches long, twisted iron and steel, browned. Barrels marked underneath breech-ends "IH 7 Co. Fine Stub Twisted" with old Birmingham "viewed" and "proof" marks. Barrels fastened to stock by pin through fore-end. Fulminate-magazines have to be removed to remove barrels from stock.

For operation of fulminate-magazines, see preceding description. Magazines stamped on inner faces "L" and "R" respectively. Capacity of each magazine, enough fulminate for forty primings.

Locks very finely made, all inner working parts highly polished. Lock-plates originally blued, hammers originally case-hardened in colors. Lock-plates marked "Forsyth & Co. Patent."

Stock walnut, probably originally polished. Checkering probably not original, as it is poorly executed. Butt-plate iron, engraved with ornamental circle around main retaining-screw. Rear of trigger-guard tang curved to form a pistol-grip; broad beneath triggers; engraved underneath with figure of a pointing dog.

Rib inlaid in gold "Forsyth & Co. London" in "Old English" lettering. Gold breech-bands on barrels and gold rays inlaid on breech-end of rib and top of standing-breech. Small, low gold bead front sight. Light, slender wooden ramrod with metal tips, in ferrules under barrels and deep tube in fore-end.

Weight of piece, eight pounds, eight ounces.

Date of manufacture, 1818, or shortly thereafter. See preceding historical account. Brought from England by the late R. G. Bickford of Newport News, Virginia, from whose estate it was purchased in February, 1926.

England.

APPENDICES

APPENDIX I

ARMS PATENTS

Patents granted by the United States from 1836 to 1873 to inventors of firearms. Compiled from "Digest of Patents Relating to Breech-Loading and Magazine Small Arms," by V. D. Stockbridge, Washington, D. C., 1875.

Several discrepancies between the number on the patent office drawing and the number on the patent office specification relating to the same invention have been found in the above work. Where the correct number could not be ascertained, such fact is indicated by a question (?) mark. Names marked by an asterisk (*) show that the patent granted relates to a revolver. Those marked (†) relate to both revolvers and ordinary small arms. All other entries relate to ordinary breech-loading and magazine small arms.

Inventor's Name	*Date*	*Patent Number*
Abbey, Frederick J., and Foster, James H.	April 25, 1871	114,081
Abbey, George T.	March 16, 1869	87,814
Adams, H. W.	September 19, 1854	11,685
Adams, J.	September 30, 1856	15,797*
	November 6, 1860	30,602*
	December 29, 1868	85,350*
Adams, John S.	August 11, 1863	39,455
	September 27, 1864	44,377
Adams, R.	May 3, 1853	9,694*
Adams, Samuel	October 3, 1838	960
Albright, Louis	May 5, 1863	38,366
Aldrich, Wales	May 12, 1863	38,455
Alexander, Charles W.	May 25, 1858	20,315
Allen, Ethan	April 16, 1845	3,998*
	July 3, 1855	13,154
	January 13, 1857	16,367*
	December 15, 1857	18,836*
Reis. 1737 and 1738, Aug. 16, 1864)	September 7, 1858	21,400*
	November 9, 1858	22,005*
(Reis. 1268, Feb. 4, 1862)	July 3, 1860	28,951*
	September 18, 1860	30,033
	September 24, 1861	33,328*
	October 22, 1861	33,509*
	April 29, 1862	35,067*
	March 7, 1865	46,617
	August 22, 1865	49,491
	December 15, 1868	84,929
Allen, F. H.	October 11, 1875	168,549*
Allin, E. S.	September 19, 1865	49,959
Alsop, C. A.	March 25, 1862	34,803*
Alsop, Charles H.	November 26, 1861	33,770*
	April 7, 1868	76,374

Inventor's Name	*Date*	*Patent Number*
Alsop, C. R.	July 17, 1860	29,213*
	August 7, 1860	29,538*
	May 14, 1861	32,333*
	January 21, 1862	34,226*
Altmaier, Peter	July 12, 1859	24,774
Ambler, N. H.	August 9, 1870	106,246
Appleby, John F.	December 20, 1864	45,466
Armstrong, J. W., and Taylor, John	November 25, 1862	37,025
Arnold, William H.	November 15, 1859	26,076
Aronson, Joseph N.	November 13, 1866	59,540
Ashcroft, E. H.	May 26, 1863	38,645
Assmus, Adolph	December 16, 1873	145,578
Aston, James	May 13, 1873	138,857
Ayres and Whittaker	February 13, 1877	187,244*
Babcock, N. L.	March 20, 1860	27,509
Bacon, A. M.	July 31, 1866	56,846
Bacon, George R.	July 21, 1863	39,270
Bailey, Ripley & Smith	February 26, 1839	1,084
Bailey, Thomas	June 7, 1859	24,274*
	June 14, 1859	24,437
Baldwin, Cyrus W.	January 19, 1869	85,897
Baldwin, E. A.	July 11, 1854	11,283
Ball, Albert	June 23, 1863	38,935
	August 15, 1864	43,827
	December 6, 1864	45,307
	January 1, 1867	60,664
Ballard, C. H.	November 5, 1861	33,631
	April 9, 1867	63,605
Barber, Joseph, and Reinried, P. C.	March 15, 1859	23,224
Barnekow, Kiel V.	June 14, 1870	104,100
Bayes, S. G.	February 9, 1869	86,723
Beals, Fordyce	September 26, 1854	11,715*
	June 24, 1856	15,167*
	May 26, 1857	17,359*
	September 14, 1858	21,478*
	January 6, 1863	37,329*
	June 28, 1864	43,284
	February 7, 1865	46,207
	January 30, 1866	52,258
Bean, Joseph H.	August 12, 1873	141,624
Bearcock, T. W.	September 25, 1877	195,562*
Beaumont, F. B. E.	June 3, 1856	15,032*
Beck, J.	May 16, 1854	10,930*
Belden, S., and Crabtree, J. F.	December 29, 1868	85,268
Bell, W. H.	March 20, 1860	27,518*
Bennett & Haviland	February 15, 1838	603*
Berdan, Hiram	January 9, 1861	51,992
	January 10, 1865	45,899
	January 9, 1866	51,991
	February 27, 1866	52,925
	December 22, 1868	85,162
	March 30, 1869	88,436
	April 5, 1870	101,418
	November 1, 1870	108,869
Berg, Henry	March 25, 1862	34,729
Bergen, Alexander J., and Williamson, D.	November 22, 1864	45,202
Bergen, Alexander J.	February 26, 1867	62,465

Inventor's Name	Date	Patent Number
Billings, Charles E.	April 24, 1866	54,100
	September 22, 1868	82,279
Blaker, John D.	August 10, 1869	93,403
Blittkowski, G. A.	April 28, 1857	17,136
Blittkowski & Hoffman	March 25, 1856	14,488*
	April 22, 1856	14,710*
Boardman and Peavey	January 18, 1876	172,243*
Bostwick, S., and Sargent, C. G.	November 11, 1862	36,891
Bourdereaux, Peter	July 29, 1873	141,198
Bourdereaux, Pierre	November 13, 1866	59,706
Bowlby, G. W.	May 21, 1867	64,941
Bowness, James	August 2, 1864	43,733
Boyd, F. E., and Tyler, P. S.	January 21, 1868	73,494
	April 6, 1869	88,540
Boynton, Paul	March 15, 1859	23,226
	January 3, 1860	26,646
	November 27, 1860	30,714
Bradley, Isaac	August 7, 1866	56,890
Brady, Freeman, Jr., and Noble, J. C.	January 14, 1862	34,126
Brand, Christopher C.	July 29, 1862	35,989
	September 23, 1862	36,505*
	April 28, 1863	38,279*
	April 28, 1863	38,280
	June 23, 1863	38,943
Brettell and Frisbie	February 10, 1857	16,575*
Briggs and Hopkins	January 5, 1864	41,117*
Briggs, George W.	October 16, 1866	58,937
Briggs, William	April 6, 1869	88,605
Broadwell, L. W.	August 22, 1865	49,583
Brooks and Bearcock	May 8, 1877	190,543*
Brooks, E., and Walter G.	July 6, 1858	20,776
Broughton, John	April 14, 1868	76,595
Brown, I. W.	August 8, 1854	11,470*
Brown, W. H.	March 4, 1862	34,561
Brugmann, Heinrich	July 16, 1872	129,312
Buchel, C. W.	February 20, 1849	6,136
Buchner, Heinrich	May 2, 1871	114,259
Buckman, I., Jr.	August 4, 1851	17,915
Bunsen, G. C.	December 26, 1865	51,690*
Burgess, Andrew	September 19, 1871	119,115
	September 26, 1871	119,218
	June 11, 1872	127,737
	June 25, 1872	128,208
	July 16, 1872	129,523
	January 7, 1873	134,589
Burghart, W.	January 12, 1858	19,068
Burke, John	May 15, 1866	54,680
	June 19, 1866	55,613
Burnside, A. E.	March 25, 1856	14,491
Burton, Bethel	December 20, 1859	26,475
	August 11, 1868	81,059
	June 29, 1869	92,013
	October 14, 1873	143,614
Buss, C.	April 25, 1854	10,821*
Callaghon, Cornelius	February 25, 1868	74,888
Calver, G. W. H.	May 17, 1870	103,013*
Campbell, J. C.	June 30, 1863	39,032*

Inventor's Name	*Date*	*Patent Number*
Carle, J. F. C.	February 27, 1866	52,938
Carter, Henry, and Edwards, George Henry	January 19, 1869	85,999
Cass, M. M.	September 26, 1848	5,814
Castle, Horace A.	November 4, 1873	144,190
Chabot, Cyprien	April 4, 1865	47,163
	September 5, 1865	49,718
Chamberlain, D. H.	April 23, 1850	7,300*
Chamberlain, Martin J. (Spelled "Chamberlin" on Pat. drawing)	February 14, 1871	111,814
	July 16, 1872	129,393
Chamberlin, M. J.	February 4, 1873	135,405
Chamberlin, Martin J., and H. M.	January 8, 1867	60,998
Chapin, Linus N.	May 17, 1864	42,748
Chassepôt, A. A.	January 1, 1867	60,832
	November 23, 1869	97,167
Christ, A.	September 11, 1866	57,864*
Christy, William J.	September 18, 1866	58,064
Churchill, R. W.	May 15, 1877	190,825*
Clark, Francis	July 19, 1864	43,571
	January 3, 1865	45,701
	March 27, 1866	53,522
Claude, E.	December 21, 1858	22,348*
Clements, Nathan S.	May 27, 1856	14,949
	October 10, 1865	50,334
Clews, W.	February 25, 1873	136,134*
Cloes, John Joseph	April 12, 1870	101,826
Cochran, F. G.	July 4, 1871	116,559*
Cochran, John Webster	April 29, 1837	183*
	April 29, 1857	188*
	December 28, 1858	22,412*
	July 7, 1863	39,120
(Reis. 2211, Mar. 27, 1866)	November 10, 1863	40,553*
	December 22, 1863	40,992
	April 4, 1865	47,088
	April 25, 1865	47,396
	February 20, 1866	52,679
	March 26, 1867	63,217
	March 17, 1868	75,627
	January 5, 1869	85,645
	May 7, 1872	126,446
	November 7, 1876	184,145*
Colborne, D. G.	June 29, 1833	None given*
Cole, O. F.	December 28, 1875	171,506*
Coleman, C. C.	May 13, 1862	35,217
	November 6, 1866	59,500
Colt, Samuel	February 25, 1836	None given*
	August 29, 1839	1,304*
	May 20, 1856	14,905*
	February 24, 1857	16,683*
	March 3, 1857	16,716*
	November 24, 1857	18,678*
	May 4, 1858	20,144*
	September 3, 1850	7,613*
	September 10, 1850	7,629*
Colvin, R. J.	March 25, 1862	34,740*
Conant, H.	April 1, 1856	14,554
Conklin, Isaiah B.	February 16, 1869	86,971
Conover, Jacob A.	July 24, 1866	56,669

Inventor's Name	Date	Patent Number
Conroy, Loughlin	December 31, 1867	72,803
	June 15, 1869	91,421
	December 2, 1873	145,154
Converse, C. A.	August 28, 1866	57,622*
Cook, Roswell F.	July 24, 1860	29,340
	March 10, 1863	37,854
Cook, Thomas	February 14, 1854	10,520
Cooke, James C.	June 3, 1862	35,488
Cooper, J. M.	March 20, 1860	27,526*
	September 4, 1860	29,864*
	September 22, 1863	40,021*
Cooper, Joseph Rock	December 15, 1868	84,938
Courter, D. A.	March 11, 1862	34,625
Cox, Calvin	April 27, 1858	20,041
	April 10, 1860	27,778
Crispin, Silas (Reis. 2238, May 8, 1866)	October 3, 1865	50,224*
	January 1, 1867	60,698
	February 5, 1867	61,722
Crocker, G. R.	October 20, 1857	18,486*
Crocker, J. A.	August 28, 1877	194,653*
Crowell, G. G.	April 17, 1866	53,955*
Cullen, Thomas	April 13, 1869	88,853
Curtiss, Frederick	February 15, 1859	22,940
	(July 15, 1859, on patent drawing)	
	September 17, 1861	33,317
	January 19, 1864	41,281
	February 9, 1864	41,489
Dangerfield, F. S.	September 3, 1872	130,984
Daniels, Henry & Charles	February 15, 1838	610
	April 3, 1838	677*
Darling, D. and B.	April 13, 1836	None given*
Davis, Jarvis	January 27, 1863	37,544
	July 7, 1863	39,198
	April 26, 1864	42,529
	November 28, 1865	51,258
	May 17, 1870	103,154
	February 28, 1871	112,127*†
	October 22, 1872	132,357*
	September 26, 1876	182,646*
Day, Joseph C.	August 8, 1854	11,477
	December 18, 1855	13,941
Day, Silas	August 31, 1837	364*
	October 8, 1840	1,810
Day and Hall	December 31, 1839	1,461
DeBrame, J. A.	July 2, 1861	32,685*
DeDartein, C. F., and J. E.	May 10, 1870	102,782*
Deeley, John and Edge, James S., Jr	July 1, 1873	140,482
Delassize, L. T.	July 20, 1869	92,799
Denzler, F.	April 29, 1862	35,086
Deprez, Jean M.	February 16, 1869	86,739
Descontures, M. L. M.	July 25, 1865	49,057
Dexter, Smith, and Marshall, Joseph C.	August 5, 1873	141,603
Diaz, Faustino Valdes	September 7, 1869	94,577
Dodge, William C. (Reis. 4483, July 25, 1871)	January 17, 1865	45,912*
	September 20, 1864	44,290

Inventor's Name	Date	Patent Number
Dodge, William C.	January 24, 1865	45,983*
	February 13, 1866	52,547
	October 16, 1866	58,790
	March 14, 1871	112,763
	April 4, 1871	113,408
	May 9, 1871	114,653*†
Dodge, William C., and Dodge, Phillip Tell	March 14, 1871	112,894
	August 22, 1871	118,350
	June 11, 1872	127,683
Doolittle, J. B.	July 29, 1862	35,996*†
	April 17, 1866	54,065
Dorwart, B. K.	September 2, 1873	142,376*
Dougall, James Dalziel	September 5, 1865	49,844
Draeger, Charles	April 8, 1862	34,922*†
Drew, Henry J.	March 14, 1871	112,563
Drew, R. W.	April 2, 1867	63,450*
Duval, Joseph	March 14, 1871	112,565
	January 30, 1872	123,159
	December 16, 1873	145,494
Earnest, George H.	July 16, 1872	129,115
Eddy, G. W.	April 11, 1876	175,951*
Edwards, D.	April 27, 1839	1,134*
Elliot, William H.	August 17, 1858	21,188*
	May 29, 1860	28,460*
(Reis. 2650, June 18, 1867)	May 29, 1860	28,461*
(Reis, 1936, 1937 and 1938, April 18, 1865)	October 1, 1861	33,382*
(Reis. 1715, July 5, 1864)	December 17, 1861	33,932*
	May 13, 1862	35,284
	July 7, 1863	39,136
	May 10, 1864	42,648*†
	May 10, 1864	42,649
	August 16, 1864	43,840
	February 7, 1865	46,225*
	April 18, 1865	47,372
	May 16, 1865	47,707*
	May 23, 1865	57,809
	October 3, 1865	50,232*†
	December 12, 1865	51,440*†
	August 27, 1867	68,292
	December 13, 1870	110,024
	May 9, 1871	114,540
	December 5, 1871	121,499
	April 2, 1872	125,127
	October 11, 1875	168,562*
Ellis, Darwin	April 12, 1870	101,845
Ellis, Willard C.	April 26, 1859	23,762
Ellis and White		
(Reis. 1528 and 1529, Aug. 25, 1863)	July 12, 1859	24,726*
	July 21, 1863	39,318*
Ells, J. (Reis. 625, Feb. 1, 1859)	April 25, 1854	10,812*
(Reis. 806, Sept. 6, 1859; add imp't. 265, Feb. 21, 1860)	August 1, 1854	11,419*
	April 14, 1857	17,032*
	April 28, 1857	17,143*
Elson, Julius	May 14, 1867	64,650
	July 23, 1867	67,933
	November 19, 1867	71,149

Inventor's Name	Date	Patent Number
Elson, J., and Schaefer, W. R.	February 2, 1869	86,378
Ely, A. B., and Clay, E. C.	July 5, 1870	105,058
Erskine, H. M.	August 20, 1878	207,168*
Escherich, Francis H.	June 2, 1868	78,519
Evans, Warren R.	September 19, 1871	119,020
Fairbanks, Louis T.	June 22, 1869	91,616
Faivre, A. C.	March 9, 1858	19,553
Fay, Henry C.	May 22, 1837	203
Ferriss, George H. (or Ferris)	October 10, 1871	119,834
Fisher, E., and Chamberlain, D. H.	April 17, 1837	168
Fitzgerald, Walter	January 17, 1865	45,919
Foehl, C.	June 3, 1873	139,461*
Forbis, Harbert K.	March 21, 1871	112,795
Forehand, S.	May 30, 1876	178,133*
Forehand and Wadsworth	June 27, 1871	116,422*
	October 14, 1873	143,566*
	April 20, 1875	162,162*
	July 24, 1877	193,367*
Forgerty, Valentine	October 23, 1866	59,126
	February 2, 1869	86,520
	July 25, 1871	117,398
Foster, George P.	April 10, 1860	27,874
Foster, G. P., and G. F.	September 19, 1865	49,994
	July 17, 1866	56,399
Fox, George H.	January 4, 1870	98,579
Freeman, Austin T.	December 9, 1862	37,091*
	January 16, 1872	122,717
	December 10, 1872	133,770
Freund, F. W.	October 17, 1876	183,389*
Frost, E. J.	June 11, 1867	65,742*
Gabel, L.	April 29, 1862	35,093*
Galand, C. F.	June 17, 1873	140,028*
Gallager, Mahlon J.	July 17, 1860	29,152
Gallager, M. I., and Gladding, W. H.	July 12, 1859	24,730
Gardner, G. H.	May 16, 1865	47,712*
Gardner, H. L.	January 23, 1877	186,470*
Gardner, William	February 16, 1869	87,038
Gedney, G. W. B.	July 29, 1862	35,999*
Geiger, Leonard	January 27, 1863	37,501
Genhart, H.	January 13, 1857	16,477*
Geraghty, M. F.	August 25, 1863	39,642*
Gerngross, Stephen	December 20, 1870	110,353
Gibbs, L. H.	October 2, 1847	5,316*
	January 8, 1856	14,057
Gibson, A. J.	May 22, 1860	28,437*
(Reis. 5,813, March 31, 1874)	July 10, 1860	29,126*
	October 9, 1860	30,309*
	October 9, 1860	30,309*
Gibson, T.	April 19, 1864	42,435*
Golcher, William	March 30, 1869	88,470
(Sometimes misspelled Goulcher)	October 19, 1869	95,998
Gordon, Charles	April 22, 1873	138,145
Gordon, J.	December 31, 1867	72,844*
	July 16, 1872	129,334*
Goshen, Ruth (Spelled "Goshan" on Pat. drawing)	February 27, 1872	124,056

Inventor's Name	*Date*	*Patent Number*
Gould, Theodore P.	January 3, 1860	26,734
Goulding, John	May 3, 1864	42,573
Graham, E. H.	January 7, 1851	10,944*
	October 4, 1853	10,084
	January 16, 1855	12,235*
	September 16, 1856	15,734*
	November 24, 1863	40,687*
Gray, Gardner B., and Romans, Joseph H.	March 21, 1871	112,803
Gray, Joshua	January 26, 1864	41,375
	November 8, 1864	44,995
	December 20, 1864	45,560
	June 20, 1865	48,337
	July 4, 1865	48,622
	April 17, 1866	54,068
Green, Charles	December 6, 1870	109,890
Greene, J. Durell (10,371 on pat. spec.)	January 3, 1854	10,391
	June 27, 1854	11,157
	November 17, 1857	18,634
	February 18, 1862	34,422
	March 23, 1869	88,161
Grillet, Alexander	November 22, 1864	45,152
Gross, Henry	May 22, 1855	12,906
	June 10, 1856	15,072
	August 30, 1859	25,259
	December 3, 1861	33,836*
	August 11, 1863	39,479
	August 25, 1863	39,645*
	August 25, 1863	39,646
	May 31, 1864	42,941
Guerriero, A.	April 11, 1865	47,252*
Gueury, F.	December 24, 1872	134,200
Guilbert, S.	September 20, 1864	44,303*
Gundersen, Gunder	December 30, 1873	145,998
Gunn, Edwinn F.	December 29, 1868	85,442
	September 10, 1867	68,736
Gywn, E., and Campbell, A. C.	October 21, 1862	36,709
Hall, A.	June 10, 1856	15,110*
	March 24, 1863	37,961*
Hamilton, Arnold	November 19, 1861	33,769
Hammond, Henry	October 25, 1864	44,798
	January 23, 1866	52,165*
	December 31, 1867	72,849
	March 14, 1871	112,589
Hancock, George	April 26, 1864	42,471
Hanson, John	November 29, 1870	109,731
Harrington, F. H.	June 15, 1858	20,607*
	February 7, 1871	111,534*
Harrington, Henry	July 29, 1837	297
Harris, C. W.	September 1, 1863	39,771*
Hartshorn, Isaac	March 31, 1863	38,042
Hartung, Charles	November 13, 1849	6,871
Haughian, P.	February 38, 1865	46,562*
Hay, Randall D.	November 22, 1870	109,514
Hayden, Hiram W.	December 20, 1864	45,495
	August 7, 1866	56,939
Haynes, W. T.	March 1, 1859	23,087*
Heath, John Sidney	May 13, 1873	138,887

Inventor's Name	*Date*	*Patent Number*
Heckenbach, John Adam	June 22, 1869	91,624
Henry, Alexander	February 13, 1866	52,654
	September 19, 1871	119,145
	December 30, 1873	145,944
Henry, B. T.	October 16, 1860	30,446
Hicks, William C.	March 10, 1857	16,797
	March 1, 1864	**41,814**
Hill, Albert V.	August 2, 1859	24,936*†
	May 28, 1861	32,421
Hill, R. B.	February 15, 1870	99,893*
Hillis, W. D.	September 20, 1864	44,312
Hinden, Mathias J.	June 29, 1869	92,048
Hoffman, F. W.	August 12, 1856	15,516
Holden, Cyrus B.	April 1, 1862	34,859
	March 29, 1864	42,139
Hollingsworth & Mershon	February 27, 1855	12,470*
	February 27, 1855	12,471*
Holman, G.	March 3, 1868	75,016*
Holt, G. L., and Marshall, J. C.	April 22, 1873	138,157
Hood, F. W.	November 8, 1864	44,953*
	July 4, 1871	116,593*
	February 23, 1875	160,192*
	April 6, 1875	161,615*
	March 14, 1876	174,731*
Hopkins, C. W.	May 27, 1862	35,419*
(Ries. 6,715, Oct. 26, 1875)	June 29, 1875	165,098*
Hopkins, Henry H.	September 23, 1873	143,012
	April 27, 1875	162,475*
Hopkins, S. S.	March 28, 1871	113,053*
Hoppenau, Henry	March 18, 1873	136,998
Hotchkiss, Benjamin B.	August 17, 1869	93,822
	January 2, 1872	122,465
	February 15, 1870	99,898
Howard, Charles	September 26, 1865	50,125
	October 10, 1865	50,358
	July 14, 1863	39,232
Howard, Sebre	October 28, 1862	36,779
Howe, Frederick W.	September 16, 1862	36,466
	March 7, 1865	46,671
Howe, J. C. (11,862 on patent drawing)	October 31, 1854	12,862
	February 17, 1863	37,693*
Hubbell, W. W. (3,649—July 1, 1844, given on patent office drawing)	February 1, 1844	3,649
	June 18, 1867	65,812
Hug, Daniel	September 2, 1873	142,396
Hughes, G. W.	November 15, 1864	45,043
Hughes, G. W., and Pusey, J. G.	August 15, 1865	49,409
Hulbert, W. A.	March 6, 1877	187,975
Hunt, Walter	August 21, 1849	6,663
Iversen, H.	March 26, 1850	7,218*
Jackson, Charles and Goodrem T.	March 17, 1863	37,937
Jaquith, E.	July 12, 1838	832*
Jarre, P. J.	June 24, 1862	35,685
Jarre, A. T., and P. J.	April 15, 1873	137,927
Jenkinson, J.	December 2, 1862	37,075*
Jenks, Barton H.	February 25, 1868	74,760

Inventor's Name	Date	Patent Number
Jenks, William	May 25, 1838	747
Jennings, L.	December 25, 1849	6,973
Johnston, William	May 13, 1862	35,241
	November 1, 1864	44,868
Johnson and Bye	June 4, 1878	204,438*
Jones, Frederick	October 2, 1860	30,228
Jones, O.	June 9, 1874	151,882*
	June 20, 1876	179,026*
	April 10, 1877	189,360*
	February 26, 1878	200,794*
Jones and Marston	January 1, 1878	198,745*
Joslyn, Benjamin F.	August 28, 1855	13,507
	July 1, 1856	15,240
	May 4, 1858	20,160*
	October 8, 1861	33,435
	June 24, 1862	35,688
	August 4, 1863	39,405*
	August 4, 1863	39,406*
	August 4, 1863	39,407
	March 22, 1864	42,000
	April 19, 1864	42,379*
	February 7, 1865	46,243*
	June 6, 1865	48,073
	June 20, 1865	48,287*
	June 20, 1865	48,288
	January 2, 1866	51,836*
	January 2, 1866	51,837
	November 15, 1870	109,218
	November 22, 1870	109,417*
	May 30, 1871	115,483*
	July 18, 1876	180,037*
	October 31, 1876	183,944*
	April 16, 1878	202,350*
	April 16, 1878	202,351*
	May 28, 1878	204,334* 204,335* 204,336* 204,337*
Josselyn, H. S.	January 23, 1866	52,248*
Judd, E. M.	February 25, 1862	34,504
Kay, Allan B.	November 22, 1870	109,419
Kellogg, Henry	May 20, 1862	35,356
Kerr, J. (England)	April 14, 1857	17,044*
Kerr, J.	August 4, 1863	39,409*
King, Benedikt	March 4, 1862	34,579
King, C. A.	August 24, 1869	94,003*
King, Nelson	May 22, 1866	55,012
	August 28, 1866	57,636
Kinsey, M.	June 8, 1858	20,496*
Kirk, John L.	June 20, 1871	116,066
Kirk, E. C., and Sneider, E.	July 9, 1867	66,596
Kittridge, B.	March 8, 1864	41,848*
Klein, Ferdinand	April 10, 1855	12,681
Kraffert, Julius	July 5, 1870	105,093
Laidley, T. T. S., and Emery, C. A.	May 15, 1866	54,743
Lancaster, P.	April 15, 1856	14,667

Inventor's Name	Date	Patent Number
Lane, Thomas W.	January 1, 1867	60,910
Landfear, W. R.	September 6, 1864	44,099
Lawrence, Richard S.	January 6, 1852	8,637
	December 20, 1859	26,504
	April 6, 1869	88,645
Leavitt, Daniel	April 29, 1837	182*
	June 14, 1859	24,394
Leaycraft, E. S.	March 28, 1871	112,471*
	March 28, 1871	112,472*
Lee, James	July 22, 1862	35,941
	May 15, 1866	54,744
	May 16, 1871	114,951
	June 20, 1871	116,068
	January 2, 1872	122,470
	January 16, 1872	122,772
Lee, Thomas	April 27, 1858	20,073
	November 19, 1861	33,745
	December 26, 1871	122,182*
LeFaucheux, Eugene (Same patent number, or two different patent drawings)	March 26, 1861	31,809
Le Mat, A. F.	October 21, 1856	15,925*
Le Mat, F. A.	December 14, 1869	97,780*
Leonard, G., Jr.	September 18, 1849	6,723*
	July 9, 1850	7,493*
	August 9, 1853	9,922*
	May 6, 1856	14,820*
Letort, James, and Mathews, H. S.	April 3, 1860	27,723
Lewis and Pfleghar	August 2, 1859	24,942*
Linberg and Phillips	December 6, 1870	109,914*
Lindner, Edward	June 27, 1854	11,197*
(Reis. 415 and 416, Dec. 23, 1856)	May 6, 1856	14,819
	May 26, 1857	17,382*
	March 29, 1859	23,378
Loomis, B. T.	February 13, 1866	52,582*
Lord, Horace	January 14, 1868	73,351
	February 11, 1868	74,387
Luce, George D.	March 11, 1873	136,660
Lull, Orin D.	June 16, 1863	38,903
Manton, Joseph	August 1, 1871	117,552
Marelli, Agostino	May 27, 1873	139,323
Marlin, John M.	February 8, 1870	99,690*
	April 5, 1870	101,637
	July 1, 1873	140,516*
Marsh, Samuel W.	December 6, 1859	26,362
	November 5, 1861	33,655
Marshall, J. P.	October 4, 1859	25,661
	April 29, 1862	35,107
Marston, W. W.	June 18, 1850	7,443
(Reis, 783, July 25, 1859; Reis. 1,029, Aug. 21, 1860)	January 7, 1851	7,887*
	September 18, 1855	13,581*
	May 26, 1857	17,386*
Mason, James M.	March 7, 1871	112,523
	August 8, 1871	117,906
Mason, W.	November 21, 1865	51,117*
	March 27, 1866	53,539*
	July 2, 1872	128,644*

Inventor's Name	Date	Patent Number
Mason, W.	September 15, 1874	155,095*
	January 19, 1875	158,957*
Maton, Francis	November 14, 1854	11,938
Mayall, T. J.	November 25, 1862	37,004*
Maynard, Edward	May 27, 1851	8,126
	December 6, 1859	26,364
	October 30, 1860	30,537
	June 27, 1865	48,423
	July 25, 1865	48,966
	August 1, 1865	49,130
	February 2, 1869	86,566
	February 18, 1873	135,928
McBeth, James E.	January 14, 1868	73,357
	August 11, 1868	80,985
McCarty, Thomas	March 11, 1837	147
McChesney, Reuben	October 2, 1866	58,444
	May 28, 1867	65,103
McGovern, John	April 13, 1869	88,890
Meigs, J. F.	October 21, 1862	35,721
	May 22, 1866	54,934
	August 18, 1868	81,100
Mellen, Dustin F.	October 4, 1864	44,545
Merlett, John	August 18, 1868	81,283
Merriam, Lincoln A.	February 16, 1869	87,058
	January 19, 1869	86,091
Merrill, George	October 17, 1871	119,939
	October 17, 1871	119,940
Merrill, James H.	January 8, 1856	14,077
	July 26, 1858	20,954
	April 9, 1861	32,032
	April 9, 1861	32,033
	May 28, 1861	32,450
	May 28, 1861	32,451
	October 22, 1861	33,536
	December 8, 1863	40,884
Mershon and Hollingsworth	September 8, 1863	39,825*
Merwin, Joseph, and Bray, Edward P.	January 5, 1864	41,166
Milbank, Isaac M.	December 2, 1862	37,048
	January 31, 1865	46,125
	February 20, 1866	52,734
	June 12, 1866	55,520
	January 8, 1867	61,082
	February 5, 1867	61,751
	June 11, 1867	65,585
	December 1, 1868	84,566
	April 16, 1872	125,829
	March 18, 1873	136,850
Miller, J. C.	February 8, 1870	99,693*
Miller, William H.	November 13, 1866	59,723
Miller, William H., and Geo. W.	May 23, 1865	47,902
	December 26, 1865	51,739
	May 14, 1867	64,786
	August 27, 1867	68,099
Millner, John Keene (or Milner)	February 17, 1863	37,723
Minesinger, David (D. Minessinger 6,139, Feb. 20, 1849, given on pat. office drawing)	July 27, 1849	6,139
Mix, Eugene M., and Horton, Henry B.	January 19, 1864	41,343
Mont-storm, William	October 9, 1855	13,660*

Inventor's Name	Date	Patent Number
Mont-Storm, William	March 11, 1856	14,420*
(15,397 in pat. spec.)	July 8, 1856	15,307
	June 14, 1859	24,414
	November 5, 1872	132,740
Moore, D	September 18, 1860	30,079*
	February 19, 1861	31,473
	December 3, 1861	33,847
	January 7, 1862	34,067*
(Reis. 1693 and 1694, June 7, 1864)	April 28, 1863	38,321*
(Reis. 7610, Apr. 17, 1877)	December 15, 1874	157,860*
	March 7, 1877	187,980*
	May 1, 1877	190,240*
	July 17, 1877	193,269*
Moore, James D	March 6, 1860	27,374
Morgenstern, William	November 29, 1864	45,262
	June 6, 1865	48,133
	December 24, 1867	72,526
	February 18, 1868	74,712
	June 23, 1868	79,291
	February 2, 1869	86,434
	February 23, 1869	87,190
	August 3, 1869	93,330
Morgenstern, W., and Morwitz, E	November 10, 1863	40,572
Morris, W. H., and Brown, C. L.		
(Reis. 4496, Aug. 1, 1871)	January 24, 1860	26,919*†
Morrison, W.H	September 19, 1854	11,698*
Morse, George W	June 27, 1856	15,995
	June 8, 1858	20,503
Moses, Myron	September 30, 1862	36,571
Moss and Johnson	June 20, 1871	116,078*
Mouncie, A. T. D	April 16, 1878	202,418*
Muller, Aloyse	May 24, 1870	103,488
Muller, Florent	February 4, 1868	74,119
Munger, Alfred S	March 12, 1867	62,873
Munson, A. L	November 13, 1866	59,629*
Needham, J., and GG. H	May 21, 1867	64,999
Nelson, C	July 28, 1874	153,597*
Nenninger, Robert	March 28, 1871	113,194
Newbury, Frederick (add. Imp't. 126, July 31, 1855)	March 20, 1855	12,555*
	June 12, 1855	13,039*
	September 18, 1855	13,582*
	March 11, 1856	14,406*†
	April 29, 1856	14,774
	August 12, 1856	15,521
	February 9, 1858	19,387*
	March 23, 1858	19,739*
(Add. Imp't. 204, Sept. 28, 1858)	June 29, 1858	20,765*
	April 10, 1860	27,868*
	October 23, 1860	30,494*
	January 31, 1865	46,131*
	January 9, 1866	51,959
Newell, J. D. S	April 6, 1869	88,730
	May 25, 1869	90,381
Newton, Abner N	June 29, 1854	11,198
	August 12, 1856	15,522
Nichols, J	September 2, 1862	36,358*

Inventor's Name	Date	Patent Number
Nichols and Childs	April 24, 1838	707*
Nickerson, C. V.	January 27, 1852	8,690
Nillus, Ferdinand E. M.	May 7, 1872	126,568
Norris, S., Mauser, W., and Mauser, P.	June 2, 1868	78,603
North, Henry S.	January 5, 1847	5,141
	June 17, 1856	15,164*
	April 6, 1858	19,868*
North and Savage	January 18, 1859	22,666*
North and Skinner	June 1, 1852	8,982*
Nutting, M.	April 25, 1838	713*
Nye, John C.	November 4, 1862	36,852
	January 6, 1863	37,356
Oliphant, John	January 13, 1863	37,407
Orr, W.	March 17, 1874	148,742*
Page, W. I.	June 21, 1870	104,636*
Palmer, William	September 28, 1858	21,623*
	July 23, 1861	32,887
	December 22, 1863	41,017
	March 8, 1864	41,857*
Palmie, G., and A. H.	October 24, 1854	11,835
Pape, William R.	November 5, 1867	70,463
	November 24, 1868	84,373
Parkhurst, C.	September 25, 1837	409
Parkinson, B. F.	June 13, 1865	48,201*
Payne, Chas. E.	May 10, 1864	42,685
Peabody, H. O. (Reis. March 13, 1866)	July 22, 1862	35,947
	December 10, 1867	72,076
	April 14, 1868	76,805
Peavey, A. J.	March 27, 1866	53,473
Pecare and Smith (Reis. August 13, 1850)	December 4, 1849	6,925*
Percival, O. B., and Smith, A.	July 9, 1850	7,496
Percy, John	August 11, 1863	39,494
Perley, Charles	February 24, 1863	37,764
Perry, A. D.	December 11, 1849	6,945
	November 28, 1854	12,001
	January 17, 1855	12,244
Perry, Samuel M.	June 21, 1864	43,259
	June 21, 1864	43,260
Perry, S. M., and Goddard, E.	April 26, 1870	102,429
Pettengill, C. S.	July 22, 1856	15,388*
	January 4, 1859	22,511*
Philips, W. H.	August 26, 1873	142,175*
Pierce, George R.	October 3, 1871	119,474
Piper, Edwin S.	December 5, 1865	51,391
Pitt, W. J.	January 7, 1862	34,093*
Plass, R. H.	January 24, 1865	46,023*
Polain, P.	March 27, 1866	53,548*
Pond, L. W.	June 17, 1862	35,623*
	June 16, 1863	38,934*
Porter, P. W.	July 18, 1851	8,210*
Post, J.	May 15, 1849	66,453*
Poultney, Thomas (Given as Poultney and Crispin on the Pat. Office drawings)	May 14, 1867	64,701
Prescott, E. A.	October 2, 1860	30,245*
Preston, James W.	February 5, 1867	61,865
Prindle, F. B (Aug. 10 on pat. office drawings)	April 10, 1858	21,149
	November 28, 1865	51,213

Inventor's Name	Date	Patent Number
Raub, A.	February 6, 1866	52,504*
Raymond and Robitaille	July 27, 1858	21,054*
Reed, J.	December 26, 1865	51,752*
Reeder, Joseph S.	November 27, 1860	30,760
Reid, J.	April 28, 1863	38,336*
Rembert, S. S.	October 29, 1867	70,264
	February 18, 1868	74,594
Remington, S.	March 17, 1863	37,921*
	February 9, 1869	86,690
Renwick, E. S.	April 26, 1870	102,434
Restell, Thomas	March 26, 1867	63,303
(Dec. 10, 1872, is date given on pat. drawing, the correction being made in pencil)	May 20, 1873	139,190
Reynolds, Henry	May 10, 1864	42,688*
(Reis. 2234, May 1, 1866)	November 22, 1864	45,176*
	May 8, 1866	54,600
Rice, Wayne H.	May 19, 1863	38,604
Richards, C. B.	July 25, 1871	117,461*
	September 19, 1871	119,048*
Richards, Westley	July 14, 1863	39,246
	October 10, 1865	50,432
	May 11, 1869	89,889
	June 22, 1869	91,668
	December 3, 1872	133,665
	May 27, 1873	139,422
Richardson, C. H.	May 22, 1877	191,178*
Richardson, George J.	August 23, 1864	43,929
Richardson, W. A.	May 23, 1876	177,887*
Rider, Joseph	August 17, 1858	21,215*
	May 3, 1859	23,861*
	September 13, 1859	25,470
(Reis. May 3, 1864)	December 8, 1861	40,887
	November 15, 1864	45,123
	January 3, 1865	45,797
	February 21, 1865	46,532
	November 28, 1865	51,269*
	March 27, 1866	53,543
	February 11, 1868	74,428
	August 15, 1871	118,152
	July 29, 1873	141,383
	July 29, 1873	141,384
	August 5, 1873	141,590
Ritter von Wessely, Z.	July 13, 1869	92,673
Roberts, Benjamin S.	September 23, 1861	36,531
	February 27, 1866	52,887
	June 11, 1867	65,607
	May 11, 1869	90,024
Roberts, Robert	December 27, 1864	45,638
Robertson, W. H., and Simpson, G. W.	February 12, 1856	14,253
	March 13, 1866	53,187
Robinson, Orvill M.	May 24, 1870	103,504
	April 23, 1872	125,988
Robitaille and Davis	August 2, 1864	43,709*
Rodier, L. C.	March 25, 1862	34,776
	July 11, 1865	48,775*
Rodier, L. C., and Bates, F. G.	April 29, 1873	138,439
Rogers, H. S.	November 4, 1862	36,861*
Rood, M. L.	November 22, 1853	10,259*

Inventor's Name	*Date*	*Patent Number*
Root, Elisha K (Reis. 846, Nov. 1, 1859)	December 25, 1855	13,999*
	June 4, 1867	65,509
	June 4, 1867	65,510*
Roper, Sylvester H	April 10, 1866	53,881*†
Rowe, A. H	April 5, 1864	42,227
Rupertus, J	April 19, 1859	23,711*
	December 2, 1862	37,059*
(Reis. 5631, Oct. 28, 1873)	July 19, 1864	43,606*
	November 21, 1871	121,199*
	July 6, 1875	165,369*
	November 9, 1875	169,848*
Russell, Charles F	May 14, 1872	126,748
Rydbeck, Sven	June 28, 1870	104,775
Saez, Cosme Garcia	January 3, 1865	45,801
Sargent, Edward L	March 1, 1870	100,455
	June 21, 1870	104,502
	November 15, 1870	109,255
Savage and North	July 30, 1844	3,686
	May 15, 1860	28,331*
Sayer, Adolf	June 19, 1866	55,719
Scharffe, G.	November 11, 1856	16,070
Schenkl, John P	June 23, 1857	17,642
Schesch, Heinrich August	April 9, 1872	125,620
Schneeloch, O	December 31, 1872	134,442*
Schofield, George W	June 14, 1870	104,211
	June 20, 1871	116,225*
	April 22, 1873	138,047*
(Reis. 8354, July 30, 1878)	July 31, 1877	193,620*
Schopp, Francis	November 28, 1865	51,225
Schroeder, Herman	June 26, 1861	32,653
Schroeder, H., Salewski, L., and Schmidt, W	December 23, 1856	16,288
Schubarth, Casper D	July 23, 1861	32,895
Schuler, Peter	January 5, 1869	85,616
	November 1, 1870	108,836
Schulz, Gustave	May 11, 1869	89,947
Scott, Cornelius	October 25, 1864	44,827
Scott, J. Q. A.	August 12, 1862	36,174
Scott, Wm. Middleditch	November 1, 1870	108,942
Scott, Walter, and Matthews, W. J	November 25, 1873	144,870
Sears, J. Hunter	December 20, 1859	26,526
Seeley, S. J	December 3, 1861	33,854
Sehorn, A. O. H. P	January 30, 1855	12,328*
Selwyn, Jasper H	August 14, 1866	57,269
Seymour, J. M	May 20, 1862	35,354
Sibert, Lorenzo	May 14, 1861	32,316
Simpson, Thomas D., Gray, G. B., and Romans, J. H.	August 2, 1870	106,083
Sharps, Christian	September 12, 1848	5,763
	December 18, 1849	6,960*
	November 11, 1856	16,072
	January 25, 1859	22,752
(Reis. 1199, June 18, 1861; reis. 2480 and 2481, Feb. 12, 1867)	January 25, 1859	22,753*†
	November 27, 1860	30,765*
	July 9, 1861	32,790
	October 29, 1861	33,607
(62,067 on patent office drawing)	February 12, 1867	62,077

Inventor's Name	*Date*	*Patent Number*
Sharps, Christian	September 5, 1871	118,752*
	April 8, 1873	137,625
Shaw, Jacob	June 30, 1857	17,698*
Shaw, T. S.	December 24, 1861	34,032*
Sheckler, Peter	April 2, 1867	63,564
Sherman, Henry N.	November 25, 1873	144,872
Shinn, F. M.	April 27, 1875	162,582*
Shull, Thomas E.	April 5, 1859	23,505
Skinner, C. D., and Tyron D.	October 20, 1857	18,472
Skinner, B. F., and Plummer, A., Jr	February 18, 1862	34,449
Sloan, T. L.	March 21, 1876	175,180*
Slocum, F. P.	January 27, 1863	37,551*
(Reis. 1839, Dec. 20, 1864)	April 14, 1863	38,204*
Smiles, James	December 27, 1870	110,505
Smith, A.	December 24, 1861	34,016*
	May 28, 1872	127,377*
Smith, D.	April 22, 1873	138,207
	March 9, 1875	160,551*
	May 11, 1875	163,032*
	December 14, 1875	171,059
	August 7, 1877	193,836*
	October 23, 1877	196,491*
	July 16, 1872	129,433
Smith, Gilbert	December 25, 1855	14,001
	August 5, 1856	15,496
	June 23, 1857	17,644
Smith, Horace	August 26, 1851	8,317
Smith, Isaac	April 26, 1864	42,542
Smith, J. C.	January 1, 1856	14,034
Smith, James D	February 27, 1866	52,933
	February 27, 1866	52,934
Smith, Joseph N	June 10, 1862	35,548
	August 18, 1863	39,591
Smith, O. A.	January 28, 1873	135,377*
	January 28, 1873	135,378*
	April 15, 1873	137,968*
Smith, R. D. O.	July 12, 1864	43,529*
Smith, W. H.	December 10, 1861	33,907
	August 23, 1864	43,957
Smith, Dexter, and Chamberlin, Martin J	March 7, 1871	112,505
Smith, D., and Marshall C. C., and J. C.	April 18, 1876	176,412*
	April 25, 1876	176,448*
	May 4, 1875	162,863*
Smith, Smith and Sweeney	March 18, 1873	136,871*
Smith, H., and Wesson, D. B	February 14, 1854	10,535
	July 18, 1861	30,990*
(Reis. 1059, Oct. 9, 1860)	July 5, 1859	24,666*
	June 16, 1863	38,921*
	November 21, 1865	51,092*
Smoot, William Sydney	August 27, 1867	68,250
	June 1, 1869	90,792
	December 14, 1869	97,821
	June 20, 1871	116,106
(Oct. 17, 1871, on patent specifications)	November 7, 1871	120,788
(933,063 on patent specifications)	November 12, 1872	133,063
	October 21, 1873	143,855*
Snedden, Wm. Tait	June 27, 1871	116,363
	June 27, 1871	116,364

Inventor's Name	Date	Patent Number
Sneider, Charles Edward	March 20, 1860	27,600
	March 18, 1862	34,703*
	August 25, 1863	39,707
	January 24, 1865	46,054
	February 28, 1865	46,612*
	May 16, 1865	47,755
	December 22, 1868	85,252
Snider, Jacob, Jr	October 15, 1867	69,941
Soper, William	June 14, 1870	104,223
Soule, George H	April 3, 1855	12,655
	July 15, 1856	15,347
	July 6, 1858	20,825
Spellerberg, Anton	July 30, 1861	32,929
Spellier, A	October 2, 1860	30,260*
Spencer, Christopher M	March 6, 1860	27,393
	February 4, 1862	34,319
	July 29, 1862	36,062
	May 26, 1863	38,702
	January 17, 1865	45,952
	October 9, 1866	58,737
	October 9, 1866	58,738
	February 11, 1873	135,671
Stabler, Edward	December 6, 1864	45,356
	March 14, 1865	46,828
Stanton, Henry	August 16, 1853	9,950
Stanton, S. F	April 29, 1856	14,780*
Starr, Eben T	January 15, 1856	14,118*
	September 14, 1858	21,523
	December 4, 1861	30,843*
	May 10, 1864	42,698*†
	December 20, 1864	45,532*
	December 19, 1865	51,628*
	March 28, 1876	175,518*
Starr, Nathan (1,141 on patent drawing)	May 3, 1839	1,411
Stephens, R. E. (given as R. E. Stevens on patent office drawing)	June 11, 1867	65,704
Stetson, George R	July 4, 1871	116,642
Stevens, Abijah C	May 4, 1869	89,699
Stevens, Joshua	November 26, 1850	7,802*
	October 7, 1851	8,412*
	August 9, 1853	9,929*
	January 2, 1855	12,189*
	September 6, 1864	44,123
Stevens, W. X	January 12, 1864	41,242
Stickney, Curtis R	July 2, 1872	128,671
Stillman, James	October 17, 1865	50,507
Stith, O.	Mar. 16, 1819	No number given
Sturtevant, Edward L	July 16, 1867	66,751
Sturtevant, Thomas L	September 19, 1865	50,048
	November 7, 1865	50,854
	December 18, 1866	60,592
Stoakes, John T	July 6, 1869	92,393
Strong, F. F	April 21, 1838	698*
Strong, Samuel	December 16, 1862	37,208
	May 19, 1863	38,643
	May 19, 1863	38,644
Swan, J. L	April 8, 1862	34,911
Sweet, A. L	December 10, 1878	210,725*

Inventor's Name	Date	Patent Number
Sweet, A. L.	August 15, 1854	11,536
Swinburn, John F.	December 17, 1872	134,014
Swingle, Alfred	April 1, 1873	137,392
Swingle, Alfred, and Huntington, Frank A.	February 18, 1873	135,947
Swyney, John	August 21, 1855	13,474
Sutvan, Isaac	March 14, 1865	46,866
Symmes, John C.	November 16, 1858	22,094
Taylor, James P.	May 6, 1873	138,711
Terrell, Charles C.	February 16, 1858	19,387
Terrill, LaFayette Z.	May 14, 1869	89,705
Terry, William	October 14, 1862	36,681
Thistle, H. L.	August 1, 1838	865
Thomas, James R.	May 19, 1840	1,611
Thomas, John F.	April 2, 1872	125,229
	May 28, 1872	127,386
Thrasher, D. C., and Aiken, B. F.	July 16, 1867	66,913
Thuer, F. A.	September 15, 1868	82,258*
	July 12, 1870	105,388
Tibbals, William	November 28, 1865	51,243
	June 19, 1866	55,743*
	July 17, 1866	56,466*
Tibbets, George H.	February 13, 1872	123,595
Tiebel, Louis B.	May 11, 1869	89,955
Tiesing, Frank, and Gerner, Chas.	April 4, 1871	113,470
	April 25, 1871	114,230
Tileston, W.	September 6, 1864	44,126*
Todd, George H.	July 27, 1869	93,023
Tonk, A.	January 27, 1857	16,411*
Townsend, Frederick	January 29, 1861	31,268
	September 6, 1864	44,127
Triplett, Louis	December 6, 1864	45,361
Trulender, Frederick	May 10, 1864	42,702
Twickeler, Theodore	March, 1862	34,706
Tyler, C. N.	May 3, 1853	9,701*
Underwood, Henry	June 2, 1863	38,772
Updegraff, Horace	September 19, 1871	119,098
	August 6, 1872	130,165
Van Choate, Silvanus Frederick	May 11, 1869	89,902
	August 24, 1869	94,047
	June 13, 1871	115,911
	October 22, 1872	132,505
Varney, A. L.	March 30, 1869	88,530
	March 30, 1869	88,531
	September 28, 1860	95,395
Vaughan, A. C.	May 27, 1862	35,404*
Vetterlin, Frederick	December 29, 1868	85,494
	November 15, 1870	109,277
Vickers, J. H.	June 17, 1862	35,657*
	May 16, 1865	47,775*
	August 21, 1866	57,448*
Vittum, Francis J., and Stevens, Edgar M.	October 22, 1861	33,560
Von Jeinsen, Ernest	December 15, 1868	84,922
Von Martini, Frederich	May 25, 1869	90,614
(115,446 given on patent office drawing)	May 30, 1871	115,546
	November 7, 1871	120,800
	October 15, 1872	132,222

Inventor's Name	*Date*	*Patent Number*
Von der Poppenburg, John	October 24, 1865	50,670
Walch, J.	February 8, 1859	22,905*
Walker, Henry	September 17, 1872	131,484
Wallis, T. M.	October 31, 1876	183,993*
	August 21, 1877	194,489*
Wampler, J. M.	March 6, 1860	27,399
Ward, H. D.	September 8, 1863	39,850*
Ward, William G.	June 29, 1869	92,129
	August 31, 1869	94,458
	December 7, 1869	97,734
	February 1, 1870	99,504
	February 21, 1871	111,994
Warner, James	January 7, 1851	7,894*
	July 15, 1851	8,229*
(Reis. 2223, Apr. 10, 1866)	June 24, 1856	15,202*
	July 28, 1857	17,904*
	February 23, 1864	41,732
	December 27, 1864	45,660
Washington, T. A.	October 28, 1856	15,990
Watson, A. T. (11,567 on patent office drawing)	March 20, 1855	12,567
Wayne, Joseph B.	August 15, 1871	118,171
Weaver, H. B.	October 16, 1855	13,691
Webley, Thomas W.	June 11, 1867	65,783
Wells, C. S.	December 10, 1872	133,732*
Werndl, Joseph	February 18, 1868	74,737
Werner, Daniel B.	October 6, 1868	82,908
Wesson, Daniel B.	December 17, 1867	72,434
	June 9, 1868	78,847
	May 2, 1871	114,374
	February 25, 1873	136,348*
	January 19, 1875	158,874*
	May 11, 1875	163,036*
	January 23, 1877	186,509*
Wesson, Edwin	June 5, 1847	5,146
	August 28, 1849	6,669*
Wesson, Franklin	November 11, 1862	36,925
	December 15, 1868	84,976*
	July 20, 1869	92,918
	May 31, 1870	103,694
	June 13, 1871	115,916*
Wesson and Bullard	February 20, 1877	187,689*
	December 18, 1877	198,228*
Wesson, F., and Harrington, N. S	October 25, 1859	25,926
Wesson and King	July 16, 1872	128,991*
Wesson, D. B., and Smith, H	February 14, 1854	10,535
Wheeler, Henry F.	February 7, 1865	46,286
	October 31, 1865	50,760
	June 19, 1866	55,752
	June 25, 1867	66,110
Wheelock, Luke	October 22, 1867	70,141
	December 1, 1868	84,598
	January 31, 1871	111,500
White, Albert M.	April 18, 1865	47,350
White, George W.	February 4, 1862	34,325
	January 6, 1863	37,369
White, Le Roy S.	January 6, 1863	37,376
White, Rollin	March 13, 1855	12,528

Inventor's Name	Date	Patent Number
White, Rollin	March 13, 1855	12,529
	April 3, 1855	12,638*†
(Reis. 1557, 1558 and 1559, Oct. 27, 1863)	April 3, 1855	12,649*
(Reis. 1802, Oct. 25, 1864; reis. 1928, Apr. 4, 1865)	April 13, 1858	19,961*
	November 29, 1864	45,290*
	July 9, 1867	66,542*
	August 10, 1869	93,572*
(Reis. 4958, June 25, 1872)	August 10, 1869	93,653*
	February 1, 1870	99,505*
	February 22, 1870	100,227*†
(Reis. 5691, Dec. 16, 1873)	September 30, 1873	143,394*
	April 20, 1875	162,208*
	April 20, 1875	162,208*
	July 27, 1875	166,173*
	June 20, 1876	179,084*
	July 4, 1876	179,633*
	March 26, 1878	201,855*
	April 16, 1878	202,613*
Whiting, E.	May 23, 1871	115,258*
Whitney, Eli	August 1, 1854	11,447*
	November 8, 1864	44,991
	January 9, 1866	51,985*
	November 26, 1867	71,349
	March 21, 1871	112,997
	June 13, 1871	115,997
	March 26, 1872	124,994
Whitney, E., Gerner, C., and Tiessing, F.	July 27, 1869	93,149
Whitney, E., and Tiessing, Frank	July 16, 1872	129,637
Whitney, James A.	July 30, 1867	67,242
Whitmore, Andrew E.	August 8, 1871	117,843
	April 16, 1872	122,775
	January 2, 1877	185,881*
Whittemore, James M.	June 14, 1870	104,387
	September 17, 1872	131,487
	October 1, 1872	131,921
Whittier, O. W.	May 30, 1837	216*
Wichmann, John H.	September 29, 1863	40,151
Wilkinson, John D.	August 29, 1871	118,569
Williams, B. H. (Reis. 6569, Aug. 3, 1875)	April 21, 1874	150,120*
Williams, G. E.	November 16, 1875	170,038*
Williams, R. D.	November 27, 1877	197,708*
Williamson, David	January 5, 1864	41,184*
	May 17, 1864	42,823*
	March 21, 1865	46,977
	October 2, 1866	58,525
	March 16, 1869	87,997
	March 18, 1873	137,043*
	November 18, 1873	144,814*
	November 18, 1873	144,815*
Wilson, Robert	November 15, 1864	45,105
Wilson, Thomas	July 14, 1868	80,043
Wilson, W. T., and Flather, Henry	August 15, 1865	49,463
Winchester, O. F.	September 4, 1866	57,808
Wohlgemuth, Friedrich	May 18, 1869	90,214
Wolcott, H. H.	June 13, 1865	48,227
	November 27, 1866	60,106

Inventor's Name	*Date*	*Patent Number*
Wood, Corbin O.	October 9, 1860	30,362
	January 1, 1861	31,050
Wood, Stephen W.	April 1, 1862	34,854
	November 18, 1862	36,984*
	August 18, 1863	39,619*
	March 1, 1864	41,803*
	September 20, 1864	44,363*
	January 16, 1866	52,105*
	June 13, 1876	178,824*
	January 23, 1877	186,445*
Woodward, F. G.	January 7, 1862	34,084
Wright, Edward S.	November 15, 1864	45,126
Wright, W.	November 7, 1854	11,917*
Wurfflein, John	April 30, 1850	7,334
Wyley, Andrew	November 24, 1868	84,459
Yates, Theodore	December 18, 1866	60,607
Ybarra, L.	April 23, 1878	202,915*
Yglesias, Jose	May 9, 1871	114,742
Young, Lewis V.	June 21, 1870	104,682
Zeller, Gottlieb	March 18, 1873	136,894

APPENDIX II

AMERICAN GUN-MAKERS, 1630-1928

The following list of 1,059 names has been compiled from every source available today, particularly the Arms Collection of the United States War Department, Old Smithsonian Building, Washington, D. C.; the Rudolph J. Nunnemacher Collection; Mr. Charles Winthrop Sawyer's "Firearms In American History," and Capt. J. G. W. Dillin's "The Kentucky Rifle." While many of the names listed are of notable and even dramatic interest, for reasons of personal distinction or genius, or contribution to the science and evolution of firearms, or for intimate connection with their country's history, it is obvious that in such a list no more comment can be given than is called for by the immediate purposes of its presentation. It has accordingly been designed for the assistance of museum-authorities, historians, collectors and students of firearms, with a view to enabling them to locate in date and place of origin pieces under their consideration. With these objects in view, no more mention is made of any name than is required to indicate the time and place of any maker's activity, when these are of record. In the cases of not a few names, either or both of these data are completely lost, when these are marked "unidentified" or "unlocated," or both.

P. B. J.

A. A., unidentified, unlocated early rifle-maker, about 1760.

Abbey, George T., Chicago, Ill.; before and after 1860.

Adam, Daniel; unlocated, flint-lock period.

Adams, W., unlocated, flint-lock period.

Adirondack Arms Co., Plattsburg, N. Y.; about 1870-1875.

Adolph, Fred, Genoa, N. Y.; present day.

Ager, A., Rumley, O.; flint-lock period.

Agy, Pennsylvania; flint-lock period.

Albertson, Douglas & Co., New London, Conn.; about 1840-1860.

Aldenderfer, M., Lancaster, Pa.; 1763-1817.

Allbright, Henry, Lancaster Co., Pa.; Durham Iron Works, Pa.; before and after 1744.

Allbright, J., Mannheim, Pa.; flint-lock period.

Allen, C. B., Springfield, Mass.; from about 1830 to Civil War.

Allen (C. B.) & Falls, Springfield, Mass.; from after 1830 to Civil War period; revolvers prior to 1865.

Allen, E. & Co., Worcester, Mass.; in business from after the Civil War up to about 1880.

Allen (Ethan), Brown & Luther; one of Ethan Allen's manufacturing affiliations, Worcester, Mass.; between 1847 and 1865.

Allen (Ethan), & Thurber, Grafton, Mass.; 1837-1842; Norwich, Conn., 1842-1847; Worcester, Mass., 1847-1856; revolvers.

Allen (Ethan) & Wheelock, Worcester, Mass., 1856-1865; revolvers.

Allen, G. F., Utica, N. Y.; before and after 1850.

Allen, Henry, New York, N. Y.; percussion period.

Allen, Oliver, Norwich, Conn.; percussion period.

Allen, Silas, Shrewsbury, Mass.; b. 1775; d. about 1850.

Allen, William, New York, N. Y.; percussion period.

Allison, T., Pennsylvania; flint-lock period.

Alsop, C. H. or C. R. (both initials are on arms in the U. S. Collection); Middletown, Conn.; revolvers, Civil War period.

American Arms Co., Chicopee Falls, Mass.; Civil War arms.

American Machine Works, Springfield, Mass.; Civil War arms.

American Repeating Rifle Co., Boston, Mass.; 1869.

American Standard Tool Co., Newark, N. J.; revolvers, prior to 1865.

Ames Arms Co., Chicopee Falls, Mass.; revolvers, prior to 1865.

Ames, David, Bridgewater, Mass.; 1790. (Supt. Springfield Armory, 1795.)

Ames, Nathaniel, Boston, Mass.; 1800.

Ames, N. P., Springfield, Mass.; Mexican and Civil War arms.

Amoskeag Mfg. Co., Manchester, N. H.; Civil War arms.

Amsden, B. W., Saratoga Springs, N. Y.; before and after 1850.

Andrus & Osborn, Canton, Conn.; Civil War arms.

Angstadt, Peter, Pa.; flint-lock period.

Annely, John, New York, N. Y.; percussion period.

Anschutz, E., Philadelphia, Pa.; percussion period.

Antis, William, Frederick Township, Pa.; Revolutionary War.

Armstrong, John, Md.; flint-lock period. Possibly the same as the following:

Armstrong, John, Pa.; d. 1827.

Armstrong, John, Jr., Gettysburg, Pa.; before and after 1855.

Armstrong, S. F., Adamsville, Mich.; percussion period.

Ashton, H. & Co., Middletown, Conn.; middle 19th century.

Astol, J. & W., New Orleans, La.; before and after 1805.

Avery, Willis, Salisbury, N. Y.; percussion period.

Babcock, Moses, Charlestown, Mass.; percussion period.
Backhouse, Richard, Easton, Pa.; prior to 1783.
Bacon Arms Co., Norwich, Conn.; revolvers, prior to 1865.
Baer, J., Lancaster, Pa.; flint-lock period.
Bailey, Thomas, New Orleans, La.; revolvers, prior to 1865.
Baird, S. S., Chittenden, Vt.; percussion period.
Baker, Clyde, 2100 East 59th St., Kansas City, Mo.; present.
Baker, John, Lancaster, Pa.; flint-lock period.
Baker, John, Providence Township, Pa.; Revolutionary War.
Ballard Arms Co., Fall River, Mass.; Civil War arms.
Bandle (J.) Gun Co., Cincinnati, O.; revolvers (perhaps long-arms).
Barent, Govert, New Amsterdam, N. Y.; 1648.
Barker, Cyrus, Providence, R. I.; percussion period.
Barlow, J., Moscow, Ind.; about 1840.
Barnes, Thomas, Brookfield (also North Brookfield), Mass.; b. 1764;
    in business up to around 1800.
Barnhardt, W., Pa.; about 1780.
Bartlett, name only; found on flint-lock arms; may be any one of the
    following:
Bartlett, Chenango Point (now Binghamton), N. Y.; about 1800-1825;
    predecessor of Bartlett Bros., below.
Bartlett, Pa.; flint-lock period.
Bartlett, A. & P., Mass.; Civil War arms.
Bartlett Bros. Binghamton, N. Y.; about 1825-1850; successors to
    Bartlett, Chenango Point, above.
Barret, "Deacon," Concord, Mass.; 1774.
Barstow, I. & C. C., N. H.; Civil War arms.
Bauer, George, Lancaster, Pa.; prior to 1783.
Bauer, J., unlocated; flint-lock period.
Bay State Arms Co., Uxbridge, Mass.; about 1870-1875.
Beadle, Indian Trail, O.; about 1840-1890.
Bean, Baxter, East Tennessee; flint-lock period.
Beauvais, R. St. Louis, Mo.; percussion period.
Beck, C., Pa.; flint-lock period.
Beck, Gideon, Pa.; flint-lock period.
Beck, Isaac, Mifflinburg, Union Co., Pa.; flint-lock period.
Beck, J., Pa.; flint-lock period. May or not be the same as the following:
Beck, J. P., Union Co., Pa.; flint-lock period.
Beech & Rigdon (also Leech & Rigdon?), Augusta, Ga.; Civil War
    Confederacy.

Bekeart, San Francisco, Cal.; percussion period to later.

Bell, Conder, unlocated; flint-lock period.

Bell, John, Boston, Mass.; 1746.

Bellis, Lancaster, Pa.; early flint-lock period.

Bemis, Edmund, Boston, Mass.; 1746.

Benfer, Amos, Snyder Co., Pa.; flint-lock and perhaps percussion periods.

Benfer, Arnig, Beaverstown, Snyder Co., Pa.; flint-lock period.

Berlin, Abraham, Easton, Pa.; prior to 1783.

Berlin, Isaac, Pa.; prior to 1783.

Berry, A. P., unlocated; flint-lock period.

Berry, W., Poughkeepsie, N. Y.; before and after 1840. Later, Albany, N. Y.

Berstro, J. H., Buffalo, N. Y.; before and after 1835.

Berstrow, H. T., Buffalo, N. Y.; before and after 1835.

Bery, R. B., unlocated; flint-lock period.

Best, unlocated, unidentified; early flint-locks. May or may not be same as either of the following:

Best, Lancaster, Pa.; flint-locks. May be same as the following:

Best M., Pa.; early 19th century. May be same as either above.

Beutter Bros., New Haven, Conn.; then Meriden, Conn.; before and after 1850.

Beyers, N., Pa.; flint-locks.

Biddle, T. & W. C., Philadelphia, Pa.; percussion period.

Bidwell, Oliver, Conn.; 1800.

Billinghurst, William, Rochester, N. Y.; about 1830-1880.

Bird Bros., Philadelphia, Pa.; flint-lock period.

Bird, John, Oskaloosa, Iowa; from about 1860 to about 1900; d. 1918, aged 86.

Bisbee, D. H., Norway, Me.; about 1835-1860.

Bisbing, Pennsylvania; percussion period.

Bishop, Henry H., Boston, Mass.; before and after 1847.

Bishop, William, Boston, Mass.; 1818 to about 1860.

Blackwood, Marmaduke, Philadelphia, Pa.; Revolutionary War.

Blake, E., unidentified, unlocated; percussion period.

Bliss & Goodyear, New Haven, Conn.; revolvers, prior to 1865.

Bloodgood, North Carolina; flint-lock period.

Blunt, Orison, New York, N. Y.; percussion period; later as below:

Blunt (Orison) & Sims, New York, N. Y.; 1837 to about 1865.

Boardlear, Samuel, Boston, Mass.; 1796.

Bolton, Enoch, Charlestown, Mass.; 1665.

Boone, E., Oley Valley, Pa.; 1818, etc., second cousin of Daniel Boone.

Boone, Samuel, Berks Co., Pa.; after 1768; nephew of Daniel Boone.

Boone, Squire, Rowan Co., N. C.; before 1800; brother of Daniel Boone.

Bosworth (also Bossworth), Lancaster, Pa.; flint-lock period.

Boulon, W. S., Kentucky; about 1800-1840.

Bourne, William, Savannah, Ga.; Confederate imitation Colts & Remingtons during Civil War, marked "W.B., C.S.A."

Bouron, P., New Orleans, La.; met. cart. period.

Bowie, James, Natchez, Miss.; killed at the Alamo, 1836.

Boyer, D., Orwigsburg, Pa.; flint-lock period; son of M. Boyer, below.

Boyer, H., Pennsylvania; flint-lock period.

Boyer, M., Lehigh District, Pa.; flint-lock period.

Boyer, N., Lehigh District, Pa.; flint-lock period.

Boyington, John, South Coventry, Conn.; flint-lock to percussion periods.

Brand Arms Co., Norwich, Conn.; about 1840-1860.

Brasirus, Joseph, Pennsylvania; flint-lock period.

Brey, Elids, unlocated; percussion period, possibly flint-lock also.

Brockway, Norman S., Bellows Falls, Vt., and West Brookfield, Mass.; b. about 1855.

Brong, Joseph, Lancaster, Pa.; flint-lock period.

Brong, Peter, Lancaster, Pa.; before and after 1800.

Brooke, I. I. & N., Pennsylvania; Civil War arms.

Brooklyn Arms Co., Brooklyn, N. Y.; revolvers, prior to 1865.

Brooks, William F., New York, N. Y.; Civil War arms.

Brooks or Brookes, Richard, Boston, Mass.; 1675.

Brown, Andrew (son of John, below); Fremont & Poplin, N. H.; percussion period.

Brown, C. E., unidentified, unlocated; percussion period.

Brown, F. B. (or F. P.), Lancaster, Pa.; flint-lock and percussion periods.

Brown, John, Poplin, N. H., 1840-1857; Fremont, N. H., 1857-1870.

Brown, J. F., Haverhill, Mass.; percussion period.

Brown Mfg. Co., Newburyport, Mass.; Civil War period and later.

Brown (William) & Sons, Pittsburgh, Pa.; percussion period.

Buck, H. A. & Co., West Stafford, Conn.; percussion period.

Buell, E., Marlborough, Conn.; 1812.

Bullard Repeating Arms Co., Springfield, Mass., and Boston, Mass.; metallic cartridge period.

Bulow, Charles, Lancaster, Pa.; about 1797.

Burd, C., Philadelphia, Pa.; flint-lock period.

Burdick, S., unidentified, unlocated; percussion period.

Burkhard, William R., St. Paul, Minn.; percussion period.

Burnett, F. L., unlocated; probably 18th century.

Burnside Rifle Co., Providence, R. I.; Civil War arms.

Burt, A. M., unlocated; Civil War arms.

Busch, Lancaster, Pa.; flint-lock period.

Buswell, J., Glens Falls, N. Y.; percussion period.

Butler, John, Lancaster, Pa.; Revolutionary War.

Butterfield, J. S., Philadelphia, Pa.; revolvers, prior to 1865.

Byers, N., Pennsylvania; 1780 to after 1800.

Calderwood, Philadelphia, Pa.; flint-lock period.

Cardis, Thomas, Pennsylvania; flint-lock period.

Carlisle, H., Pennsylvania; flint-lock period.

Cartwright, John, Ottowah, Ohio; before and after 1865.

Caup, Levi, West Buffalo, Snyder Co., Pa.; 19th century.

Carver, James W., Pawlet, Vt.; before and after 1885.

Chapman, C., unidentified, unlocated; Civil War Confederate arms.

Chapman, James, Bucks Co., Pa.; Revolutionary War.

Chapple, Thomas, New York, N. Y.; percussion period.

Charlottesville Rifle Works, Charlottesville, N. C.; from about 1740 to
     through the Revolutionary War.

Chase, Anson, Enfield, Mass., before 1830; Hartford, Conn., 1830-
     1834; later, New London, Conn.

Chase, William, Pandora, Ohio; before and after 1860.

Cherington, Sr., Pennsylvania, unlocated; father of Thomas P. Cher-
     ington, probably early 19th century.

Cherington, Thomas P., Cattawissa, Pa.; also associated with George
     Schalk, at Pottsville, Pa.; middle 19th century.

Chilcote, Pennsylvania; percussion period.

Chnader, J.; Kentucky rifles.

Chrisky, L., Philadelphia, Pa.; Kentucky rifles.

Christ, D., Lancaster, Pa.; Kentucky rifles.

Christ, Jacob, unlocated; Kentucky rifles.

Clark, Carlos, Windsor, Vt. Claimed invention of false-muzzle in 1836;
     patented by Alvan Clark (brother?), celebrated telescope-maker,
     in 1840.

Clarke, John, Pennsylvania; flint-lock period.

Clarke, N., Pennsylvania; percussion period.

Clause, Nathan, Pennsylvania; flint-lock period.

Clewflin, W., unlocated; Kentucky rifles.

Cline, C., Pennsylvania; flint-lock period.

Cochran, J. W., New York, N. Y.; percussion period.

Colderwood, Philadelphia, Pa,; Kentucky rifles.

Coleman, H., Boston, Mass.; before and after 1847.

Collier, Elisha H., Boston, Mass., 1800-1817; flint-lock revolving arms.

Colt Patent Fire Arms Co., Paterson, N. J., 1836-1842; New York City, 1847-1848; Hartford, Conn., 1848—; revolvers chiefly; also revolving muskets, rifles, carbines, shotguns; also non-revolving arms; also automatic pistols.

Cone, D. D., Washington, D. C.; revolvers, prior to 1865.

Connecticut Arms Co., Norfolk, Conn.; revolvers, prior to 1865.

Connecticut Arms & Mfg. Co., Naubec, Conn.; metallic cartridge period, Hammond rifles.

Constable, Richard, Philadelphia, Pa.; before and after 1847.

Cook & Bro., New Orleans, La., before 1863; Athens, Ga., 1863-1866, Civil War Confederate arms.

Cookson, John, Boston, Mass., 1727.

Cooley, D., unlocated; Kentucky rifles.

Coons; some unidentified and unlocated gunsmiths of this name (relatives of E. Coons of Philadelphia, Pa.), made early American shotguns.

Coons, E., Philadelphia, Pa.; Kentucky rifles.

Cooper Firearms Co., Frankford, Philadelphia, Pa., revolvers, prior to 1865.

Cooper, Henry T., New York, N. Y.; before and after 1845.

Cope, Jacob, unlocated; Kentucky rifles.

Cosmopolitan Arms Co., Hamilton, Ohio; Civil War arms.

Cowell, Ebenezer, Allentown, Pa.; Revolutionary War period.

Cowell, Joseph, Boston, Mass., 1745.

Cowell, P., Pennsylvania; prior to 1783.

Craig, Robert, Philadelphia, Pa.; Revolutionary War.

Craig, William, Alleghany, Pa.; before and after 1850.

Cramer, Pennsylvania; Kentucky rifles.

Crandall, W. F., Gowanda, N. Y.; percussion period.

Crissey, Elias, Pennsylvania; percussion period.

Crosman Arms Co., Rochester, N. Y.; airguns only, present.

Crow, C. A., Lima, Ohio; before and after 1870.

Cryth, John, Lancaster, Pa.; Kentucky rifles.

Cushing, A. B., Troy, N. Y.; about 1840-1870.

Dallam, Richard, Maryland; American Revolution.

Davenport. W. H., Fire Arms Co., Norwich, Conn.; about 1855-1910.

Davidson, F. & Co., Cincinnati, Ohio; before and after 1850.

Daub, J., Berks Co., Pa.; flint-lock period.

Decherd (also spelled Dechard, Dechert, Deckard, Descherd, Dickert),
    Jacob, Philadelphia, before and after 1732; Lancaster, Pa., before
    and after 1753.

Deeds, W., Pennsylvania; Kentucky rifles, "smooth-bores."

De Haven, Hugh, Philadelphia, Pa.; Revolutionary War.

De Haven, Peter, Philadelphia, Pa., 1776. (French Creek, Chester Co.,

DeHuff, Henry, Lancaster, Pa., 1800.

DeLaney, Sussex Co., N. J.; percussion period.

Delaney, Nelson, Reading, Pa.; flint-lock period.

DeReiner, Michael, Lancaster, Pa.; Kentucky rifles.

Deringer, Henry, Sr., Richmond, Va., latter 19th century to 1806;
    probably Philadelphia, Pa., 1806—? (Also Easton, Pa.?); father
    of maker of "Deringer" pistols.

Deringer, Henry, Jr., Philadelphia, Pa.; b. 1786, d. 1868; maker of
    "Deringer" pistols. (Also Easton, Pa.?)

Derr, John, Lancaster and Oley Valley, Berks Co., Pa.; flint-lock and
    percussion periods.

Deterer, Adam, Lancaster, Pa.; Revolutionary War.

Dewarson, R., Boston, Mass.; before and after 1847.

D. G. & Co., Cincinnati, Ohio; percussion period.

Dickson, Nelson & Co., Alabama; Civil War Confederate arms.

Diets, Pennsylvania; Kentucky rifles.

Dike, Bridgewater, Mass.; American Revolution.

Dimmick, H. E., St. Louis, Mo.; before and after 1850.

Disboch, Pennsylvania; Kentucky rifles.

Dish, R., New York, N. Y.; percussion period.

Doell, Frederick, Boston, Mass., 1884-1909.

Doll (or Dull), Jacob, Lancaster, Pa., 1802.

Dooley, Scranton, Pa.; percussion period.

Doplier, Robt., Wheeling, W. Va.; percussion period.

Dorn (or Lorn), unidentified, unlocated; Kentucky rifles.

Douglass, John, Huntingdon, Pa., 1830.

Doyle, John, Lancaster, Pa., 1784.

Drepperd (also Drippard and Dreppard), Henry, Lancaster, Pa.; flint-lock period.

Dreppert (perhaps son of F. Drippard), Lancaster, Pa.; flint-lock period.

Driggs-Seabury Ordnance Co., Utica, N. Y.; related to Savage Arms Co.; World War arms.

Drisbach, G., unidentified, unlocated; Kentucky rifles.

Dull (or Doll), Jacob, Lancaster, Pa., 1802.

Dunkle, G., unidentified, unlocated; Kentucky rifles.

Dunkle, G., Path Valley, Pa.; flint-lock period.

Dunlap, G., Pennsylvania; percussion period. (Proprietor Pennsylvania Rifle Works.)

Dunwicke, Chester Co., Pa.; Revolutionary War.

Durham Iron Works, Easton, Pa.; prior to 1783.

Dwight, H. D., Belchertown, Mass.; before and after 1847.

Eagle Arms Co., Rock Falls, N. Y.; revolvers, prior to 1865.

Eagle Mfg. Co., Mansfield, Conn.; Civil War arms.

Earl, Thomas, Leicester, Mass., 1776.

Eaton, J., Boston, Mass.; before and after 1847.   (Connection with the following is not known.)

Eaton, J., Concord, N. H.; about 1870 to about 1916.

Eberly, Lancaster, Pa.; Revolutionary War.

Eckles, H., Pennsylvania, 1820.

Edgerton, H. S., German, N. Y.; percussion period.

Eggers, Samuel, New Bedford, Mass.; about 1840 to 1865.

Ehlers, unidentified, unlocated; Kentucky rifles.

Ehrmon, H., Pennsylvania; Kentucky rifles.

Eicholtz & Bro., Lancaster, Pa.; flint-lock period to 1888.

Ellis, Reuben, Albany, N. Y.; flint-lock period; made Hall rifles in 1829.

Ells, Josiah, Pittsburgh, Pa.; revolvers, prior to 1865.

Ely, Martin, Springfield, Mass.; American Revolution.

Emmes, Nathaniel, Boston, Mass., 1796-1825.

Ensley, M., unidentified, unlocated; percussion period.

Ernst, J., Pennsylvania; flint-lock period.

Ernst, Maryland; Kentucky rifles.

Estabrook, J. M., Milford, Mass.; percussion period.

Estabrook, William, Armada, Mich.; percussion period.

Evans Repeating Rifle Co., Mechanic Falls, Me.; about 1871 to about 1880.

Evans, Stephen, partner with Daniel Walker & Joseph Williams, Mount
    Joy Armory, later Valley Forge, Pa., established 1742.
Evans, W. L., Valley Forge, Pa., 1822.

Fairbanks, A. B., Boston, Mass., d. 1841.
Falley (also Foley), Richard, Westfield, Mass., 1774.
Farnot, Frank, Lancaster, Pa.; prior to 1783.
Farnot, Jacob, Pennsylvania; prior to 1783.
Farrington, William B., Concord, N. H.; percussion period.
Farrow Arms Co., Holyoke, Mass.; about 1885-1890; later Mason,
    Tenn., defunct.
Farrow, William Milton, Holyoke, Mass.; about 1880-1885.
Fayetteville Arsenal, Fayetteville, N. C.; Civil War Confederate arms.
Feder, G., unlocated; "straight-cut" Kentucky rifles.
Fehr, J., Nazareth, Pa., 1835.
Ferree, Jacob, Lancaster, Pa.; prior to 1783; with son Joel moved to
    24 miles north of Pittsburgh on Monongahela River.
Ferree, Joel (not son of Jacob), Leacock Township, Lancaster Co., Pa.,
    on Pequa Creek before 1750 to Revolutionary War.
Ferris, Geo. H., Utica, N. Y.; about 1850-1875.
Finch, Joseph, New York, N. Y., before 1828.
Fischer, Gustav, New York, N. Y.; before and after 1860.
Fish, New York, N. Y.; before and after 1845.
Fitch, John, Trenton, N. J., 1771-1775. Gun-maker for N. J. in Revo-
    lution. Invented steamboat.
Fogerty Repeating Rifle Co., Boston, Mass.; defunct about 1868.
Fogg, G. E., Manchester, N. H.; percussion period.
Fohrer, Ludwig, Pennsylvania; Revolutionary War.
Foilke, A., Lehigh District, Pa.; associated with John Young, 1776.
Follecht, Lancaster, Pa.; prior to 1783.
Folsom, Henry & Co., St. Louis, Mo.; percussion period.
Fondegrift, Pennsylvania; prior to 1783. So of record; possibly same
    as John Vondegrift, Bucks Co., Pa., during Revolutionary War.
Fondersmith & Son, John, Lancaster, Pa.; also Stroudsburg, Pa.; prob-
    ably 1749-1802.
Ford, J., Virginia; Kentucky rifles.
Fordney, C., Maryland; Kentucky rifles.
Fordney, I., unlocated; Kentucky rifles.
Fordney, Melchior, Lancaster, Pa.; flint-lock period.
Fortune, Thomas L., Mt. Pleasant, Kansas; metallic cartridge period.

Foster, George P., Taunton, Mass.; percussion period.

Foster, Joseph, Pennsylvania; Revolutionary War.

Foulke, Adam, Easton, Pa., with John Young, from about 1770 to after 1776; later Allentown, Pa., and Philadelphia, Pa.

Fox, A. H., Gun Co., Philadelphia, Pa., present.

Franck, Lancaster Pa.; prior to 1783.

French, Thomas, Canton, Mass.; b. 1778; d. 1862.

French, Blake and Kingsley, Mass.; Civil War arms.

Frock, unidentified, unlocated; Kentucky rifles.

Frost, Gideon, Massachusetts; American Revolution.

Fry, Francis, Doniphan Co., Kansas; before and after 1855.

G. D. & Co., Cincinnati, Ohio; middle 19th century.

Gable & Son, Henry, Williamsport, Pa., 19th century.

Gardner, J. N., Scranton, Pa.; percussion period.

Garret, Herman, Boston, Mass., 1650.

Gaspard, Lancaster, Pa., prior to 1783.

Gemmer, John P., St. Louis, Mo., 1862-1915; ex-Hawken-employee; d. 1919.

Georg, Jacob, Pennsylvania, 1817.

Getz, Philadelphia, Pa., before 1811.

Gerrish, John, Boston, Mass., 1709.

Gibbs, H., Lancaster, Pa., 1824.

Gibbs, Tiffany & Co., Sturbridge, Mass.; about 1820-1850.

Gilbert, Daniel, Brookfield, Mass., b. 1729, d. 1824.

Gilbert, Daniel, Massachusetts; Civil War arms.

Gilbert, E., Rochester, N. Y.; percussion period.

Giles, Richards & Co., Boston, Mass.; flint-lock period.

Gillespie, unidentified, unlocated; maker of fine "Deringer"-style pistols in first half of 19th century.

Gingerich, Henry, Lancaster, Pa.; Revolutionary War.

Glassbremer, G., Pennsylvania; Kentucky rifles.

Glassbrenner, H., unidentified, unlocated; percussion period.

Glaze & Co., William, Columbia, S. C.; Civil War Confederate arms. (Sometimes marked his work "Palmetto Armory.")

Goetz & Westphall, Pennsylvania; Civil War arms.

Golcher (often Goulcher), George, New York, N. Y.; percussion period.

Golcher (often Goulcher), James, Philadelphia, Pa., d. 1805.

Golcher (often Goulcher), Joseph, Philadelphia, Pa., later on Pacific Coast; late flint-lock to percussion period.

Golcher (often Goulcher), John, Easton, Pa., prior to 1783. At one
    time in Philadelphia, Pa., and probably New York, N. Y.
Golcher (often Goulcher), Manuel, Philadelphia, Pa., 1824.
Golcher (often Goulcher), Philadelphia, Pa., unidentified descendant of
    James Golcher; made shotguns in 1877.
Gompf, Pennsylvania; percussion period.
Gonter, Peter, Jr. (also spelled Gonder and possibly Gontee, and pos-
    sibly the same as Gontec), Lancaster, Pa., d. 1818.
Gontec, Peter, Sr. (possibly the same as Gonter), Lancaster, Pa.; Rev-
    olutionary War period.
Goodling, P., unlocated; Kentucky rifles.
Goodrich, worked with Hyde, New Orleans, La.; percussion period.
Gorsage, Thomas, Mt. Pleasant, O.; percussion period.
Gouger, Pennsylvania; Revolutionary War.
Gove, Carlos, Denver, Colo.; metallic cartridge period.
Graeff, John, Lancaster, Pa., 1798.
Graeff, William, Lancaster, Pa.; before and after 1751.
Graeff, William, Reading, Pa.; before and after 1761.
Graham, J., unlocated; Kentucky rifles.
Grandstatt, J., unlocated; Kentucky rifles.
Great Western Gun Works, Pittsburgh, Pa.; percussion to metallic car-
    tridge period. See Johnston, James H.
Greene (J. D.), Millbury, Mass., 1857.
Gregory, Richard, Boston, Mass., 1727.
Gresheim, Lancaster, Pa.; prior to 1783.
Griffin & Howe, New York, N. Y., present.
Grimes, Pennsylvania; Kentucky rifles.
Groff, H. S., unlocated; Kentucky rifles.
Groot, Henry, Pittsfield, Mass.; percussion period.
Gross Arms Co., Tiffin, Ohio; about 1840-1880.
Grover & Lovell (John P.), Boston, Mass., 1841-1844. See Lovell.
Grubb, George, New York, N. Y.; percussion period.
Grubb, T., Philadelphia, Pa., 1820.
Guest, J., Pennsylvania; Kentucky rifles.
Guger, J. P., Pennsylvania; Kentucky rifles.
Guillam, Benjamin, Massachusetts; American Revolution.
Gumph, A., unlocated; Kentucky rifles.
Gumph, Christopher, Lancaster, Pa.; Kentucky rifles up to 1888.
Gumph, James, Lancaster, Pa., d. 1887.
Gwyn & Campbell, Hamilton, Ohio; Civil War arms.

Haeffer, J., Berks Co., Pa.; Kentucky rifles.

Haeffer, P. B., unlocated; Kentucky rifles.

Hagi, Pennsylvania; flint-lock period.

Hahn, W., New York, N. Y.; percussion period.

Haines, Isaac, Pennsylvania; about 1730.

Halbern, Caspar, Lancaster Pa.; Revolutionary War.

Haldeman, F., Heidelberg Township, Berks Co., Pa.; Kentucky rifles.

Halk, I., unlocated; Kentucky rifles.

Hall, Alexander, New York; before and after 1850.

Hall, Charles, Lancaster, Pa.; d. about 1887.

Hall, E. L., Springfield, Mass.; percussion period.

Hall, John H., Yarmouth, Mass., now Me., until 1816; Harper's Ferry Armory, 1816-1840; d. in Missouri, 1841. Inventor and maker of Hall breech-loading rifles and percussion arms.

Hapgood, Joel, Shrewsbury, Mass.; b. about 1805; d. about 1890.

Harden, J., Lock Haven, Pa.; percussion period.

Harpers Ferry, U. S. Arsenal, Va., 1796.

Harrington, H. B., Lebanon, N. H.; percussion period.

Harrington, Luke, Sutton, Mass.; before and after 1832.

Harrington & Richardson, Worcester, Mass., present.

Harris, Isaac, Maryland; American Revolution.

Harris, Jason L., location unknown; flint-lock period.

Hart (B. J.) & Bro., New York, N. Y.; percussion period.

Harter, Lock Haven, Pa.; 19th century.

Harwood, Nathaniel H., Brookfield, Mass.; about 1825-1840.

Harwood, Reuben, Somerville, Mass.; metallic cartridge period.

Hatch, Warren, Plattsburg, N. Y.; before and after 1850.

Haven, see De Haven.

Hawk, Nicholas, Monroe Co., Pa., 19th century.

Hawkins, Pennsylvania, unidentified; possibly same as Henry Hawkins.

Hawkins, Henry. (Hawkins appears to have been the ancestral form of the name, his sons changing it to Hawken, in which form it became famous throughout the West. It often appears in either form.) Probably originally at Lancaster, Pa.; thence Harper's Ferry Arsenal, Va., and Hagerstown, Md., at unrecorded periods. Believed to have removed to St. Louis, Mo., about 1808, whither his sons later followed him.

Hawken, Jacob, son of Henry Hawkins; b. Hagerstown, Md., 1786; d. St. Louis, Mo., May, 1849; business, St. Louis, 1820-1849. Alleged also to have been in business in Louisville, Ky., about 1822.

Hawken, Samuel T., son of Henry Hawkins; b. Hagerstown, Md., October, 1792; business, St. Louis, Mo., 1822-1860; said also to have been in business in Independence, Iowa, and Denver, Colo.

Haynes, Joshua, Waltham, Mass.; percussion period.

Hench (also Hennch), Peter, Lancaster, Pa.; flint-lock period.

Hench, Pottsville, Pa.; Kentucky rifles.

Henry, Abraham, Lancaster, Pa.; with John Graeff made rifles for Commonwealth of Pennsylvania in 1798 and thereafter. Probably relative of William Henry, 1st, as lived in family, but relationship unknown.

Henry, Granville, son of James; partner with father; later alone until in the '80's; Philadelphia, Pa., and Boulton, Pa. B. 1835; d. 1912.

Henry, James, son of John Joseph; b. Philadelphia, Pa., 1809; proprietor Boulton Iron Works, 1836—? Later took as partner son, Granville.

Henry, John, Lancaster, Pa., 1762 and thereafter; not known what relationship, if any, with William Henry, 1st.

Henry, J. & Son; a partnership name of James and Granville Henry, Boulton, Pa., and Philadelphia, Pa.

Henry, John Joseph, son of William Henry, 3rd; proprietor Boulton Iron Works, Boulton, Pa., Nov., 1822, to death, Dec., 1836.

Henry, W. & I. I., Pennsylvania; Civil War arms.

Henry, William (1st); b. Chester Co., Pa., 1729; d. 1786; apprenticed to (Martin or Peter?) Roeser, 1750. Armorer of Braddock's Expedition, 1755. In business with Joseph Simons, firm known as Simons & Henry; partnership dissolved, 1759.

Henry, William (2nd), known as William Henry, Jr., son of William Henry, 1st; began at Nazareth, Northampton Co., Pa., 1778. In 1800 at Jacobsburg, same neighborhood; in 1812-1813 in the Henry Rifle Works, Boulton, Pa., 3 miles N. E. of Nazareth, Pa.

Henry, William (3rd), son of Wm. Henry (2nd); in 1810 at Philadelphia, Pa., in partnership with Bro., John Joseph Henry; in 1822 sold to latter, who organized the Boulton Iron Works.

Henry Repeating Arms Co., New Haven, Conn.; ceased 1866.

Hep, Philip, Jr., unlocated; Kentucky rifles.

Hepburn, Lewis L., Colton, Vt.; percussion period; inventor Remington-Hepburn system.

Herthe, August, Pennsylvania; Kentucky rifles.

Hess, J., unlocated; Kentucky rifles.

Hetrick (I. or J.) & Co., unidentified, unlocated; percussion period.

Hill, Thomas, Carlotta, Vermont, 1800.
Hillegas, H., unlocated; Kentucky rifles.
Hillegas, J., Pottsville, Pa.; about 1810-1830.
Hilliard, D. H., Cornish, N. H.; percussion period.
Hinds, John, Boston, Mass., 1745.
Hoake, J., unlocated; Kentucky rifles.
Hoard, C. B., Watertown, N. Y.; Civil War arms.
Hoard's Armory, Watertown, N. Y.; revolvers, prior to 1865.
Hobbs, P., Monterey, Mass.; percussion period.
Hodge, J. T., unlocated; Civil War arms.
Hoffman Arms Co., formerly Cleveland, Ohio, now Ardmore, Okla.,
    present.
Holburn, Casper L., unlocated; Kentucky rifles.
Holden, C. B., Worcester, Mass.; metallic cartridge period.
Hollingsworth, Henry, Elkton, Md.; American Revolution.
Holmes, George H., Defiance, Ohio; before and after 1870.
Hopkins & Allen, Norwich, Conn.; metallic cartridge period and mod-
    ern arms.
Horn, Conrad, Hazleton, Pa.; about 1820-1855; brother of William.
Horn, Stephen, Lancaster, Pa., and Easton, Pa., 18th century.
Horn, William, Hazelton, Pa., brother of Conrad.
Houghton, Richard W., Norway, Me.; percussion period.
Howard Bros., Whitneyville, Conn., metallic cartridge period.
Howland, Rufus J., Binghamton, N. Y., about 1840-1870.
Hudson, W. L., Cincinnati, Ohio; percussion period.
Humbarger, Adam, Somerset, Ohio; percussion period.
Humble, Michael, Louisville, Kentucky; Kentucky rifles.
Hunter Arms Co., Fulton, N. Y., present. (Makers of L. C. Smith
    guns.)
Huntingdon, V., Allensville, Pa.; Kentucky rifles.
Hurd, Jacob, Boston, Mass., 1816-1825.
Hutchinson, R. J., Williamsport, Pa.; Kentucky rifles.
Hutz, Lancaster, Pa., 1803.
Huyslop (also Hyslop), R., New York, N. Y.; flint-lock period.
Hyde & Goodrich, New Orleans, La.; Civil War Confederate arms.

Ingall, Brown; Portland, Bucksport, Andover and Blue Hill, Me.; per-
    cussion period.
Isch, Christian, Lancaster, Pa.; prior to 1783.
Ithaca Gun Co., Ithaca, N. Y., present.

Iver Johnson's Arms & Cycle Works, New York, N. Y.; Fitchburg, Mass.

James, M., Pennsylvania; Kentucky rifles.

James, Morgan, Utica, N. Y., about 1820-1860.

Jenison & Co., J., Southbridge, Mass.; Civil War arms.

Jenks, Alfred & Son, Bridesburg, Pa.; Civil War arms.

Jenks, Stephen, North Providence and Pawtucket, R. I.; American Revolution.

Johnson's Arms & Cycle Works, New York City, N. Y.; Fitchburg, Mass.

Johnson, Robert, Middletown, Conn., U. S. Model 1819 rifles.

Johnson, Seth, Old Rutland, Mass.; American Revolution.

Johnson, William, Worcester, Mass., 1787.

Johnston, James H., Pittsburgh, Pa., son of John, below; b. 1836; d. about 1916; proprietor Great Western Gun Works, Pittsburgh, Pa.; in business 1866-1915.

Johnston, John H., Waynesboro, Pa.; b. 1811; d. 1889.

Jones, Charles, Lancaster, Pa.; prior to 1783.

Jones, Robert, Lancaster, Pa.; prior to 1783.

Jordan, L., South Adams, Mass.; percussion period.

Josan, Pennsylvania; flint-lock period.

Joslyn, B. F., Firearms Co., Stonington, Conn.; Civil War arms.

Jost, White Plains Township, Pa.; Revolutionary War.

Kantz, F., unlocated; Kentucky rifles.

Kascheline, Peter, Northampton Co., Pa.; Revolutionary War.

Kaup, Levi, Union Co., Pa., 19th century.

Keely, Matthias, Pennsylvania; Revolutionary War.

Keely, Sebastian, Pennsylvania; Revolutionary War.

Keffer, Jacob, Lancaster, Pa., 1802.

Keller, I., Cumberland Valley, Pa.; Kentucky rifles.

Kellogg Bros., New Haven, Conn.; about 1850-1890.

Kelly, Samuel, unlocated; Kentucky rifles.

Kemmerer, David, Carbon Co., Pa.; percussion period.

Kemmerer, David, Jr., location uncertain; percussion period.

Kempton, Ephraim, Salem, Mass., and Boston, Mass., about 1677.

Kendall, N., Windsor, Vt.; before and after 1850.

Kennedy, E. M., unlocated; Kentucky rifles.

Kerlin, John, Pennsylvania; Revolutionary War.

Kessler, John, Weston, Mo., about 1840-1860.

Key, R., Central Pa.; Kentucky rifles.

Kile, Nathan, Raccoon Creek, Jackson Co., O.; before and after 1817.
Kinder, Pennsylvania; Revolutionary War.
Kirkwood, David, Boston, Mass.; metallic cartridge period.
Kiser, A., unlocated; Kentucky rifles.
Kitchen, Wheeler, Broadway, Luzerne Co., Pa.; Kentucky rifles.
Kittinger, J., unidentified, unlocated; percussion period.
Klein, George, Pennsylvania; before and after 1800.
Klein, P. H., New York, N. Y.; middle 19th century.
Kleist, Daniel, Bethlehem Township, Pa.; prior to 1783 up to 1810.
Kline, C., Pennsylvania; Kentucky rifles.
Kochler, P., Lewisburg, Pa.; Kentucky rifles.
Koesler, unlocated; Kentucky rifles.
Koons, Frank, Berks Co., Pa.; Kentucky rifles.
Kornman, A. D., Central Pa.; Kentucky rifles.
Krammer, Pennsylvania; percussion period.
Krider, John, Philadelphia, Pa.; about 1820-1870.
Kunz, J., Philadelphia, Pa.; flint-lock period.

Lagunbra, Pennsylvania; Kentucky rifles.
Lamberson & Furman, Windsor, Vt.; percussion period.
Lamson, E. G., Windsor, Vt.; Civil War arms.
Lamson, Goodnow & Yale, Windsor, Conn.; Civil War arms.
Lancaster Rifle Works, Lancaster, Pa.; percussion period.
Lander, C., unidentified, unlocated; percussion period.
Lane & Read, Boston, Mass., 1826 to about 1835.
Langdon, W. G., Boston, Mass.; Civil War arms.
Latil, L. A., Baton Rouge, La.; percussion period.
Lawrence, Laconia, N. H.; percussion period.
Lawrence, Philadelphia, Pa., before 1830.
Lawrence, Richard S., Hartford, Conn.; percussion period.
Laydendecker, George, Allentown, Pa., prior to 1783.
Leader, Richard, Boston, Mass., 1646.
Leavitt, Daniel, Springfield, Mass.; revolvers, prior to 1865.
Lechiler, unlocated; Kentucky rifles.
Lee Firearms Co. (James Paris Lee), Milwaukee, Wis.; metallic car-
    tridge period.
Lee, James Paris, Stevens Point, Wis., to 1861; Milwaukee, Wis.;
    Ilion, N. Y.; later lived in England. Inventor and maker of many
    gun-mechanisms, including the Lee-Hotchkiss, Remington-Lee,
    Lee-Metford (English), etc.

Lefever Arms Co., Ithaca, N. Y., present.

Lefevre, Philip, Beaver Valley, Lancaster Co., Pa., 1731-1756.

Leman, H. E., Lancaster, Pa., before and after 1740.

Leman, Henry E., son of H. E. (formerly Lehmann), Lancaster, Pa.;
  b. 1812; d. 1887; began business 1834.

Leman, Peter, Lancaster, Pa., before and after 1740.

Lennard, Lancaster, Pa., prior to 1783.

Leonard, A. & Sons, Saxons River, Vt.; about 1840-1860.

Leonard, Eliphalet, Easton, Mass.; before and after 1800.

Leonard, G., Charlestown, Mass., pepper-box pistols.

Leonard, Johnathan (son of Eliphalet), Canton, Mass., about 1800-
  1820.

Leonard, R. & C., Massachusetts; Civil War arms.

Lewis, Troy, N. Y.; percussion period.

Libeau, V. G. W., New Orleans, La.; before and after 1835.

Lindsay, John Parker, unidentified, unlocated; percussion period.

Little, D., Bellefonte, Pa., 19th century.

Livermore, E. K., New York, N. Y.; percussion period.

Lloyd, William, Snyder Co., Pa., 19th century.

Loder, Lancaster, Pa.; prior to 1783.

Long, James, Beaver Springs, Snyder Co., Pa.; Kentucky rifles.

Lorney, M., Boalsburg, Pa.; Kentucky rifles.

Losey, B., unidentified, unlocated; percussion period.

Lovell, John P.; b. 1833; d. 1915; Philadelphia, Pa., before 1872; Denver, Colo., 1872-1915.   See Grover & Lovell.

Lovell, J. P., Arms Co., Boston, Mass.; metallic cartridge period; succeeded by Iver Johnson Co.

Lowe, William V., Fitchburg, Winchester and Woburn, Mass., about
  1875-1895.

Ludington, Lancaster, Pa.; flint-lock period.

Ludwig, Paul, Pennsylvania; before 1831.

Lupus, Dover, N. H.; percussion period.

Mange, H., unlocated; Kentucky rifles.

Manhattan Arms Mfg. Co., New York, N. Y.; revolvers, prior to 1865.

Manhattan Fire Arms Co., Newark, N. J.; revolvers, prior to 1865.

Marble Arms & Mfg. Co., Gladstone, Mich., present.

Marker, Daniel, Pennsylvania; Kentucky rifles.

Marker, Jr., son of Daniel, unlocated; Kentucky rifles.

Marlin Fire Arms Co., New Haven, Conn.; metallic cartridge period; present.

Marlin-Rockwell Co., New Haven, Conn., successors to Marlin Arms Co.; World War arms.

Marsh, J., Binghamton, N. Y., about 1850-1870.

Marshall, Job, Fairmount Township, Luzerne Co., Pa.; Kentucky rifles.

Marston Firearms Co., New York, N. Y.; percussion period.

Marston, William W., New York, N. Y.; revolvers, prior to 1865.

Marston, W. W., & Knox, New York, N. Y., 1854.

Martin, M., unlocated; Kentucky rifles.

Martin, T., unlocated; Kentucky rifles.

Mason, William, Taunton, Mass.; Civil War arms.

Massachusetts Arms Co., Chicopee Falls, Mass.; revolvers (Wesson's, Leavitt's, J. Stevens' and Adams'); 1849-1880.

Matson, Thomas, Boston, Mass., before 1658 to after 1682.

Mauger, H., unlocated; 1780.

Mause, F. E., Mausedale, Montour Co., Pa., 19th century.

Maxim Munitions Co., New Haven, Conn., modern, World War arms.

Mayesch, Lancaster, Pa., prior to 1783.

Maynard Gun Co., Chicopee Falls, Mass.; percussion period.

McCartney, Robert, Boston, Mass., 1805-1818.

McComas, Alexander, Baltimore, Md.; percussion period.

McCormick, James, Pennsylvania; Revolutionary War.

McElwaine, W. S., Holly Springs, Miss.; Civil War Confederate arms.

McKim, Baltimore, Md., before 1825; also McK. & Bro.

Meacham, I. & H., Albany, N. Y.; flint-lock period.

Melchior, M., unlocated; Kentucky rifles.

Mendenhall, James and Gardner, Greensboro, N. C.; Civil War Confederate arms.

Meriden Mfg. Co., Meriden, Conn.; Civil War arms.

Merrill, J. H., Baltimore, Md.; Civil War arms.

Merrimac Arms & Mfg. Co., Merrimac Co.; metallic cartridge period.

Merritt, Allen, East Randolph, Mass.; percussion period.

Merritt, John, Boston, Mass.; before 1789 to after 1798.

Merwin & Bray, New York, N. Y.; metallic cartridge period.

Messer, W. W., Boston, Mass.; percussion period.

Messersmith, John, Maryland; American Revolution; Lancaster, Pa., 1776.

Metropolitan Arms Co., 97 Pearl St., New York, N. Y.; revolvers, prior to 1865.

Meunier, John, Milwaukee, Wis.; active about 1855-1900; d. 1919.

Meunier, Steve, Meunier Gun Co., Milwaukee, Wis., 1880-1915.

Meylan, Martin, Lancaster, Pa., began 1719.

Middletown Mfg. Co., Middletown, Conn.; revolvers, prior to 1865.

Midvale Steel & Ordnance Co., Midvale, Pa.; World War arms.

Miles, D. D., Aurora, Ill., 1892.

Miles, John, New Jersey, 1800 to Civil War period.

Miles, Thomas, Lancaster, Pa., prior to 1783.

Miles, Thomas, location uncertain; probably Philadelphia, Pa., 19th century.

Milholland, James, unlocated; Civil War arms.

Miller, A., unidentified, unlocated; flint-lock period.

Miller, C., Honoeye, N. Y.; percussion period.

Miller, I., unlocated; Kentucky rifles.

Miller, John, Lancaster, Pa., prior to 1783.

Miller, John, Penfield and Monroe, Mich., about 1830-1875.

Miller, Mathias, Easton, Pa., prior to 1783.

Miller, S. C., New Haven, Conn.; before and after 1850.

Miller, Simon, Hamburg, Pa.; Kentucky rifles.

Miller, Simon (Jr.?), Hamburg, Pa.; percussion period.

Miller, W. D., Pittsfield, Mass.; percussion period.

Mills, Benjamin, Harrodsburg, Ky.; before 1790 to 1815.

Moll, I., unlocated; Kentucky rifles.

Moll, John, Allentown, Pa., prior to 1783.

Moll, John, Hellertown, Pa.; flint-lock period.

Moll, J. & W. H., Hellertown, Pa.; Kentucky rifles.

Moll, N., Pennsylvania; flint-lock period.

Moll, Peter, Pennsylvania; began about 1820.

Moll, Peter and David, Allentown, Pa.; Kentucky rifles.

Moore, Abraham, Chester Co., Pa.; Revolutionary War.

Moore, J. P., Union, N. Y.; before and after 1845.

Moore, R. A., New York, N. Y.; percussion period.

Moore's Patent Fire Arms Co., Brooklyn, N. Y.; revolvers, prior to 1865.

Morrison, Milton, Pa., 19th century.

Morse, Augusta, Ga., and Columbia, S. C.; Civil War Confederate arms.

Morse, Thomas, Lancaster, N. H.; about 1866-1890.

Mossberg & Sons, O. F., New Haven, Conn.; pistols, present.

Mount Joy Forge, later Valley Forge Armory, Pa., 1742-1842.   See Valley Forge.

Mower, P., Columbia Co., Pa., 19th century.

Mowry, D. D., unlocated; Civil War arms.

Muir, William, Windsor, Conn.; Civil War arms.

Mullin, John, New York, N. Y.; before and after 1850.

Mullin, Patrick, New York, N. Y.; before and after 1850.

Murray, J. P., Columbus, Ga.; Civil War Confederate arms.

Musser, H., Mulheim, Pa.; Kentucky rifles.

Myer, Henry, Lancaster, Pa.; Revolutionary War.

Nason, C. F., Auburn, Me.; percussion period.

National Arms Co., location unknown; revolvers, prior to 1865.

Nelson & Co., unidentified, unlocated; percussion period.

Nepperan Arms Co., Yonkers, N. Y.; revolvers, prior to 1865.

Newbury Arms Co., Albany, N. Y., middle 19th century.

Newhardt, Jacob, Allentown, Pa., before 1800.

Newhardt, Peter, Allentown and Whitehall, Pa., prior to 1783.

New Haven Arms Co., New Haven, Conn., 1860-1866.

Newton Arms Co., Buffalo, N. Y., recent years; name varies at times.

Niedner, A. O., Melrose, Mass., and Dowagiac, Mich., present.

Norris, S. & W. T., Clement, Mass. (?); Civil War arms.

North, H. S., Middletown, Conn.; revolvers, pistols and other arms, prior to 1865.

North, Simeon, Middletown, Conn., and Berlin, Conn.; U. S. arms from flint-lock period to Civil War.

Norwich Arms Co., Norwich, Conn.; Civil War arms.

Oakes, Samuel, Philadelphia, Pa., about 1800.

Oberholzer, Christian, Lancaster, Pa.; Revolutionary War.

Odlin, John, Boston, Mass.; before 1671 to after 1682.

Ogden, J., Owego, N. Y.; percussion period.

Ogden, W. E., Owego, N. Y.; percussion period.

Olmstead, Morgan L., Auburn, N. Y.; percussion period.

Ong, E., Philadelphia, Pa.; Revolutionary War. (Workman in factory of Peter de Haven.)

Orr, Hugh, Bridgewater, Mass., 1737-1798.

Overly, Peter, Kentucky; flint-lock period to percussion period.

Owen, Robert, Sauquoit, N. Y., present.

Palm, Isaac, Pennsylvania; flint-lock period.

Palm, Jacob, Lancaster, Pa., prior to 1768, when moved to Esopus, N. Y.

Palm, John, Lancaster Pa.; Kentucky rifles.

Palmer, unidentified, unlocated; flint-lock period.

Palmer, W. R., New York, N. Y.; middle 19th century.

"Palmetto Armory," Columbia, S. C. See Glaze, William.

Pannabacker (also spelled Pannebecker and Pennypacker), Jefferson;
    brother of John; Hopeland, Lancaster Co., Pa., about 1800.

Pannabacker, John, brother of Jefferson, Lancaster, Pa.; percussion
    period.

Pannabacker, S., Muddy Run, Lancaster Co., Pa., about 1780.

Pannabacker, William, Sr., Mohntown, Berks Co., Pa., about 1800.

Pannabacker, William, Jr., Mohntown, Pa., 1818-1880; also Trenton,
    N. J., 1860-1865.

Pannebecker (also spelled Pannabecker and Pennypacker), Jesse,
    Adamstown, Lancaster Co., Pa., and Elizabeth Township, Pa.,
    about 1820 to percussion period.

Parker Bros., Meriden, Conn., present.

Parker, H. & Co., Trenton, N. J.; percussion period.

Parker, Snow & Co.; Civil War arms.

Parrish, W. A., Pennsylvania; flint-lock period.

Parsons, Plattsburg, N. Y.; before and after 1860.

Patent Arms Mfg. Co., Paterson, N. J.; revolvers and revolving rifles
    and carbines; Colt's first manufacturing company.

Patt, C., Alma, Wis.; metallic cartridge period.

Paul, Andrew, Pennsylvania, 1831.

Pearson, James, Pennsylvania; Revolutionary War.

Pecare & Smith, New York, N. Y.; middle 19th century.

Peck, J., New Haven, Conn.; revolvers, prior to 1865.

Peck, J. C., Atlanta, Ga.; Civil War Confederate arms.

Penhallow, Capt. John, Boston, Mass., 1726.

Pennsylvania Rifle Works, Philadelphia, Pa.; percussion period. See
    Dunlap, G.

Perkins, James, Bridgewater, Mass., 1800.

Perkins, Joseph, Philadelphia, Pa., about 1800.

Perkins, Luke, Bridgewater, Mass., 1800.

Perkins, Rufus, Bridgewater, Mass., 1800 to Civil War period.

Perry, J. V., Jamestown, N. Y.; before and after 1840.

Perry Patent Arms Co., Newark, N. J., Alonzo D. Perry, proprietor;
    about 1850-1858.

Pessota, with J. Wurfflein, Philadelphia, Pa.; Kentucky rifles.

Petersen, A. W., Denver, Colo., present.

Philips, E., New York, N. Y.; percussion period.

Philpy, J., unlocated; Kentucky rifles.

Phips, James, Kennebec River, Mass., 1650. Father of Gov. Phipps of Massachusetts.

Pickle, Henry, Lancaster, Pa., 1802.

Pim, Boston, Mass., 1722. Made 11-shot repeater.

Pitcher, Neillsville, Wis., 1892.

Plant Mfg. Co., New Haven, Conn.; revolvers, prior to 1865.

P. M., S. J. III, I. M., S. C., unidentified maker(s) of Confederate Civil War arms.

Pomeroy (also written Pumeroy), Ebenezer, Northampton, Mass.; b. 1669; d. 1754; son of Medad.

Pomeroy (also written Pumeroy), Eldad, Boston, Mass., Northampton, Mass., and Hampshire, Mass., 1630-1662; son of "Elty."

Pomeroy (also written Pumeroy), commonly called "Elty" (full name appears in records as Eltwed, Eltwud, Eltweed and Eltwood), Boston, Mass., 1630; Dorchester, Mass., 1633-1637; Hartford and Windsor, Conn., 1637-1671.

Pomeroy (also written Pumeroy), Lemuel, Northampton, Mass., and Pittsfield, Mass., 1790-1840.

Pomeroy (also written Pumeroy), Medad, Northampton, Mass., 1659-1716; son of "Elty."

Pomeroy (also written Pumeroy), Gen. Seth., Northampton, Mass.; b. 1706; d. 1777; son of Ebenezer.

Pond, Lucius W., Worcester, Mass.; revolvers, prior to 1865.

Pope, Harry M., Newark, N. J., present.

Porter, P. W., New York, N. Y.; revolvers, prior to 1865.

Potter, Daniel, Hartford, Conn.; percussion period.

Prahl, Lewis, Philadelphia Co., Pa.; Revolutionary War.

Pratt, Alvan, Concord, Mass.; at least as early as 1835.

Pratt, Henry, Roxbury, Mass.; b. 1790; d. 1880.

Prescott, E. A., Worcester, Mass.; revolvers, prior to 1865.

Pringle, John, Pennsylvania; Revolutionary War.

Prissey, Elias, Hooversville, Pa.; b. Oct. 25, 1835; specialty, ornamented m. 1. squirrel rifles.

Providence Tool Co., Providence, R. I., about 1850-1917.

Puling, J., unlocated; Kentucky rifles.

Putnam, Enoch, Granby, Mass.; American Revolution.

Quackenbush, H. M., Defriet, N. Y.; metallic cartridge period.

Quinnepaug Rifle Co., Southbridge, Mass.; metallic cartridge period.

Radcliffe, T. W., Columbia, S. C.; Civil War Confederate arms.

Radfang (also appears as Rathfong, Redfan, Redfang, and Rothfong), George, Lancaster, Pa.; Revolutionary War period.

Ramsdell, G. V., Bucksport, Me.; percussion period.

Rappahannock Forge, Fredericksburg, Va.; Revolutionary War. Established by Virginia Assembly, 1775.

Raymond & Robitaille, Brooklyn, N. Y.; revolvers, prior to 1865.

Read, Robert, Chesterton, Md.; American Revolution.

Read, William, Boston, Mass.; percussion period.

Reasor, David, Lancaster, Pa., 1749.

Reed, Joseph, Lancaster, Pa., prior to 1783.

Reid, Templeton, Milledgeville, Ga., 1824.

Reiner, Michael De, Lancaster, Pa.; Revolutionary War.

Reising Mfg. Corp'n, New York, N. Y., modern automatic pistols.

Remington Arms Co., Inc., New York, N. Y.; Ilion, N. Y.; and Bridgeport, Conn.; from Eliphalet Remington, 1816, to present.

Remington, Eliphalet (Sr.), Ilion Gorge, N. Y.; began with son, 1816; flint-lock period to percussion period; d. 1828.

Remington, Eliphalet (Jr.), Ilion, N. Y.; d. 1861; originator of Remington gun-making; father of Philo, Samuel, and Eliphalet 3rd; began in 1816 with father, Ilion Gorge, N. Y.

Remington, Eliphalet (3rd); member of firm of E. Remington & Sons, Ilion, N. Y.; son of Eliphalet, Jr.

Remington, Philo, son of Eliphalet, Jr.; member of firm of E. Remington & Sons, Ilion, N. Y.

Remington, Samuel; son of Eliphalet, Jr.; member of firm of E. Remington & Sons, Ilion, N. Y.

Remington, E. & Son; Eliphalet and Son, Philo, 1845-1856; then name changed to E. Remington & Sons, Ilion, N. Y.

Remington, E. & Sons; Eliphalet Remington and sons, Philo, Eliphalet 3rd, and Samuel, Ilion, N. Y.; later Remington Arms Co.

Remington-U. M. C. Co., name of Remington organization after taking over Union Metallic Cartridge Co., and prior to taking name of Remington Arms Co., Inc.

Remmerer, David. This name thus exists in records, but is probably an erroneous form of writing Kemmerer, David, Carbon Co., Pa., or Kemmerer, David, Jr., location uncertain; both of percussion period.

Resor, J., unlocated; Kentucky rifles.

Reynolds, Lancaster, Pa.; flint-lock period.

Rice, Ralsa C., Ohio; b. 1838; d. 1911.

Richardson, Joel, Boston, Mass., 1816-1825.

Richardson, O. A., Lowell, Mass.; percussion period.

Richardson & Overman, Philadelphia, Pa.; Civil War arms.

Richmond Armory, Richmond, Va.; Civil War Confederate arms.

Rickmers, A. F., Cape Girardeau, Mo.; d. in '50's; Kentucky rifles; a pupil of the Hawkens' of St. Louis.

Rickmers, A. F., Kansas City, Mo., 1890-present; grandson of A. F. Rickmers of Cape Girardeau, Mo.

Ricks, Thomas, Boston, Mass., 1677.

Riddel; possibly same as Riddle; Lancaster, Pa., prior to 1783.

Riddle; possibly same as Riddel; Lancaster, Pa., prior to 1783.

Rider, Joseph, Newark, O.; revolvers, prior to 1865.

Rigert, or Reigart, Peter, Lancaster, Pa.; flint-lock period.

Riggins, Thomas, McMinn Co., Tenn., and Knoxville, Tenn.; b. 1821, living in 1910; Civil War Confederate arms.

Riggs, B., Bellows Falls, Vt.; before and after 1850.

Righter, J. G., Cadiz, O., 19th century.

Riner, Michael, Lancaster, Pa.; Kentucky rifles.

Ripley Bros., Windsor, Vt.; before and after 1835.

Rittenhouse, Benjamin, Norrington, Pa.; Revolutionary War.

Robbins & Lawrence, Windsor, Vt.; Mexican and Civil War arms.

Robbins, W. E., Manesburg, Pa.; percussion period.

Robbins, W. G., unidentified, unlocated; percussion period.

Roberts, W., Dansville, N. Y.; before and after 1850.

Robertson, Philadelphia, Pa.; Kentucky rifles.

Robinson, Philadelphia, Pa.; Kentucky rifles.

Robinson, Edward, New York, N. Y.; Civil War arms.

Robinson, S. G., also Robinson Arms Co., Richmond, Va.; Civil War Confederate arms.

Rocketer, J. H., Syracuse, N. Y.; before and after 1850.

Rock Island Arsenal, Rock Island, Ill., U. S. Government arms manufactured since 1843.

Roemer, O. E., unidentified, unlocated; percussion period.

Roesser (also Roeser), Peter, Lancaster, Pa.; before and after 1780.

Roesser (also Roeser), Matthew, or Mathias, Lancaster, Pa.; before and after 1744.

Roger, J., Highland, Ill.; percussion period.

Rogers, R., California; percussion period.

Rogers, Spencer & Co., Williow Vale, N. Y.; revolvers, prior to 1865.

Roop, John, Lehigh District and Allentown, Pa.; before and after 1775.

Roper Repeating Arms Co., Amherst, Mass.; metallic cartridge period, about 1866-1870.

Roper Repeating Rifle Co.; same as Roper Repeating Arms Co., Amherst, Mass. This name said to have been occasionally used.

Roth, Charles, Wilkes-Barre, Pa.; percussion period.

Roth, Henry, Wilkes-Barre, Pa., 19th century.

Rowe, A. H., Hartford, Conn., 1864.

Rowell, Horace, Sawmill Flats, Calif.; metallic cartridge period.

Rugart, Peter, Lancaster, Pa., prior to 1783.

Ruggles, Stafford Hollow, Conn.; percussion period.

Rupp, J., Pennsylvania; flint-lock period.

Ruppert, William, Lancaster, Pa.; before and after 1777.

Rusily, Jacob, Lancaster, Pa.; Kentucky rifles.

Ruslin, Jacob, unidentified, unlocated; flint-lock period.

Rynes, Michael, Pequa Creek, Lancaster Co., Pa., 1775-1783.

Sargent & Smith, Newburyport, Mass.; percussion period.

Sarson & Roberts, unlocated; Civil War arms.

Savage Arms Co., Utica, N. Y., present.

Savage Revolving Fire Arms Co., Middletown, Conn.; revolvers, prior to 1865.

Schaeffer, unidentified, unlocated; flint-lock period.

Schaefer, William R., Boston, Mass., began business in 1853; living, 1917.

Schaefer (William) & Werner, Boston, Mass., about 1860-1870.

Schalk, Chris., Williamsport (?) Pa., about 1825-1875.

Schalk (also Schalch), George, associated with Thomas P. Cherington, Pottsville, Pa., middle 19th century.

Schenkl, J. P., Boston, Mass.; before and after 1850.

Schmelzer Arms Co., Kansas City, Mo., 1880 to present.

Schmelzer, J. H., Leavenworth, Kansas; percussion period.

Schnaut, T. G., Monmouth, N. J.; flint-lock period; d. 1838.

Schneelock, Brooklyn, N. Y.; percussion period.

Scho'b, I., Pennsylvania, 1780. (Usual mark: I. S.)

Schorer, Andrew, Bethlehem Township, Pa., prior to 1783.

Schoyen, George, New York, present.

Schridt (also Schreidt), John, Reading, Pa.; before and after 1758.

Schubarth, C. D., unlocated; Civil War arms.
Schuler, John, Liverpool, Perry Co., Pa.; Kentucky rifles.
Schull, M., unidentified, unlocated; flint-lock period.
Schultz, unidentified, unlocated; flint-lock period.
Scrivener, James A., Auburn, N. Y.; percussion period.
Seaver, Vergennes, Vermont; percussion period.
Seibert, Columbus, O.; percussion period.
Sell, Frederick, Maryland; Kentucky rifles.
Senseney, J., Chambersburg, Pa.; before and after 1855.
Sever, Joseph, Framingham, Mass.; American Revolution.
Sever, Shubabel, Framingham, Mass.; American Revolution.
Seward, Benjamin, Boston, Mass., 1796-1803.
Shafer, Joseph, unlocated; Kentucky rifles.
Shakanoosa Arms Mfg. Co., unlocated; Civil War Confederate arms.
Shannon, W. & H., Pennsylvania; Civil War arms.
Shaw, Massachusetts; American Revolution.
Shaw, Joshua, Bordentown, N. J., after 1800; also Philadelphia, Pa.
    Invented percussion cap.
Sharps, C. (Christian) & Co., Philadelphia, Pa.; rifles and revolvers,
    the latter prior to 1865.
Sharps & Hankins, Philadelphia, Pa.; Civil War arms; in business until
    1880.
Sharps (Christian) Rifle Mfg. Co., Hartford, Conn., 1848-1871;
    Bridgeport, Conn., 1871-October, 1881.
Sheesley, George, Hartley Township, Union Co., Pa., 19th century.
Sheets, M., Virginia; Kentucky rifles.
Sheffield, Jeremiah, 1775.
Shell, John, Liverpool, Pa.; flint-lock to percussion period.
Shell, John, Leslie Co., Ky., 19th century.
Shell, M., Pennsylvania; flint-lock period.
Shell, N., Pennsylvania; flint-lock period.
Shell, Samuel, Kentucky; Kentucky rifles.
Sherman, Nathaniel, Boston, Mass., 1692.
Shields, D., unidentified, unlocated; percussion period.
Shirley, Jeremiah, Cloverdale, O.; before and after 1870.
Shuler, John, Liverpool, Pa.; percussion period.
Shultz, H., Pennsylvania; Kentucky rifles.
Sieber, C. C., Columbus, O.; percussion period.
Siebert, H. L., Cincinnati, O.; percussion period.

Siner, L. K., Philadelphia, Pa. (successor to John Krider), present.

Slocum, Harding, Worcester, Mass,; began 1820.

Slocumb, Samuel D., New Orleans, La.; Kentucky rifles.

Slotterbeck, Charles, Sr., San Francisco, Cal.; percussion period.

Slotterbeck, Chas., Jr., California; metallic cartridge period.

Smart, Eugene, Dover, N. H., about 1865-1890.

Smith, Anthony, Bethlehem Township, Pa., prior to 1783.

Smith, H., Norwich, Conn.; Civil War arms.

Smith, Horace, Springfield, Mass.; b. 1808. After 1856, Smith & Wesson.

Smith, Jeremiah, Lime Rock, R. I., 1770.

Smith, J. F., Huntingdon, Pa.; flint-lock period.

Smith, Johnson, Lehigh District, Pa., and Easton, Pa., associated with John Young, 1776.

Smith, Martin, Greenfield, Mass.; before and after 1830.

Smith, M., unidentified, unlocated; flint-lock period.

Smith, Obadiah, Brunswick Co., Va.; flint-lock period up to about 1820.

Smith, P., Buffalo, N. Y.; percussion period.

Smith, Stoeffel, Pennsylvania; before and after 1794.

Smith & Wesson, Norwich, Conn., 1849-1856; Springfield, Mass., 1856 to present time; revolvers, also pistols and automatic pistols.

Sneider, Anthony, Lancaster, Pa., prior to 1783.

Sneider, T., unlocated; Kentucky rifles.

Snell, Chauncey (son of Elijah), Auburn, N. Y., about 1830-1860.

Snell, Elijah, Auburn, N. Y.; flint-lock period; d. 1834.

Snyder, Ira, Woodward, Pa., 19th century.

Snyder, John, Henry, George, Adam, Providence Township, Lancaster Co., Pa.; several generations of the same family; Kentucky rifles.

Soleil, Francis, New Amsterdam, N. Y., 1656.

Soubie, Armand, New Orleans, La.; percussion period.

Spang & Wallace, Philadelphia, Pa.; flint-lock period.

Spangler, G., Pennsylvania; Kentucky rifles.

Specht, Eley, Beavertown, Snyder Co., Pa.; Kentucky rifles.

Speed, Robert, Boston, Mass., after 1818.

Spellier, A., Philadelphia, Pa.; revolvers, prior to 1865.

Spencer, Dwight, West Hartford, Conn.; middle 19th century.

Spencer Repeating Arms Co., Boston, Mass.; ceased about 1870.

Spies, A. W., New York, N. Y.; b. about 1783; d. about 1863.

Spiller & Burr, Macon, Ga.; Civil War Confederate arms.

Spitzer, Sr. and Jr., Newmarket, Va., prior to 1783.

Sprague & Marston, New York, N. Y., pepper-boxes, 1849.

Springfield Arms Co., Springfield, Mass.; revolvers, prior to 1865.

Springfield Armory, Springfield, Mass.; U. S. Government arms manufactured since 1795.

Stahl, C., Pennsylvania; Kentucky rifles.

Stalter, William, Logan, O.; before and after 1870.

Standard Arms Co.; modern, defunct.

Starr, Lancaster, Pa.; before 1783 and after 1800.

Starr Arms Co., Yonkers, N. Y.; Civil War arms.

Starr, Nathan, Middletown, Conn., 1790 to 1850.

St. Clair, S. H., Pennsylvania; flint-lock period.

State, J. S., New Haven, Conn.; Civil War arms.

State Rifle Works, Greenville, S. C.; Civil War Confederate arms.

Stenzel, or Stengel, Lancaster, Pa.; flint-lock period.

Sterewith, Maryland; American Revolution.

Stevens, J., Arms & Tool Co., Chicopee Falls, Mass., present; now part of the Savage Arms Co., Utica, N. Y.

Stillman, Ethan, Brookfield, Conn., 1800 to 1812.

Stinger, Thomas, Antony Township, Lycoming Co., Pa., and Jersey Shore, Pa.; before and after 1840.

Stuart, Charles, Binghamton, N. Y., about 1850-1870.

Sunderland (also Sunderland & Blair), Boulton, Bethlehem District, Pa.; Kentucky rifles.

Sutherland, Samuel, Richmond, Va.; Civil War Confederate arms.

Sweet, E. S., Kalamazoo, Mich.; percussion period.

Sweet, Jenks & Sons, Rhode Island; Civil War arms.

Switzer, Daniel & Co., Lancaster Co., Pa.; Kentucky rifles.

Taylor, Argulus, Ira, N. Y.; percussion period.

Taylor, George, Easton, Pa., prior to 1783. With Durham Iron Works.

Teaff, Nimrod, Steubenville, O.; Kentucky rifles, before and after 1847.

Teff, George, Rhode Island, 1775.

Tell, Frederick, Pennsylvania; flint-lock period.

Thomas, Benjamin, Hingham, Mass.; before and after 1845.

Thompson, John, Philadelphia, Pa., about 1800.

Thorsen, Thomas M., Philadelphia, Pa.; metallic cartridge period.

Three-Barrel Gun Co., Moundville, W. Va., recent years; defunct.

Tidd, Marshall, Woburn, Mass.; d. about 1890.

Todd, George, Austin, Texas; Civil War Confederate arms.

Tomes, Henry & Co., New York, N. Y.; before and after 1845.

Tomlinson, Pennsylvania; Revolutionary War.

Tonks, Joseph, 49 Union St., Boston, Mass.; before and after 1860.

Town, Benjamin, Pennsylvania; Revolutionary War.

Trant, Geo. B., Thornville, O.; before and after 1880.

Trenton Arms Co., Trenton, N. J.; Civil War arms.

Tryon (Geo.) & Getz, Philadelphia, Pa., 1811 only.

Tryon, Geo. W., Philadelphia, Pa., 1836-1843; also & Co.; also Son & Co. Predecessor of E. K. Tryon & Co.

Trout, John, Williamsport, Pa., 19th century.

Tyler Arsenal, Tyler, Texas; Civil War Confederate arms.

Tyler, John, Easton and Allentown, Pa., prior to 1783; Kentucky rifles.

Union Arms Co., location unknown; revolvers, prior to 1865.

Union Mfg. Co., Richmond, Va.; Civil War Confederate arms.

Underwood, Thomas, Lafayette, Ind.; percussion period.

Updegraph, Jacob, Schuylkill Co., Pa.; Kentucky rifles.

Vallee, Prosper, Philadelphia, Pa.; before and after 1840.

Valley Forge Armory, Pennsylvania, first called Mount Joy, established by Stephen Evans, Daniel Walker and Joseph Williams, 1742; Revolutionary War; operated until 1842.

Valsey & Son, location unknown; revolvers, prior to 1865.

Vanderheyden, John, Auburn, N. Y.; percussion period.

Vander Poel, Albany, N. Y., 1740.

Vanderslice, T., Pennsylvania; Kentucky rifles.

Varney, David M., Burlington, Vt.; before and after 1850.

Velee, Philadelphia, Pa., 1826.

Virginia Manufactory, Richmond, Va.; Civil War Confederate arms.

Virginia Manufactory, Richmond, Va., 1797-1865; Provincial, then State.

Volcanic Repeating Arms Co., New Haven, Conn., 1855-1857; later absorbed into the Winchester Repeating Arms Co.

Volvert, or Volkert, Lancaster, Pa.; flint-lock period.

Vondegrift, John, Bucks Co., Pa., during Revolutionary War. Possibly same as Fondegrift.

Vondersmith, Lancaster, Pa., prior to 1783. So of record; possibly same as John Fondersmith & Son, Lancaster, Pa., probably 1749-1802.

Wagonhorst, unlocated; 19th century.

Walch Fire Arms Co., New York, N. Y.; revolvers, prior to 1865.
Waldren, Alexsander, Piscataqua River, Mass., 1672.
Waldren, William, Boston, Mass., 1671.
Walker, Daniel, partner with Stephen Evans and Joseph Williams, Mount Joy Armory, later called Valley Forge, Pa., established 1742.
Wallace (& Spang), Philadelphia, Pa.; flint-lock period.
Wallach, Moses A., Boston, Mass., 1800 to about 1825.
Walters, A., New York, N. Y.; before and after 1822.
Ware, J. S., Worcester, Mass.; percussion period.
Ware, Orlando, Worcester, Mass.; percussion period.
Ware & Morse (Jos. S., John R.), Worcester, Mass., began 1833.
Ware & Wheelock, Worcester, Mass., began 1825.
Warner, Horace, Williamsport, Pa., about 1890.
Warner, Horace, Ridgeway, Pa.; percussion period.
Warner (Horace) & Lowe (Wm. V.), Syracuse, N. Y.; before and after 1880.
Warner, James, Springfield, Mass.; b. about 1818; d. about 1870.
Washington Arms Co., location unknown; rifles and revolvers, prior to 1865.
Waters, Dutchess Co., N. Y.; American Revolution.
Waters, Andrus, Sutton, Mass., about 1776; d. 1778.
Waters, Asa, Sutton, Mass., about 1776.
Waters, Asa H., Sutton, Mass., 1789-1841. Son of Asa.
Waters, A. (Asa, father of Asa H.) & Co., Sutton, Mass., and Milbury, Mass.; percussion period.
Waters, A. & Son (Asa and Asa H.), Sutton, Mass., and Milbury, Mass.; percussion period.
Waters, Elijah, brother of Asa 2nd.
Waters, Richard, Salem, Mass., 1632.
Waters & Whitmore, Massachusetts; Civil War arms.
Watson, Jonathan, Chester, N. H.; before and after 1800.
Weaver, Crypret; Pennsylvania; before and after 1818.
Weaver, H. B., South Windham, Conn.; percussion period.
Webber Arms Co., Denver, Colo.; metallic cartridge period.
Weiss, G., Pennsylvania; percussion period.
Welch, W. W., unlocated; Civil War arms.
Welshens, J., unlocated; Kentucky rifles.
Welton, Lieut. Ard, Waterbury, Conn.; American Revolution.
Wesle, N., Milwaukee, Wis.; percussion period.

Wesson, Daniel B., Groton, Mass., Worcester, Mass., and Taunton, Mass., to 1856; after 1856, Springfield, Mass., firm of Smith & Wesson.

Wesson, Edward (cousin of Daniel, Edwin and Frank), Northboro, Mass.; before and after 1840.

Wesson, Edwin (brother of Frank and Daniel, cousin of Edward), Hartford, Conn., and Northboro, Mass.; percussion period.

Wesson, Frank (brother of Edwin and Daniel, cousin of Edward), Worcester, Mass.; percussion period.

Wesson (Edwin), Stevens (Joshua) and Miller (S. C.), Hartford, Conn., about 1848-1855.

Westinghouse Mfg. Co., Chicopee Falls, Mass.; World War rifles.

Wetzel, J., Pennsylvania; flint-lock period.

Whall, William, Jr., Boston, Mass., 1813-1819.

W. G. M., unidentified, unlocated; Kentucky rifles.

Wheeler & Morrison, Virginia; Civil War arms.

Whetcroft, William, Annapolis, Md.; American Revolution.

White, Horace, Springfield, Mass.; American Revolution.

White, Peter, unlocated; Kentucky rifles.

Whitesides, John M., Tennessee; Kentucky rifles.

Whitmore, Andrew E., Somerville, Mass.; before and after 1860.

Whitmore, H. G., Boston, Mass.; after about 1853.

Whitmore, Nathan, Marshfield, Mass.; about 1825-1880.

Whitney, Eli; b. 1765; Whitneyville (later part of New Haven), Conn.; flint-lock to percussion period; rifles, rifled muskets, pistols, revolvers.

Whittemore, Anos, Boston, Mass., 1775-1800—.

Whittemore, D., Cambridge, Mass.; before and after 1860.

Wickham, Marine T., Philadelphia, Pa.; flint-lock period.

Wigfal, Samuel, Philadelphia, Pa.; Revolutionary War.

Wiley, Pennsylvania; Revolutionary War.

Wilkinson, J. D., Plattsburg, N. Y., after 1866.

Williams, Abe, Covington, Ky.; percussion period.

Willis, John, Pennsylvania; Revolutionary War.

Williams, Joseph, partner with Stephen Evans and Daniel Walker, Mount Joy Armory, later called Valley Forge, Pa.; established 1742.

Wilmot, N. N., Boston, Mass.; percussion period.

Winchester Repeating Arms Co., New Haven, Conn., 1866-present.

Winner, Nippes & Co., Pennsylvania; Civil War arms.

Winters, Elisha, Maryland; American Revolution.
Withers, Michael, Lancaster, Pa., prior to 1783.
Wolf, New York, N. Y.; flint-lock period.
Wolfheimer, Philip, unlocated; flint-lock period.
Wood, B. C., Painted Post, N. Y.; percussion period.
Wood, John, Boston and Roxbury, Mass.; American Revolution.
Wood, Josiah, Norrington Township, Pa.; Revolutionary War.
Wood, Luke, Sutton, Mass.; flint-lock period.
Woods, Robert, Pennsylvania, 1800.
Workmen, J., Pennsylvania; Kentucky rifles.
Worl, H., Pennsylvania; flint-lock period.
Worly, J., unlocated; Kentucky rifles.
Worthen, Barney, San Francisco; percussion period.
Wosnek, Ulrich, Milwaukee, Wis.; now Antigo, Wis.; 1908-present.
Wright, J., unidentified, unlocated; before and after 1815.
Wrisley, Loren H., Norway, Me., after 1834.
Wundhammer, Ludwig, Los Angeles, Calif., present.
Wurfflein, Andrew, Philadelphia, Pa.; Kentucky rifles.
Wurfflein (J.) & Pessota, Philadelphia, Pa.; Kentucky rifles.
Wurfflein, William, Philadelphia, Pa., 1875-1910.

Yesley, H., unidentified, unlocated; percussion period.
Yochum, D., Pennsylvania; Kentucky rifles.
Yost, Maryland; Kentucky rifles.
Youmans, Lancaster, Pa.; flint-lock period.
Youmans, also Yomens (various members of family of this name),
    Charlotte, N. C., 18th and probably 19th centuries.
Young, D., Middleburg, Snyder Co., Pa.; Kentucky rifles.
Young, Henry, Easton, Pa., prior to 1783.
Young, John, Lehigh District, Pa., and Easton, Pa., before and after
    1776; also associated with Johnson Smith & A. Foilke.
Young, Joseph, Haynes Run, W. Va.; Kentucky rifles.
Young, P., Pennsylvania, prior to 1783.
Yost, John, Maryland; American Revolution.

Zettler Bros., New York, N. Y., about 1880-1890.
Zimmerman, Pennsylvania; Kentucky rifles.
Zischang, Syracuse, N. Y.; late metallic cartridge period.

APPENDIX III

GLOSSARY OF ARMS TERMS

AMMUNITION, see range.

ARQUEBUS, early name for military musket.

ARQUEBUSIER (Fr.), a gunsmith or gun-maker; early name for foot-soldier armed with military musket.

AUTOMATIC (1), self-operating mechanism effecting successive discharges of a firearm. (2) Common name for any self-operating repeating firearm, usually applied to an arm held and fired in one hand. *Semi-automatic,* a self-operating mechanism whereby after each shot the firearm is automatically reloaded, but requires a pull of the trigger for each individual shot. The mechanism performs only the operation of reloading and does not fire the arm.

BALLISTICS, (1) the science of the performance of projectiles; (2) characteristics of the flight or motion of a projectile, namely, its energy, velocity, range, penetration, etc. *Exterior b.,* ballistics from muzzle to target. *Interior b.,* ballistics behind projectile while it is still within the gun barrel.

BANDOLIER, a belt worn on the person and carrying a number of separate charges or cartridges for a firearm.

BANDS, strips of metal or other material encircling the barrel and stock of a firearm and holding these together.

BARLEYCORN, a block-shaped front sight, or metal support of a front sight.

BATTERY (1), a form of fortification; (2), a division in artillery; (3), a frizzen.

BAYONET, a cutting or thrusting weapon affixed to a firearm.

BEAD, a small front sight which appears round as viewed in taking aim. *"To draw a bead,"* vernacular for the act of aiming.

BLADE, the front sight of a firearm when consisting of an upright piece of thin metal.

BLUING, the coloring of the metallic portions of any firearm in various shades of blue, black, brown, etc., by a process of artificially-induced rusting, checked when the desired color is secured.

BLUNDERBUSS, early form of smooth-bore gun with round or oval, bell-muzzle designed to scatter a charge of shot.

BOLT (1), that portion of the locking-mechanism of a breech-loading firearm designed to prevent the opening of the mechanism on discharge. (2) The breech-block portion of the mechanism of a firearm having a motion to and fro in line with the bore for the purpose of opening or closing the breech. It may or may not have a rotary motion also, for the purpose of locking the mechanism. *Bolt-action,* the mechanism of a firearm in which the opening and closing of the breech is effected by a bolt.

BORE, the interior of the barrel of a firearm.

BOW, a weapon using the elastic or reflex power of wood or steel to propel a projectile.

BREECH, the rear portion of the barrel and accompanying mechanism of a firearm. *Standing b.,* the upright face of the mechanism of a breech-loading firearm which supports the rear of the cartridge and away from which the barrel or barrels of the piece are moved for loading.

BREECH-BLOCK, the portion of the mechanism of a firearm supporting the charge at the rear and designed to receive the rearward effect of the discharge.

BREECH-BOLT, a bolt-shaped breech-block.

BREECH-PRESSURE, the pressure against the walls of the chamber produced by the expansion of the gases of an exploding charge. Usually measured in tons per square inch of chamber surface.

BROWNING, same as bluing.

BULL'S-EYE (1), the central portion of a target, usually indicated by being of a different color or shape from the portions surrounding it. (2) Vernacular, a direct hit.

BUTT, (1) the rearward portion of the stock of a firearm; that portion of the stock of a gun which is placed against the shoulder. (2) the object, material, etc., used to support a target and to stop projectiles.

BUTT-PLATE, the metal or other material used to cover and protect the butt.

CALIBER, CALIBRE, the diameter of the bore of a firearm. On the next page we have inserted a table for converting calibres and gauges into millimeters.

Caliber (Cont.)

TABLE OF FIREARM BORE DIAMETERS

Caliber	Gauge	Millimeters	Caliber	Gauge	Millimeters
.835	8	21.8	.453	50	11.45
.805	9	20.9	.450		11.43
.775	10	20.0	.440		11.17
.752	11	19.3	.433		11.00
.729	12	18.6	.410		10.41
.711	13	18.2	.393		10.00
.693	14	17.8	.380		9.65
.677	15	17.3	.360		9.14
.662	16	16.8	.354		9.00
.650	17	16.5	.320		8.12
.638	18	16.2	.314		8.00
.626	19	15.9	.310		7.87
.615	20	15.6	.303		7.69
.577	24	14.6	.301		7.65
.571	25	14.4	.300		7.62
.537	30	13.6	.297		7.54
.526	32	13.2	.275		6.98
.500	35	12.7	.236		5.99
.488	40	12.4	.220		5.58

(In the Gauge column of the right-hand group, spanning the rows: "Shotguns of smaller bore are very unusual.")

(1)   Rifle calibres are expressed in decimals of an inch.

(2)   The so-called "gauge" of a firearm describes the diameter of the bore as stated in terms of the number of balls of pure lead which, if of a diameter equal to that of the bore, would weigh one pound. Thus, a "12-gauge" gun is one of whose bore-diameter balls, 12 would weigh one pound. This is the earliest standard description of the bore or caliber of firearms, dating at least to the 16th century, and probably to the 15th.

Caliber (Cont.)

Exact conformity by any firearm of its projectile to the *nominal* caliber or gauge claimed for either, is rare. Many stated calibers or gauges of either bores or projectiles are designated only by the numerical name of the popularly-known "standard" figure most nearly approximated by the actual measurements of the bore, ball, bullet, etc.

This is equally true, though for different reasons, of arms both ancient and modern. Of modern and semi-modern arms it is true because the actual diameter of the bullet has been determined by (1) the quantity or weight of lead or other projectile-metal, and (2) the shape of the bullet, found to combine to give the best results at given or desired ranges or velocities, with (a) the predetermined, or (b) the required amount of explosive. These factors having determined the actual diameter of the projectile—and, of course, that of the bore of the weapon in which it is to be fired—it is announced, stated, advertised, etc., as nominally of that standard, accepted and more-or-less-generally-understood "caliber" or "gauge" which its actual size most nearly approximates; this for the purpose of avoiding, as far as possible, the confusing of inexpert minds concerning it. Thus, any two rifles, revolvers, pistols, etc., of nominally the same "caliber" may yet differ in actual size of bullets and bores by many thousandths of an inch, so that the bullet used in the one may perhaps not even enter the bore of the other; or, again, may drop through it without touching the sides of the tube.

Of ancient, "antique" and similar arms of the muzzle-loading period the same is true, for two reasons. The first is naturally that accuracy in barrel-boring was formerly vastly less; less possible with the ruder tools and methods used, and even sometimes hardly considered by ancient makers—certainly in any modern sense and appreciation of the fit of a barrel and its bullet. A second reason is that in many arms—for instance, the American "Kentucky" rifles—the bore of the arm, though nominally of a certain specified or required size, "caliber" or "gauge" (to which size the ball itself was made), was actually bored several hundredths of an inch larger in order to allow for the use of the "patch" of linen, paper or thin leather in which the ball was wrapped when placed in the barrel. This explains the fact that 99 in every 100 rifles of the mentioned muzzle-loading type, though nominally of certain calibers, or of bore-diameter designed to accommodate a ball of a certain number-per-pound (i.e., "gauge" measurement), each will be found on measurement to be actually several hundredths of an inch larger in

Caliber (Cont.)

bore-diameter than required for the stipulated ball-size; the explanation being, as mentioned, in the allowance for the use of the "patch."

A similar discrepancy is almost invariably found between the nominal "gauge" or bore-diameter of shotguns, and the actual diameter of the barrel as ascertained by measurement. Here it is largely due to the presence or absence, and the degree, of the "choke" given the interior of the bore. Thus, two nominally "twelve-gauge" shotguns may be found to have markedly different muzzle-diameters, or other differences of interior bore-diameter at other points within the barrel, according as one or the other is more or less "choked." This is so generally the case that makers of round balls for use in smooth-bores, or of moulds for making such balls, always make these in sizes up to as much as eight one-hundredths of an inch smaller than their nominal gauge-diameter, to allow for such variations in the bore of guns in which they may be used.

P. B. J.

CANNELURE (Fr.), (1) Sunken rings around a bullet, produced in casting or by swaging, designed to hold the lubricant or to receive the compressed mouth of a metallic cartridge and so to retain the bullet. (2) A channel or groove.

CANNON, a firearm of such size as requires transportation other than on the person. *Hand-cannon,* early name for the first guns.

CAP, (1) a small device of metal, paper, etc., containing a percussion-ignited compound for firing the explosive of a firearm and designed to retain such compound in position for firing. See percussion-cap. (2) a cover or protection of any portion of a firearm, as the grip-cap, butt-cap, etc.

CARBINE, a shortened form of long arm, originally designed for mounted men.

CARTRIDGE, container for a charge of explosive (may or may not include ball), and designed for placing in the chamber of the piece. N.B. The early form was not so designed but was broken and the contents emptied into the chamber.

CATCH, a portion of the mechanism of a firearm designed to hold any other portion in a desired position, as a bayonet-catch, locking-catch, etc.

CHAMBER, the portion of the bore of a firearm in which the charge is placed for firing.

CHAMBER-PRESSURE, the pressure against the sides of the chamber produced by the expansion of the gases of an exploding charge.

CHECKERING, also CHECKING, roughening of the wood of the stock or grip for ornamental purposes, and making the grip by the hand or hands more secure.

CHEEK-PIECE, a portion of the stock of a gun designed to support the cheek and face, and, therefore, the eyes, of the user in a desired position for aiming.

CHOKE, a narrowing or constriction of the bore of a gun for the purpose of keeping the discharged projectiles closer together in flight.

CHOKE-BORE, the bore of a firearm when narrowed or constricted for any portion of its length.

CLIP, a device for holding a number of cartridges together for placing simultaneously in the magazine of a firearm.

COCK, that external movable portion of the firing-mechanism of a firearm whose function it is to ignite the priming of the explosive. (verb), the action of placing the movable ignition-producing portion of the mechanism of a firearm in position for firing the piece.

COMB, the top of the stock of a firearm when in the firing-position at the shoulder and against which the face of the user rests in aiming.

CONE, the bore of a shotgun immediately in front of the chamber.

CORD, same as match.

CORDITE, a form of smokeless powder, so called from being made in strips instead of grains.

CORNED POWDER, early name for gunpowder when in the form of larger grains than the earlier dust-like product afforded.

CRIMP, the constriction of the mouth of a cartridge for retaining the contained charge. (verb), the act of constricting the cartridge of a breech-loading firearm to retain the charge.

CROSS-BOW, a bow which is affixed to a stock when in use.

CYLINDER, a magazine of a firearm holding a number of cartridges and having a rotary motion about an axis, designed to present its contained cartridges successively in position for firing.

DAG, an early name for a short pistol.

DAMASCUS, a combination of strips of iron and steel welded together and twisted so as to afford a "pattern" of their contrasting-colored metals on the surface, used in gun-barrels. Differs from "laminated" and "twist" materials in the proportions of the metals and the fineness and degree of twist.

DETONATING, an early name for percussion ignition.

DETONATOR, an early name for a firearm using the percussion-method of ignition.

DOUBLE-BARRELED, a firearm having two parallel barrels.

DOUBLE-TRIGGER, see trigger.

EJECTOR (1), a mechanism which withdraws from the chamber and throws the fired cartridge clear of the piece. (2) A term or name for a firearm employing the above mechanism.

ELLIPTICAL, any portion of the bore of a firearm whose cross-section at that point is oval instead of round.

EXPRESS, used to describe a rifle, bullet, etc. This word is the surviving portion of the name "Express Train," devised in 1856 by a London gunmaker, James Purdey, as a trade-mark or trade-name, to characterize a model of a rifle brought out by him at that time, having an unusually long so-called "point-blank" range and low trajectory. The arm meeting with popular favor, the name shortened to "Express," has since then been frequently used to describe any rifle possessing similar features in a marked degree. *Express bullet,* a projectile used in an express rifle.

FIRING-PIN, that part of the mechanism of a firearm which, by delivering a blow upon the cap or primer, ignites the contents of the same, effecting the discharge. It may or may not be a part of the hammer, but is usually used to describe the rod intervening between the hammer and the primer, delivering the blow of the former upon the latter.

FINE SIGHT, TAKING, the act of aiming a firearm, in which the user sees only the smallest discernible portion of the front sight in the notch, or over the bar, of a notched or bar-shaped rear sight.

FLINT, a form of silica, possessing the properties of hardness, of fracturing to a sharp edge, and of producing sparks when such an edge is struck on iron or steel, used accordingly to effect the discharge of the firearm known as the flint-lock by delivering such sparks into the powder.

FLINT-AND-STEEL devices, of remotely ancient origin and use and
continuing to modern times, for effecting the sudden contact of
flint with iron or steel, the resultant sparks igniting combustible
materials.

FLINT-LOCK, (1) a device for igniting the charge of a firearm by
means of the sparks produced by the contact of a piece of flint
upon iron or steel. (2) A firearm equipped with a flint-lock for
the purpose of igniting its charge.

FORE-ARM, or FORE-END, that portion of the wooden stock of a
firearm extending forwardly under the barrel in front of the trigger-guard.

FORE-STOCK, is usually used to describe the forward extension of
the wooden stock of a musket-type of gun, i.e., one having a wooden
stock extending beneath the entire length, or nearly so, of the barrel.

FORK, a support used in holding, for aiming, firing, etc., the early
military muskets and other guns, whose weight and the slow operation of their ignition-mechanism, rendered desirable such assistance
in holding them in the desired firing-position. It took the form
of a rod, stick or other material, one end of which was often
pointed for placing on the ground, the other end, at a length suitable for holding the weapon in a horizontal position, had a notch
or crotch in which the forward portion of the weapon was placed
for support in the acts of aiming and firing. Also called a rest.

FRIZZEN, the upright iron or steel portion of a flint-lock mechanism
against which the flint strikes to produce the sparks.

FRONT SIGHT, the device affixed to the muzzle-end of a firearm for
use in aligning the arm with an object aimed at.

FULMINATE, an explosive substance or compound having the
property of igniting on receiving a blow.

FULL SIGHT, TAKING, the act of aiming a firearm, in which the
user sees a considerable portion of the front sight in the notch, or
over the bar, of a notched or bar-shaped rear sight.

GAUGE, the diameter of the bore of a firearm expressed in the number of balls of a corresponding size required to make a pound;
thus, a "12-gauge" arm is one, the diameter of whose bore is such
that 12 balls of lead, each fitting the bore, will weigh one pound.
An early form of indicating the size of the bore of a firearm. Still
in use in England, etc., for describing the bore-size of large-caliber

rifles as well as shotguns, though in America and elsewhere the use of the term is usually restricted to describing the size of the bore of shotguns. The word "bore" is often used as synonymous with "gauge," as a "12-bore" rifle or shotgun, etc. See note under caliber.

GILDING-METAL, a metallic compound or alloy having a lustrous yellow "gilt" or "gilded" appearance and coming into recent use for the "jacket" of modern bullets by way of securing the various advantages in avoiding direct contact of the lead of a projectile with the iron or steel of the bore of a rifled arm.

GRIP, that portion of the stock of a gun held by the right hand in aiming the weapon; or, in a weapon held in one hand in use, that portion so held. *Pistol g.,* that portion of the stock of a gun held by the right hand when in use, when given somewhat the shape of the handle or butt of a pistol.

GRIP-CAP, a cover, protection, etc., of metal or other material, often applied to the "grip" of a firearm when such grip assumes a "pistol-grip" shape.

GROOVES, the continuous lengthwise parallel hollows cut in the bore of a firearm for the purpose of imparting a rotary motion to a projectile coming in contact with such grooving in its passage through the barrel. In certain early firearms such grooves were straight, instead of turning around the axis of the bore, it being (apparently) believed at the time that such grooving tended to impart a direct line of flight to the projectile.

GUARD, that which protects or encloses. *Hand g.,* that portion of a firearm designed to protect the hand of the user from heat generated in the barrel, as by rapid firing. *Trigger g.,* that portion of a firearm protecting the firing-mechanism from being actuated otherwise than by the contact of the finger of the user with the trigger.

GUN, GONNE and GUNNE, early forms, (1) a firearm of any kind, (2) usually restricted in England to describe a shotgun as distinct from a rifle, etc., (3) occasionally, but rarely, used in mediaeval literature to describe the projectile as well as the arm from which it was discharged.

HAIR-TRIGGER, a device capable of being so set or adjusted, as to permit of the operation of the firing-mechanism of a firearm by

an exceedingly light touch of the finger on the trigger. Such a
device usually consists of a second trigger, pressure upon which
"sets" the firing-mechanism in position, and the firing-trigger,
which does the actual firing, upon a subsequent light touch. It
may, however, consist of a screw or other mechanism for effecting
a similar adjustment, and may be so arranged that its operation
is either requisite or optional for the firing of the arm. The name
is derived from the old popular idea that the degree of touch upon
the firing-trigger required to operate it is so slight as to allow of
its operation by the contact of a "hair."

HAMMER, that portion of the mechanism of a firearm whose fall or
blow either directly or indirectly effects the discharge of the piece.

HAND-GUN, early name for a firearm held in the hands in use.

HARQUEBUS, see arquebus.

HARQUEBUSIER, see arquebusier.

HEEL, that part of the butt or shoulder-portion of the stock of a gun
which, when in position at the shoulder, is uppermost.

HIGH-POWER, a name coming into popular use during the decade
1890-1900 to describe the then new military and sporting rifles
characterized by using a long, small-caliber bullet driven by a large
charge of smokeless powder, as contrasted with the larger-bore arms
using black powder and lead bullets, which the new weapons superseded.

HOLLOW-POINT, of a bullet; it has in the forward end or point a
hollow, usually cylindrically-shaped, of greater or less depth, for
the purpose of expanding or "mushrooming" the forward portion
of the bullet upon impact, by the accompanying expansion or resistance of the air in the hollow.

IGNITION, (1) the kindling or setting on fire of the explosive portion
of the charge of a firearm. (2) The means of igniting the charge
of a firearm, as in the phrase, "cartridges containing their own
ignition."

JACKET, *of a bullet;* an outer complete or partial covering of the
(usually) lead core or interior-portion of a bullet, for the purpose
of avoiding the several disadvantages of direct contact between
the iron or steel of the bore of a barrel and the lead surface of a
bullet in motion at high speed. *Of a gun-barrel;* a cover, protection, or other accessory more or less entirely surrounding the

length of the barrel of a firearm, provided for any one or more of various purposes, as for protecting the hands of the user from heat generated by rapid firing, for containing water for cooling the piece when fired rapidly, or, as in automatic and semi-automatic arms, to provide a frame in which the moving parts of the mechanism may be actuated, etc.

JAWS, of the "hammer" or "cock" of a flint-lock mechanism; the opposable faces of the mechanism, one of them movable, whereby the flint is held firmly for the purpose of effecting its blow upon the iron or steel "battery" or "frizzen" of the mechanism.

LAMINATED, the arrangement of alternate layers of iron and steel and their welding together into a strip suitable for use as the barrel of a firearm, resulting in a barrel upon whose proper treatment by acids, etc., there is visible a resultant "pattern" of the different colors of the welded metals.

LANDS, the portions of the original inner surface of the bore of a rifled firearm left between the grooves of the rifling.

LEAD (1) [*lēd*] verb, the distance in front of a moving object the firearm must be aimed in order that the projectile may cover the intervening space and meet the object on its arrival at the point aimed at; (2), a word used in England to describe that portion of the bore of a rifle immediately in front of the chamber, and leading into the grooved bore, which in American usage is commonly called the "throat" of the bore.

LEAD [*lĕd*] verb, to deposit minute portions of the outer surface of a projectile on the inner surface of the bore of a firearm as a result of heat, friction, etc.

LEVER-ACTION, a repeating-mechanism operated by an outwardly extending or extensible pivoted or hinged arm or limb.

LOCK, (1) the mechanism of a firearm whereby the explosive is ignited; (2) that portion of the mechanism of a firearm whereby the breech, chamber and related parts of the arm are held together at the moment of firing; (3) verb, to set the safety so that the arm cannot be fired.

LOCK-PLATE, an iron plate on the outside of the stock of certain firearms upon which the mechanism of the igniting device is mounted for the mutual interaction of the parts.

LONG-ARM, a classification-name used of all longer firearms—guns, muskets, rifles, etc.—as distinguished from shorter weapons—pistols, revolvers, etc.

LUBALOY, a "trade-name" for a yellow metallic alloy or compound used by an American ammunition-making company for covering rifle bullets.

MACHINE-GUN, an automatically self-operating repeating firearm. In military phraseology confined to weapons fired from some form of mount, fixed or portable. In trade-usage the name is also applied to at least one make of automatic pistol held in both hands for firing.

MAGAZINE, that portion of the mechanism of a repeating firearm holding ammunition which, by appropriate mechanism, is made available for successive discharges of the piece.

MAKER'S MARK, any form of indication upon any part of a firearm, as to the maker of the part so marked or of the entire arm. Most frequently used of such indications when placed upon the barrel or barrels of the piece.

MANDREL, an iron or steel rod, around which an iron or steel or composite strip of metal is wrapped and hammered or welded into a barrel with a bore the diameter of such rod.

MATCH or MATCH-CORD, a cord of combustible material used for igniting the priming-charge of a match-lock firearm. In use, one end was placed in the jaws of the depressible portion of the ("cock" or "serpentine") of the mechanism and ignited. The depression of the cock brought the lighted end of the match into the powder in the pan.

MATCH-LOCK, a firearm, the firing of whose charge was effected by the use of a match.

MILLIMETER, MILLIMETRE, a measure of length or distance equal to one-thousandth of a metre (or meter), or .03937 of an inch. One meter equals 39.37 inches.

MINIÉ, originally the name of a French infantry officer (a captain of infantry in the Chasseur d'Orleans, a rifle-brigade) who, about 1840, invented a rifle-bullet having a cup-shaped hollow in the base, the sides of which hollow were by the gases of the explosion expanded into the grooving of the rifle-barrel, effecting a practically gas-tight fit of the bullet to the bore and giving greater energy, velocity and accuracy than had formerly been procured in rifles with non-expanding projectiles. It is probable that others than Captain Minié invented and used somewhat similar bullets, but his name has ever since been associated with the device. It was immediately adopted throughout the world, and bullets designated by his name became among the most famous of those in use in the middle of the nineteenth century.

MUSHROOMING, the expansion of a bullet on impact or penetration to a shape somewhat like that of a mushroom, the point, nose or front portion spreading broadly to a shape effecting an increased diameter of the path of travel.

MUSKET (Fr. mousquet, musquet), originally any arm fired from the shoulder, but later generally used of such an arm having a wooden stock extending underneath the full length of the barrel.

MUSKETEER (Fr. mousquetaire, musquetaire), the user of a musket-style firearm.

MUZZLE, the end of a firearm from which the projectile makes its exit. *Bell-muzzle,* a muzzle flaring outwardly to a diameter larger than that of the rest of the bore, for the purpose of more widely scattering the discharged projectiles. See blunderbuss.

NEEDLE-GUN, a firearm in which the ignition of the fulminate is effected by means of a slender or needle-like rod or firing-pin. Originally the Anglicization of the German name (Zündnadelgewehr) for the military rifle and its mechanism invented by Dreyse of Sommerda in 1831 and improved from 1844 to 1870 as the regulation German military musket of that period. In the paper cartridge used in this arm the fulminate was placed in the center of the cartridge, in a wad immediately behind the bullet or in the base of the bullet itself, both methods having been used, where it was ignited by the blow of a long, slender, needle-like firing-pin attached to or operated by a hammer or striker, which pin was driven through the powder-charge from the rear of the cartridge to reach the fulminate. The first "needle-guns" were muzzle-loaders, the mechanism being later altered to breech-loading. The name "needle-gun" was later popularly though incorrectly applied to various breech-loading arms in which a rod or firing-pin of some length intervened between the hammer or striker, among these the United States original "Springfield" breech-loading rifles being occasionally so called.

NIPPLE, the external opening or tube of a percussion arm on which the cap containing the priming-compound or fulminate was placed.

OPEN SIGHT, any device for sighting or aiming a firearm, no part of which device consists of an aperture or tube through which the aim is taken. Chiefly used to describe sights consisting of a block, bar or bead front sight, and a notch, bar or notched-bar rear sight.

OVAL-BORE RIFLING, a device for imparting to a projectile a rapid revolution about its longer axis. The bore of the barrel was oval in cross-section, the bullet being similarly shaped to fit the bore. The tube of the barrel was so bored that its oval cross-section was made to revolve about the axis of the bore a given number of times in the length of the barrel, the tight-fitting bullet was correspondingly revolved in such passage and received a spinning or revolving motion about its long axis maintained in its flight. Oval-bore rifling was first experimented with by a Capt. Berners of England in 1835. Later it was reintroduced and a considerable number of rifles of various calibers on this system manufactured by Mr. Charles Lancaster of London, about 1850. The interior of such a barrel presents to the eye a perfectly smooth surface, like that of a shotgun.

PAN, in a match-lock, wheel-lock or flint-lock arm, the receptacle provided for holding the priming-powder for ignition.

PAN-COVER, the device on a match-lock, wheel-lock or flint-lock for covering the deposited priming-powder in the pan, retaining it in position for use and protecting it against jarring out, against wet, etc.

PATCH, a wrapping paper (linen, leather, metal, etc.), for a bullet for the purpose of avoiding a direct contact between the lead of the bullet or its core and the sides of the bore. Originally adopted for facilitating loading of the bullet in a muzzle-loading rifle by wrapping in such a greased "patch," modern usage continues the covering of high-velocity bullets with thin metal patches or "jackets," for the purpose of avoiding the several disadvantages of direct contact between the lead core of the projectile and the iron or steel of the bore. Such modern bullets are often described as "metal-patched," designated by the abbreviation "M.P.," "full metal-patched," "F.M.P.," "metal-jacketed," "M.J.," etc.

PATCH-BOX, the receptacle often provided in the stock of flint-lock or percussion muzzle-loading rifles for carrying such patches ready for use, or the material and grease for making and using them.

PATTERN, the spread or concentration of the pellets of a charge of shot in flight, secured by firing at a target, and shown by the imprint or perforations of the pellets.

PEEP-SIGHT, a device for aiming a projectile weapon, consisting of an aperture through which the eye looks at the object aimed at. Not used of "tube," "tubular" or "telescope" sights, which are preferably distinguished by these more fully descriptive names.

PEPPER-BOX, a vernacular name common in the United States and elsewhere about the middle of the nineteenth century for a type of revolving percussion pistol having a number of parallel barrels revolvable about a central axis for firing in succession by a single fixed hammer.

PERCUSSION-CAP, a small metal cup-shaped device for holding a charge of fulminate and designed to be placed on the tube or nipple of a percussion firearm, in which position the ignition by the blow of the hammer forced the resultant flash through the ignition-tube —"vent," "flash-hole," "touch-hole," etc.—and ignited the charge within the barrel. Also, a double piece or strip of paper between the two surfaces of which such a charge of fulminate was placed for the same purpose, as in the "Maynard primer," in modern "cap-pistols," etc. Sometimes also used of the modern "primer" of central-fire fixed ammunition.

PERCUSSION-LOCK, a device for firing the charge of a firearm by the use of suitably held small amounts of fulminate, ignited by the blow of the corresponding portion of the mechanism of the piece.

PINS, appropriately-shaped pieces of iron or steel placed in openings in the wooden stock of a firearm and passing through similarly-shaped holes in metal studs affixed to the barrel, thus holding the barrel and stock together.

PISTOL, a firearm primarily designed for holding in one hand in firing.

PISTOL-GRIP, that portion of the wooden stock of a gun shaped somewhat like the handle or butt of a pistol, for more convenient holding by one of the hands in firing the same.

PITCH, (1) of the grooving of the barrel of a rifle, the angle made by the grooving and any straight line of the barrel; i.e., exterior line or, preferably, the axis of the bore. Also often called the *twist*, or *"degree of twist,"* of the grooving of a rifle. Commonly measured by (1) the number of turns of the grooving in the length of the barrel, or (2) the number of inches of barrel-length required for one complete turn of the grooving, or (3), as in the preferred technical modern method, the number of calibers (bore-diameters) required to make the distance in which the grooving makes a complete turn. (2) Of the external shape of a firearm, the angle made by the intersection of the line of sight of the barrel projected to the rear of the piece, at its intersection with a projected line of the chord of the arc of the butt-plate. Commonly measured and stated, however, by standing a gun against a wall, with the top of the barrel toward the wall and the heel and toe of the butt touching the floor, when the "pitch" will be illustrated, and is often thus stated or measured, by the distance of the muzzle of the piece away from the wall.

POINT-BLANK, in common parlance the distance to which the projectile of a firearm will travel in a direct line before descending below the line of aim. There is, of course, actually no such distance or phenomenon, as such a projectile actually begins to fall at the instant of its leaving the muzzle, but the phrase having become a convenient and popular one to describe the first almost-straight-in-the-line-of-sight travel of a projectile, it has been retained in popular use.

PRESSURE, BREECH, see breech-pressure.

PRESSURE-GUN, a firearm equipped with a device for measuring the breech-pressure of a given charge.

PRIMER, the device used in a firearm for holding the igniting charge of fulminate in position for the firing of the charge by the flash of its ignition. Most commonly, the "cap" of a cartridge, though also used of the "friction-primer" or similar device for firing certain pieces of artillery, etc.

PRIMING, the charge of gunpowder or fulminate, as the case may be, whose ignition in turn ignites the charge of a firearm.

PROPELLANT, in a firearm, the charge of powder the expanding gases of whose ignition or explosion propels the projectile.

PRIMING-PAN. Same as PAN.

PROTECTED POINT, a part of the bullet or its construction, which holds together and supports the metal of the point against undesirably great shattering, etc., on impact with an object. Usually a wire, rod, or other stiff central part or inner core inserted within the bullet for the purpose, and which may or may not be connected with a cap, point or nose of the projectile.

PYRITES, a mineral substance (sulphid of iron) characterized for firearms purposes by its quality of producing sparks when suddenly brought in contact with iron or steel. Used in wheel-lock and pyrites-lock arms. *P.—lock,* the first form of the flint-lock, characterized by its use of a piece of pyrites for producing sparks for igniting the priming-charge, whereas the later flint-lock proper used a piece of flint.

QUARREL, an early common name for the projectile of a cross-bow.

RAMROD, (1) a rod of wood, iron or steel, used for pushing down into the barrel of a muzzle-loading firearm the charge of powder and the projectile; (2) the name is still sometimes popularly used for the cleaning-rod of an arm.

RANGE, the distance of the flight of a projectile. *Effective r.,* the distance at which the projectile of a firearm is capable of the desired effect—i.e., whether for hunting, target-shooting, military purposes, etc. *Extreme r.,* the utmost distance to which the projectile of a given firearm may be made to travel before striking the ground.

RANGES OF THE AMERICAN ARMY RIFLES AND AMMUNITION.

The following notes on the American Army rifles and their ammunition are inserted because of popular interest in the subjects, a certain absence of generally available accurate information thereon, and the fact that the ammunition in use is in a stage of transition from the older variety to a more modern type.

There are at present two forms of "regulation" military ammunition available for use in the United States Army rifles officially designated as "Model 1903" and "Model 1917." The Model 1903 rifle is popularly known as the "Springfield," from the location of the arsenal (Springfield, Massachusetts), where it was developed and first manufactured; it is often called the "New Springfield," to distinguish it from the old and now obsolete single-shot army rifle of .45-70 caliber, which in its day (1868-1892) was also popularly termed the "Springfield." The Model 1917 rifle, brought out for emergency-use during the World

Ranges (Cont.)

War, is often spoken of as the "Enfield," from the name of the British army rifle which it closely resembled.

Of the mentioned two forms of regulation ammunition the older, in use with various improvements and slight changes since its adoption in the year 1906, is accordingly known as the "Model 1906," or "M '06" or "'06." Its ballistic data are given under the description of the Springfield rifle in the text of this Bulletin. While it is the cartridge now issued and in use for all present-day military purposes by the War Department, its manufacture by the Government has ceased, being supplanted by that of the second form, and it is announced as the official intention to use up all the Model 1906 ammunition now on hand and then to supersede it by the second form, now in production at the Frankford arsenal.

The new ammunition is known as the "Mark I" cartridge, whose ballistic data will also be found under mention of the Springfield rifle in the text.

The maximum range of the Model 1906 ammunition, when fired in the Model 1903 rifles, averages 3,450 yards, or 2.04 miles.[1] This distance is ascertained by actual firing[2] and is 797 yards less than was, previous to actual firing-tests, officially considered and announced[3] as the maximum range of the Model 1906 cartridge; such greater range having been only theoretically computed, and never having been tested for actual determination, so far as is of record, prior to the U. S. War Department firing-tests of 1919 at Daytona Beach, Florida.

This maximum range of 3,450 yards for the Model 1906 cartridge is secured by firing the rifle at an angle of elevation of 30 degrees.[4]

There is prevalent to some extent a popular impression that the maximum range of the projectile from a firearm is secured on firing the piece at an angle of elevation of 45 degrees. This impression, while capable of some purely theoretical support, is found erroneous as demonstrated by actual firing of modern arms. Modern small-arms ammunition is found by test to attain its maximum range when the piece is fired at an angle of elevation, varying in various arms and ammunition, between 25 and 32 degrees.

The time of the flight of the bullet of the Model 1906 cartridge,

---

[1]War Department Technical Regulations No. 1350-A, June 1st, 1927, p. 22.
[2]Ibid.
[3]Description and Rules for the Management of the U. S. Magazine Rifle, Model of 1903; Revised April 18, 1906; U. S. Ordnance publication No. 1923, p. 48.
[4]War Department T. R. No. 1350-A, op. cit., p. 22.

Ranges (Cont.)

when fired in the Model 103 rifle at an angle of elevation of 30 degrees and to its maximum range, is 26 seconds.[5]

It must be understood that the *maximum range* of a bullet and the *maximum efficiency range* are different. The *maximum range* is the utmost linear distance to which a bullet can be made to travel, when fired in a given arm, before striking the ground. The *maximum efficiency range* is the distance at which the bullet retains sufficient accuracy and energy for military purposes against an enemy. This maximum efficiency range of the Model 1906 bullet in the Model 1903 rifle is computed at 2,800 yards; which distance is accordingly that of the extreme range for which the Model 1903 rifle is at present sighted, as indicated on the rear sight of the arm.

With the use of the mentioned newer Mark I cartridge, both the maximum range and the corresponding maximum efficiency range of the Model 1903 rifle are found by actual firing-test to be greatly increased. Its maximum range averages approximately 6,000 yards,[6] or 3.46 miles; though official tests of both maximum range and maximum efficiency range of this cartridge are at present (February, 1928) in progress and uncompleted. When completed, their definitely ascertained results will be officially published.[7] The maximum range of the Mark I cartridge, as at present announced (above), are secured by firing the rifle at an angle of elevation of 30 degrees.[8]

The Model 1903 rifle, with the Model 1906 ammunition, was generally recognized during the World War as the most accurate military rifle in existence;[9] while, as mentioned, the adoption of the new Mark I ammunition has greatly increased both its range and its accuracy.

The U. S. Model 1917 rifle, adopted as an emergency arm during the World War, will use both the Model 1906 and the Mark I ammunition. It has, however, a shorter efficiency range than has the Model 1903 rifle, as it is sighted for such a range of but 1,600 yards, and no automatic adjustment for drift of the bullet, no wind-gauge adjustment, and no provision for finer sighting than by hundreds of yards, are provided on its rear sight as at present issued.

P. B. J.

---

[5] Ibid.

[6] Ibid. The latest tests with the 172-grain "boat-tail" bullet give 5,800 yards.

[7] Ibid.

[8] In view of the greater range and accuracy of the Mark I ammunition as compared with the Model 1906 ammunition, it is probable that on the complete supersession of the latter by the former, as mentioned, new rear sights for the Model 1903 rifle will be issued, better adapted to the use of the Mark I cartridge than the present rear sight, which was designed, as indicated, for the use of the Model 1906 ammunition.—Editor.

[9] "Rifles And Ammunition," H. Ommundsen and E. H. Robinson, Funk & Wagnalls Co., New York and London, pub. 1915, Chap. 14, p. 324.

REAR SIGHT, the device on a firearm between the front sight and the eye, whereby the piece may be aimed at or aligned with an object.

RECOIL, the rearward motion of a firearm upon its discharge. *r.-pad,* anything affixed to the butt of a gun for the purpose of lessening the violence of the blow or push of the recoil against the shoulder of the user.

REST, any support for the piece, the barrel, etc., or of the hand or hands of the user, for the purpose of steadying the piece during the acts of aiming and firing. In early times, a support for the barrel of a firearm was carried by the user or an assistant. See fork.

REPEATER, an arm capable of firing successive shots without reloading. Usually used of single-barreled firearms other than revolvers or automatic arms.

REVOLVER, a firearm with (1) several parallel barrels revolving about a common axis, or (2) a revolving magazine containing chambers for the presentation of successive charges for firing through a single barrel.

RIFLE, a firearm in whose bore grooves are cut for the purpose of imparting a desired revolving or spinning motion to a projectile. (Such is the common meaning of the word as universally used today, though in earlier days straight rifling or grooving was not uncommon, and was believed to assist in maintaining the projectile in a direct line of flight.)

RIFLING, n. (1) the grooves cut in the sides of the bore of a rifle. (2) verb, the act or process of cutting such grooves.

ROD-BAYONET, a form of bayonet for use on the barrel of a firearm, placed in a groove or tube in the stock under the barrel and capable of being drawn out to a desired point or extent and locked in such position for use, being sharpened at its point, as a thrusting-weapon. Sometimes combined with design and used as a rod, cleaning-rod, etc., as well, one end being appropriately formed for such service. Such bayonets have been somewhat experimented with by the United States and other military forces, but have invariably been discarded as impractical.

SAFETY, a part of the mechanism of a firearm capable of such manipulation as to prevent its accidental discharge.

SALTPETER, a chemical salt occurring in nature, known in chemistry as Potassium Nitrate or Nitrate of Potash, also capable of artificial manufacture, one of the three essential components of common or black gunpowder (the other two being sulphur and charcoal).

SEAR, in England often written SCEAR, a part of the lock of many firearms, operating between the trigger and hammer or striker.

SELF-LOADING, a firearm mechanism which, upon the discharge of the piece, automatically withdraws and ejects the fired cartridge and places another in position for firing.

SERPENTINE, originally the early name of the S-shaped metal portion or limb of a match-lock which held the lighted match-cord in readiness for igniting the priming. It was presently commonly applied to the powder, and at times was used as a name of the gun itself.

SET-SCREW, a screw so set in the mechanism of a firearm as by its motion, adjusts some other portion of the mechanism for a desired degree of motion. Usually, a screw whereby the degree of pressure upon the trigger required to discharge the firing-mechanism can be regulated.

SET-TRIGGER, an auxiliary trigger on a firearm by whose operation the firing-mechanism is adjusted for effecting the discharge of the arm by very light pressure on the firing-trigger.

SHELL, the case of metal or stiff paper containing the charge for a breech-loading firearm.

SHOT, (1) pellets of metal, usually lead, composing the charge of a gun. Also commonly used in the vernacular to allude to (2) any discharge of a firearm, e.g., "He fired a shot," and (3) for any projectile so fired, e.g., "I heard the shot strike."

SHOTGUN, a firearm designed primarily for the use of a multiple charge rather than a single projectile.

SIGHT, a device affixed to or part of a firearm designed for use in aiming or aligning a firearm. See front sight, open sight, rear sight, telescope sight, peep sight, etc.

SLIDE-ACTION, used to describe the mechanism of a repeating firearm when so constructed as to be operated by the reciprocal movement of an actuating-portion movable by the hand to and fro underneath the barrel and parallel thereto.

SLING, (1) a strap attached to a firearm for the purpose of carrying; (2) chiefly of a strap fastened to and underneath the stock of a firearm for the purpose of use by the hand and arm of the shooter to steady the piece in the acts of aiming and firing.

SMALL-ARMS, firearms capable of being carried on the person and fired from the hand or hands. Chiefly a military term for all such arms, as the official British army "Text-Book of Small-Arms," etc.

SMOKELESS, a form of propellant having the quality of a more or less complete absence of smoke accompanying its explosion.

SMOOTH-BORE, a firearm the interior of whose barrel is not grooved as in a rifle.

SNAPHAUNCE, an early name for a firearm whose discharge was effected by a "flint-and-steel" mechanism.

SPADE-BAYONET, a bayonet whose blade is so shaped as to be capable of use as a spade.

SPANNER, (1) a flat wrench for operating a portion of a firearm mechanism, (2) the auxiliary device used for winding up the spring of a wheel-lock arm.

STANDING BREECH, the upright portion of the mechanism of a breech-loading firearm which supports the rear (base or head) of the cartridge when fired and away from which the barrel or barrels of such firearm are moved for removal of the cartridge or cartridges and for reloading. Chiefly only of the upright face of the frame of a single or double shotgun or rifle of the "tipping" form of opening, against which the head of the cartridge rests for support when the gun is ready for firing.

STOCK, (1) the wooden portion of a firearm; also (2) the portion of the wood of a firearm which in the firing position is placed against the shoulder.

STRIKER, that portion of the mechanism of a firearm whose movement delivers either directly or indirectly the blow requisite for ignition of the primer-charge.

STRIPPING, in a rifled firearm, the forcing of the projectile through the barrel without its fitting against the grooves of the rifling and receiving therefrom a rotary motion. So called from the fact that such failure to "take" the grooves of the rifling effects a "stripping" off of portions of the lead or other metal of the projectile in its passage through the barrel.

STUD, a projection upon any portion of a firearm designed by contact with some other portion to retain such other portion in a desired position or to effect a desired operation of the same; as "the bayonet-stud" for holding a bayonet in position on the muzzle of a gun; a "sight-stud," a projection for supporting a sight, etc., etc.

SWIVELS, portions of the mechanism of a firearm or affixed thereto, for the purpose of allowing a desired motion of a part or attachment while at the same time effecting the retaining of such part in connection with the piece. E.g., "swivels" for holding a rifle-sling to the arm while permitting any necessary or desired motion of the same; also a "swivel-ramrod" or ramrod permanently attached to the piece while yet capable of all necessary motion for use in loading the same.

TANG, used of certain portions of the construction of firearms formed into more-or-less flat or curved strips of metal; chiefly of such strips when projecting before or behind the trigger-guard, and of a projection of the metal of, or affixed to, the barrel when projecting toward the butt of the piece and let into or affixed upon the wood of the stock. *Tang-sight,* a rear sight mounted upon such rearward strip of metal in or upon the top of the stock behind the breech-end of the barrel. *Tang-screw,* a screw through a tang and the wood of the stock and holding these together.

TELESCOPE-SIGHT, a telescope affixed or capable of being affixed to a firearm for the purpose of aiding in the aiming or alignment of the arm upon an object by the magnifying of such object.

THROAT, that portion of the bore of a rifle forming the front of the chamber and conducting the moving bullet into contact with the grooving of the barrel. In writings, military parlance, etc., in England often called the "lead" of the bore of a rifle.

THROATED, of a rifle, the forming of the front end of the chamber in such slightly funnel-shaped dimensions as to lead the bullet into desired contact with the grooving of the barrel.

THROATING, the process of forming the front end of the chamber into a throat. See throat.

TOE, of the butt of a gun-stock, the narrower or lower end of that part of the stock which in the firing position abuts against the shoulder.

TOUCH-HOLE (ancient form TUTCH-HOLE), the "vent" opening

or tube of a firearm communicating between the priming and the powder of the charge in the barrel. Hence, also *touch-powder* and *tutch-powder,* as names for the priming-powder, which in the earliest firearms was touched by the gunner, or an assistant, with a lighted "match" or combustible, and later by the lighted "match-cord" of the match-lock type of firearm.

TRAJECTORY, the path of the projectile from the muzzle of the piece to its contact with an object or, in the case no object intervenes, with the ground at the end of its flight. Also used of projectiles from weapons other than firearms, arrows, bolts, stones, etc.

TRIGGER, that portion of the mechanism of a firearm pressure upon which actuates a part or all of the firing-mechanism and directly or indirectly effects the discharge of the piece. Usually of the projecting limb of the mechanism within the trigger-guard, pressure upon which by the finger of the user effects the discharge of the piece; but triggers of firearms may be, and not infrequently are, of other location, operation, etc., as "thumb-trigger," "trigger-button," "set-trigger," etc. *Double-trigger,* as usually used, describes the construction of the firing-mechanism of a firearm in such fashion that pressure on one of two triggers places the mechanism in position for release by a very slight pressure upon the second. See hair-trigger, set trigger. *Single-trigger,* a firearms mechanism in which one trigger effects the successive discharges, at the intent and operation of the user, of more than one barrel, chamber, charge, etc., of the arm.

TRIGGER-PLATE, that portion of the frame or receiver of a firearms mechanism through which the trigger or triggers project externally.

TRIGGER-GUARD, that portion of the construction of a firearm so placed and shaped as to protect the trigger or triggers from being moved by accident or by other than the intentionally-operating finger of the user.

TWIST, (1) the angle of the grooving with the axis of the bore. Measured, and commonly stated, in terms of the linear distance in which such grooving makes one complete turn; or, in the most modern manner, as in military and technical descriptions, etc., in terms of the number of times the caliber (diameter) of the barrel equaling such linear distance. As, "one turn in 10 calibers," etc. (2) Formed by welding iron, steel, or iron and steel, into a strip which is then (a) either twisted upon itself for the purpose of greater toughness and strength, or (b) around a mandrel or rod against and around which it is welded and hammered into a tube, or (c) both such operations combined. All the mentioned processes are technically known as various forms of twist in barrel-making.

VENT, a small hole or tube communicating with the chamber of the bore of a gun and the exterior of the piece, either (a) for the purpose of conveying the flame of the ignited priming to the powder of the charge, as in match-lock, wheel-lock and percussion arms, or (b) for the purpose of permitting the escape of such gas of the explosion as might result from the giving way of a side or head of a fired cartridge, around the primer, etc. Such latter form of vent is provided on many modern arms for the escape of gas in case of such occurrence, to avoid its penetrating into the mechanism or escaping otherwise to the annoyance or injury of the shooter, etc.

VERNIER, an adjustable portion of the sight of a gun, movable parallel to and in connection with a fixed scale, for the purpose of so elevating the path of the projectile, in the act of aiming the piece, as to hit a more or less distant object. Usually of the sliding portion of an adjustable upright rear sight of a gun, though such vernier sights have also been made for use as front sights.

WAD, or WADDING, anything used to retain in place any part of the charge of a firearm, either in the barrel of the piece or in a cartridge. *Base-w.*, material placed inside a cartridge between the head or base of the cartridge and the charge of explosive, perforated to allow the flame from the primer to reach such explosive.

WALL-PIECE, a firearm, usually of early form and large size, bore and weight, primarily designed for firing from a wall or battlement, and accordingly frequently equipped with a stud or projection beneath the piece, designed to be placed in contact with such wall for the purpose of effecting the push or blow of the recoil against such wall rather than against the shoulder of the shooter.

WHEEL-LOCK, an early form of firearm, chronologically developed between the match-lock and the flint-lock, in which the ignition of the priming was effected by the contact of a piece of pyrites (q.v.) held by a portion of the mechanism against the rapidly revolving serrated rim of a small wheel driven by a spring, such contact producing sparks for the ignition of the immediately adjacent priming.

WORM, a twisted metal rod or wire affixed, or capable of being
affixed, to one end of the ramrod of a muzzle-loading firearm for
the purpose of withdrawing a charge from the barrel of such arm.

INDEX*

Rifle (Cont.)

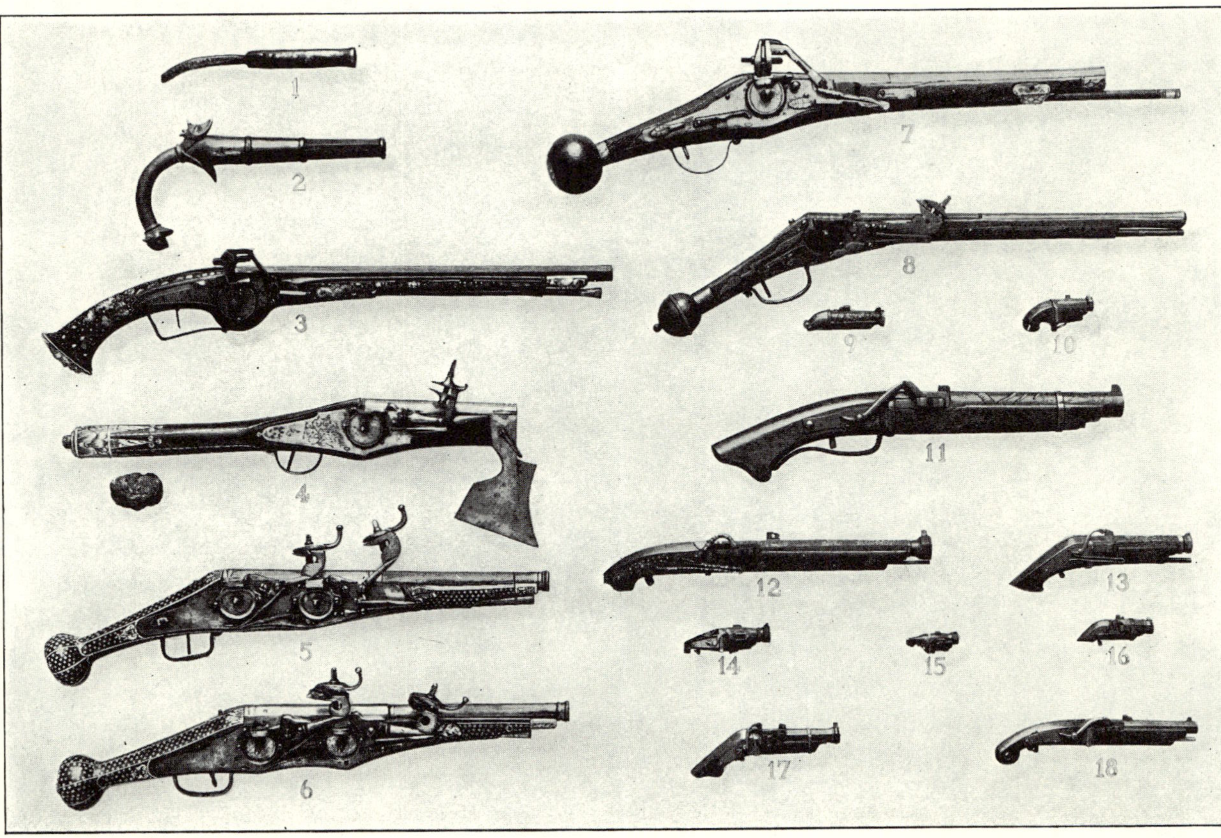
1
2
3
4
5
6
7
8
9
10
11
12
13
14
15
16
17
18

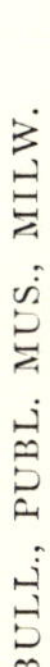

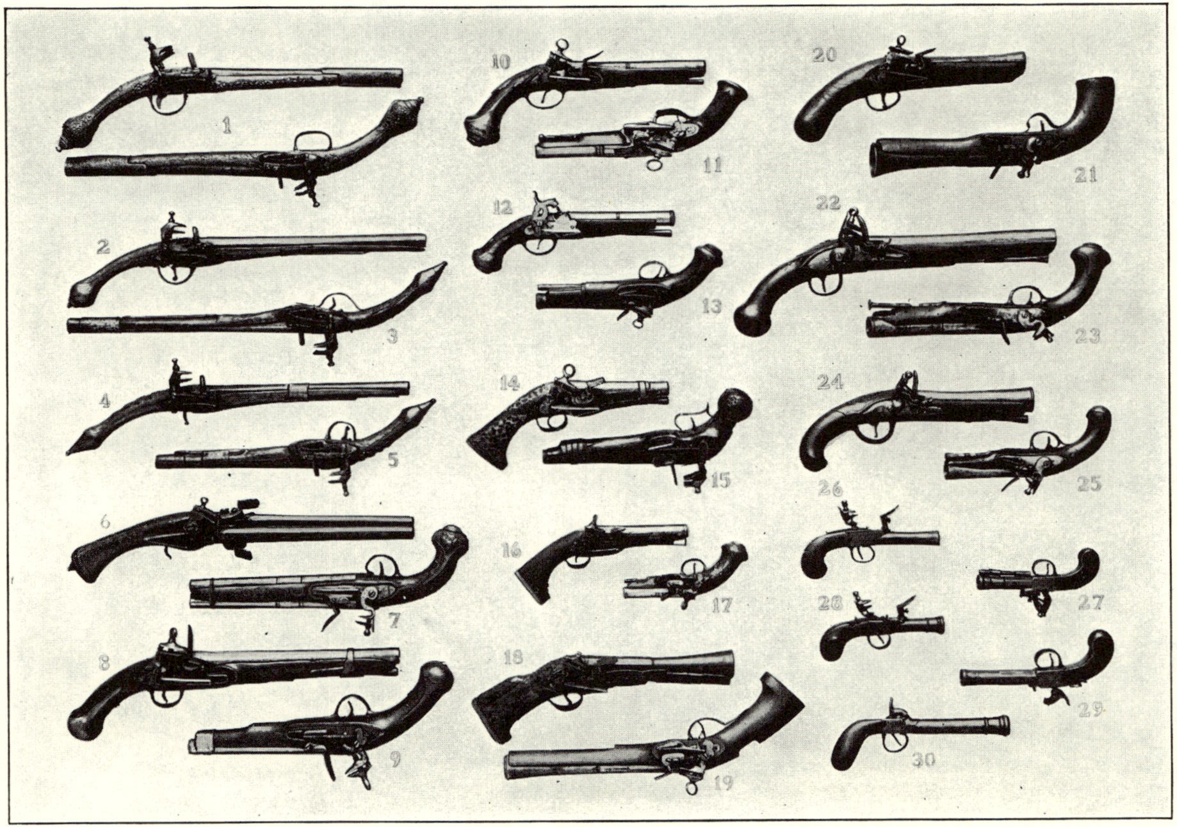
1
2
3
4
5
6
7
8
9
10
11
12
13
14
15
16
17
18
19
20
21
22
23
24
25
26
27
28
29
30

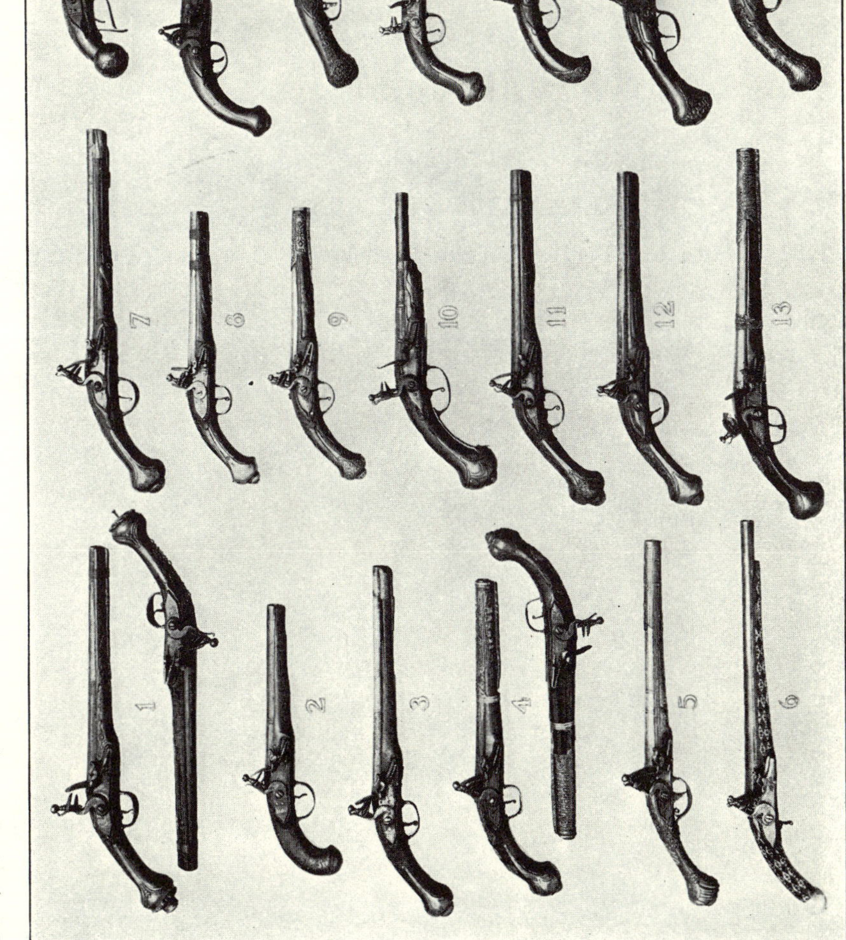

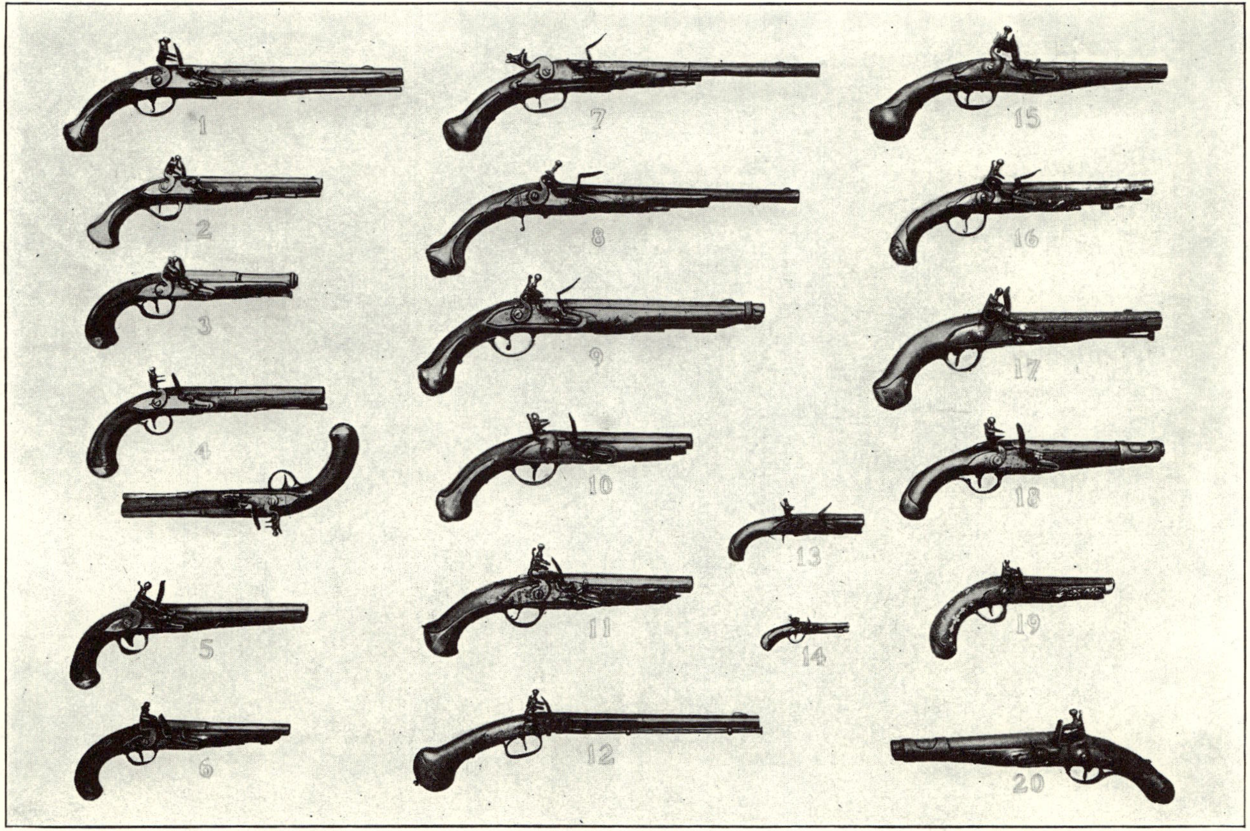
1
2
3
4
5
6
7
8
9
10
11
12
13
14
15
16
17
18
19
20

EXPLANATION OF PLATE 68.

English Flint-Lock Pistols (page 435).

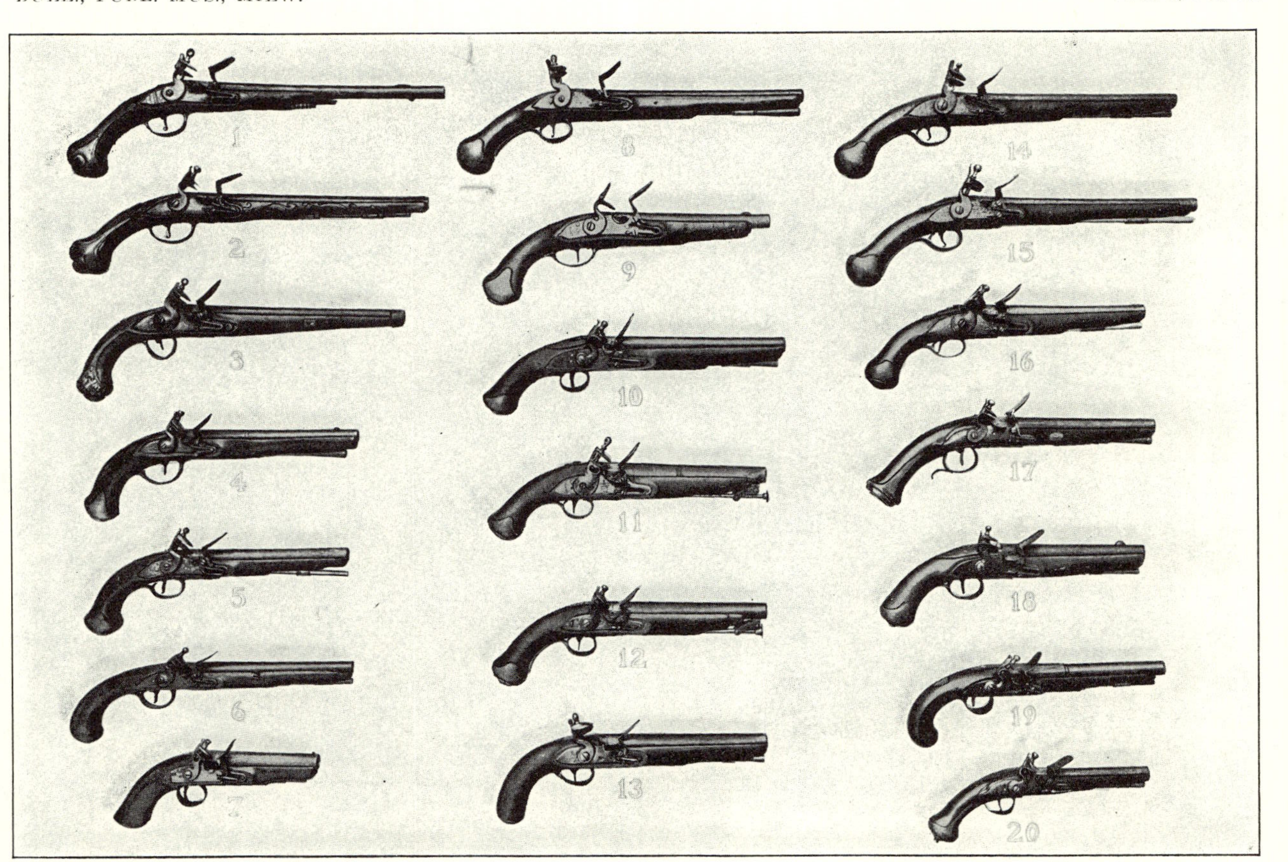

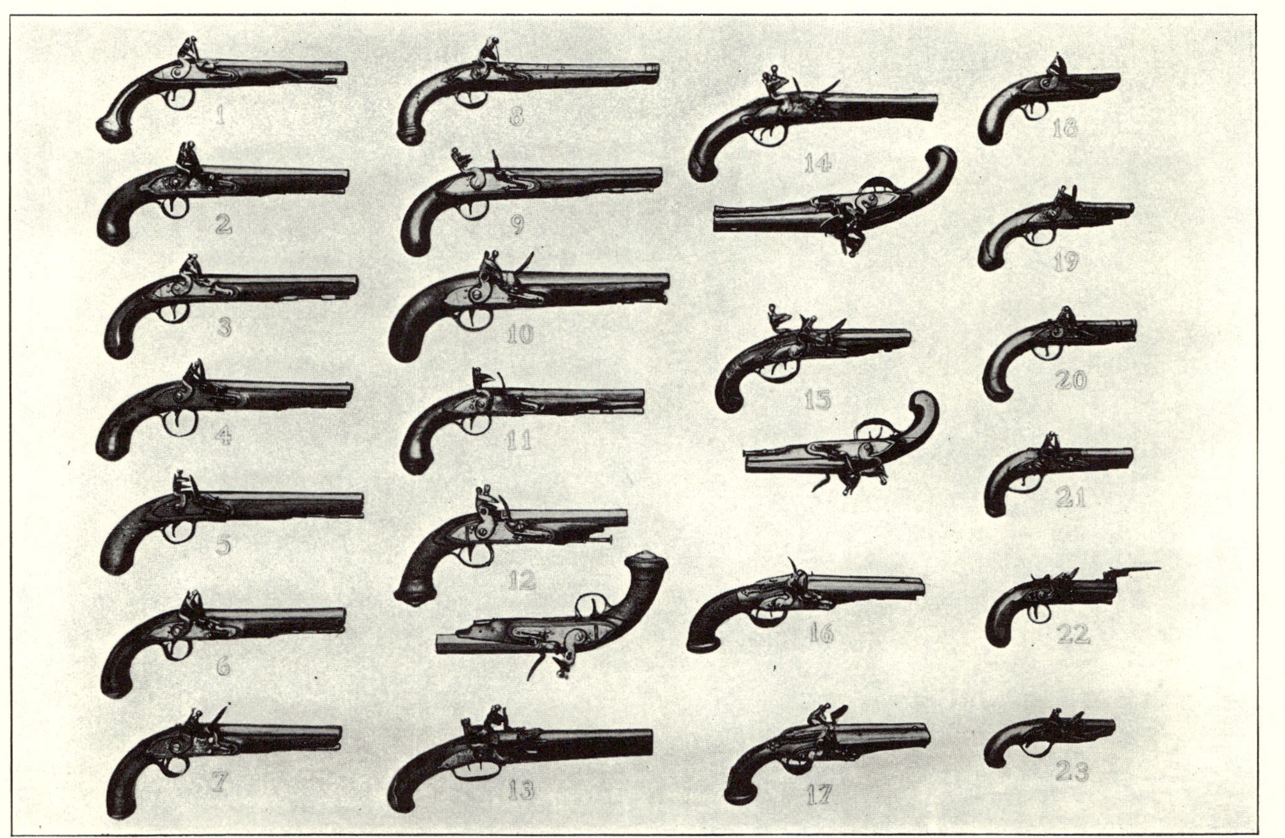

EXPLANATION OF PLATE 70.

Flint-Lock Pistols, Chiefly French Military (page 442).

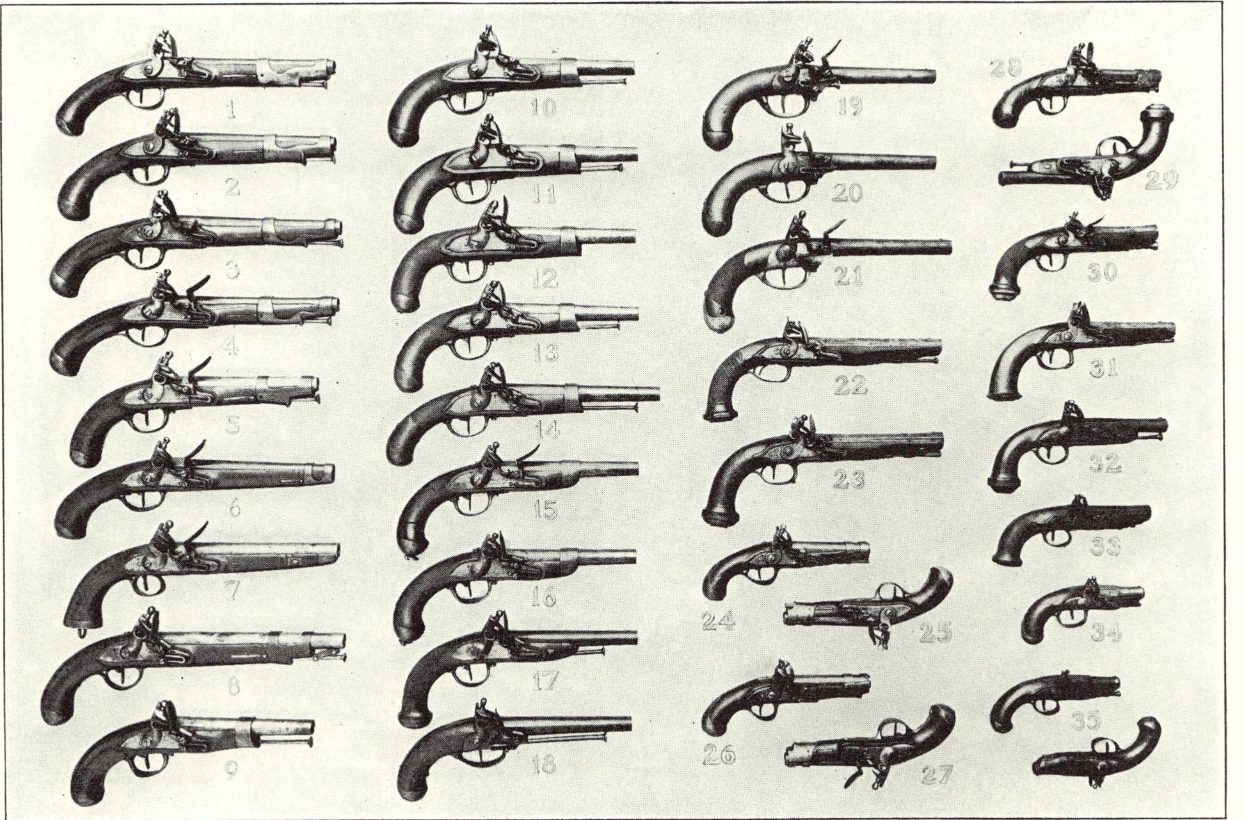

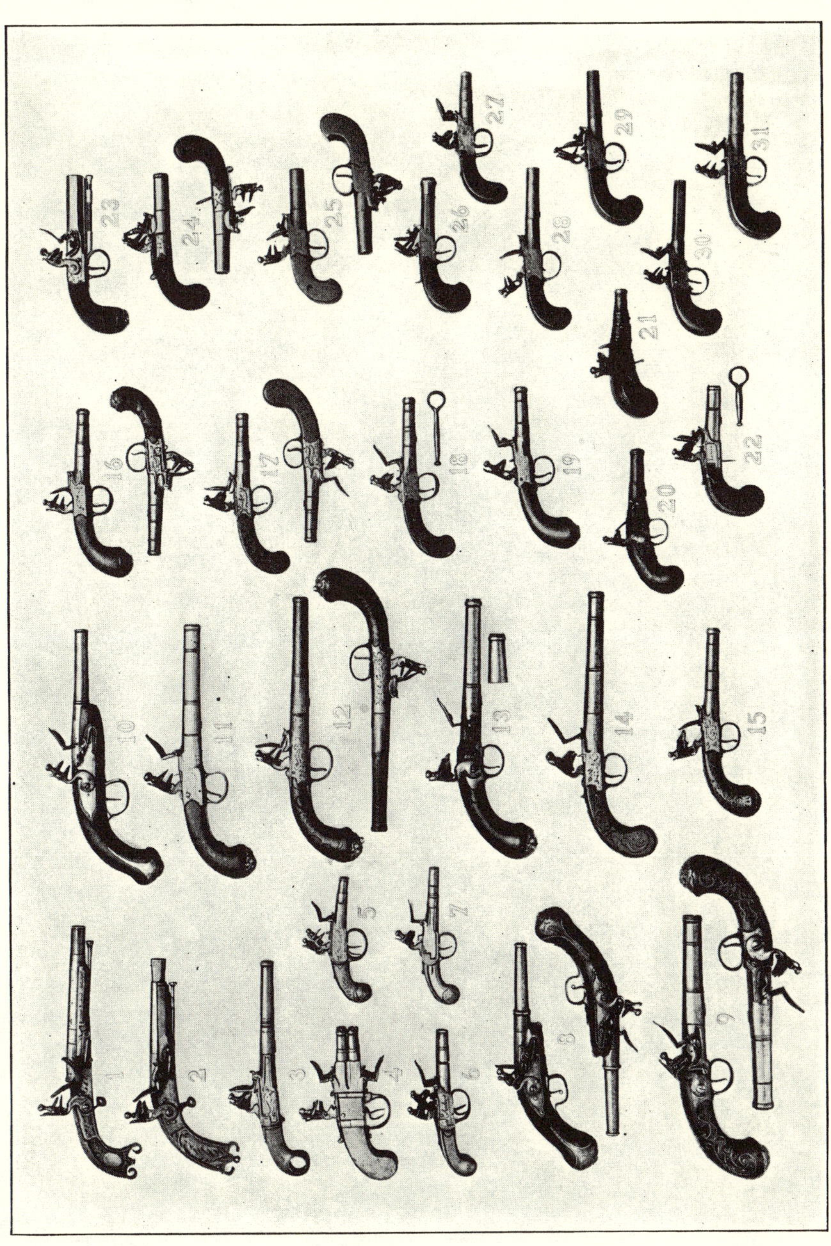

EXPLANATION OF PLATE 72.

Flint-Lock Pocket Pistols, Mostly English (page 457).

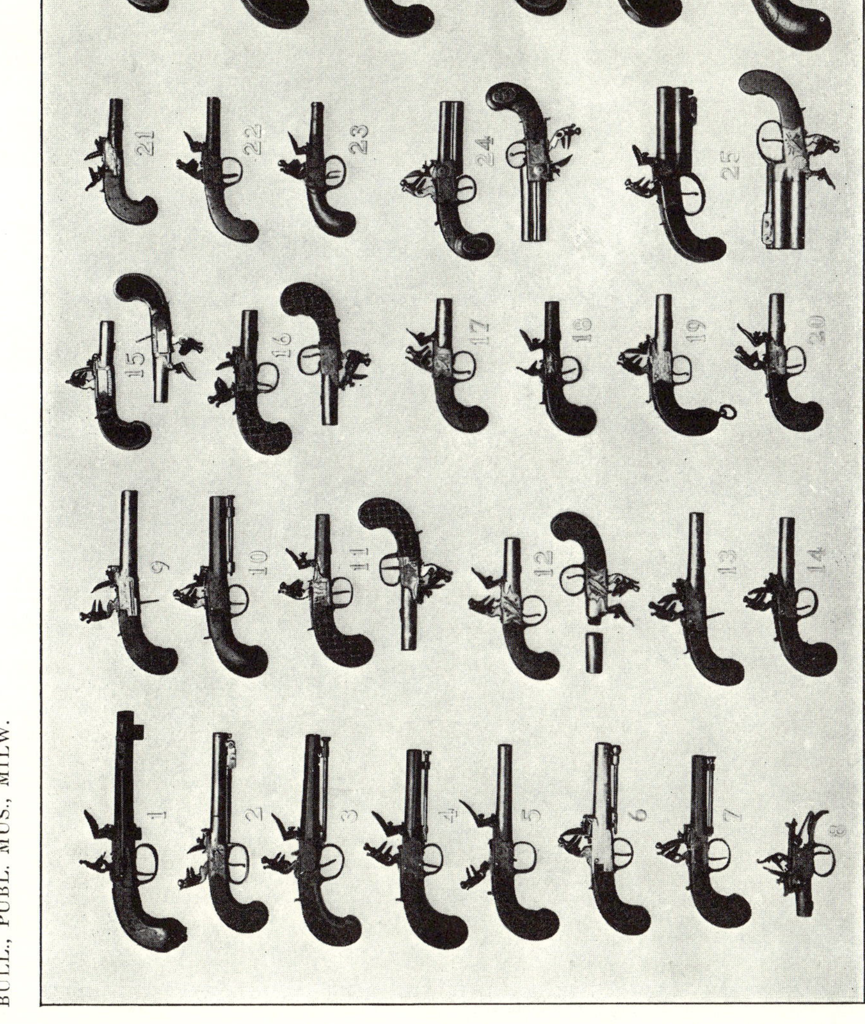

EXPLANATION OF PLATE 73.

Pistols in Cases (page 464).

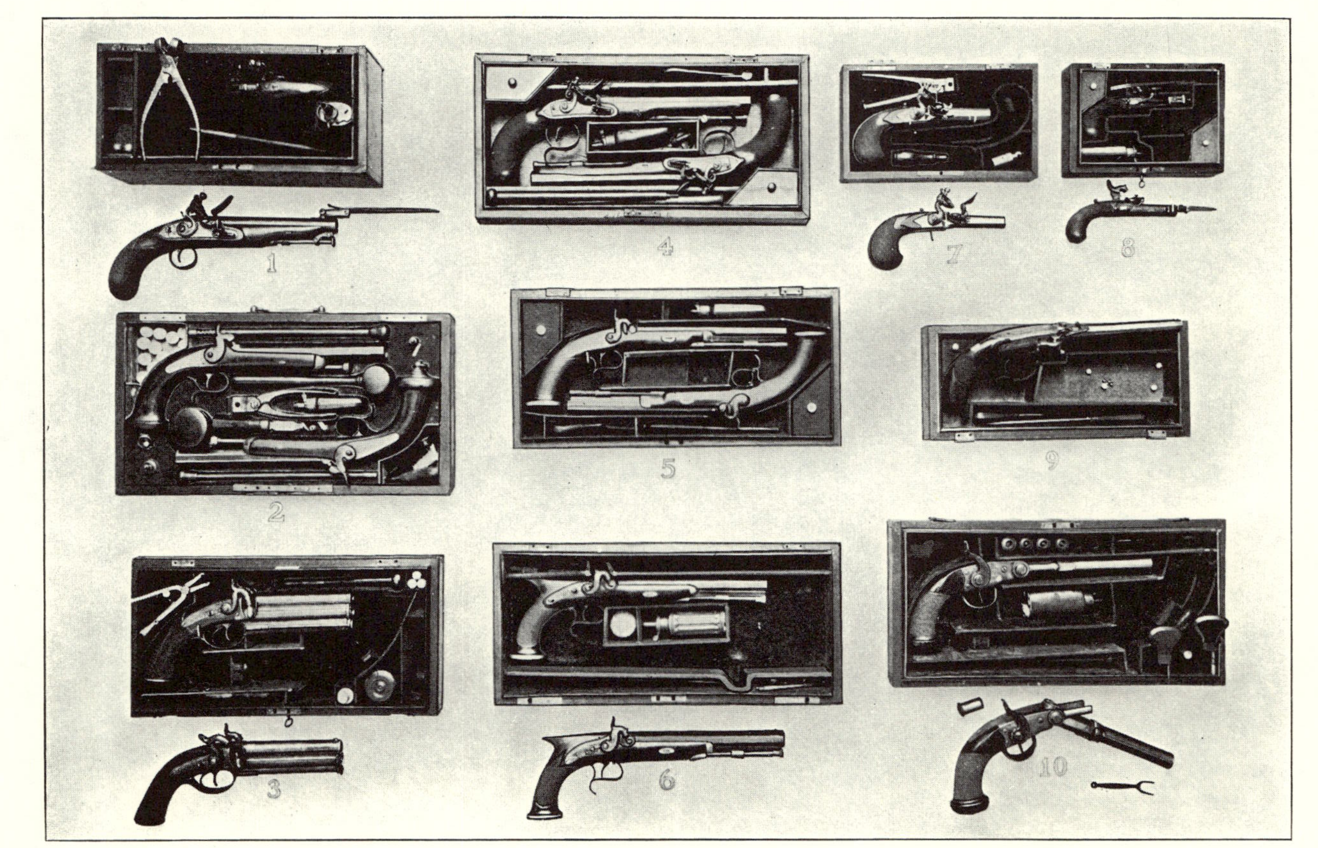

EXPLANATION OF PLATE 74.

Pistols in Pairs, Flint-Lock and Percussion (page 470).

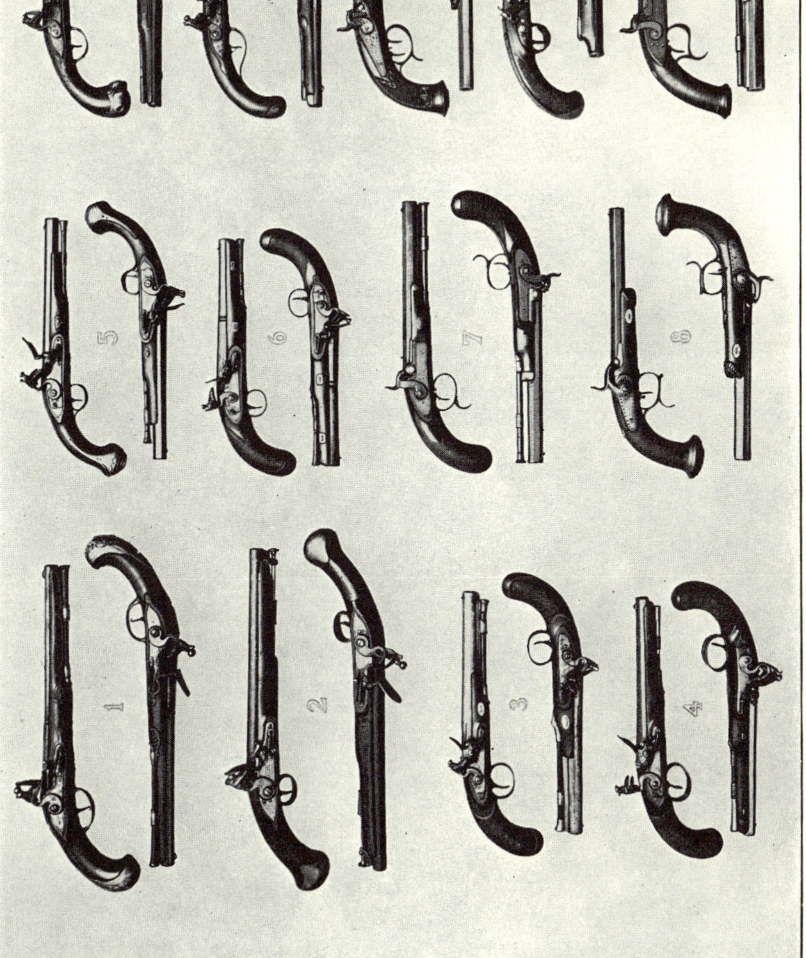

EXPLANATION OF PLATE 75.

Flint-Locks, Converted Flint-Locks, and Percussion Arms of
Various Countries (page 473).

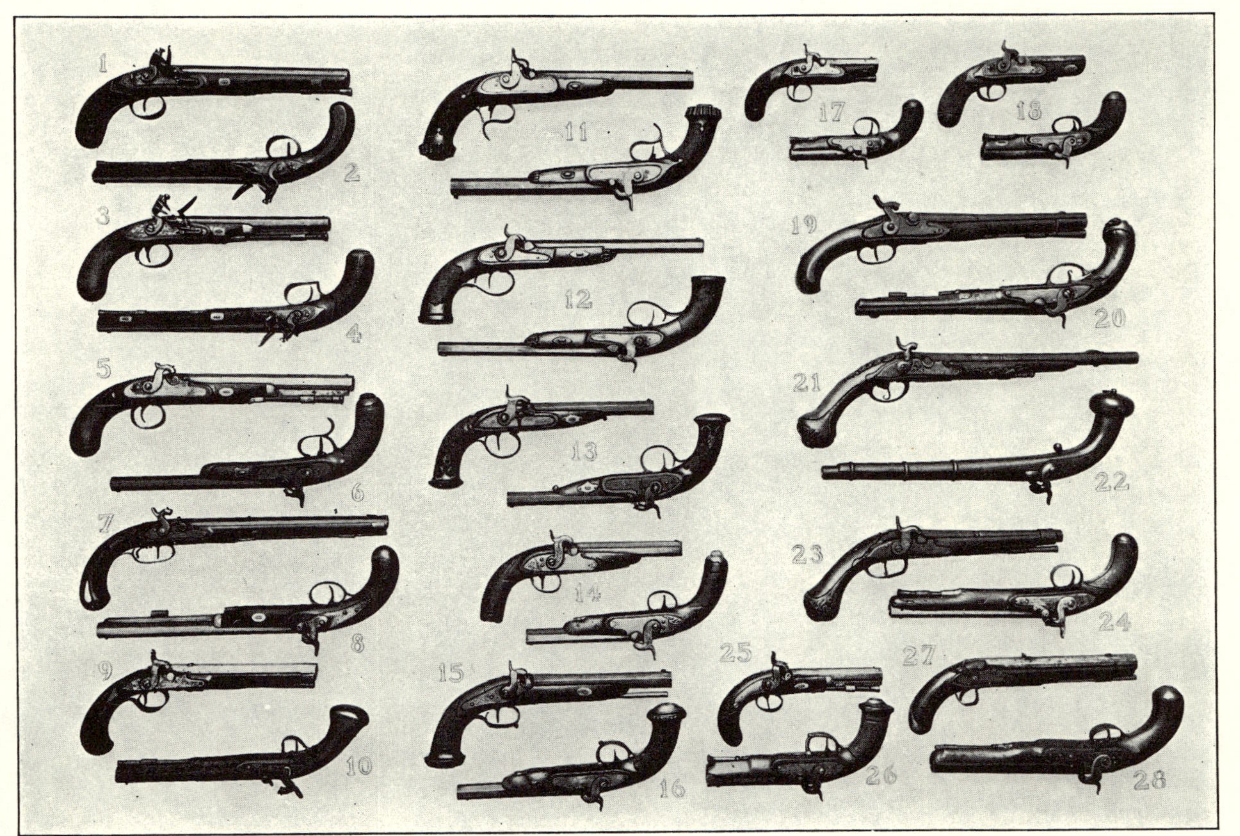

EXPLANATION OF PLATE 76.

United States Single-Shot Military Pistols, Flint-Lock (page 479).

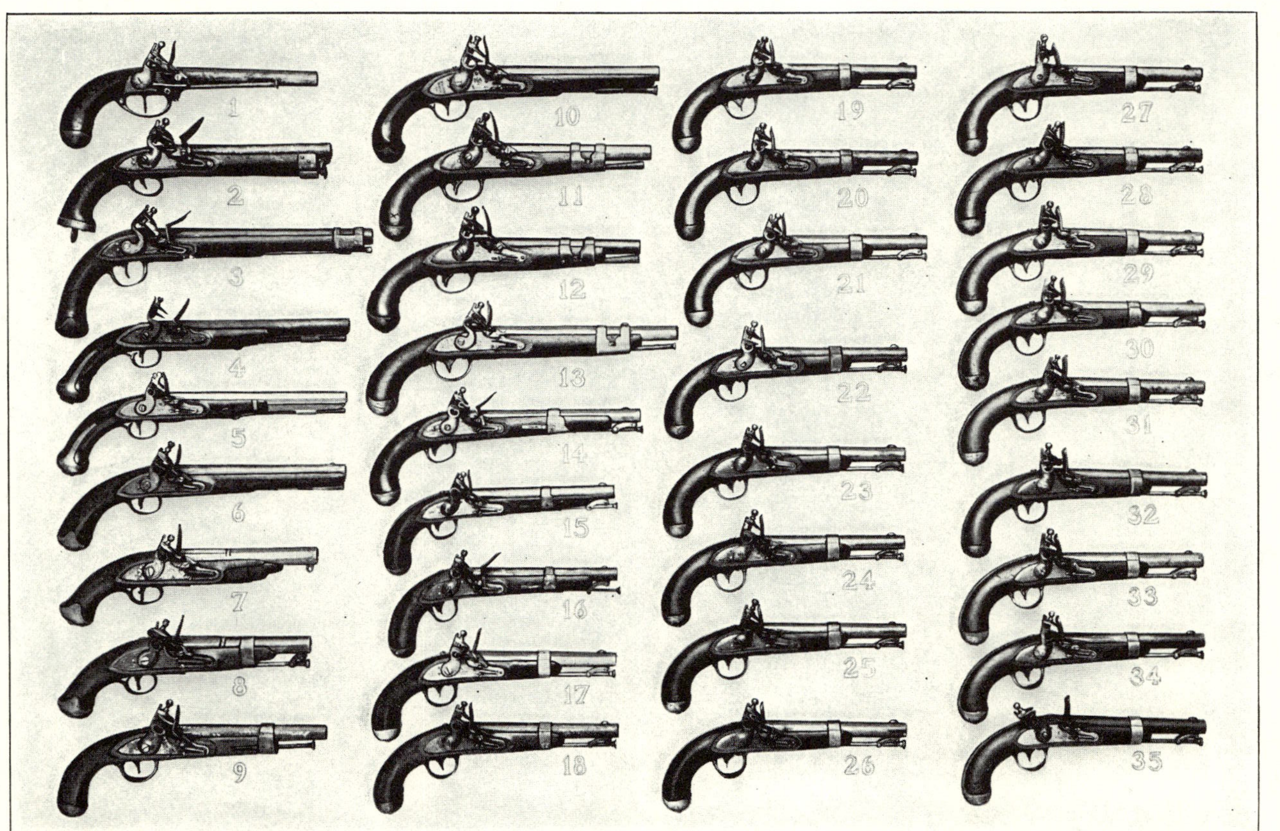

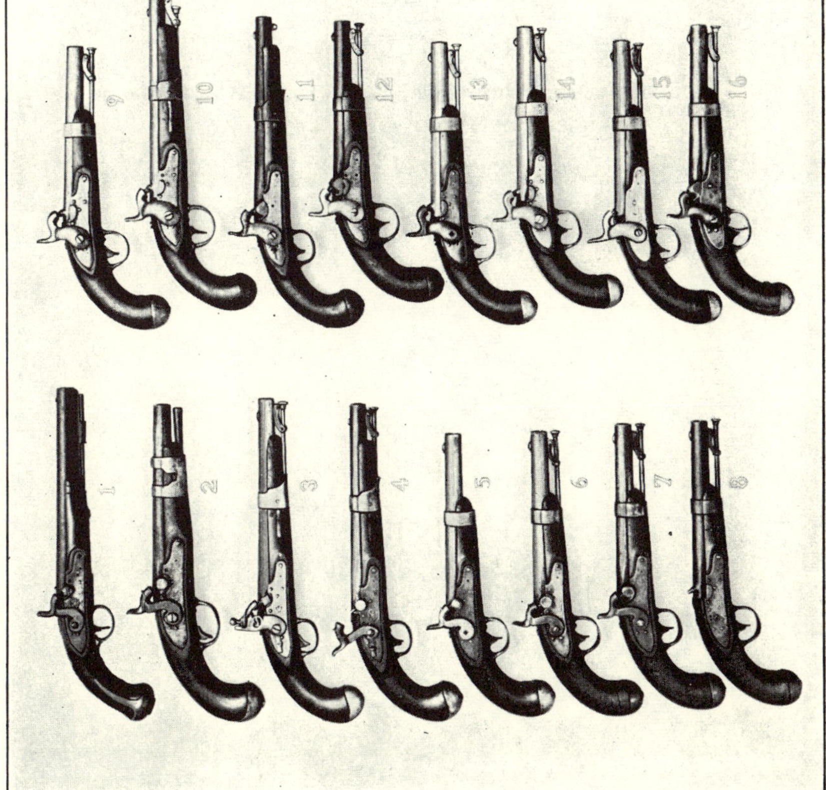

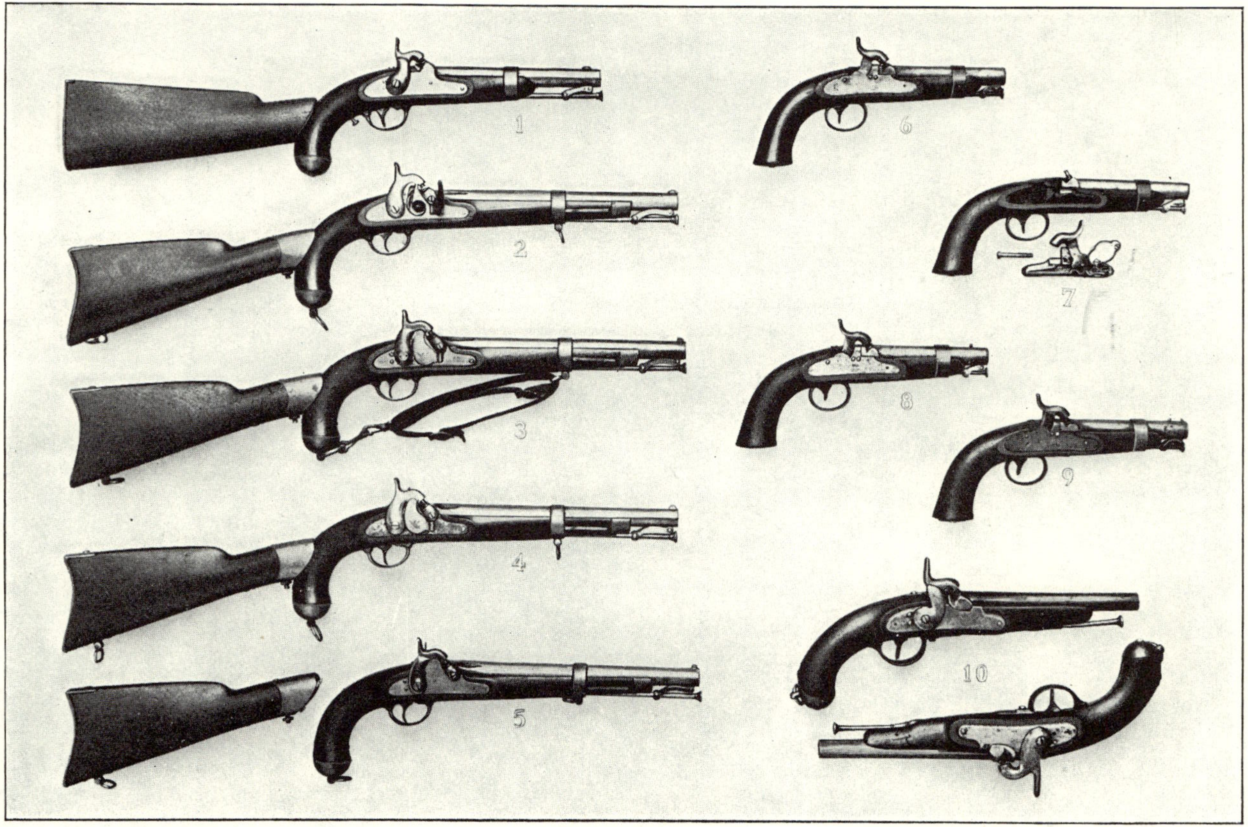

EXPLANATION OF PLATE 79.

Chiefly European Percussion Military Pistols (page 498).

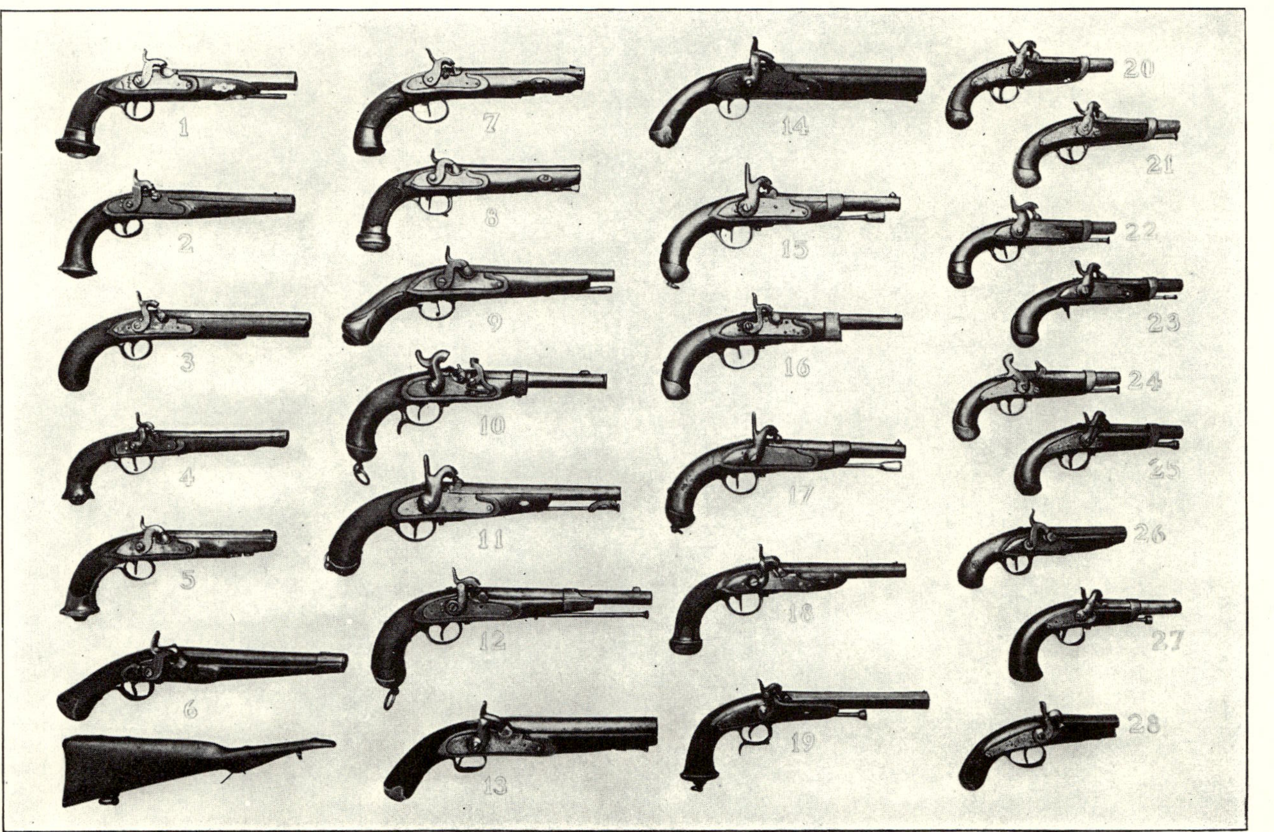

EXPLANATION OF PLATE 80.

Deringers and Similar Pistols (page 505).

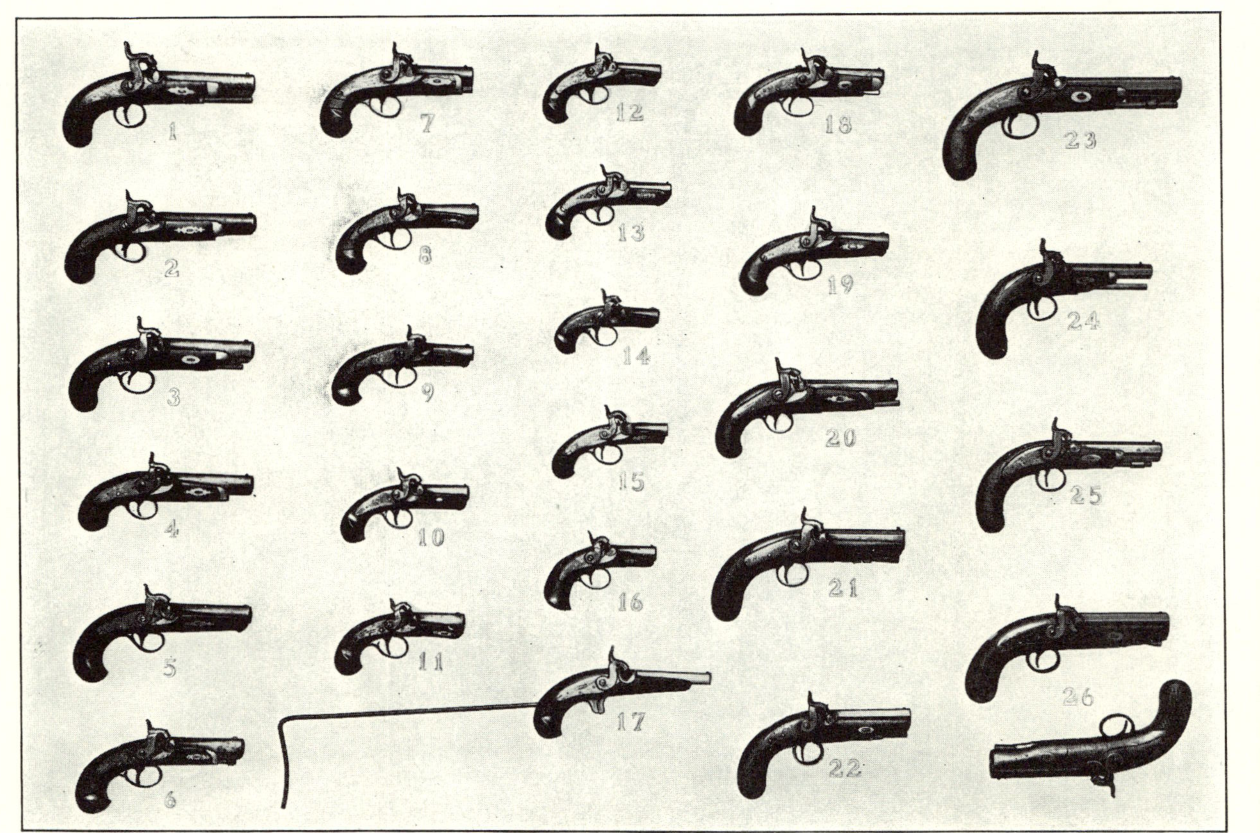

EXPLANATION OF PLATE 81.
Percussion Pistols, Top or Center-Hammer Types (page 511).

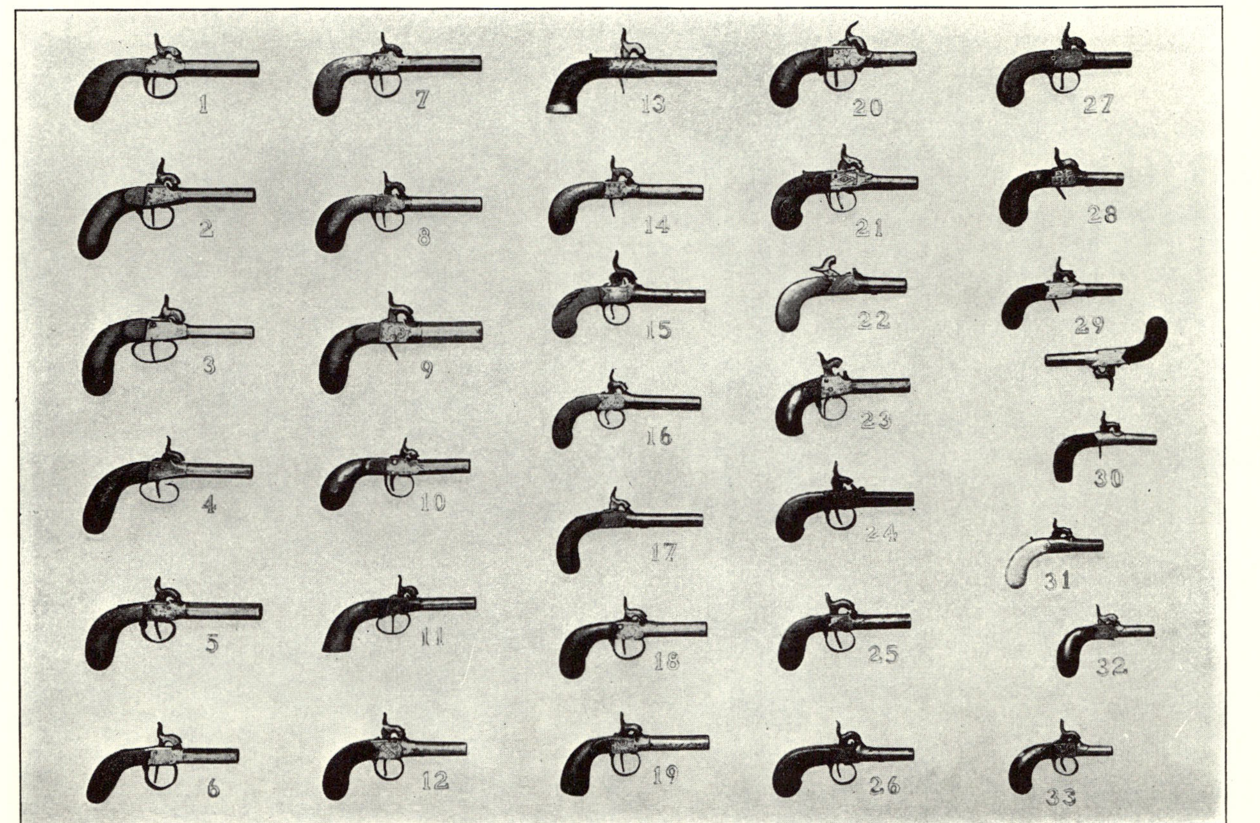

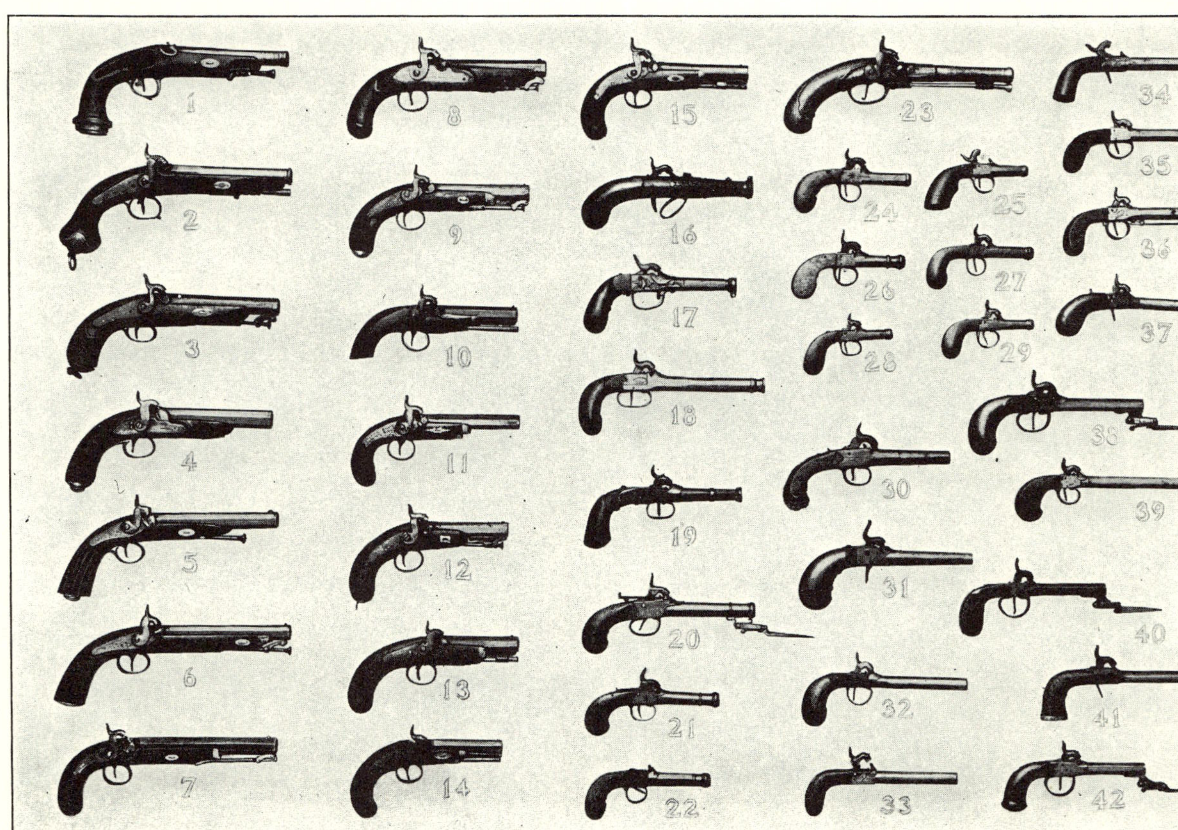

EXPLANATION OF PLATE 83.

Various Kinds of Double-Barreled Percussion Pistols (page 525).

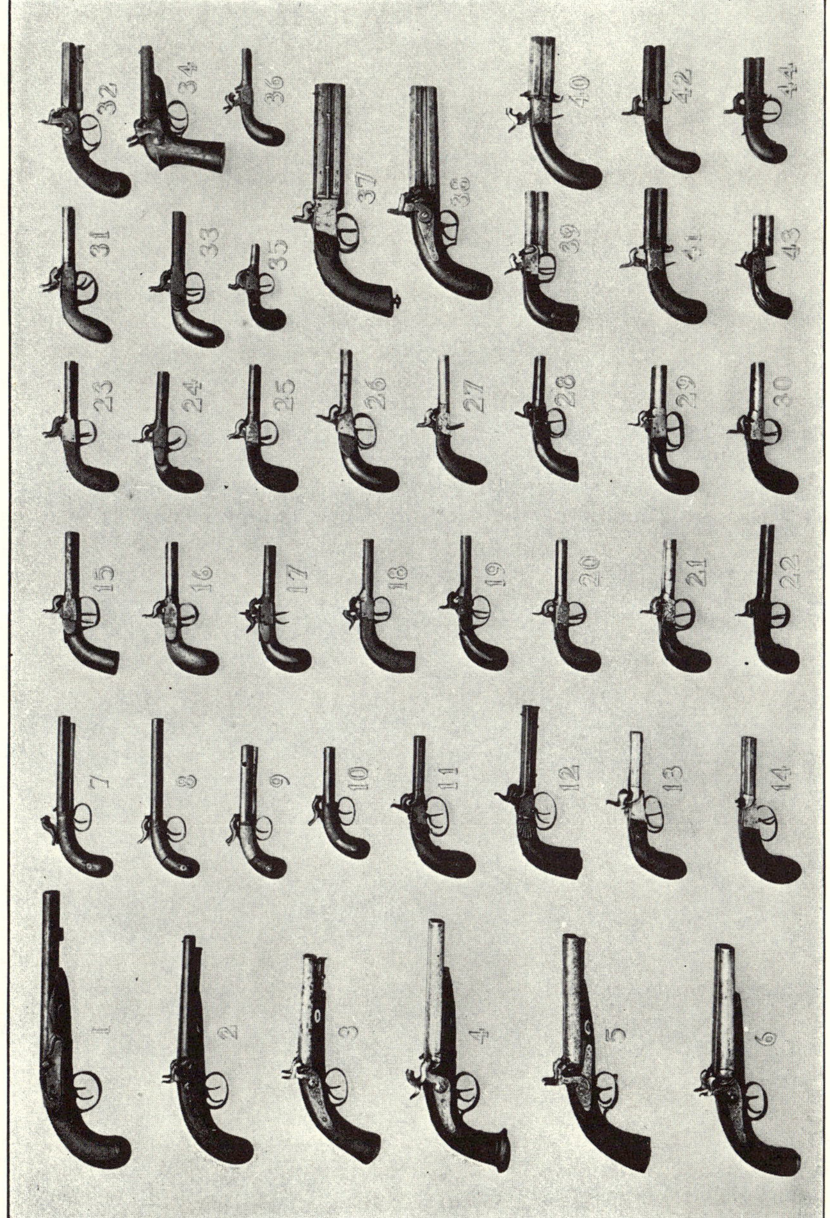

EXPLANATION OF PLATE 84.

Under-Hammers, Self-Cocking Top-Hammers, and Other
Percussion Pistols (page 534).

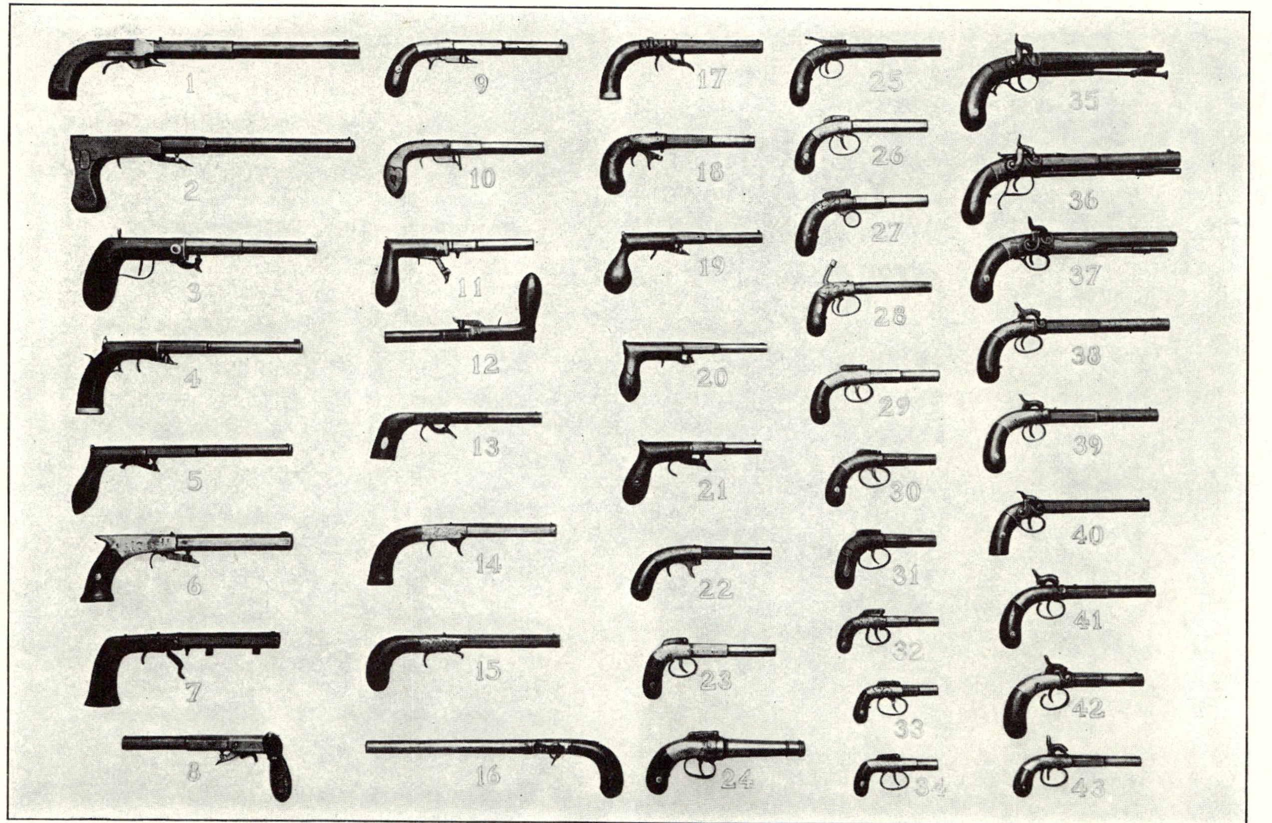

EXPLANATION OF PLATE 85.

Miscellaneous Percussion Pistols (page 543).

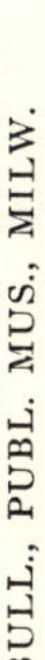

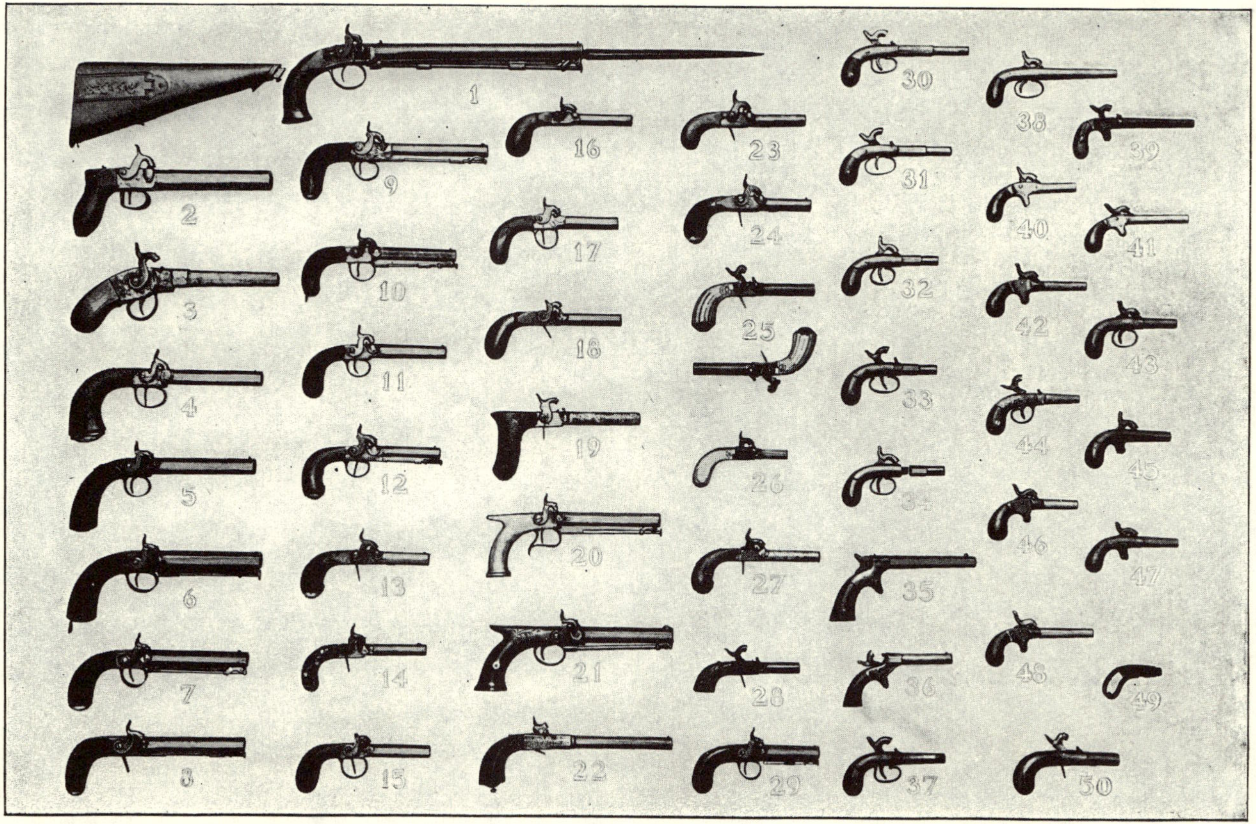
1
2
3
4
5
6
7
8
9
10
11
12
13
14
15
16
17
18
19
20
21
22
23
24
25
26
27
28
29
30
31
32
33
34
35
36
37
38
39
40
41
42
43
44
45
46
47
48
49
50

EXPLANATION OF PLATE 86.

Percussion Pepper-Boxes, Foreign and United States (page 553).

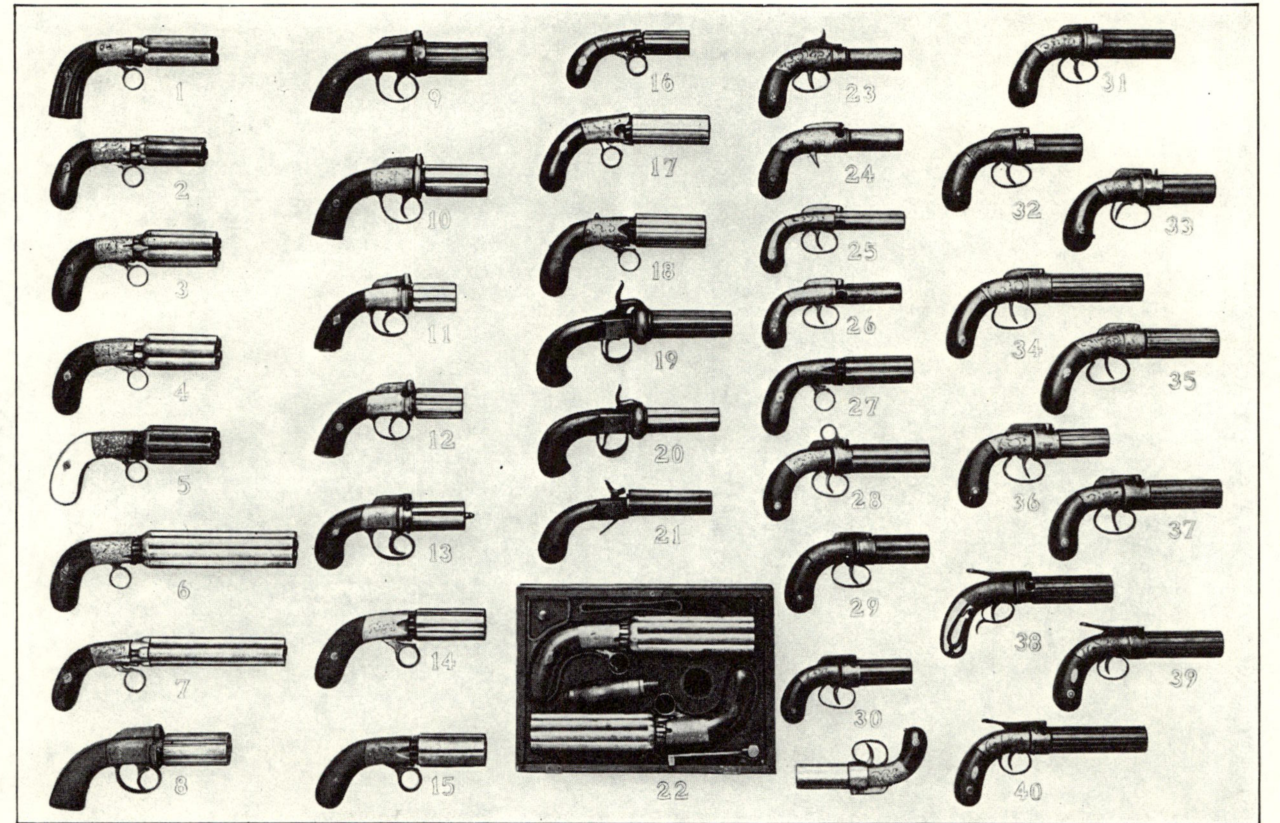

EXPLANATION OF PLATE 87.

Percussion Pepper-Boxes of United States Manufacture (page 561).

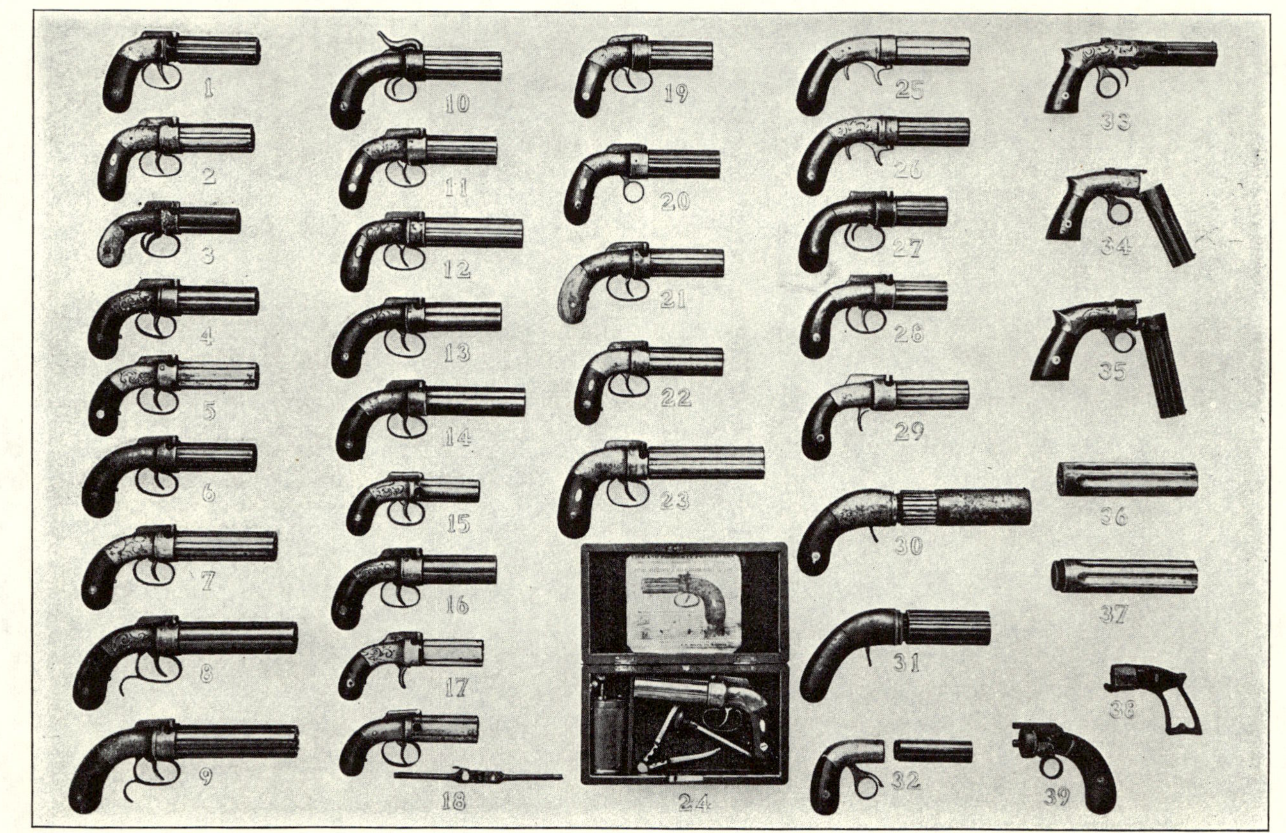

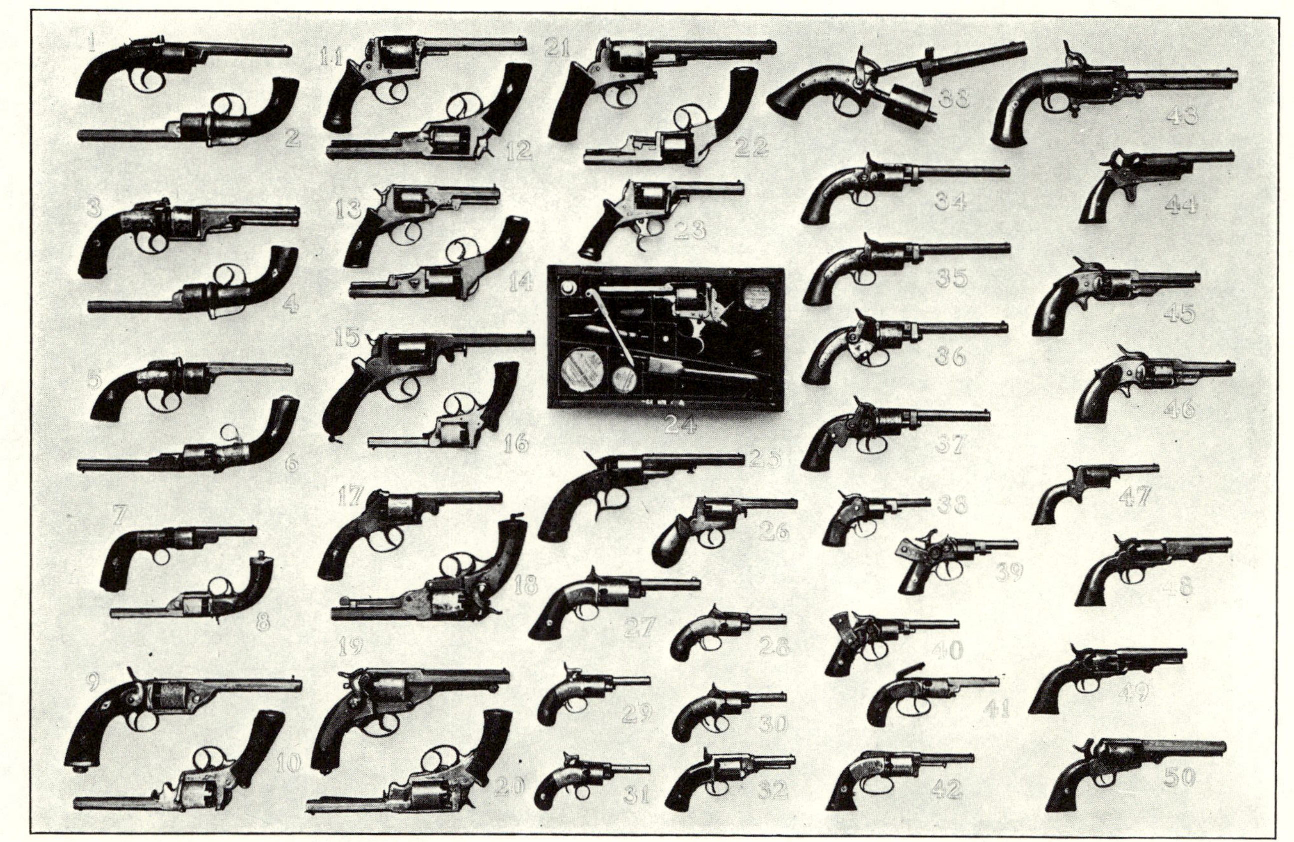

EXPLANATION OF PLATE 89.
Percussion Revolvers, Chiefly United States (page 582).

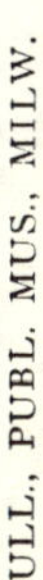

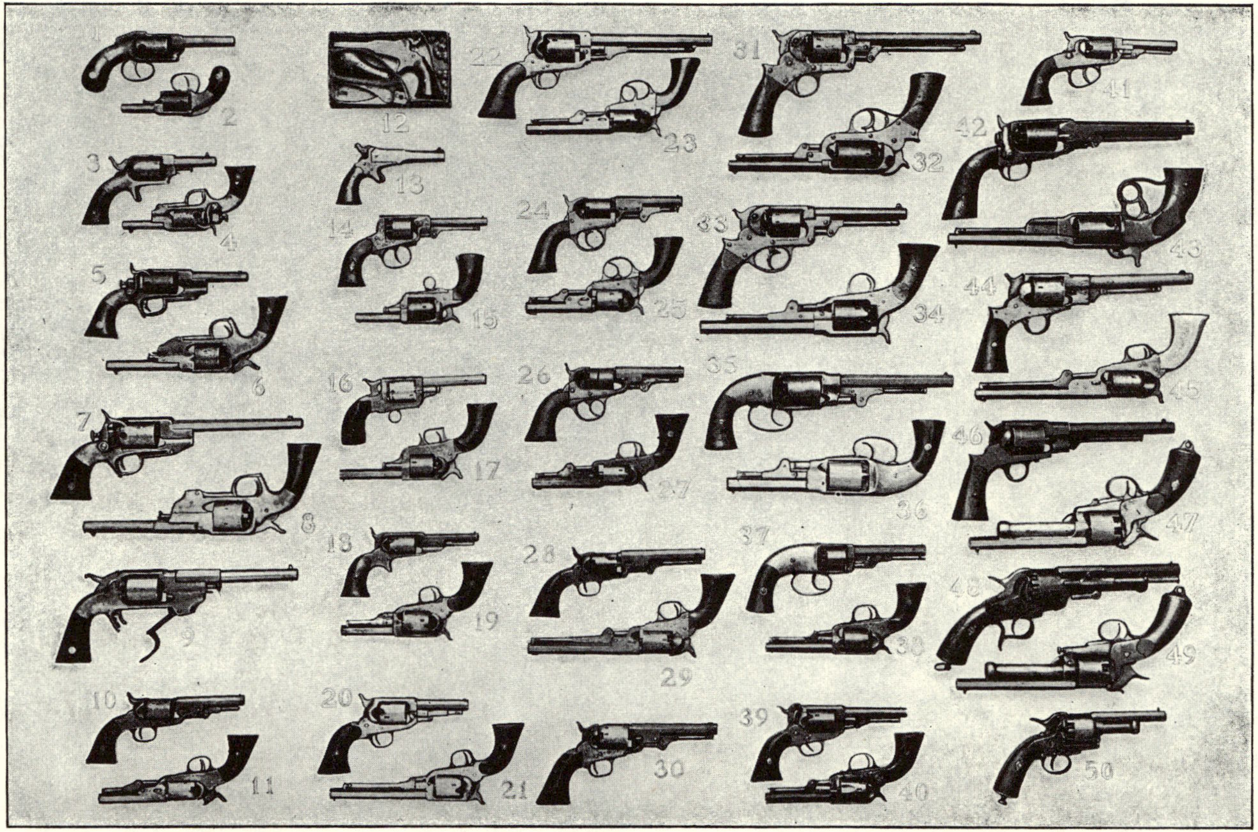
1
2
3
4
5
6
7
8
9
10
11
12
13
14
15
16
17
18
19
20
21
22
23
24
25
26
27
28
29
30
31
32
33
34
35
36
37
38
39
40
41
42
43
44
45
46
47
48
49
50

EXPLANATION OF PLATE 90.

Rim-Fire Breech-Loading Pistols (page 594).

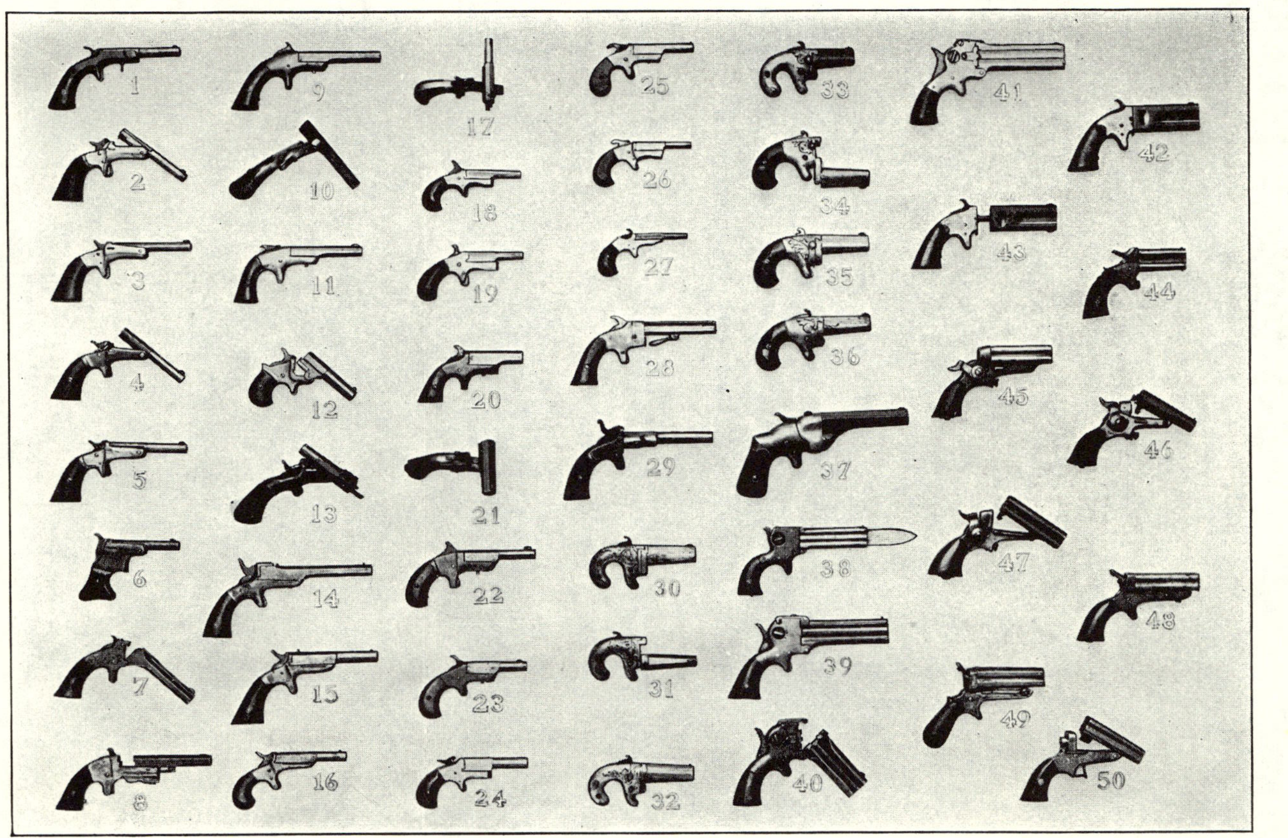

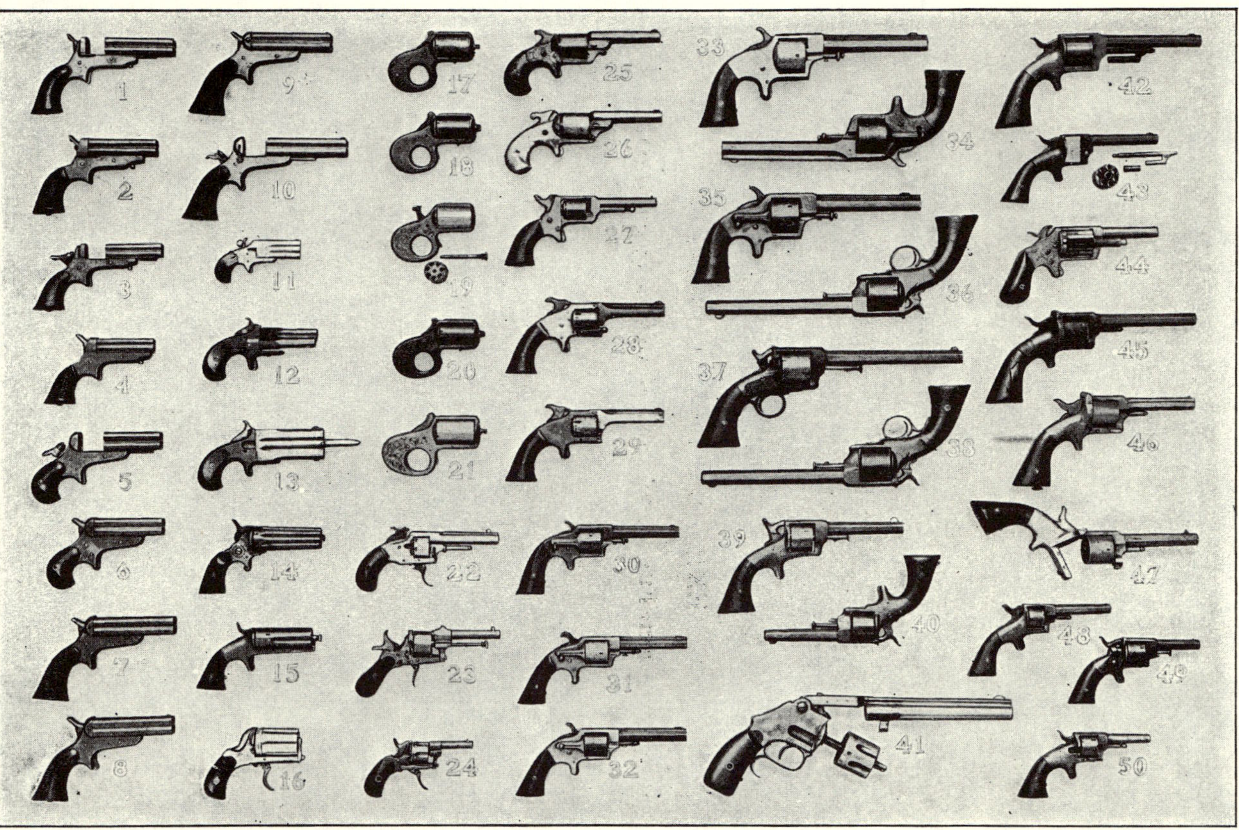

EXPLANATION OF PLATE 92.
Rim-Fire Revolvers (page 619).

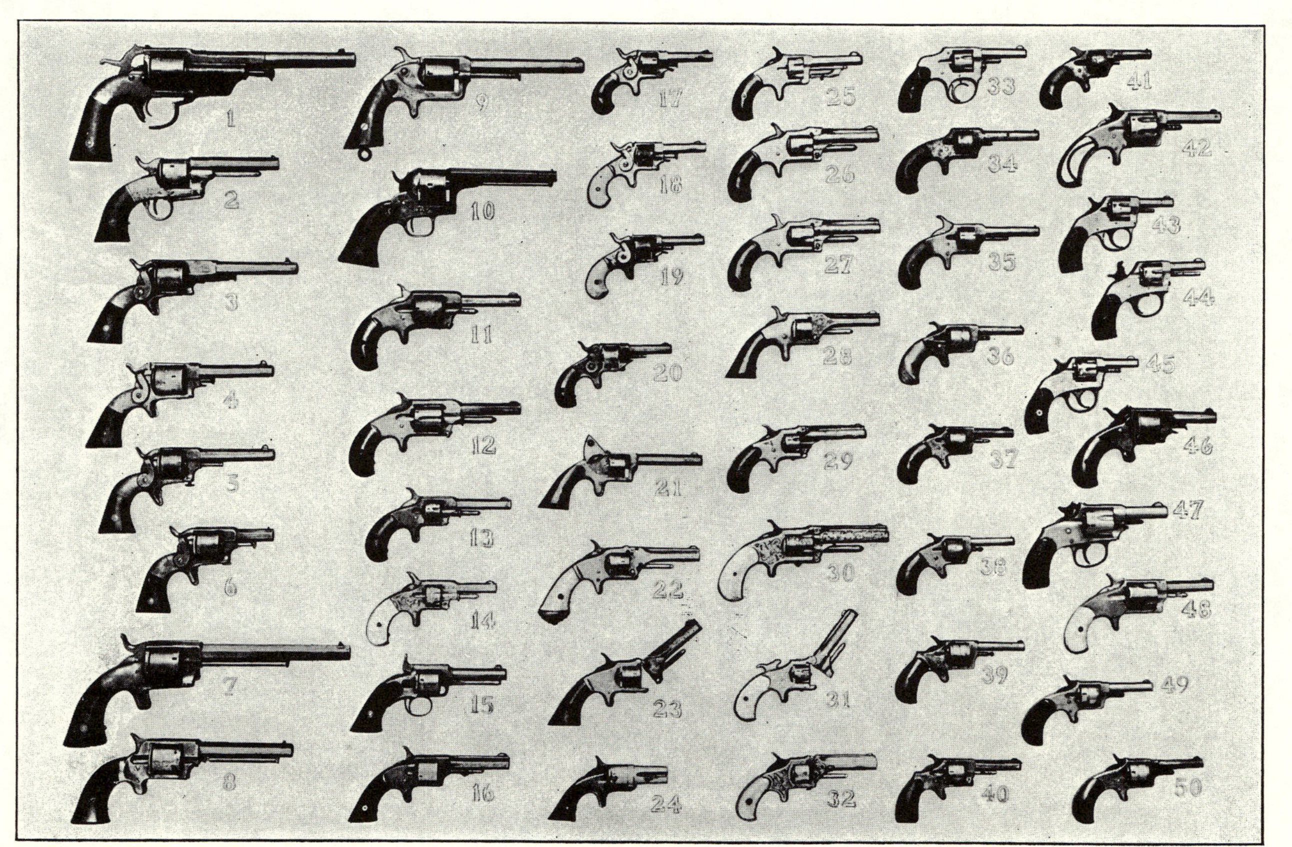

EXPLANATION OF PLATE 93.
American Rim-Fire Revolvers (page 632).

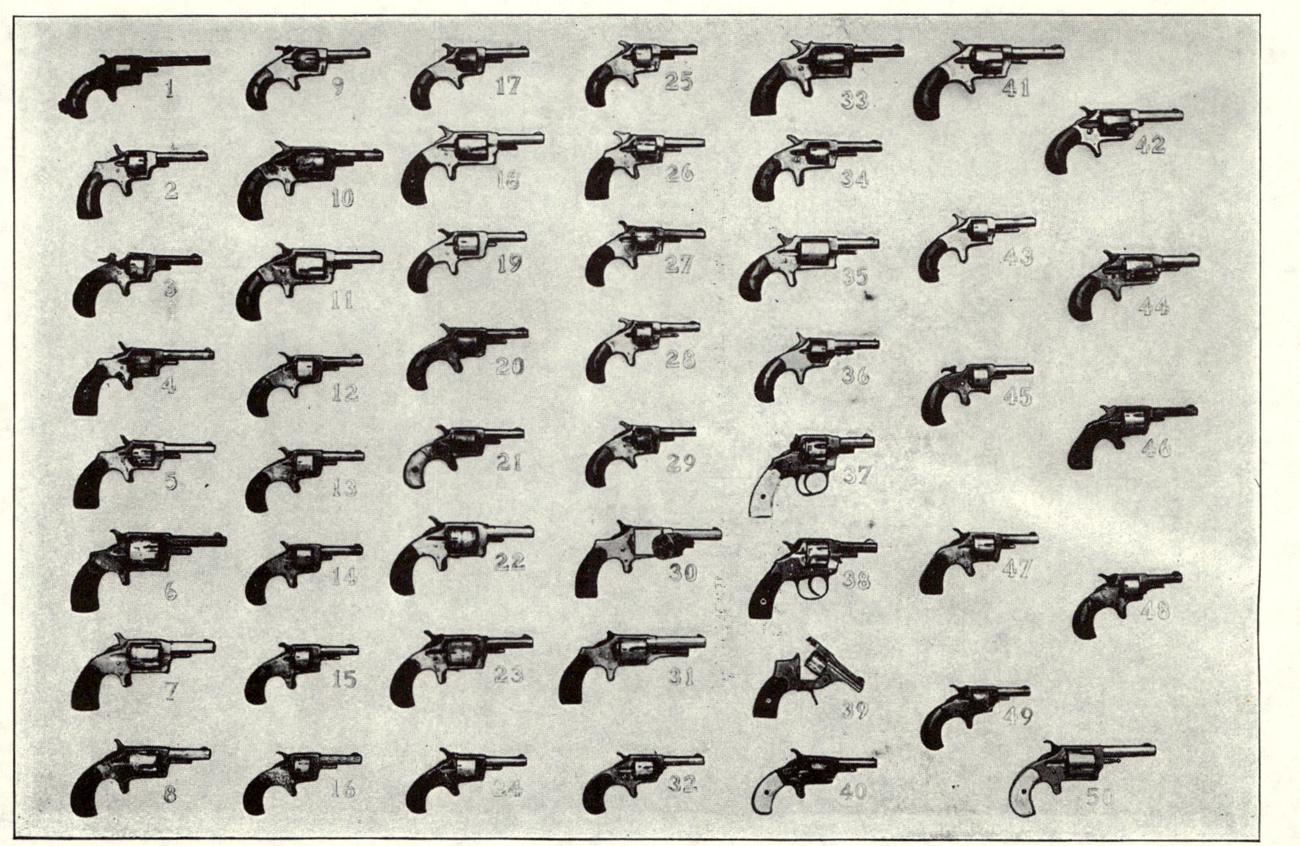

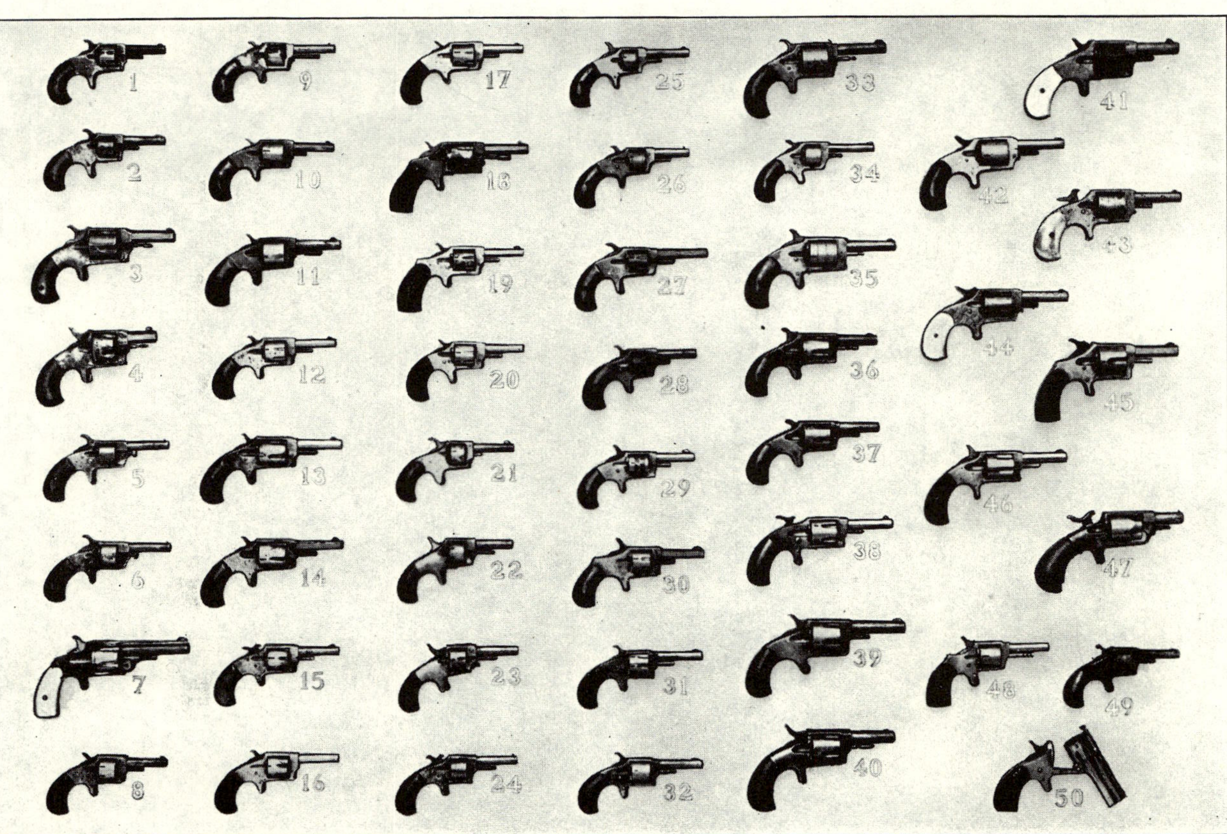

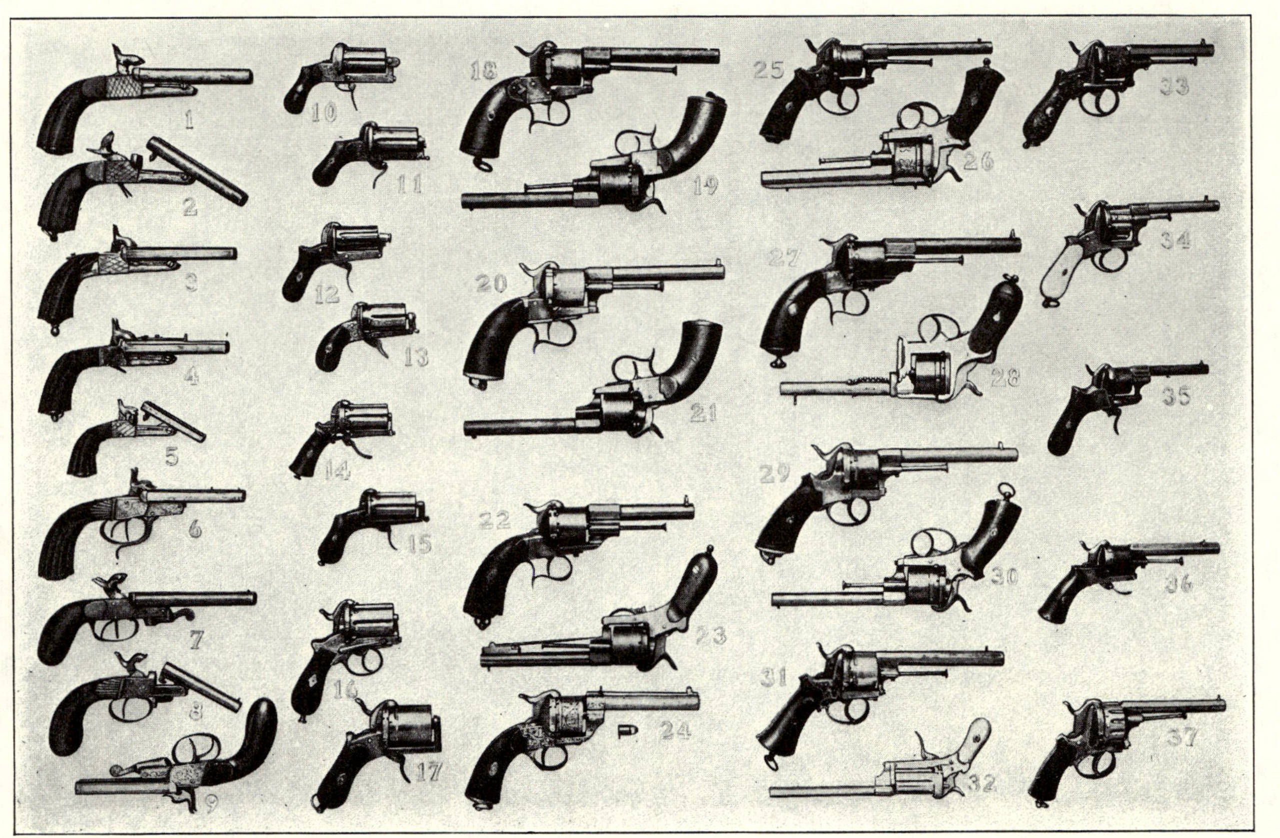

EXPLANATION OF PLATE 96.

Pin-Fire, Needle, and Center-Fire Pistols and Revolvers of Foreign Manufacture (page 663).

Foreign and United States Center-Fire Revolvers (page 672).

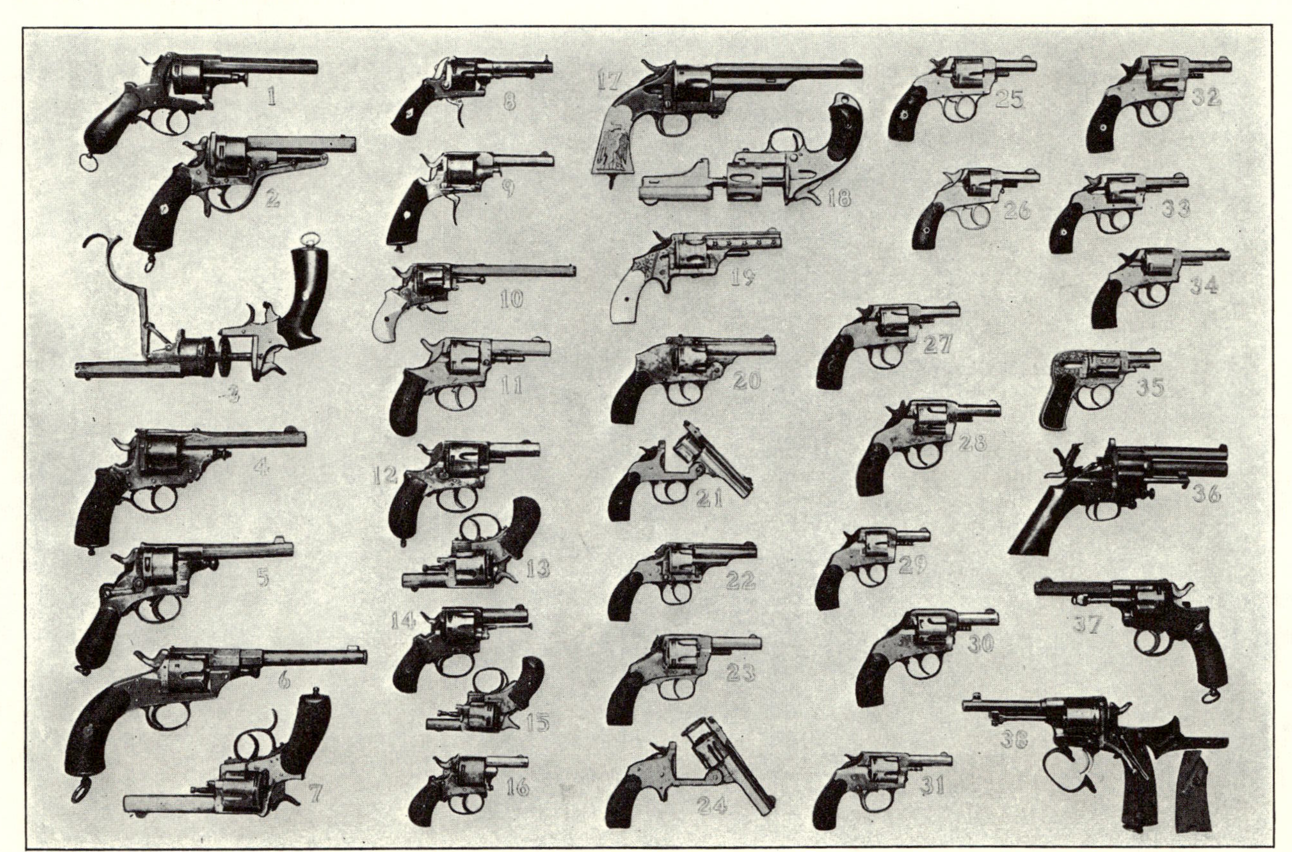

EXPLANATION OF PLATE 98.

Colt-Patersons and Colt Pocket Pistols (page 684).

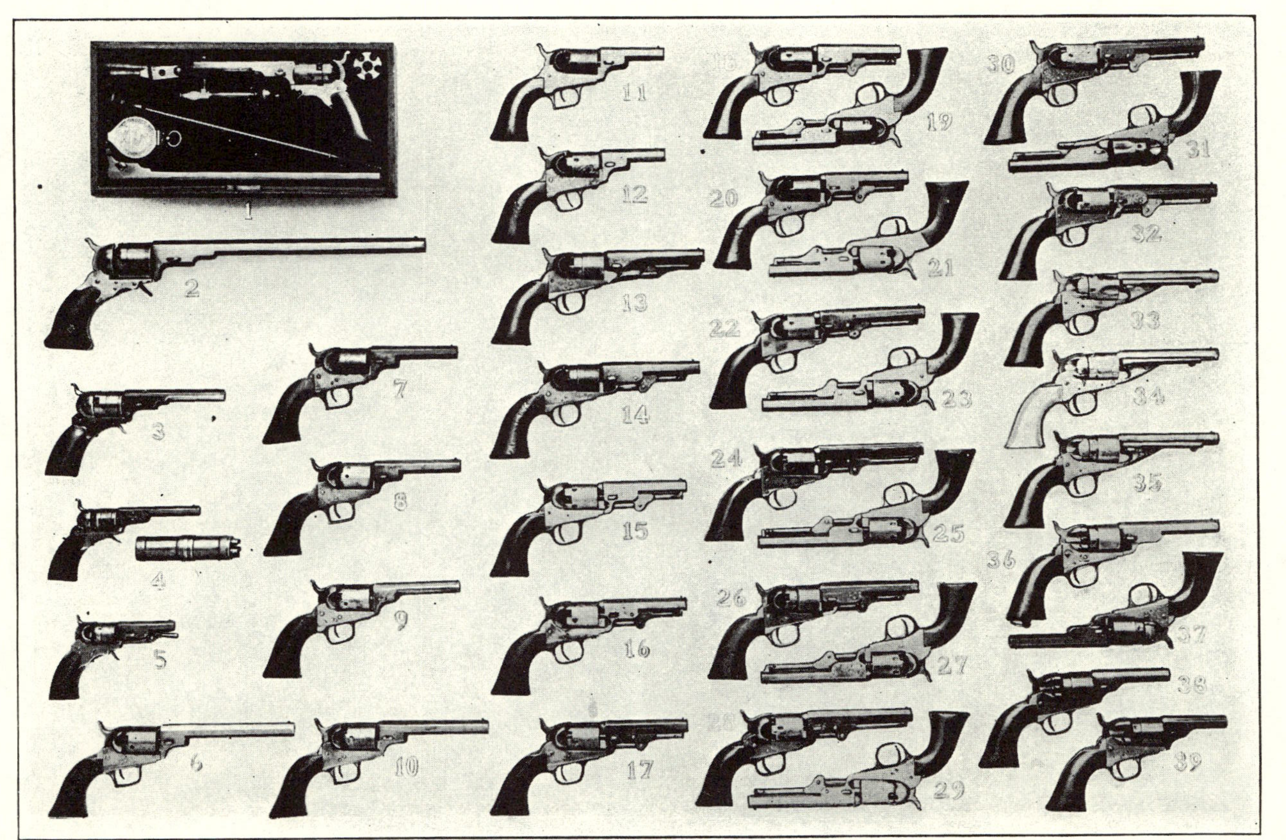

EXPLANATION OF PLATE 99.

Colt Dragoon Revolvers and Colt Revolvers with Shoulder Stocks
(page 694).

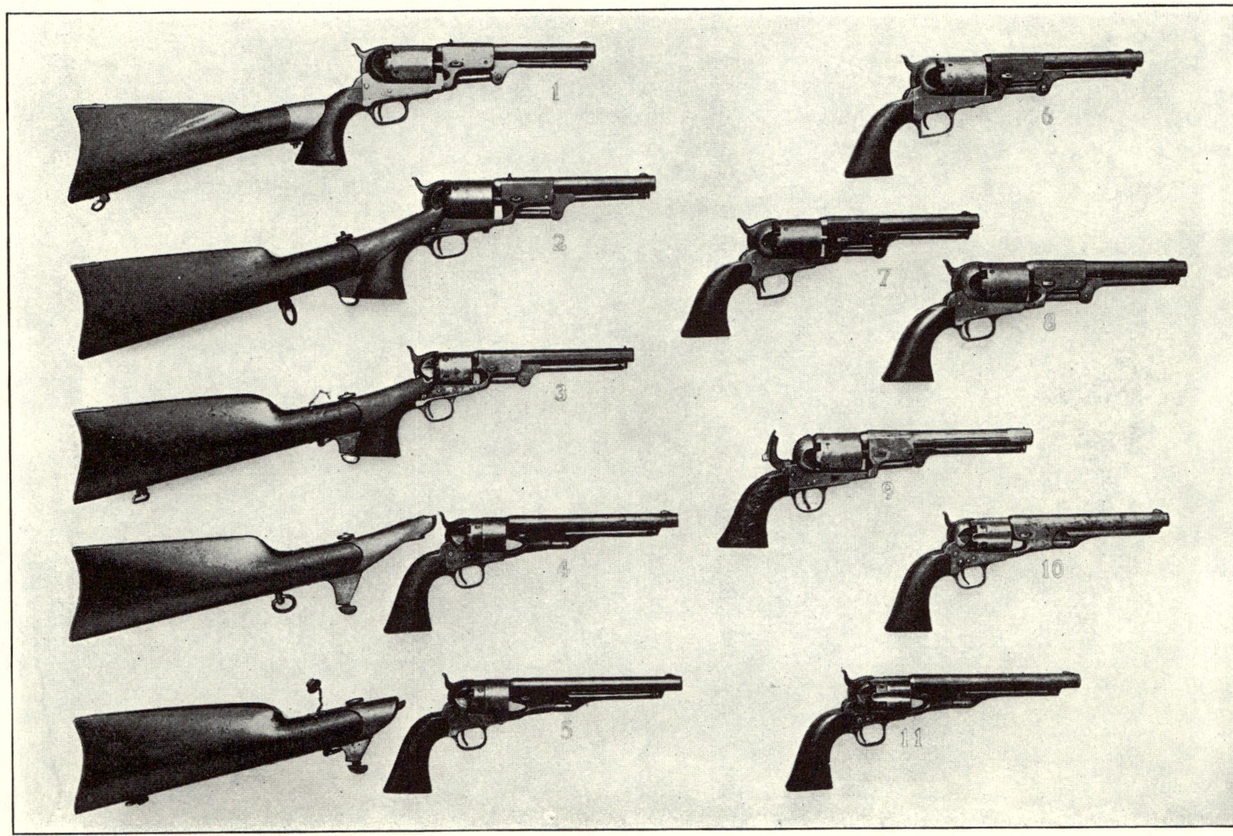

Colt Army and Navy Percussion Revolvers and Alterations of the
Same (page 698).

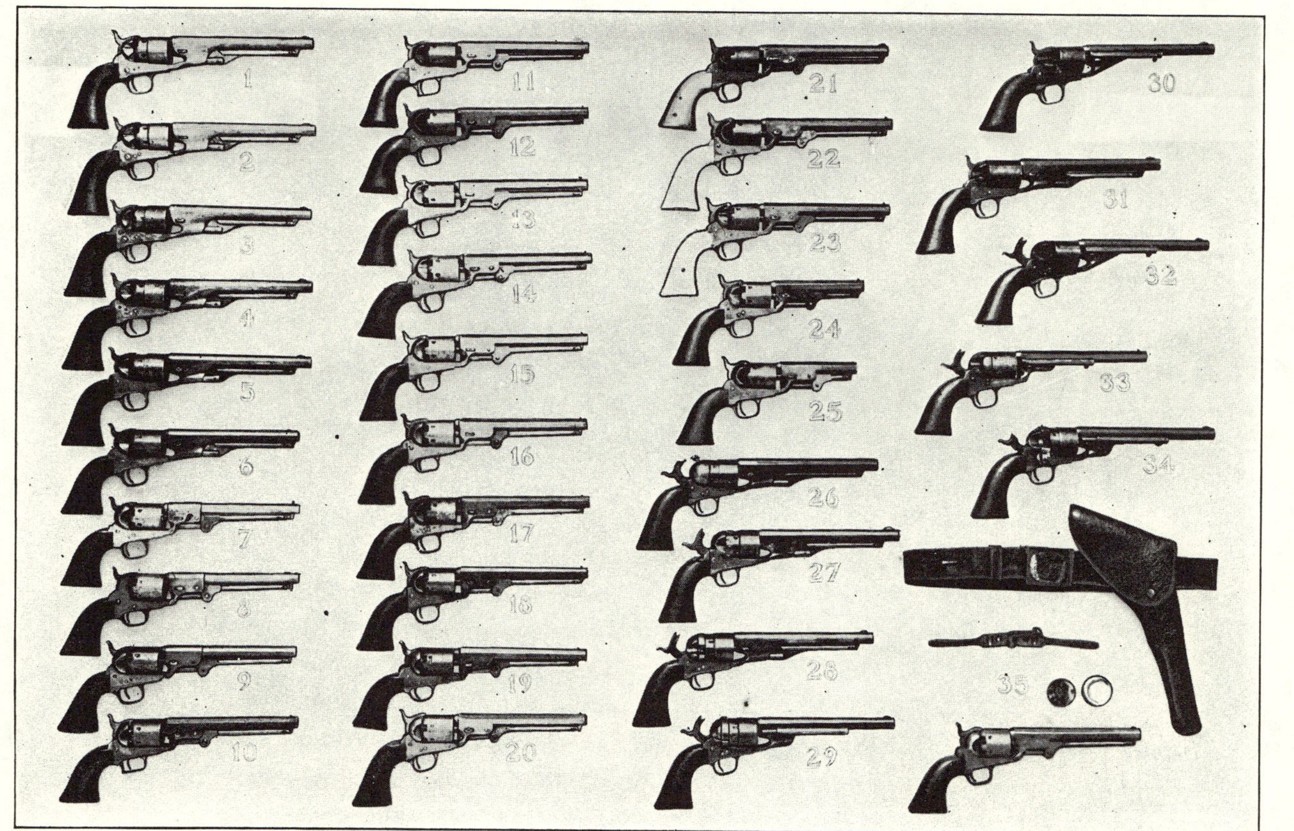

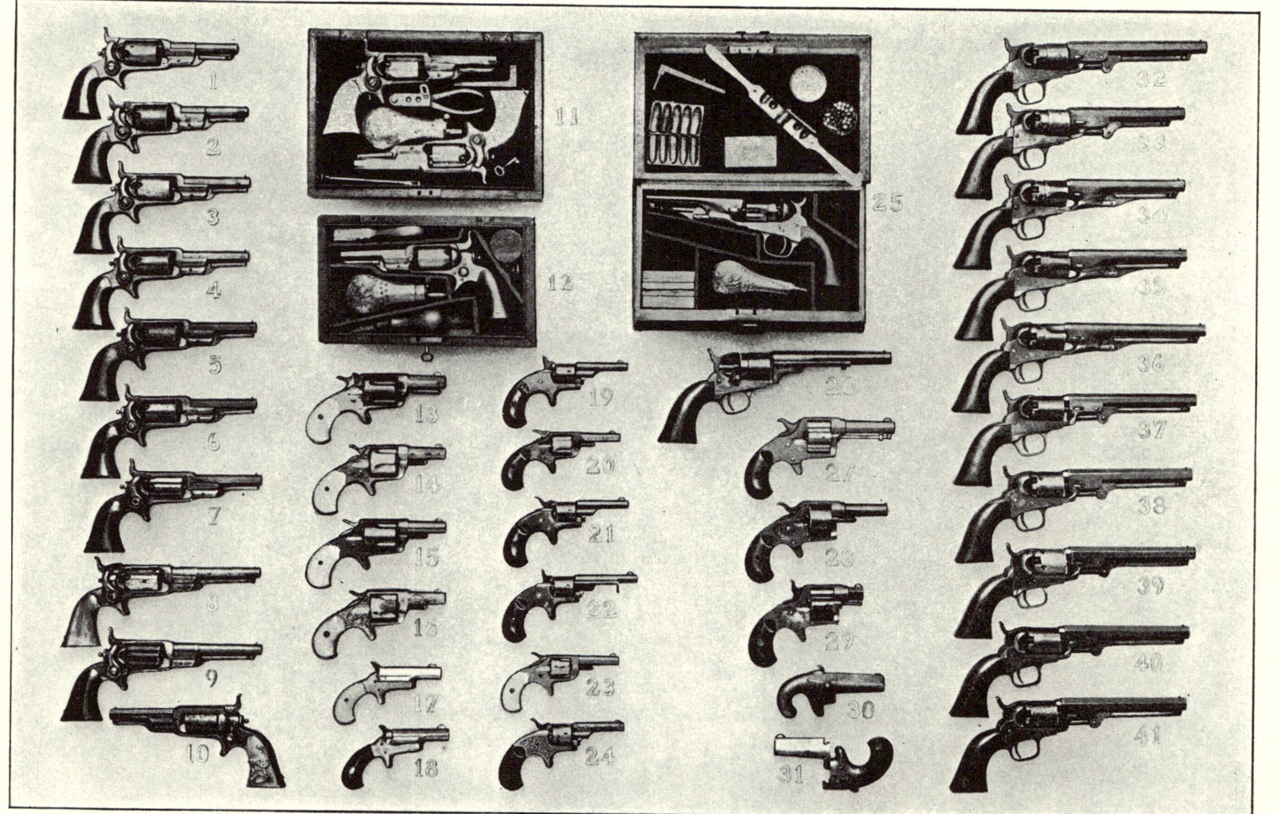

EXPLANATION OF PLATE 102.

Colt Metallic Cartridge Revolvers (page 717).

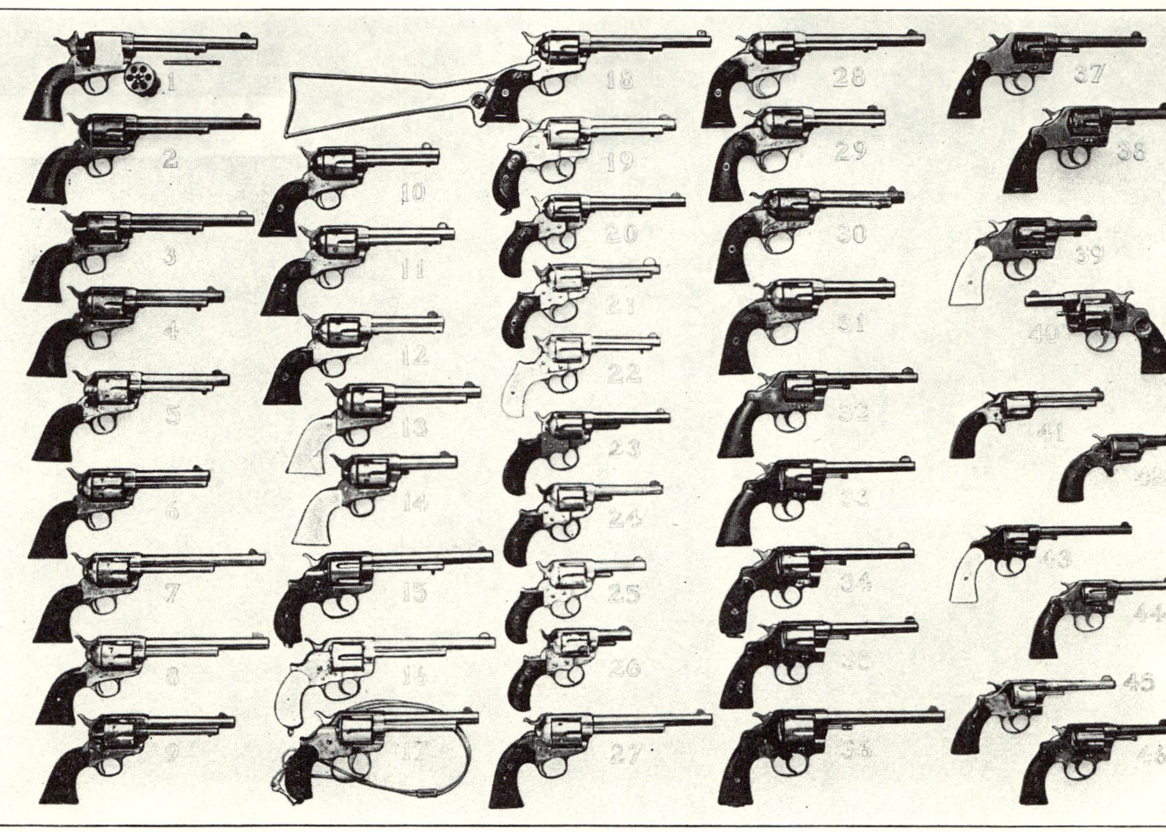

EXPLANATION OF PLATE 103.

Colt Automatic Pistols (page 728).

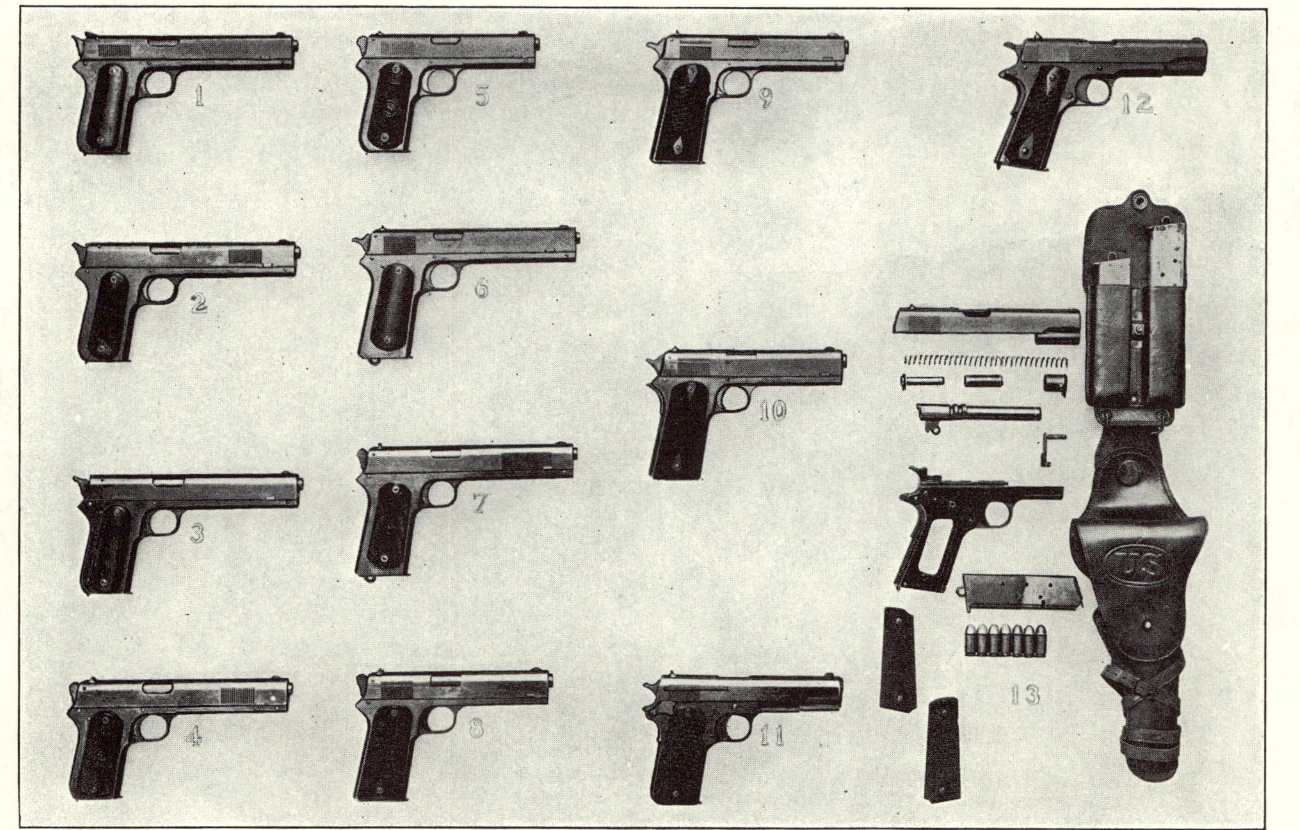

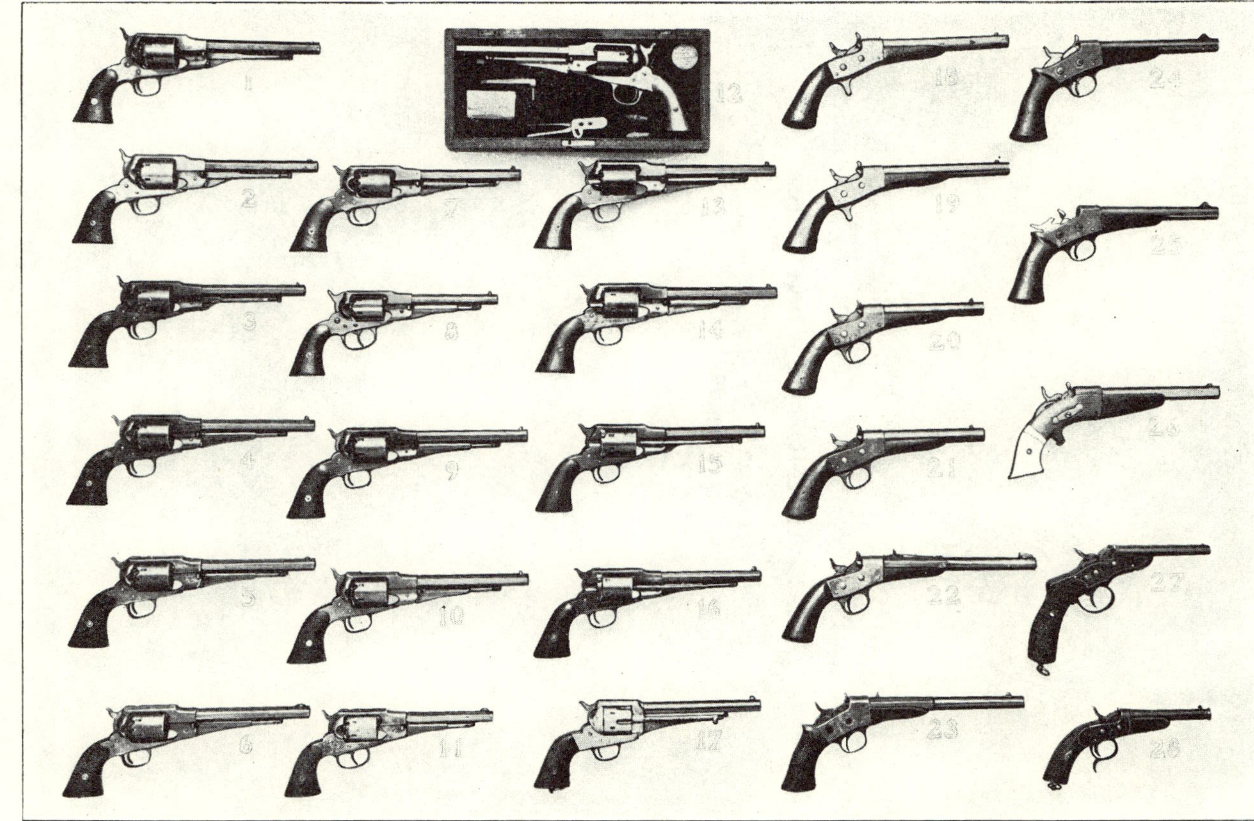

EXPLANATION OF PLATE 105.

Remington, and Smith and Wesson Revolvers and Pistols (page 738).

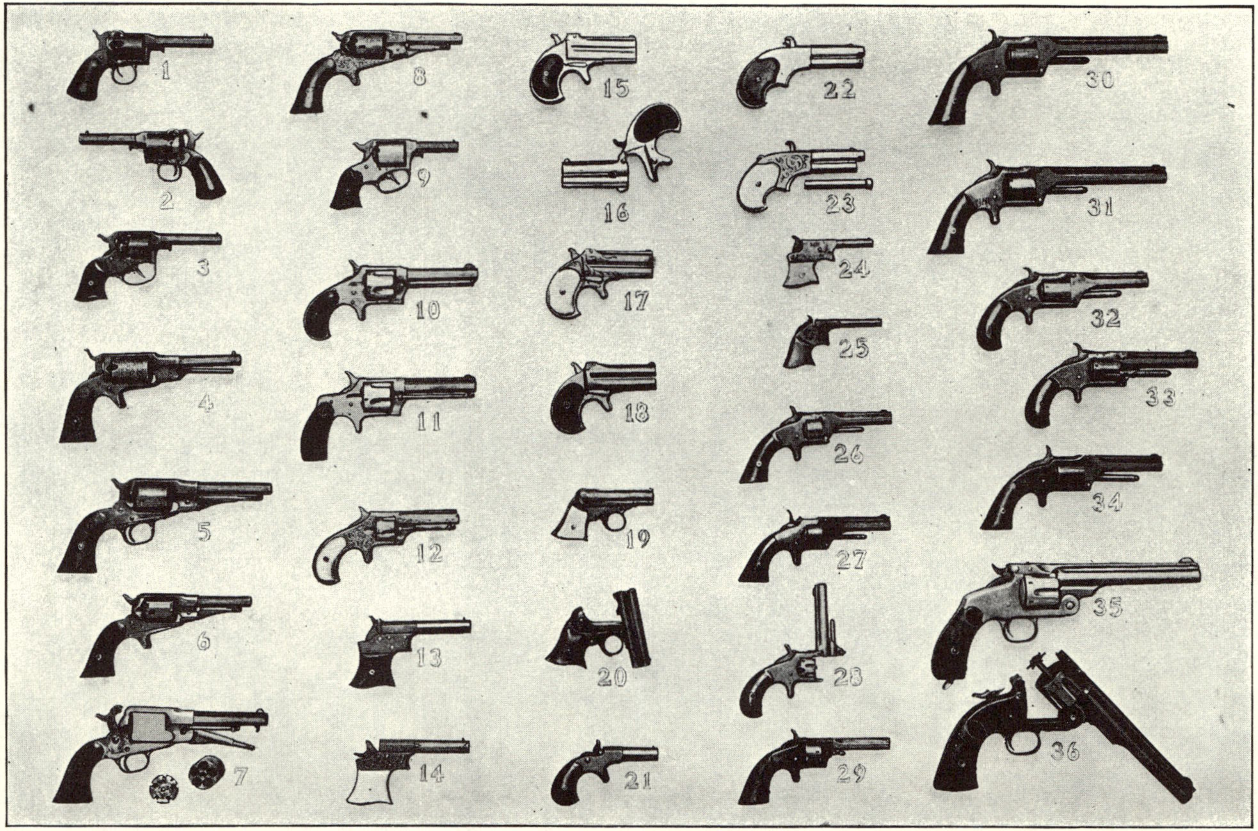
1
2
3
4
5
6
7
8
9
10
11
12
13
14
15
16
17
18
19
20
21
22
23
24
25
26
27
28
29
30
31
32
33
34
35
36

EXPLANATION OF PLATE 106.

Volcanic Pistols and Smith and Wesson Revolvers (page 747).

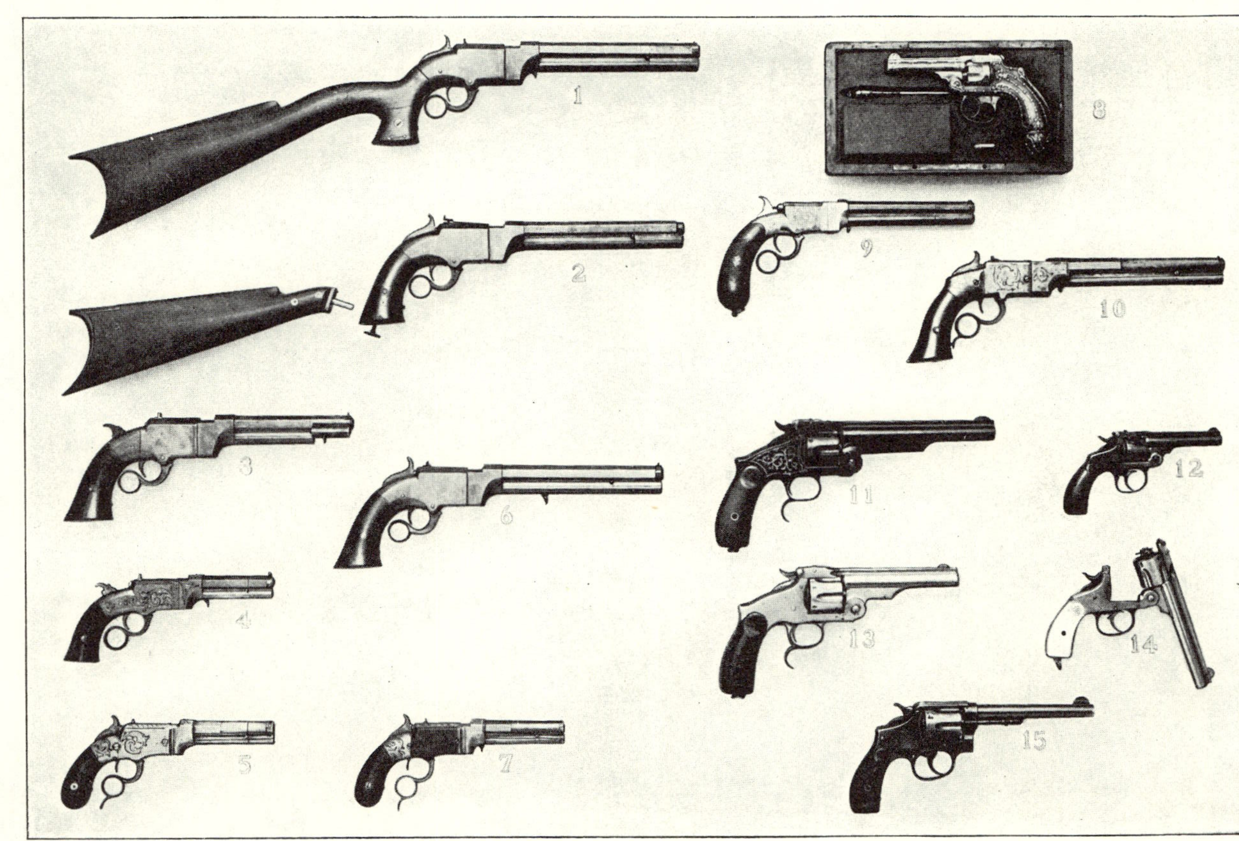

EXPLANATION OF PLATE 107.

Miscellaneous Pistols and Revolvers (page 747).

Powder Testers, Fire Lighters, Signal Pistols, Cap Testers, and Trap Guns (page 759).

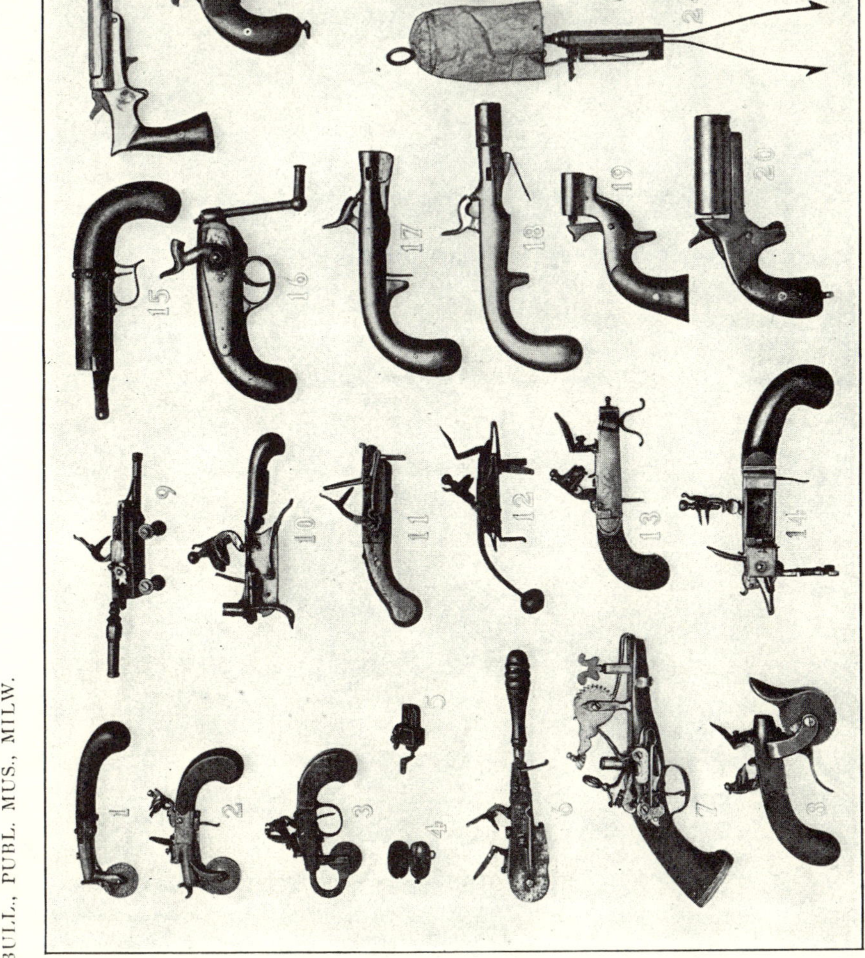

EXPLANATION OF PLATE 109.

Miscellaneous Guns and Freaks (page 765).

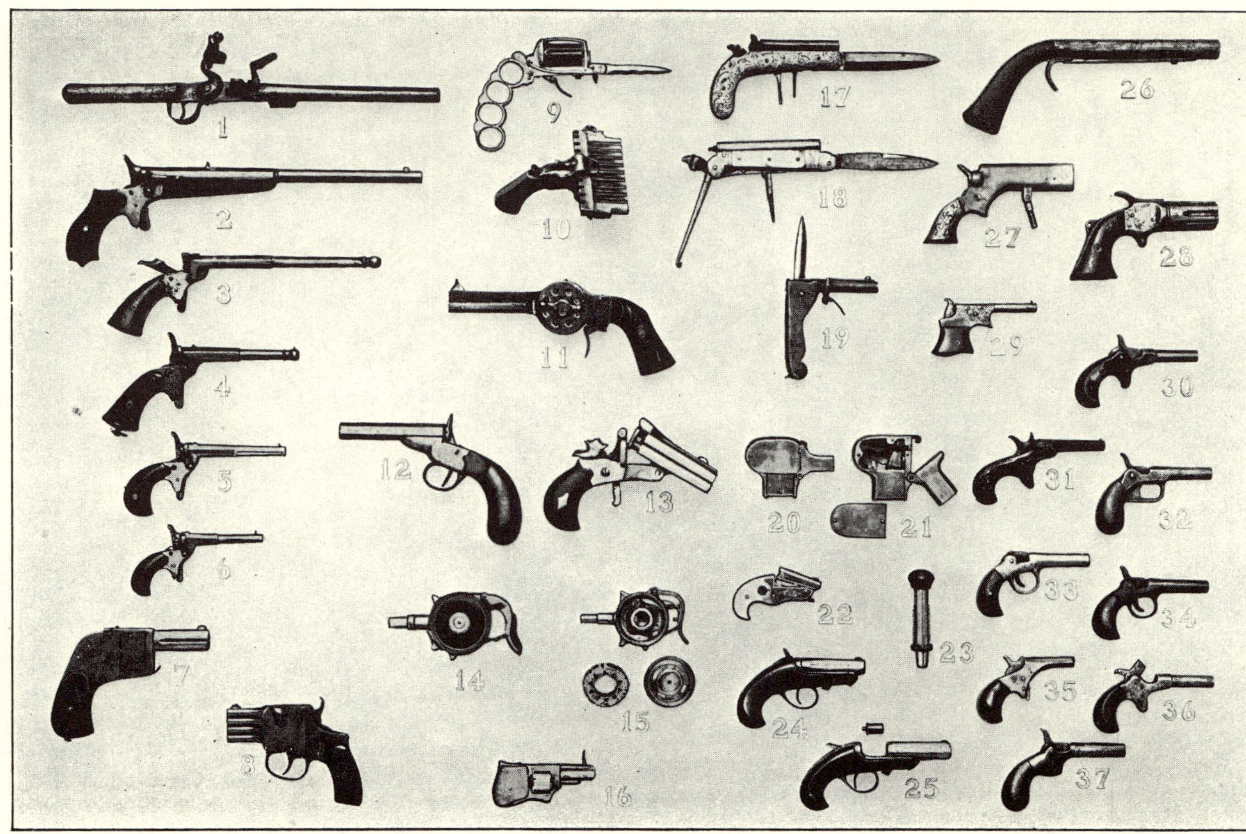

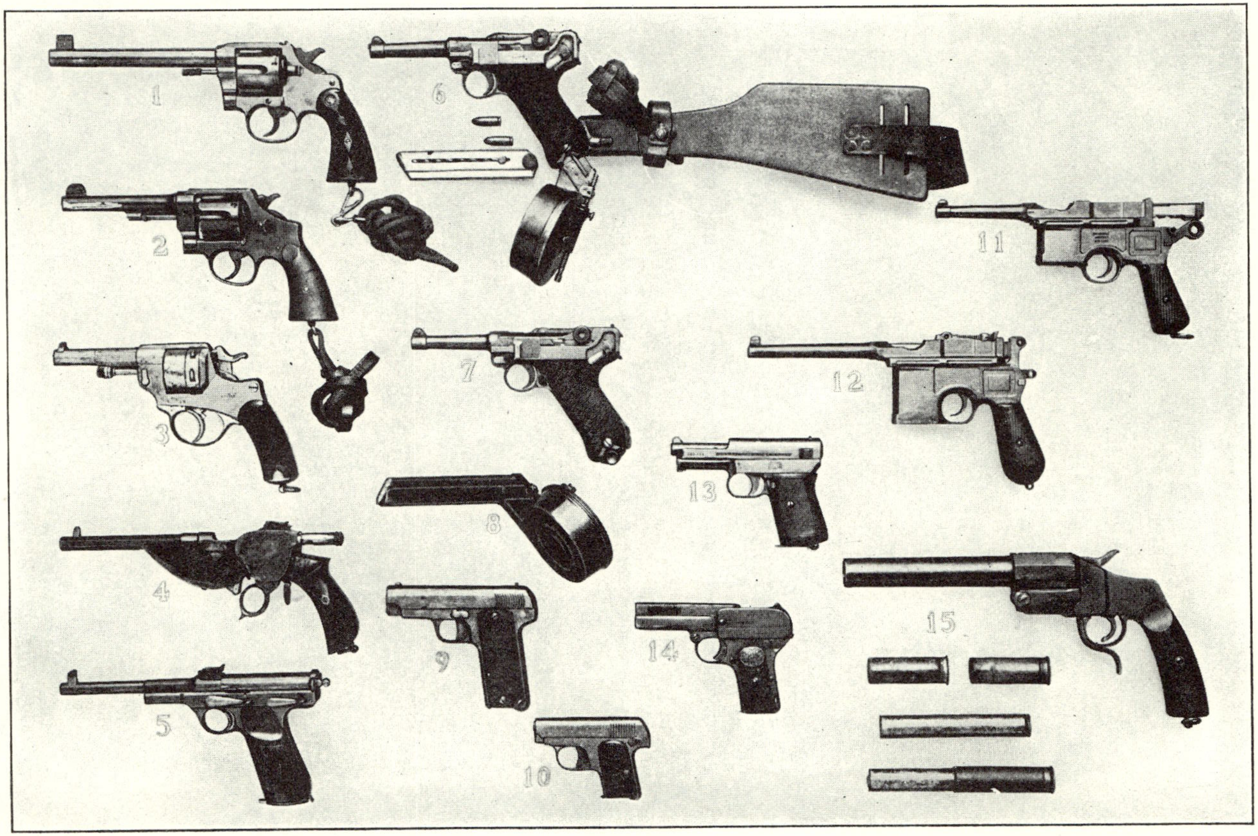
1
2
3
4
5
6
7
8
9
10
11
12
13
14
15

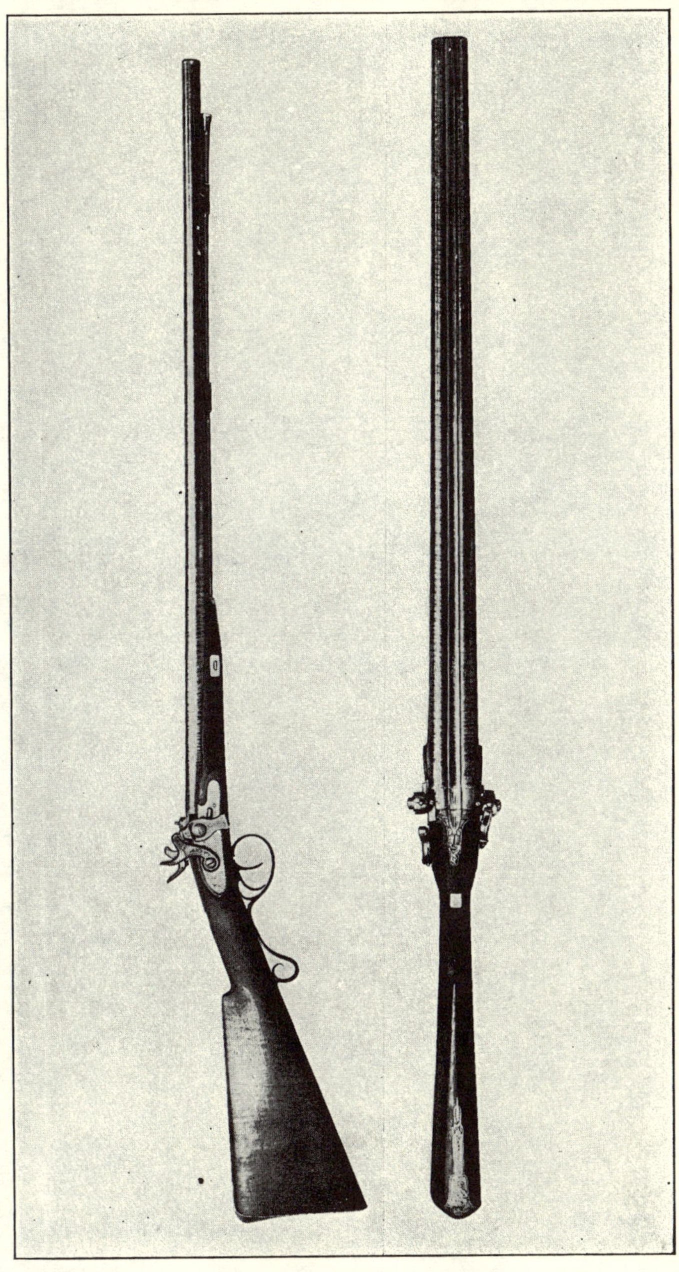

EXPLANATION OF PLATE 112.

Forsyth Percussion or "Detonating" Shotgun (page 783).

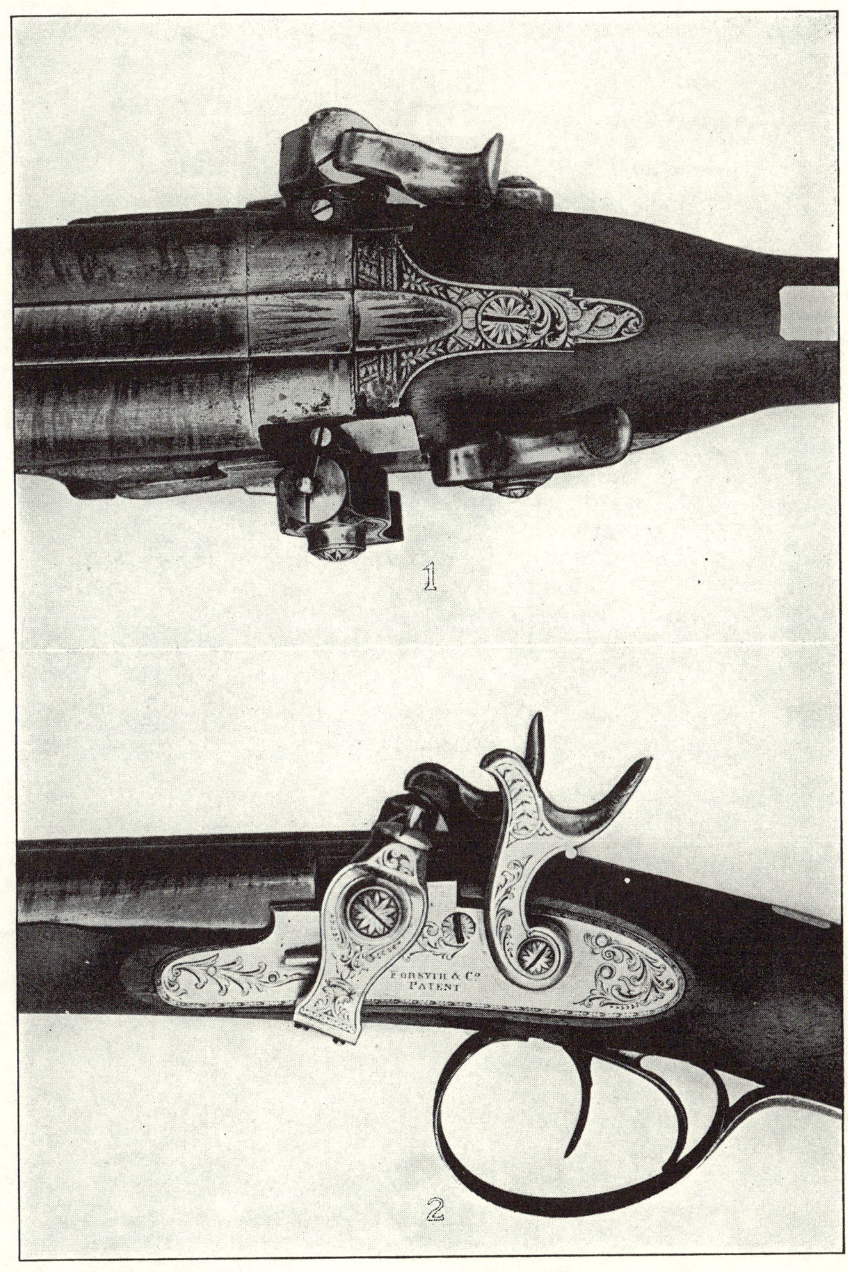

1

2